T0236065

Lecture Notes in Computer Science   10111

Commenced Publication in 1973
Founding and Former Series Editors:
Gerhard Goos, Juris Hartmanis, and Jan van Leeuwen

More information about this series at http://www.springer.com/series/7412

Shang-Hong Lai · Vincent Lepetit
Ko Nishino · Yoichi Sato (Eds.)

Computer Vision – ACCV 2016

13th Asian Conference on Computer Vision
Taipei, Taiwan, November 20–24, 2016
Revised Selected Papers, Part I

 Springer

Editors
Shang-Hong Lai
National Tsing Hua University
Hsinchu
Taiwan

Vincent Lepetit
Graz University of Technology
Graz
Austria

Ko Nishino
Drexel University
Philadelphia, PA
USA

Yoichi Sato
The University of Tokyo
Tokyo
Japan

ISSN 0302-9743   ISSN 1611-3349   (electronic)
Lecture Notes in Computer Science
ISBN 978-3-319-54180-8   ISBN 978-3-319-54181-5   (eBook)
DOI 10.1007/978-3-319-54181-5

Library of Congress Control Number: 2017932642

LNCS Sublibrary: SL6 – Image Processing, Computer Vision, Pattern Recognition, and Graphics

Printed on acid-free paper

This Springer imprint is published by Springer Nature
The registered company is Springer International Publishing AG
The registered company address is: Gewerbestrasse 11, 6330 Cham, Switzerland

Preface

Welcome to the 2016 edition of the Asian Conference on Computer Vision in Taipei. ACCV 2016 received a total number of 590 submissions, of which 479 papers went through a review process after excluding papers rejected without review because of violation of the ACCV submission guidelines or being withdrawn before review. The papers were submitted from diverse regions with 69% from Asia, 19% from Europe, and 12% from North America.

The program chairs assembled a geographically diverse team of 39 area chairs who handled nine to 15 papers each. Area chairs were selected to provide a broad range of expertise, to balance junior and senior members, and to represent a variety of geographical locations. Area chairs recommended reviewers for papers, and each paper received at least three reviews from the 631 reviewers who participated in the process. Paper decisions were finalized at an area chair meeting held in Taipei during August 13–14, 2016. At this meeting, the area chairs worked in threes to reach collective decisions about acceptance, and in panels of nine or 12 to decide on the oral/poster distinction. The total number of papers accepted was 143 (an overall acceptance rate of 24%). Of these, 33 were selected for oral presentations and 110 were selected for poster presentations.

We wish to thank all members of the local arrangements team for helping us run the area chair meeting smoothly. We also wish to extend our immense gratitude to the area chairs and reviewers for their generous participation in the process. The conference would not have been possible without this huge voluntary investment of time and effort. We acknowledge particularly the contribution of 29 reviewers designated as "Outstanding Reviewers" who were nominated by the area chairs and program chairs for having provided a large number of helpful, high-quality reviews. Last but not the least, we would like to show our deepest gratitude to all of the emergency reviewers who kindly responded to our last-minute request and provided thorough reviews for papers with missing reviews. Finally, we wish all the attendees a highly simulating, informative, and enjoyable conference.

January 2017

Shang-Hong Lai
Vincent Lepetit
Ko Nishino
Yoichi Sato

Organization

ACCV 2016 Organizers

Steering Committee

Michael Brown	National University of Singapore, Singapore
Katsu Ikeuchi	University of Tokyo, Japan
In-So Kweon	KAIST, Korea
Tieniu Tan	Chinese Academy of Sciences, China
Yasushi Yagi	Osaka University, Japan

Honorary Chairs

Thomas Huang	University of Illinois at Urbana-Champaign, USA
Wen-Hsiang Tsai	National Chiao Tung University, Taiwan, ROC

General Chairs

Yi-Ping Hung	National Taiwan University, Taiwan, ROC
Ming-Hsuan Yang	University of California at Merced, USA
Hongbin Zha	Peking University, China

Program Chairs

Shang-Hong Lai	National Tsing Hua University, Taiwan, ROC
Vincent Lepetit	TU Graz, Austria
Ko Nishino	Drexel University, USA
Yoichi Sato	University of Tokyo, Japan

Publicity Chairs

Ming-Ming Cheng	Nankai University, China
Jen-Hui Chuang	National Chiao Tung University, Taiwan, ROC
Seon Joo Kim	Yonsei University, Korea

Local Arrangements Chairs

Yung-Yu Chuang	National Taiwan University, Taiwan, ROC
Yen-Yu Lin	Academia Sinica, Taiwan, ROC
Sheng-Wen Shih	National Chi Nan University, Taiwan, ROC
Yu-Chiang Frank Wang	Academia Sinica, Taiwan, ROC

Workshops Chairs

Chu-Song Chen	Academia Sinica, Taiwan, ROC
Jiwen Lu	Tsinghua University, China
Kai-Kuang Ma	Nanyang Technological University, Singapore

Tutorial Chairs

Bernard Ghanem	King Abdullah University of Science and Technology, Saudi Arabia
Fay Huang	National Ilan University, Taiwan, ROC
Yukiko Kenmochi	Université Paris-Est, France

Exhibition and Demo Chairs

Gee-Sern Hsu	National Taiwan University of Science and Technology, Taiwan, ROC
Xue Mei	Toyota Research Institute, USA

Publication Chairs

Chih-Yi Chiu	National Chiayi University, Taiwan, ROC
Jenn-Jier (James) Lien	National Cheng Kung University, Taiwan, ROC
Huei-Yung Lin	National Chung Cheng University, Taiwan, ROC

Industry Chairs

Winston Hsu	National Taiwan University, Taiwan, ROC
Fatih Porikli	Australian National University, Australia
Li Xu	SenseTime Group Limited, Hong Kong, SAR China

Finance Chairs

Yong-Sheng Chen	National Chiao Tung University, Taiwan, ROC
Ming-Sui Lee	National Taiwan University, Taiwan, ROC

Registration Chairs

Kuan-Wen Chen	National Chiao Tung University, Taiwan, ROC
Wen-Huang Cheng	Academia Sinica, Taiwan, ROC
Min Sun	National Tsing Hua University, Taiwan, ROC

Web Chairs

Hwann-Tzong Chen	National Tsing Hua University, Taiwan, ROC
Ju-Chun Ko	National Taipei University of Technology, Taiwan, ROC
Neng-Hao Yu	National Chengchi University, Taiwan, ROC

Area Chairs

Narendra Ahuja	UIUC
Michael Brown	National University of Singapore
Yung-Yu Chuang	National Taiwan University, Taiwan, ROC
Pau-Choo Chung	National Cheng Kung University, Taiwan, ROC
Larry Davis	University of Maryland, USA

Contents – Part I

Dictionary Learning, Retrieval, and Clustering

Segmentation and Classification

Realtime Hierarchical Clustering
Based on Boundary and Surface Statistics

Dominik Alexander Klein[1]([⊠]), Dirk Schulz[1], and Armin Bernd Cremers[2]

[1] Department of Cognitive Mobile Systems, Fraunhofer FKIE, Wachtberg, Germany
dominik.klein@fkie.fraunhofer.de
[2] Bonn-Aachen International Center for Information Technology (B-IT),
Bonn, Germany

Abstract. Visual grouping is a key mechanism in human scene perception. There, it belongs to the subconscious, early processing and is key prerequisite for other high level tasks such as recognition. In this paper, we introduce an efficient, realtime capable algorithm which likewise agglomerates a valuable hierarchical clustering of a scene, while using purely local appearance statistics.

To speed up the processing, first we subdivide the image into meaningful, atomic segments using a fast Watershed transform. Starting from there, our rapid, agglomerative clustering algorithm prunes and maintains the connectivity graph between clusters to contain only such pairs, which directly touch in the image domain and are reciprocal nearest neighbors (RNN) wrt. a distance metric. The core of this approach is our novel cluster distance: it combines boundary and surface statistics both in terms of appearance as well as spatial linkage. This yields state-of-the-art performance, as we demonstrate in conclusive experiments conducted on BSDS500 and Pascal-Context datasets.

1 Introduction

One of the major challenges in computer vision is the question of semantic image partitioning. While today's cameras do record impressive numbers of pixels per image, the amount of possible segmentations and subdivisions of such pictures is even incredibly much higher. Therefore, generation of coherent parts and object candidates is a crucial step in every vision processing pipeline prior to higher level semantic interpretation.

There are several algorithmic variants how to break down the data for semantic classification and object detection: For instance, the family of plain window scanning methods has become popular in classification of certain object types [1–3]. Usually, a vast amount of candidates is sampled regularly without respect to the image content itself. Still, such methods have issues generating the proper candidates, e.g. with objects that do not fit well in rectangular shapes. A second group of candidate generating methods is based on interest point operators, which first locate prominent points before aligning known shapes with respect

© Springer International Publishing AG 2017
S.-H. Lai et al. (Eds.): ACCV 2016, Part I, LNCS 10111, pp. 3–19, 2017.
DOI: 10.1007/978-3-319-54181-5_1

to matched ones [6–8]. This usually generates less and high quality object candidates. However, it requires an object model to transform the set of localized interest points into object candidates. This is a significant drawback when dealing with a large number of different part- and object classes or priorly unknown ones to be found. Unsupervised clustering is a third line in this taxonomy of approaches and the most biologically inspired one, since it resembles the mechanism of visual grouping, which has been explored by psychologists and neurophysiologists for decades [9,10]. Such methods can provide a complete partitioning of the scene arising from the data itself. Even more, some approaches yield a hierarchy of nested segmentations, which naturally corresponds to object-part relations. In essence, these methods rely on a proper distance metric to measure differences between image parts. The general hypothesis is that a semantically meaningful entity is of consistent appearance and in contrast to its surroundings in some way. A comprehensive research about what defines this *consistency* in human perception was performed by the Gestalt school of psychology [11] and influenced many technical approaches.

The algorithm introduced in this paper belongs to the unsupervised clustering approaches and is named RaDiG (**Ra**pid **Di**stributions **G**rouping). At its core, it follows the well known agglomerative clustering paradigm, greedily merging the pair of *closest* segments in each step up to total unification, but innovating in many details. Our main contributions comprise

- an innovative, threefold distance metric incorporating boundary contrast with surface dissimilarity as well as spatial linkage,
- the texture-aware enhancement of Ward's minimum variance criterion with the meaningful and efficient Wasserstein distance in the space of normal distributions of features,
- a considerable reduction of runtime due to several algorithmic tweaks with pruning and maintaining the cluster connectivity graph as well as an economic feature processing.

Fig. 1. Exemplary results from Pascal-Context dataset [4]. From left to right per row: ultrametric contour map (UCM) and optimal image scale (OIS) segmentation of our approach RaDiG (blue), ground-truth boundaries (green), OIS and UCM of the leading approach MCG-UCM [5](red) (Color figure online).

In total, these improvements result in an expeditious algorithm which meets strict realtime requirements. Still, the quality of results is on level with the best, globally optimized and highly complex approaches (cf. Fig. 1). Therefore, we claim that it is the method of choice in mobile use cases such as robotics. This is supported by our experimental findings presented in Sects. 4 and 5.

2   Related Work

There is a huge history of research concerned with segmentation in general settings as well as more specific in computer vision. Here, we restrict our review to the most related, seminal and/or up-to-date methods, and in addition referring the reader to appropriate survey literature. Unsupervised clustering is a field at least as old as computer science itself. We recommend the textbook of Hastie, Tibshirani and Friedman [12] as a comprehensive work about algorithms and statistical background. More specialized on agglomerative hierarchical clustering and thus related to this paper is the overview written by Murtagh and Contreras [13].

Image segmentation in computer vision is a wide subject with many applications. In this paper, we narrow our interest down to methods which segment natural images into semantically meaningful entities in a fully unsupervised fashion, i.e. without any prior knowledge about the image content or objects.

There are "planar" methods, which aim to find an optimal partitioning of the input image. An example is the famous mean shift approach [14], in which pixels are assigned to the next stationary point of an underlying density in the joint domain of color and image position. Some planar methods do integrate a notion of a best segmentation across all scales of structures in the image [15,16]. Recently, the Superpixel paradigm became popular, which focuses on segmenting the basic building blocks of images [17,18]. This usually attains an oversegmentation of semantic entities, therefore some approaches add a further accumulation step of Superpixels into larger clusters [19,20].

Besides planar clustering techniques, there are methods providing a hierarchical tree (dendrogram) of nested segmentations. This additionally determined subset-relation can be useful for further processing. With such hierarchy, the scale of results is not fixed: the nodes close to the root represent a coarse segmentation, which becomes a fine oversegmentation traversing the tree to the leaves [21]. Most approaches can be categorized into either divisive or agglomerative type. Divisive ones build the hierarchy in a top-down manner: iteratively, the current partitioning is split into a finer one. For each step, one could apply a planar segmentation algorithm on each remaining cluster, until they show a uniform appearance. An advantage of divisive methods is that they can naturally exploit the statistics of an entire image/region to find an optimal splitting. Among such methods, there is the family of normalized cuts based approaches [15,22]. In contrast, agglomerative approaches merge initial clusters in a bottom-up way until complete unification. Region based approaches maintain and match feature statistics of a segment's surface, such as the mean color or a histogram of texture [23–25]. As Arbeláez [26] pointed out, a hierarchy tree is the dual representation of an ultrametric contour map (UCM), if the underlying metric holds the *strong triangle inequality*. With this insight, the way of thinking image segmentation shifted to concentrate on contour detection. This fruitful ansatz led to a family of algorithms which globalize the results of sophisticated contour detectors using spectral methods, before a greedy agglomerative clustering constructs the hierarchy based on average boundary strengths [5,27,28].

Since a planar segmentation is achievable at lower computational complexity than hierarchical approaches, for the sake of processing speed it can be advantageous to first compute an initial oversegmentation, before constructing a hierarchy on top [29–31]. In a similar way, contour based approaches benefit from an atomic oversegmentation since it can significantly accelerate the globalization of boundary strengths [28]. However, the spectral contour globalization process can be speed up alike by iterated decimation and information propagation of the affinity matrix [5].

Our algorithm RaDiG also provides a hierarchy of segments for further processing and is designed to be realtime capable. Thus, we have chosen the greedy framework of agglomerative clustering starting from a topological Watershed transform. While this basis is similar to several other approaches, ours is considerably different in its components, such as the performant processing of the image graph structure, the efficient representation of feature distributions, and the profitable combination of novel boundary contrast, region dissimilarity as well as spatial linkage within an innovative cluster-distance.

3   Rapid Distributions Grouping

The execution of our method comprises four main steps. At first, we convert the colorspace of the input image to CIE-Lab. In the following sections, we will name the luminance dimension L and the chromatic dimensions with a resp. b. Next, we compute an oversegmentation of the image (Sect. 3.1). This is used to initiate the finest layer of the hierarchical clustering (Sect. 3.2). Finally, our agglomerative approach greedily merges the pair of *closest* segments in each step

Fig. 2. RaDiG processing steps. From left to right: input image, gradient magnitude, watershed segments, initial cluster distances, final ultrametric contour map, optimal image scale (OIS) segmentation.

up to total unification (Sect. 3.3). Intermediate results of the approach are shown in Fig. 2. All gained subset relations are kept in a tree structure to be available for further steps of a processing pipeline.

3.1 Atomic Subdivision Using Watershed Transform

Since the number of clusters is doubling with each additional level of the cluster hierarchy, it can potentially save a lot of processing time to prune the finest layers from the pixel level and replace them with a planar, atomic subdivision. An established method producing such oversegmentation based on irregularities in the image function is the watershed transform. We apply a variant of the topographical watershed transform based on hill climbing [32] which features $\mathcal{O}(n)$ runtime complexity in the number of pixels and is very fast in practice. It operates on the gradient magnitude of the input image, which we determine as follows: the partial image derivatives in x and y directions are calculated by convolution with the optimized, derivative 5-tap filters of Farid and Simoncelli [33] for each layer L, a, and b. Then, our Lab gradient magnitude is made of independent luminance and chromaticity parts

$$|\nabla| = \sqrt{L_x^2 + L_y^2} + \sqrt{2(a_x^2 + a_y^2 + b_x^2 + b_y^2)}. \tag{1}$$

There is a complete field of research on edge strength calculations by itself. While recently machine learning based methods have shown to improve bounding edge detections [34,35], that sophisticated and costly quantification is not crucial or often not even beneficial when initializing an oversegmentation.

3.2 Founding of the Hierarchy Tree Structure

The watershed transform uniquely labels each pixels such that it results in a set of connected components each denoting an atomic building block of the image. In order to setup the graph based hierarchical clustering, we need to gather region adjacencies and feature statistics. Our graph structure is made of indexed sets of cluster (node) and boundary (edge) objects, each containing the information listed in Table 1. With a single scan through the image and labels, one could assign this information. The hierarchical connectivity is initialized by a constant NOT_CONNECTED. Each pixel joins its corresponding cluster's size and

Table 1. Data stored per cluster and boundary element.

cluster (node)	*hierar. connectivity*: parent, left-, and right child cluster-IDs
	planar connectivity: adj.-list of $\begin{pmatrix} \text{neighbor-ID} \\ \text{boundary-ID} \\ \text{cl.-distance} \end{pmatrix}$; near.-neighbor index
	statistics: area A in number of pixels; color distribution $\left\{ \begin{pmatrix} \mu_L \\ \mu_a \\ \mu_b \end{pmatrix}, \begin{pmatrix} \sigma_L \\ \sigma_a \\ \sigma_b \end{pmatrix} \right\}$
boundary (edge)	*hierar. connectivity*: parent boundary-ID
	statistics: boundary length l; average contrast $\bar{\delta}$

color statistics. If the label to the right or bottom is different from the current, we have to initiate/update the corresponding boundary element:

– Let i and j be the labels and w.l.o.g. $i < j$, we look for the first neighbor-ID $\geq i$ in the (sorted) adjacency list of cluster j. If we found i, we increase the border length of b_{ij} by 1, else we initiate a new boundary of length 1 and add the link ahead of the found position in j's adjacency list.
– We check the labels below/right for presence of a diagonal boundary (cf. Fig. 3). If found, we shorten the border length of b_{ij} by $\sqrt{2}/2$ in order to approximate $\mathbf{L}^2$-norm (cf. Fig. 4).

Fig. 3. The boundary segment between i and j is a diagonal one, if the labels of i with l or j with k are equal.

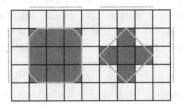

Fig. 4. Both blue and red clusters have a boundary length of 12 counting in $\mathbf{L}^1$-norm. With shortening, $\mathbf{L}^2$-norm is approximated to length 10.82 for the blue and 8.49 for the red cluster (Color figure online).

Then, we iterate the clusters once: for each neighbor in the current cluster's adjacency list, we fill in the value of our cluster-distance and insert the backlink at the end of the neighbor's adjacency list. Note that this operation retains the correct sorting of adjacency lists, because all neighbor-IDs were bigger and the clusters are processed by increasing index. Furthermore, we bookmark the nearest neighbor per cluster with respect to cluster-distances.

The runtime complexity of this tree foundation is $\mathcal{O}(n)$, since we visit each pixel only once and the individual operations are of constant complexity. Please note that the degree of a node with respect to neighborhood edges (its adjacency list size) is constant on average (< 6) as a conclusion from "Euler's planar graph formula", since only segments touching in the image domain become connected by an edge in the graph.

3.3 Agglomerative Clustering

We impose a graph structure of nodes and edges (cf. Table 1) on the image. A node represents a cluster (set of connected pixels). There are two kinds of edges between clusters: the first are planar edges, which represent a common boundary between some pair of clusters with respect to 4-connectedness of comprised pixels and when existing at the same *time* in the hierarchical tree. In this sense, a node's lifetime spans from its creation until it is merged. Thus, the timeline is made of the chrono-logical order of merge events in the hierarchy tree. The others are hierarchical edges that connect a pair of merged segments with their parent. These hierarchy edges form a binary tree with the root node representing the whole scene down to the leaves being atomic clusters resulting from the preceding watershed transform.

At each step, agglomerative clustering chooses the cluster pair of lowest dis-tance to be merged. Therefore, all current candidate pairs are maintained in a priority queue ordered by their distance. For this, we employ a Fibonacci heap implementation. Here, the DELETE operation is most costly with $\mathcal{O}(\log m)$ run-time complexity in the number of heap entries. Since every merged pair has to be deleted from the heap once, the overall complexity is dominated by $\mathcal{O}(n \log m)$. Thus, to optimize the runtime, we have to keep the number m of candidates in the heap as low as possible. From Sect. 3.2 we already know that $m < 3n$ even if we consider all planarly connected clusters. Pushing only the nearest-neighbor connected to each cluster to the heap would guarantee $m \leq n$. We can improve this further to $m \ll \frac{n}{2}$ pushing only reciprocal nearest neigh-bor (RNN) pairs to the heap, since this is a necessary condition for the pair of *globally lowest* distance. Unfortunately, a further improvement which caches nearest-neighbor-chains[1] (NNC) is not applicable, since the local connectivity violates the *reducibility* prerequisite [36].

Each time two clusters are merged, a new parent cluster is initialized with the joined statistics and adjacencies. Merging of two sorted adjacency lists into one is of linear time in the number of entries, thus amortized constant in this case. Furthermore, this way the case where a cluster was neighboring both children can be easily identified and treated in special way: the two boundaries between the cluster and both children are concatenated, in other words a new boundary object with added lengths and weighted harmonic mean of contrasts is created to replace them. The cluster-distances between the parent and its neighbors are calculated, nearest-neighbor edges identified and finally corresponding RNNs updated in the heap. When only a single active cluster is left over, it represents the whole scene and is the root of the finalized hierarchy tree.

3.4 Cluster-Distance from Boundary Contrast and Surface Dissimilarity

While handling of data structures is important for efficiency, the quality of the hierarchical segmentation is determined by the design of a proper cluster-distance. Hence, this is one of our main contributions. Our distance function $\mathcal{D}$ is threefold, comprising the following parts:

$$\begin{aligned} \mathcal{D}(P,Q) = {}& \log\left(\omega\left(P,Q\right)\right) & (surface\,dissimilarity) \\ & + \log\left(\overline{\delta}\left(P,Q\right)\right) & (boundary\,contrast) \\ & + \log\left(\eta\left(P,Q\right)\right) & (spatial\,linkage). \end{aligned} \quad (2)$$

---

[1] https://en.wikipedia.org/wiki/Nearest-neighbor_chain_algorithm.

Here, P and Q denote clusters with a common boundary. All three parts are innovative in some aspect. In the following subsections, we will explain each term in more detail. Please note that adding the logarithms is a geometric fusion of terms (equivalent to a product), hence one does not need to normalize the scales of individual parts to a common range. Nonetheless, it is suggestive to train weights for an optimized combination of parts. This could further improve results at virtually no costs, but was not yet exploited for experiments in this paper.

Surface Dissimilarity Term. Bucking the trend, our approach is very thrifty in computing different kinds of features. Indeed, we only use a normal distribution of colors to describe the appearance of a cluster (cf. Table 1). These statistics are gathered in the form of ML-estimates from the individual colors of all pixels belonging to a certain cluster. We primarily came to this decision, because the joining of Gaussian statistics is a very efficient, constant time operation [37]. The second good reason is that the Wasserstein distance between normal distributions is meaningful and fast to compute. The Wasserstein distance is a transport metric between probability distributions. It accounts how much (probability-) mass needs to be carried how far wrt. an underlying metric space. Here, the underlying space is CIE-Lab with $\mathbf{L}^1$-norm[2], which mimics human texture discrimination in a simplified way when working on top of the perceptually normalized CIE-Lab colorspace [38]. Using normal distributions, the 1st Wasserstein distance solves to the expression

$$\mathcal{W}_1(\mathcal{N}_P, \mathcal{N}_Q) = |\mu_P - \mu_Q| + \left| \operatorname{tr}\left(\sqrt{\Sigma_P}\right) - \operatorname{tr}\left(\sqrt{\Sigma_Q}\right) \right|. \tag{3}$$

Ward's clustering criterion [39] is a distance which estimates the growth in data variance when joining clusters,

$$\operatorname{Ward}(P, Q) = \frac{d(\mu_P, \mu_Q)^2}{1/A_P + 1/A_Q} \tag{4}$$

where A refers to the clusters' surface areas in pixels. If you think of this in terms of color means, the intuition is that mixing more and more colors always ends up in some grayish tone, but then subtle differences between such mashes can still make a clear difference for large surfaces. This measure is known for developing clusters of more balanced sizes. As a novelty, we propose to lift this concept from color to texture by replacing the distance between means with the 1st Wasserstein distance, which yields

$$\omega(P, Q) = \frac{\mathcal{W}_1(\mathcal{N}_P, \mathcal{N}_Q)^2}{1/A_P + 1/A_Q}. \tag{5}$$

Boundary Contrast Term. Each boundary element carries information about its average (weighted harmonic mean) boundary contrast $\bar{\delta}(P, Q)$ between the

---

[2] This is similar to the well known earth mover's distance (EMD) on histograms, which is in fact the discretized $\mathcal{W}_1$ distance.

clusters on either side. This value is kept up-to-date if two boundary elements are concatenated during a merge event (cf. Sect. 3.3). The remaining question is how it should be initialized. Instead of using the gradient magnitude or related measures on the contour at pixel resolution, we decided to plug in the Wasserstein distance between P and Q. Obviously, this is most economic since it is already computed as a part of the surface dissimilarity. But there are more good reasons for this decision: a measure incorporating the area alongside the contour-line better assesses the strength of blurry edges often emerging from motion or defocussing. Furthermore, we smartly avoid cross-talk artifacts at the crossings of edges, a problem Arbeláez et al. [5,27] explicitly addressed by a procedure called Oriented Watershed Transform (OWT). Please note that despite re-using the Wasserstein distance, the boundary contrast is essentially different to the surface dissimilarity: from the way boundaries and clusters are merged, the former is the average difference between the opposing, atomic clusters along a boundary, while the latter includes the difference between both entire surfaces.

Spatial Linkage Term. From graph topology, it is a binary decision if two segments are connected by an edge or not. However, it is reasonable to consider a certain factor of how close they actually are. We introduce the simple yet effective connectivity term

$$\eta(P,Q) = \left(\frac{\sqrt{A_P}}{l_{PQ}} \cdot \frac{\sqrt{A_Q}}{l_{PQ}} \right)^{1/2} = \frac{(A_P \cdot A_Q)^{1/4}}{l_{PQ}}, \tag{6}$$

which combines the extents of the common boundary and both surfaces. This way, the linkage between clusters is rated independently of concrete positions and thus is very flexible in shape. Relating the radical of the cluster surface to the common boundary's length is a normalized measure of connectivity, but advances versus simply using the perimeter by preferring more convex, smooth shapes versus elongated or jagged ones. We put the geometric mean of both surfaces' score, since it devaluates a cluster pair more towards the inferior result and thus better avoids residues staying alive.

4   Evaluation of Key Contributions

Since we introduced several contributions in this paper, it is reasonable to evaluate how much the overall result depends on every single innovation. Therefore, we compare different variants of our system each with one enhancement deactivated or replaced. The experiments are conducted on the trainval images of the BSDS500 benchmark [27]. This dataset provides boundary annotations of several human subjects, which where interpreted to define a segmentation.

Segmentation Quality Measure. It is not straightforward to rate the quality of a segmentation with respect to a ground truth. Pont-Tuset and Marques [40] defined the F_{op} measure in the context of generating object or part candidates for recognition. Basically, this allows a segment to be fragmented into several parts.

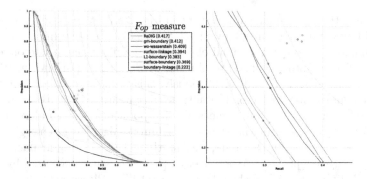

Fig. 5. Precision/recall curve of RaDiG and its reduced variants on BSDS500 trainval set using the object-and-parts F_{op} measure. The optimal dataset scale (ODS) is drawn with an asterisk on each curve, while the optimal per image scale (OIS) is drawn with a circle and numbers are given in the legend.

Regions need to exceed some relative overlaps to be accepted as valid object resp. part matches. Those matches are weighted by their fraction of overlap, thus, parts are only partially counted as true positive matches. On this basis, a precision and recall for object-and-part matches is defined. Varying a threshold on the cluster-distance of the cluster hierarchy (or UCM) produces different levels of segmentations from coarse to fine. Coarse segmentations are likely of low precision, but high recall, and vice versa comparing fine oversegmentations. The F_{op} measure is the maximally achievable value of the harmonic mean of these precision and recall pairs.

Boundary Contrast Evaluation. The RaDiG boundary contrast is initialized with the Wasserstein distance between neighboring, atomic clusters and further on averaged with respect to approximated $\mathbf{L}^2$-norm length when two parts are concatenated. To assess if our novelties really contribute to the overall results, we conducted two experiments: Fig. 5 shows the original algorithm's result (RaDiG) in comparison to a variant where we averaged the gradient magnitude along the boundary in order to initialize the boundary contrasts (gm-boundary), which is more expensive to compute but slightly inferior. We also tried $\mathbf{L}^1$-norm lengths (Fig. 5, L1-boundary) for averaging the boundary contrast, but here the little additional effort for approximation pays off well.

Appearance Distance Evaluation. Next, we assess the benefit of integrating covariance information using the Wasserstein distance. Thus, we replace it with usual squared Euclidean metric between the means, yielding the classic Ward clustering criterion and a simpler boundary contrast (Fig. 5, wo-wasserstein). We deduce that the Wasserstein distance is a worthwhile enhancement, raising the F_{op} from 0.409 to 0.417 and showing a consistently better trend on the precision/recall curves. From our observation, the difference is pronounced with larger segments, when textures are more often of non-uniform color.

Table 2. Runtime of RaDiG on a single core of an Intel Xeon X5680 @ 3.33 GHz for different resolutions.

Resolution	#pixel	Time in ms
135 × 90	12150	11
270 × 180	48600	41
320 × 240	76800	64
481 × 321	154401	107
540 × 360	194400	151
1080 × 720	777600	604

Table 3. Runtime of different components of our RaDiG approach at QVGA resolution (320 × 240).

Component	Time in ms	Time in %
Colorspace	10	16
Watershed	7	10
Founding tree	20	32
Agg. clustering	27	42
$\sum$serial	64	100
$\sum$parallel	27	42

Threefold Cluster-Distance Evaluation. Our agglomerative clustering approach efficiently maintains statistics from both the boundaries between and the surfaces of clusters, which gives rise to our sophisticated, threefold cluster-distance term (cf. Eq. (3)). In this experiment, we figure out if all three components contribute to the fine overall results, by omitting one of the parts at a time. We named the twofold cluster-distances by the remaining components: surface-linkage, surface-boundary, and boundary-linkage (cf. Fig. 5). Dropping one of the parts severely worsens the results. Particularly important is the Ward-based term of surface dissimilarity, probably because it controls the clusters' sizes not to diverge too much during agglomeration.

Runtime Evaluation. We claimed that our implementation is realtime capable and therefore the best choice for applications on mobile platforms such as autonomous robots. Table 2 shows the overall runtime of our approach for different resolutions, estimated as an average from six cluttered, natural images. Empirically, it seems as if, thanks to pruning of candidate pairs in the heap data structure (cf. Sect. 3.3), we achieved an amortized linear behavior in the number of pixels. For QVGA resolution, Table 3 breaks down how much time each component of our approach consumes. Since our implementation is based on the ROS-framework [41] and every component is implemented as a processing node using its own thread, only the slowest component restricts the framerate when processing video data. Thus we are able handle up to 37 fps in QVGA on our machine. On BSDS500 image size (481 × 321), the single thread variant of RaDiG takes on average 107 ms to run the whole algorithm. Thereby, it outperforms the approaches of Arbeláez et al. [5] by one resp. two orders of magnitudes. Testing their code of version 2.0 on the same machine, we measured runtimes of 17.6 s for MCG-UCM (165× slower) and 2.8 s for SCG-UCM (26× slower) on BSDS500 data.

5   Comparative Experiments on BSDS500 and Pascal-Context

We compared absolute performances and ranked our approach versus state-of-the-art ones on widely used datasets and appropriate measures. Pont-Tuset and Marques [40] provide a number of results from different algorithms based on their F_{op} measure. In addition, we wrote a tool to convert data from the result formats of Arbeláez et al. [5] plus Ren and Shakhnarovich [42] to update their comparison with recent approaches. We are convinced that F_{op} is the most adequate performance metric for our kind of application, since we aim to perceive semantically meaningful entities. Our method is not an edge detector in the first place, nevertheless we include the F_b measure, which quantifies how well contours are reproduced, as additional assessment. Figure 6 first row shows the precision/recall

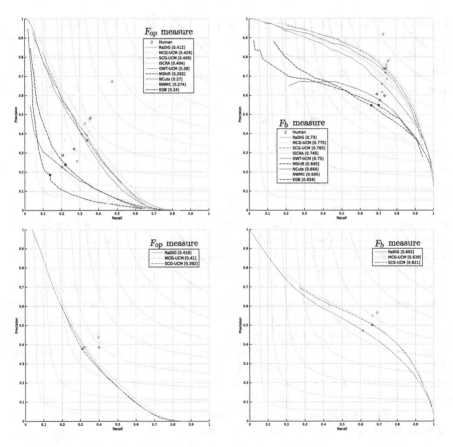

Fig. 6. Precision/recall curves of RaDiG and leading approaches using F_{op} as well as F_b measure. Top row: results on BSDS500 test images; second row: results on Pascal-Context dataset. The optimal per image scale (OIS) values (circles) are given in the legend.

curves for F_{op} and F_b measures of all tested approaches on the test images of BSDS500. There is a noticeable gap between EGB (efficient graph based method of Felzenszwalb and Huttenlocher [16]), NCUTs (normalized cuts [15]), MShift (mean shift [14]), and NWMC (binary partition tree [23]) on one hand, but the recent approaches OWT-UCM (global probability of boundary with oriented watershed transform followed by agglomerative merging [27]), ISCRA (image segmentation by cascaded region agglomeration [42]), SCG-UCM (single-scale combinatorial grouping [5]) MCG-UCM (multi-scale combinatorial grouping [5]) and our proposed RaDiG method on the other hand. A clear difference between OWT-UCM, ISCRA, SCG-UCM and MCG-UCM versus the others is that they tuned certain aspects of their algorithms using machine learning techniques. Also, they are rather costly to compute and do not target realtime applications. This said, results of RaDiG are very encouraging and our algorithm is presumably the best choice for realtime tasks in unconstrained environments, such as often present in autonomous robotics. Figure 7 shows some characteristic outcome of RaDiG on BSDS500 images.

While the BSDS500 dataset is the popularly accepted benchmark in this kind of unsupervised image segmentation, we felt that the ground truth is not entirely appropriate for evaluation of object and part candidates. The Pascal-Context dataset [4] contains pixel-precise labels for more than 400 object categories on

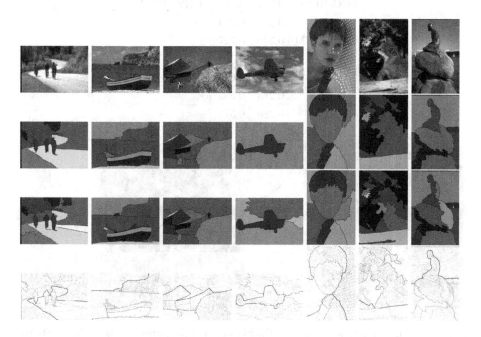

Fig. 7. Exemplary segmentations of RaDiG on BSDS500 images. From top to bottom per column: original, optimal image scale (OIS), optimal dataset scale (ODS), and ultrametric contour map (UCM) of the corresponding hierarchy.

the train/val Pascal VOC 2010 data [43] (10103 images), which is appealing also for comparison of unsupervised approaches due to this variety and size. We wrote a tool to convert this ground truth labels into segmentations by separating labels of the same category into spatially connected components, numbering them increasingly and convert them into the ground truth format of BSDS500. Fortunately, two of the best performing approaches on BSDS500, SCG-UCM and MCG-UCM, published their ultrametric contour maps online[3] for the main Pascal VOC 2012 images, which is a superset of 2010, so that we are able to compare results. All three approaches retain their very good performance on this large dataset (cf. Fig. 6, second row). However, RaDiG supersedes SCG-UCM on the precision/recall curve and the gap between MCG-UCM and RaDiG is even smaller than on BSDS500. Furthermore, looking at F_{op} from optimal image scale, RaDiG took the lead. However, RaDiG is only moderately suited for precise

Table 4. Optimal dataset scale (ODS) results for both comparative experiments.

Method	BSDS500		Pascal-C	
	F_{op}	F_b	F_{op}	F_b
RaDiG	0.363	0.688	0.351	0.533
MCG-UCM	0.379	0.744	0.356	0.575
SCG-UCM	0.352	0.737	0.342	0.571
ISCRA	0.363	0.714	–	–
OWT-UCM	0.349	0.727	–	–
MShift	0.229	0.598	–	–
NCuts	0.213	0.634	–	–
NWMC	0.215	0.552	–	–
IIDKL	0.186	0.575	–	–

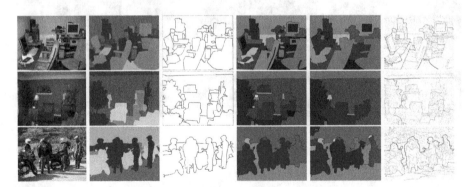

Fig. 8. Exemplary segmentations of RaDiG on Pascal-Context images. From left to right per row: original, ground truth labels, g.t. segmentation, optimal image scale (OIS), optimal dataset scale (ODS), and ultrametric contour map (UCM).

---

[3] http://www.eecs.berkeley.edu/Research/Projects/CS/vision/grouping/mcg/.

contour detection, as F_b results suggest. Table 4 reports numbers for the optimal dataset scale results of all approaches, which are also marked in the Figs. 5 and 6 by asterisks. Figure 8 shows some qualitative results of RaDiG on Pascal-Context dataset.

6   Conclusion and Future Work

We proposed RaDiG (Rapid Distributions Grouping), a fast and greedy algorithm to construct a tree of nested image segmentations starting from an atomic subdivision. Whereas its overall clustering framework is approved in awhile, we contributed several valuable findings. The feature computations and handling of the graph structure were optimized to enable a very low runtime needed for real-time applications. At the core of the approach, a novel, threefold cluster-distance was introduced and stepwise shown to advance the quality of clustering. Overall, our method plays in the same league wrt. precision/recall as current state-of-the-art approaches, but is at least an order of magnitude faster. By now, RaDiG is a parameter free approach. We are confident that involving machine learning techniques to a weighted balancing of components could further improve results. At the moment, there are several projects under development, where we deploy RaDiG on mobile robots, e.g. in the area of object manipulation.

References

1. Viola, P., Jones, M.J.: Robust real-time face detection. Int. J. Comput. Vis. **57**, 137–154 (2004)
2. Dalal, N., Triggs, B.: Histograms of oriented gradients for human detection. In: Computer Vision and Pattern Recognition (CVPR), pp. 886–893 (2005)
3. Felzenszwalb, P., Girshick, R., McAllester, D., Ramanan, D.: Object detection with discriminatively trained part-based models. Trans. Pattern Anal. Mach. Intell. **32**, 1627–1645 (2010)
4. Mottaghi, R., Chen, X., Liu, X., Cho, N.G., Lee, S.W., Fidler, S., Urtasun, R., et al.: The role of context for object detection and semantic segmentation in the wild. In: 2014 IEEE Conference on Computer Vision and Pattern Recognition (CVPR), pp. 891–898. IEEE (2014)
5. Arbelaez, P., Pont-Tuset, J., Barron, J., Marques, F., Malik, J.: Multiscale combinatorial grouping. In: 2014 IEEE Conference on Computer Vision and Pattern Recognition (CVPR), pp. 328–335. IEEE (2014)
6. Lowe, D.G.: Object recognition from local scale-invariant features. In: International Conference on Computer Vision (ICCV), pp. 1150–1157 (1999)
7. Sivic, J., Zisserman, A.: Video Google: a text retrieval approach to object matching in videos. In: International Conference on Computer Vision (ICCV), pp. 1470–1477 (2003)
8. Leibe, B., Leonardis, A., Schiele, B.: Combined object categorization and segmentation with an implicit shape model. In: Workshop on Statistical Learning in Computer Vision, ECCV, p. 7 (2004)
9. Bruce Goldstein, E.: Perceiving objects and scenes. In: Sensation and Perception, 8th edn., pp. 99–130. Wadsworth Cengage Learning, Belmont, USA (2009). ISBN-13: 978-0-495-60149-4

10. Wagemans, J., Elder, J.H., Kubovy, M., Palmer, S.E., Peterson, M.A., Singh, M., von der Heydt, R.: A century of gestalt psychology in visual perception I. perceptual grouping and figure-ground organization. Psychol. Bull. **138**, 1172–1217 (2012)
11. Wertheimer, M., Spillmann, L., Wertheimer, M.: On Perceived Motion and Figural Organization. MIT Press, Cambridge (2012)
12. Hastie, T., Tibshirani, R., Friedman, J.: The Elements of Statistical Learning. Springer Series in Statistics. Springer, Heidelberg (2009)
13. Murtagh, F., Contreras, P.: Algorithms for hierarchical clustering: an overview. Interdisc. Rev. Data Min. Knowl. Disc. **2**, 86–97 (2012)
14. Comaniciu, D., Meer, P.: Mean shift: a robust approach toward feature space analysis. IEEE Trans. Pattern Anal. Mach. Intell. **24**, 603–619 (2002)
15. Cour, T., Benezit, F., Shi, J.: Spectral segmentation with multiscale graph decomposition. In: Computer Vision and Pattern Recognition (CVPR), vol. 2, pp. 1124–1131 (2005)
16. Felzenszwalb, P.F., Huttenlocher, D.P.: Efficient graph-based image segmentation. Int. J. Comput. Vis. **59**, 167–181 (2004)
17. Achanta, R., Shaji, A., Smith, K., Lucchi, A., Fua, P., Susstrunk, S.: SLIC superpixels compared to state-of-the-art superpixel methods. IEEE Trans. Pattern Anal. Mach. Intell. **34**, 2274–2282 (2012)
18. Van den Bergh, M., Boix, X., Roig, G., de Capitani, B., Van Gool, L.: SEEDS: superpixels extracted via energy-driven sampling. In: Fitzgibbon, A., Lazebnik, S., Perona, P., Sato, Y., Schmid, C. (eds.) ECCV 2012. LNCS, vol. 7578, pp. 13–26. Springer, Heidelberg (2012). doi:10.1007/978-3-642-33786-4_2
19. Kovesi, P.: Image segmentation using SLIC superpixels and DBSCAN clustering (2013). http://www.peterkovesi.com/projects/segmentation
20. Zhou, B.: Image segmentation using SLIC superpixels and affinity propagation clustering. Int. J. Sci. Res. **4**(4), 1525–1529 (2015)
21. Peng, B., Zhang, L., Zhang, D.: A survey of graph theoretical approaches to image segmentation. Pattern Recogn. **46**, 1020–1038 (2013)
22. Shi, J., Malik, J.: Normalized cuts and image segmentation. IEEE Trans. Pattern Anal. Mach. Intell. **22**, 888–905 (2000)
23. Vilaplana, V., Marques, F., Salembier, P.: Binary partition trees for object detection. IEEE Trans. Image Process. **17**, 2201–2216 (2008)
24. Calderero, F., Marques, F.: Region merging techniques using information theory statistical measures. IEEE Trans. Image Process. **19**, 1567–1586 (2010)
25. Alpert, S., Galun, M., Brandt, A., Basri, R.: Image segmentation by probabilistic bottom-up aggregation and cue integration. IEEE Trans. Pattern Anal. Mach. Intell. **34**, 315–327 (2012)
26. Arbelaez, P.: Boundary extraction in natural images using ultrametric contour maps. In: 2006 Conference on Computer Vision and Pattern Recognition Workshop (CVPRW 2006), p. 182 (2006)
27. Arbelaez, P., Maire, M., Fowlkes, C., Malik, J.: Contour detection and hierarchical image segmentation. IEEE Trans. Pattern Anal. Mach. Intell. **33**, 898–916 (2011)
28. Taylor, C.J.: Towards fast and accurate segmentation. In: 2013 IEEE Conference on Computer Vision and Pattern Recognition (CVPR), pp. 1916–1922 (2013)
29. Haris, K., Efstratiadis, S.N., Maglaveras, N., Katsaggelos, A.K.: Hybrid image segmentation using watersheds and fast region merging. IEEE Trans. Image Process. **7**, 1684–1699 (1998)

30. Marcotegui, B., Beucher, S.: Fast implementation of waterfall based on graphs. In: Ronse, C., Najman, L., Decencière, E. (eds.) Mathematical Morphology: 40 Years On. Computational Imaging and Vision, vol. 30, pp. 177–186. Springer, Heidelberg (2005)

31. Jain, V., Turaga, S.C., Briggman, K., Helmstaedter, M.N., Denk, W., Seung, H.S.: Learning to agglomerate superpixel hierarchies. In: Shawe-Taylor, J., Zemel, R.S., Bartlett, P.L., Pereira, F., Weinberger, K.Q. (eds.) Advances in Neural Information Processing Systems 24, pp. 648–656. Curran Associates, Inc., New York (2011)

32. Roerdink, J.B., Meijster, A.: The watershed transform: definitions, algorithms and parallelization strategies. Fundamenta Informaticae **41**, 187–228 (2001)

33. Farid, H., Simoncelli, E.P.: Differentiation of discrete multidimensional signals. IEEE Trans. Image Process. **13**, 496–508 (2004)

34. Xie, S., Tu, Z.: Holistically-nested edge detection. In: International Conference on Computer Vision (ICCV), pp. 1395–1403 (2015)

35. Li, Y., Paluri, M., Rehg, J.M., Dollár, P.: Unsupervised learning of edges. In: International Conference on Computer Vision and Pattern Recognition (CVPR) (2016)

36. Bruynooghe, M.: Methodes nouvelles en classification automatique de donnees taxinomiqes nombreuses. Statistique et Anal. des Donnes **3**, 24–42 (1977)

37. Chan, T.F., Golub, G.H., LeVeque, R.J.: Algorithms for computing the sample variance: analysis and recommendations. Am. Stat. **37**, 242–247 (1983)

38. Rubner, Y., Tomasi, C., Guibas, L.J.: The earth mover's distance as a metric for image retrieval. Int. J. Comput. Vis. **40**, 99–121 (2000)

39. Ward Jr., J.H.: Hierarchical grouping to optimize an objective function. J. Am. Stat. Assoc. **58**, 236–244 (1963)

40. Pont-Tuset, J., Marques, F.: Measures and meta-measures for the supervised evaluation of image segmentation. In: Computer Vision and Pattern Recognition (CVPR), pp. 2131–2138. IEEE (2013)

41. Robotics Foundation: ROS - Robot Operating System (2016). http://www.ros.org

42. Ren, Z., Shakhnarovich, G.: Image segmentation by cascaded region agglomeration. In: 2013 IEEE Conference on Computer Vision and Pattern Recognition (CVPR), pp. 2011–2018. IEEE (2013)

43. Everingham, M., Van Gool, L., Williams, C.K.I., Winn, J., Zisserman, A.: The PASCAL Visual Object Classes Challenge 2010 (VOC2010) Results (2010). http://host.robots.ox.ac.uk/pascal/VOC/voc2010/workshop/index.html

Weakly-Supervised Video Scene Co-parsing

Guangyu Zhong[1,2]($\boxtimes$), Yi-Hsuan Tsai[1], and Ming-Hsuan Yang[1]

[1] UC Merced, Merced, USA
{gzhong,ytsai2,mhyang}@ucmerced.edu
[2] Dalian University of Technology, Dalian, China

Abstract. In this paper, we propose a scene co-parsing framework to assign pixel-wise semantic labels in weakly-labeled videos, i.e., only video-level category labels are given. To exploit rich semantic information, we first collect all videos that share the same video-level labels and segment them into supervoxels. We then select representative supervoxels for each category via a supervoxel ranking process. This ranking problem is formulated with a submodular objective function and a scene-object classifier is incorporated to distinguish scenes and objects. To assign each supervoxel a semantic label, we match each supervoxel to these selected representatives in the feature domain. Each supervoxel is then associated with a series of category potentials and assigned to a semantic label with the maximum one. The proposed co-parsing framework extends scene parsing from single images to videos and exploits mutual information among a video collection. Experimental results on the Wild-8 and SUNY-24 datasets show that the proposed algorithm performs favorably against the state-of-the-art approaches.

1 Introduction

Scene parsing, the task to assign labels for every pixel in images or videos [1–3], has attracted much attention in recent years. Many applications such as 3D layout estimation [4] and auto-driving [5] benefit from the results of scene parsing. However, existing scene parsing methods typically require large scale pixel-level annotated training images with fixed semantic categories and fully supervised [6] or retrieval-based [3,7] methods. Due to the restriction of category numbers and labor-intensive pixel-level annotations, it is not easy to directly apply these methods for videos with complex and dynamic scenes.

To relax the dependence on pixel-level annotations, several weakly-supervised methods [8–10] for video object segmentation have been proposed, in which only video-level semantic labels are given for each video. In these methods, segment-based classifiers are learned to distinguish objects from the background by extracting information with relevant and irrelevant frames from videos [8,9]. Object

---

Both authors contribute equally to this work.

Electronic supplementary material The online version of this chapter (doi:10.1007/978-3-319-54181-5_2) contains supplementary material, which is available to authorized users.

© Springer International Publishing AG 2017
S.-H. Lai et al. (Eds.): ACCV 2016, Part I, LNCS 10111, pp. 20–36, 2017.
DOI: 10.1007/978-3-319-54181-5_2

detectors are also used to help locate potential objects with specific semantics [10]. However, segment-based classifiers are suspecting to ambiguous training instances, and object detectors are not effective for parsing both scenes and objects in videos.

To address the above-mentioned issues, we extend scene parsing from single images to videos and propose a co-parsing framework to assign semantic labels in weakly-labeled videos. The proposed weakly-supervised method relaxes the constraints of large-scale annotations with fixed category numbers, while the co-parsing framework exploits information among a video collection to alleviate the ambiguity of individual segments. Considering the temporal consistency, we first segment each video into supervoxels [11,12]. Here, each supervoxel belongs to one or more parts of objects or scenes, which are quite different in terms of the contents (e.g., usually skies are smooth and objects are textured). Hence we develop a scene-object classifier to understand the contents of each supervoxel. Compared to previous methods using segment-based classifiers [8,13,14], our approach aims to learn a generalized scene-object classifier without the need to know specific semantic categories.

Since using the information within only one video is limited, we develop a co-parsing method to include videos that share the same labels. For each semantic category, we first collect all the supervoxels in videos with such label. We then select representative supervoxels through a submodular optimization problem guided by the scene-object classifier. These representative supervoxels selected in each category are further utilized in a matching process, in which we assign each supervoxel a potential to be a specific category by considering the similarities between the supervoxel and representative ones. Finally, a category is assigned to each supervoxel according to the maximum potential obtained from the matching process.

We demonstrate the effectiveness of the proposed weakly-supervised video co-parsing algorithm on the Wild-8 [13] and SUNY-24 [15] benchmark datasets. We first show the effectiveness of the proposed scene-object classifier incorporated in the submodular function for scene labeling. In addition, we extend the scene-object classifier to a detailed scene classifier with multiple categories, and show that the performance can be further improved. Overall, our experimental results show that the proposed algorithm performs favorably against the state-of-the-art methods in terms of visually quality and accuracy.

The main contributions of the proposed algorithm are summarized as follows. First, we propose a scene co-parsing framework for weakly-labeled videos, in which relations of supervoxels between different videos are exploited. Second, we formulate a submodular objective function to select representative supervoxels in semantics guided by a scene-object classifier from a collection of videos. Third, we propose an effective matching process to re-rank supervoxels in semantics and obtain final semantic labels in videos.

2   Related Work

Video Object Segmentation/Co-segmentation. In general, video object segmentation methods aim to detect and extract one or more dominant objects from a number of categories in image sequences [16–22]. These approaches achieve state-of-the-art performance by tracking segments [17], exploiting motion cues [18,23], propagating labels [19] or using spatial-temporal graph-based models [22]. However, these methods are not exploited to extract common objects from a video collection. Several video object co-segmentation methods have been proposed to separate common foreground objects from the background [24–29]. By analyzing the coherent motion and similar appearance in different videos, foreground objects from a collection of videos can be identified. However, these approaches usually assume that common objects appear in all the input videos, which is rarely true in real-world scenarios. In addition, these methods have limited capability in separating scenes in videos due to less motion information and large appearance variations of scenes. In contrast, the proposed method is able to parse scenes and objects with large appearance changes.

Object Segmentation in Weakly-labeled Videos. Weakly-supervised methods have recently been proposed for multi-class video segmentation [8,10,13,14]. Given the videos with video-level category labels, several learning-based approaches [8,14] use a large set of training samples to learn segment-based classifiers to distinguish foreground objects. To minimize the effect of ambiguous instances in the learning-based methods, pre-trained object detectors are incorporated to help locate objects [10]. However, object detectors can only locate instances from a number of known categories, rather than extract scenes in videos. Liu et al. [13] extend previous methods to a more challenging multi-class setting including objects and scenes. Based on the assumption that common objects in multiple videos should be similar in appearance, this method transfers video-level labels to each supervoxel via a nearest neighbor-based scheme. In contrast, the proposed algorithm introduces a more general and robust scene-object classifier without the need of training for specific categories, and considers the relations between different videos to facilitate the co-parsing task.

Image/Video Scene Parsing. Numerous approaches have been proposed for scene parsing [1,3,7,30–33]. These methods address the problem in single images via dense scene alignment [3], superpixel retrieval [7], neural networks [1,30] or context information [32]. However, directly applying single image parsing approaches to each video frame does not exploit temporal information and performs poorly in complex and dynamic scenes. Liu et al. [2] construct a conditional random field model to extract spatial-temporal information for video scene parsing, in which dense connections on supervoxel level and sparse object-level potentials are used for labeling. However, this method performs on single videos and requires manually pixel-wise labeled exemplars for initialization and propagation. In contrast, the proposed algorithm focuses on the co-parsing task from a video collection and requires only video-level labels. Recently, Chen et al. [34] propose a co-labeling task to parse scenes in multiple images, but without considering temporal connections. In addition, this method requires pixel-wise training samples and has a limitation that the training and test images should contain similar scenes. In contrast, the proposed algorithm exploits temporal consistency via supervoxels in a weakly-supervised fashion from a video collection.

3   Proposed Algorithm

3.1   Overview

Given a collection of videos with video-level labels, we aim to assign a semantic label to each pixel in image sequences. We formulate the labeling problem as a co-parsing task by simultaneously considering a video collection. To this end, our approach consists of two stages: (1) semantic supervoxel ranking via a submodular function: In this stage, we aim to discover representative semantic supervoxels for each category. We first segment weakly-labeled videos into supervoxels that maintain spatial-temporal consistency. For each semantic category, we construct a graph to connect the supervoxels collected from videos labeled with such category. To model the relations between supervoxels, we formulate it as a submodular optimization problem guided by a scene-object classifier based on the appearances and semantic information. The most representative supervoxels for each category are then extracted by solving this proposed submodular function; (2) scene co-parsing via region-based matching: In this stage, we aim to assign each supervoxel a semantic label by computing its category potentials. For each supervoxel, we compute similarities between it and representative ones for each category as its corresponding category potential. Each supervoxel is then assigned to a semantic label according to its maximum potential. Figure 1 shows the main steps of the proposed algorithm.

3.2   Supervoxel Ranking via Submodular Function

Weakly-supervised video segmentation methods [2,13] usually transfer semantic labels to nearby regions spatially [13] or temporally [2] based on appearance features. However, globally searching the neighbors from all videos is likely to introduce redundant information and cause ambiguity when the videos share multiple semantic labels. In contrast, we start from each semantic label and aim to select representatives for each category. For each semantic category, we collect all the videos that share the same label and segment them into supervoxels. We then construct a graph where supervoxels are considered as nodes. We formulate a submodular optimization problem to select nodes that can represent each semantics.

Graph Construction. Given a collection of weakly-labeled videos V, we denote the full semantic label set as $\mathcal{L} = \{1, 2, \ldots, L\}$. For each category $l \in \mathcal{L}$, we collect videos containing l and segment them into supervoxels, which are denoted as $\mathcal{O}$. We construct a graph $G = (\mathcal{V}, \mathcal{E})$, in which each element $v \in \mathcal{V}$ is a supervoxel from $\mathcal{O}$ and the edge $e \in \mathcal{E}$ represents the pairwise relation between two supervoxels. To exploit the supervoxels that best represent a target category, we aim to select a subset $\mathcal{A}$ from $\mathcal{O}$.

Submodular Function. We model the supervoxel selection task as a facility location problem which can be solved by submodularity [35,36]. We design the submodular objective function to find representative supervoxels that meet two

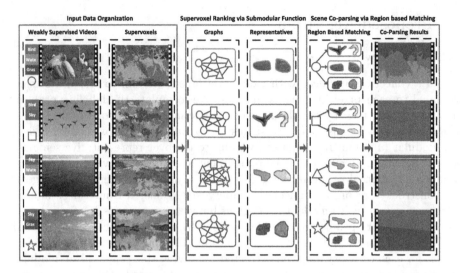

Fig. 1. Overview of the proposed algorithm. Given a collection of weakly-labeled videos, we aim to assign a semantic label to each pixel in image sequences. First, we segment each video into supervoxels. The supervoxels are illustrated by different patterns (e.g., circle represents all the supervoxels in the first video and rectangle represents the ones in the second video). We collect the supervoxels in videos that share the same semantic category and construct a graph. Each category is associated with a unique color (e.g., dark blue represents sky and green represents grass). We then formulate a submodular optimization problem to discover representative supervoxels for each category. Next, we match each supervoxel to the corresponding representatives and compute their similarities as the category potentials. Finally, a category is assigned to each supervoxel according to the maximum potential calculated during the matching process (Color figure online).

criteria: (1) sharing high mutual similarities; (2) maintaining high probability to match the target category. To this end, we formulate the objective function with two terms, i.e., a facility-location term to show similarities among all the elements [24,37] and a semantic sensitive term to represent the potential of each element belongs to the target category. The formulation of facility-location (FL) term is defined by:

$$\mathcal{F}(\mathcal{A}) = \frac{1}{N_\mathcal{V}} \sum_{i \in \mathcal{A}} \sum_{j \in \mathcal{V}} w_{ij} - \sum_{i \in \mathcal{A}} \phi_i, \tag{1}$$

where ω_{ij} is the pairwise relation between a potential facility v_i and an element v_j. The cost of opening a facility is defined as ϕ_i and fixed to a constant σ (i.e., 1 in this work).

We represent the supervoxel v_i by a hierarchical convolutional neural network (CNN) feature vector f_i. For each supervoxel, we first extract CNN features in each frame. The CNN features are computed by combining the first three convolutional layers [6] (i.e., 448-dimensional features). We then apply an average pooling method on all the frames and generate a feature vector for each super-pixel. As f_i is extracted from hierarchical CNN layers which represent both visually fine-gained details and semantic information, features that share similar appearance and semantics should have higher mutual similarities.

To meet the first criteria, we define the pairwise relations ω_{ij} as the similarity between facilities and elements in the feature domain. This strategy encourages to select the node that well presents or is similar to its group elements so that the selected facilities in $\mathcal{A}$ are representative. We define the weight ω_{ij} of each edge e_{ij} in (1) as:

$$\omega_{ij} = S(v_i, v_j), \tag{2}$$

where $S(v_i, v_j)$ is the inner product of the features f_i and f_j, i.e., $\langle f_i, f_j \rangle$. The second term ϕ_i with a constant in (1) is to penalize excessive facilities. With the growth of $\mathcal{A}$, the cost of opening facilities becomes higher and thus it avoids selecting all the nodes.

However, videos within the same category usually contain various objects and scenes, and hence introduce large appearance variations and ambiguities between different categories. Therefore, it is not sufficient to use only facility-location term to discriminate different semantic information. To this end, we propose a unary term to represent the category sensitivity of each supervoxel. We define the proposed semantic sensitive (SS) term by:

$$\mathcal{U}(A) = \sum_{i \in \mathcal{A}} \psi_i, \tag{3}$$

where ψ_i denotes the potential of supervoxel v_i belonging to the target cate-gory. In the proposed algorithm, we apply classifiers to estimate these potentials. Considering large variations of semantic labels, learning classifiers for each cate-gory as [37] is time-consuming and labor-intensive. Hence we learn a generalized scene-object classifier based on the fully convolutional network (FCN) [6]. For each supervoxel with unknown category, we predict its category probabilities from the FCN output layer, i.e., scene and object. Then ψ_i for each supervoxel is computed in a way similar to the feature generation step, where the probability is first extracted from each frame and then averaged through all the frames.

Optimization for Supervoxel Ranking. To ensure that the selected facility set $\mathcal{A}$ shares more similarities with group elements and maintains high semantic sensitivities, we formulate the proposed submodular objective function of both facility-location term $\mathcal{F}(\mathcal{A})$ of (1) and semantic sensitive term $\mathcal{U}(\mathcal{A})$ of (3) by:

$$\max_{\mathcal{A}} \mathcal{C}(\mathcal{A}) = \max_{\mathcal{A}} \mathcal{F}(\mathcal{A}) + \lambda \mathcal{U}(\mathcal{A}),$$
$$\text{s.t. } \mathcal{A} \subseteq \mathcal{O} \subseteq \mathcal{V}, \mathcal{N}_\mathcal{A} \leq \mathcal{N},$$
$$\mathcal{J}(\mathcal{A}^i) \geq 0, \tag{4}$$

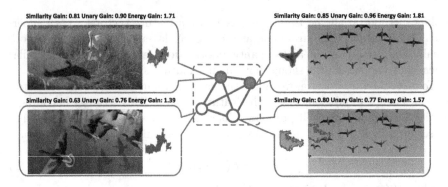

Fig. 2. Illustration of the proposed submodular function for selecting representatives. For the bird category, we show four supervoxels collected from three videos. The top two supervoxels are selected as the representatives as birds (denoted as circles with solid brown color). For each supervoxel, we show energy gain, similarity gain (FL term) and unary gain (SS term). As the video containing birds are usually accompanied with large sky or water regions, the supervoxels (e.g., the bottom two) in the scenes are likely to be similar to other regions and have high similarity gains. Owning to our scene-object classifier, the supervoxels containing the scenes (e.g., sky and water) are associated with lower unary gain in the object category (e.g., bird), and hence provide lower energy gains (Color figure online).

where $\mathcal{N}_\mathcal{A}$ is the number of selected facilities in $\mathcal{A}$, and $\mathcal{N}$ is the maximum number of $\mathcal{A}$. We set the energy gain $\mathcal{J}(\mathcal{A}^i)$ at the i-th iteration during the optimization as: $\mathcal{C}(\mathcal{A}^i) - \mathcal{C}(\mathcal{A}^{i-1})$. In addition, λ is the parameter to balance the contribution of two terms.

As the proposed objective function in (4) is the non-negative linear combination of two submodular terms, we can maximize $\mathcal{C}(\mathcal{A})$ via a greedy algorithm similar to [24]. We first initialize the facility set $\mathcal{A}$ as an empty set $\emptyset$. Then the element $a \in \mathcal{V} \setminus \mathcal{A}$ which leads to the maximum energy gain is added into $\mathcal{A}$. We iteratively select other elements and this absorbing process stops when either one of the following conditions is satisfied: (1) the maximum facility number is reached, i.e., $\mathcal{N}_\mathcal{A} > \mathcal{N}$; (2) the cost of opening facilities is larger than the gain from elements, i.e., $\mathcal{J}(\mathcal{A}^i) < 0$. In addition, due to the submodularity of the objective function, the optimization process can be sped up by an evaluation form as proposed in [38]. The process of selecting representatives for each category is presented in Algorithm 1 and the effectiveness of our submodular function is shown in Fig. 2.

3.3 Scene Co-Parsing via Region-Based Matching

Next, we aim to assign each supervoxel a semantic label by considering its relations to representatives in each category. Previous approaches generally formulate the labeling task as markov random field (MRF) [13] or conditional random field (CRF) [2] models that require additional optimization process to estimate

Algorithm 1. Representatives Selection for Each Category

Input: $G = (\mathcal{V}, \mathcal{E}), \mathcal{N}, \lambda$
Initialization: $\mathcal{A}^0 \leftarrow \emptyset,\ \mathcal{O}^0 \leftarrow \mathcal{V},\ i \leftarrow 1$
loop
    $a^* = \underset{\{\mathcal{A}^i \in \mathcal{V}\}}{\arg\max}\ \mathcal{J}(\mathcal{A}^i)$, where $\mathcal{A}^i = \mathcal{A}^{i-1} \cup a$
    if $\mathcal{N}_A > \mathcal{N}$ or $\mathcal{J}(\mathcal{A}^i) < 0$ **then**
      break
    end if
    $\mathcal{A}^i \leftarrow \mathcal{A}^{i-1} \cup a^*,\ \mathcal{O}^i \leftarrow \mathcal{O}^{i-1} \setminus a^*$
    $i = i + 1$
end loop
Output: $\mathcal{A} \leftarrow \mathcal{A}^i,\ \mathcal{O} \leftarrow \mathcal{O}^i$

---

category posteriors. In contrast, we propose an efficient way and predict the potential of each supervoxel by matching them to category representatives. We then assign each supervoxel a semantic label according to the maximum potential computed during the matching process.

Region-based Matching. The energy gain in the proposed submodular function can be utilized to estimate how likely each supervoxel belonging to which category as proposed in [39]. However, generating energy gains for all the elements in each category is ineffective. In addition, different submodular functions for various categories may differ in the graph size, element appearance and semantic distribution, which causes incomparable results when comparing energy gains between different categories. In this work, we propose to compute the category potentials for each supervoxel by a matching process, which is efficient and can reduce confusions between categories (see Fig. 3).

For the input video collection V, we denote all the supervoxels as $B = \{b_1, \ldots, b_M\}$, where M is the number of supervoxels. The corresponding video-level labels of B are denoted as $Y = \{y_1, \ldots, y_M\}$, where y_i is a semantic label set according to the video that b_i belongs to. For instance, y_i is identical to y_j if they belong to the same video. For each supervoxel b_i, we aim to compute its category potentials with respect to all the semantic labels $\mathcal{L}$. Hence we generate a M by L matrix P, where each element p_i^l denotes the potential of a supervoxel b_i belonging to a category l. To ensure that potentials between different categories are comparable, we compute P in the feature domain. For each category l, the p_i^l is computed as the average similarity between the feature f_i of supervoxel b_i and feature f_j^l of category representatives a_j^l in $\mathcal{A}_l$, where $\mathcal{A}_l$ is the corresponding facility set:

$$p_i^l = \begin{cases} \dfrac{1}{\mathcal{N}_l} \displaystyle\sum_{a_j^l \in \mathcal{A}_l} \langle f_i, f_j^l \rangle & l \in y_i, \\[4mm] -\infty & l \notin y_i, \end{cases} \qquad (5)$$

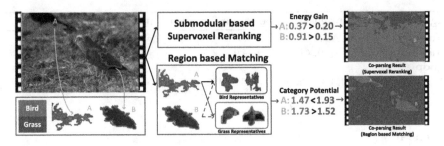

Fig. 3. Illustration of the proposed region-based matching process. Given a video with weakly-supervised labels (e.g., bird and grass), we first extract its supervoxels (e.g., A and B with red boundaries). We then compare them with representative supervoxels for each category and compute category potentials. By comparing these potentials, we can further assign a label to each supervoxel corresponding to the category with the maximum one (bottom row). In contrast, using the energy gain for comparing scores and assigning labels may produce wrong results, due to incomparable gains in different categories (upper row). Note that categories of the potential scores are associated with different colors (Color figure online).

where $\mathcal{N}_l$ denotes the number of representatives in $\mathcal{A}_l$. Note that if the video does not contain the target category l, p_i^l is assigned as the value $-\infty$, meaning that supervoxels in this video do not share any similarities with category l.

Scene Label Assignment. After matching supervoxels to representatives in each category, we assign a semantic label for each supervoxel b_i based on its category potential vector $P_i = [p_i^1, \ldots, p_i^L]$ as:

$$c_i = \underset{l \in \mathcal{L}}{\operatorname{argmax}} \ P_i(l). \tag{6}$$

Hence all the pixels within a supervoxel have the same assigned label c_i. The proposed region-based matching strategy to assign labels is illustrated in Fig. 3. Note that if there are existing multiple semantic labels with the same potential, we use their submodular energy gains for further comparisons. In addition, all the video-level labels in each video should be at least assigned to one of the supervoxels. For those labels that are not assigned to any supervoxel in the video, we ensure that at least top K (i.e., 15) supervoxels with high category potentials are re-assigned to that label.

4   Experimental Results

We evaluate the proposed scene co-parsing algorithm on the Wild-8 [13] and SUNY-24 [15] datasets with comparisons to the state-of-the-art methods, including CRANE [8], MIN [40], SVM [14], MIL [41] and WILD [13]. We use the same metrics in [13] for evaluation including average accuracy of each class, average per category accuracy (aveAcc), and mean average precision (mAP). More experimental results can be found in the supplementary material and the MATLAB codes will be made available to the public.

4.1   Experimental Settings

We use the streamGBH algorithm [12] at the fifteenth level with the default parameters to generate supervoxels in videos, For optimizing the submodular function in (4), the maximum number $\mathcal{N}$ of each category is set to 10, and the parameter λ is set to 1. For matching representatives in (5), we use the top-5 ranked supervoxels as $\mathcal{A}_l$.

To learn the scene-object classifier, we finetune a fully convolutional network (FCN) [6], which is a state-of-the-art algorithm for semantic segmentation. We follow the same setting used in [42] for collecting training images in the LMSun dataset [7], but merging all the categories into scene and object categories. The parameters are fixed in all the experiments.

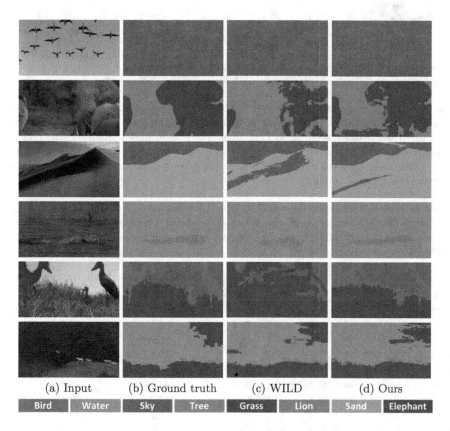

(a) Input   (b) Ground truth   (c) WILD   (d) Ours

| Bird | Water | Sky | Tree | Grass | Lion | Sand | Elephant |

Fig. 4. Sample results of the proposed method (with the generalized scene-object classifier) and the WILD [13] method on the Wild-8 dataset. The results show that the proposed algorithm generates more complete and accurate results than WILD, especially on objects (lion and elephant). Each color indicates a semantic label and the legend is shown on the bottom (Color figure online).

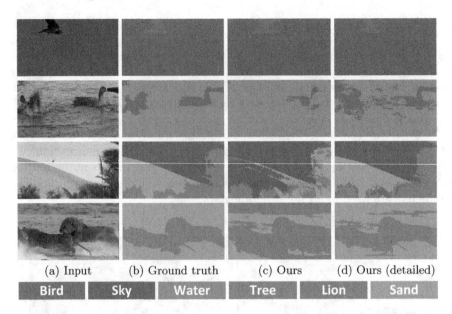

(a) Input   (b) Ground truth   (c) Ours   (d) Ours (detailed)

| Bird | Sky | Water | Tree | Lion | Sand |

Fig. 5. Sample results of the proposed method with different classifiers on the Wild-8 dataset. The results in (c) and (d) are generated by the proposed method with the generalized and detailed scene-object classifier. The results show that the detailed scene classifier further improves the performance on both object (e.g., bird) and scene (e.g., sand) categories. Each color indicates a semantic label and the legend is shown on the bottom (Color figure online).

Table 1. Video scene co-parsing results on the Wild-8 dataset. We measure the average accuracy of each class, average per category accuracy (aveAcc) and mean average precision (mAP). The highest score is marked in bold and the second highest score is marked with underlines.

Category	MIL	SVM	MIN	CRANE	WILD	Ours (similarity)	Ours (generalized)	Ours (detailed)
bird	31.5	42.5	48.1	47.8	<u>53.0</u>	35.8	41.5	**66.2**
water	<u>79.3</u>	74.5	75.2	76.5	77.3	60.5	72.5	**82.6**
sky	85.4	86.9	87.2	89.5	93.8	78.3	<u>96.0</u>	**98.2**
tree	41.1	45.5	36.7	42.8	50.1	<u>93.0</u>	<u>93.0</u>	**95.5**
grass	<u>78.3</u>	74.0	74.1	73.7	76.5	**81.0**	72.4	68.3
lion	2.1	16.6	15.4	19.3	21.3	47.1	**95.8**	<u>91.5</u>
sand	55.2	42.1	43.3	43.2	<u>60.1</u>	35.0	44.7	**75.3**
elephant	5.5	12.3	13.2	16.8	28.1	51.8	**87.1**	<u>73.4</u>
aveAcc	47.3	49.3	49.2	51.2	57.5	60.3	<u>75.4</u>	81.4
mAP	41.8	41.0	42.1	43.9	52.4	59.6	<u>68.6</u>	**78.6**

4.2   Wild-8 Dataset

The Wild-8 dataset consists of 100 weakly-labeled videos, and 33 of them are with pixel-level annotations. It contains 8 categories including scenes (sky, tree,

Table 2. Comparisons of using energy gains and region-based matching process for assigning final semantic labels on the Wild-8 dataset. We show two sets of results using different classifiers and measure the average per category accuracy (aveAcc) and mean average precision (mAP). The results consistently show the effectiveness of the proposed matching strategy. The highest scores are marked in bold.

Indicator	Energy gain (generalized)	Matching (generalized)	Energy gain (detailed)	Matching (detailed)
aveAcc	70.2	**75.4**	78.0	**81.4**
mAP	64.0	**68.6**	75.3	**78.6**

grass, sand and water) and objects (bird, lion and elephant). Each video is associated with multiple video-level labels and contains 30 frames with 640 × 480 resolution.

We first evaluate the contribution of the scene-object classifier in the submodular function. Table 1 shows that the proposed algorithm with the generalized scene-object classifier (second column from the right) significantly improves the performance of the state-of-the-art methods (e.g., more than 15% gain in terms of aveAcc and mAP). In addition, the scene-object classifier improves results in most categories compared with only using the similarity term (third column from the right).

Overall, the proposed algorithm performs favorably on both object (e.g., lion and elephant) and scene (e.g., sky and tree) categories. The MIL method [40] achieves high accuracy in categories of water and grass since it uses the max-margin strategy, which contributes more to categories with larger regions while usually ignoring small objects (e.g., lion and elephant). The WILD scheme [13] performs well in the bird and sand categories. However, when objects and scenes have similar appearances, the smoothness assumption used in [13] may introduce ambiguity and thus it leads to low accuracy (e.g., lion, elephant and tree). In contrast, the proposed algorithm not only considers similarities between supervoxels but also utilizes a scene-object classifier, which is able to guide the submodular function for separating scenes and objects. Figure 4 shows some results generated by the WILD method [13] and proposed algorithm with the generalized scene-object classifier. As the codes and results of the WILD method [13] are not available, we use the reported results from the original paper for illustration. The parsing results of the proposed algorithm are more complete and accurate, especially on object regions (lion and elephant). The results also demonstrate the usefulness of the proposed scene-object classifier as it helps analyze video contents and discriminate objects from various scenes.

Next, we extend the scene-object classifier to a detailed scene classifier with multiple categories. To achieve this, we carry out another finetuned FCN by using the same training set as mentioned before, but keeping different scene categories (without merging to a single scene category). Here, we still keep a single object category due to the fact that objects usually vary a lot in different scenes while there are common scene categories appearing in different videos.

Table 1 (rightmost column) shows that the proposed algorithm with the detailed scene classifier achieves highest accuracy in terms of aveAcc (i.e., 81.4%) and mAP (i.e., 78.6%) when compared to all the other methods. As the detailed scene classifier provides more discriminative information for each scene category, the proposed algorithm further improves results in most categories, especially in bird and sand classes, which are significantly improved from 41.5% to 66.2% and from 44.7% to 75.3%, respectively. Figure 5 shows sample results of the proposed algorithm guided by the proposed two classifiers, i.e., the generalized scene-object classifier and the detailed one.

Furthermore, we validate the effectiveness of the proposed region-based matching process by comparing it with directly using the energy gain obtained from the submodular optimization. When iteratively selecting supervoxels into the facility set in the submodular function for a target category, each supervoxel is associated with an energy gain to represent the potential to be the target category (see Fig. 3 for an illustration). Table 2 shows that compared with the method using energy gain, the matching process consistently improves the results with both generalized and detailed scene-object classifiers.

4.3 SUNY-24 Dataset

We also evaluate the proposed method on the challenging SUNY-24 dataset [15], which contains 24 categories and 8 videos. Each video is taken in one scene with motion of camera or objects, and contains 70 to 88 frames with pixel-level annotations. Each single video usually contains multiple categories with small semantic regions and complex scenes (see Fig. 6). The mutual information among the video collection is insufficient and ambitious. These factors make this dataset even challenging. We compare our method with the generalized

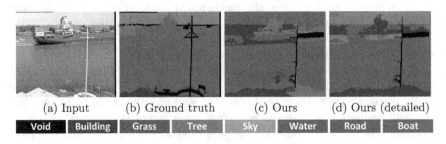

(a) Input   (b) Ground truth   (c) Ours   (d) Ours (detailed)

Void   Building   Grass   Tree   Sky   Water   Road   Boat

Fig. 6. Sample results of the proposed method with different classifiers on the SUNY-24 dataset. The results in (c) and (d) are generated by the proposed method with the generalized and detailed scene-object classifiers. The results show the challenges in the SUNY-24 dataset, i.e., some semantics in the video are small (e.g., building, grass and road) and the scenes are various and complex (e.g., the sample video contains 6 different scene labels). The proposed method with the generalized scene-object classifier successfully labels some of the semantics (e.g., void and water), and the performance is further improved by the detailed one (e.g., boat, water and sky). Each color indicates a semantic label and the legend is shown on the bottom (Color figure online).

Table 3. Video scene co-parsing in terms of the average per category accuracy (aveAcc) on SUNY-24.

Indicator	CRANE	WILD	Ours (generalized)	Ours (detailed)
aveAcc	13.8	14.1	21.6	35.42

and detailed scene-object classifiers to the CRANE [8] and WILD [13] methods. Figure 6 shows the challenge in the SUNY-24 dataset and demonstrates the effectiveness of the proposed method with the generalized and detailed scene-object classifiers. Table 3 shows that the proposed method performs favorably on this challenging dataset. The results are improved with more than 7% gains in terms of aveAcc using the proposed generalized model, and more than 20% gains using the detailed scene classifier.

5   Concluding Remarks

In this paper, we propose a scene co-parsing framework to assign semantic pixel-wise labels in weakly-labeled videos. We first extract representative supervoxels for different categories by a supervoxel ranking scheme. To relax the constraints of large-scale annotations with fixed category numbers, we incorporate the generalized scene-object classifier into the submodular objective function which provides guidance in distinguishing the contents of objects and scenes. By iteratively optimizing the submodular function, we select representatives that share mutual similarities and maintain higher probabilities to match the target category. To exploit semantic information among the video collection, we predict category potentials and assign semantic labels for each supervoxel by matching it to the selected representatives. Experimental results on the Wild-8 and SUNY-24 datasets show that the proposed algorithm performs favorably against the state-of-the-art approaches in terms of visual quality and accuracy.

Acknowledgments. This work is supported in part by the NSF CAREER grant #1149783, NSF IIS grant #1152576, and gifts from Adobe and Nvidia. G. Zhong is sponsored by China Scholarship Council and NSFC grant #61572099.

References

1. Farabet, C., Couprie, C., Najman, L., LeCun, Y.: Learning hierarchical features for scene labeling. IEEE Trans. Pattern Anal. Mach. Intell. **35**, 1915–1929 (2013)
2. Liu, B., He, X.: Multiclass semantic video segmentation with object-level active inference. In: Proceedings of IEEE Conference on Computer Vision and Pattern Recognition (2015)
3. Liu, C., Yuen, J., Torralba, A.: Nonparametric scene parsing via label transfer. IEEE Trans. Pattern Anal. Mach. Intell. **33**, 2368–2382 (2011)
4. Liu, X., Zhao, Y., Zhu, S.C.: Single-view 3d scene parsing by attributed grammar. In: Proceedings of IEEE Conference on Computer Vision and Pattern Recognition (2014)

5. Zhang, C., Wang, L., Yang, R.: Semantic segmentation of urban scenes using dense depth maps. In: Daniilidis, K., Maragos, P., Paragios, N. (eds.) ECCV 2010. LNCS, vol. 6314, pp. 708–721. Springer, Heidelberg (2010). doi:10.1007/978-3-642-15561-1_51

6. Long, J., Shelhamer, E., Darrell, T.: Fully convolutional networks for semantic segmentation. In: Proceedings of IEEE Conference on Computer Vision and Pattern Recognition (2015)

7. Tighe, J., Lazebnik, S.: Superparsing: scalable nonparametric image parsing with superpixels. Int. J. Comput. Vision **101**, 329–349 (2013)

8. Tang, K., Sukthankar, R., Yagnik, J., Fei-Fei, L.: Discriminative segment annotation in weakly labeled video. In: Proceedings of IEEE Conference on Computer Vision and Pattern Recognition (2013)

9. Wang, L., Hua, G., Sukthankar, R., Xue, J., Zheng, N.: Video object discovery and co-segmentation with extremely weak supervision. In: Fleet, D., Pajdla, T., Schiele, B., Tuytelaars, T. (eds.) ECCV 2014. LNCS, vol. 8692, pp. 640–655. Springer, Cham (2014). doi:10.1007/978-3-319-10593-2_42

10. Zhang, Y., Chen, X., Li, J., Wang, C., Xia, C.: Semantic object segmentation via detection in weakly labeled video. In: Proceedings of IEEE Conference on Computer Vision and Pattern Recognition (2015)

11. Grundmann, M., Kwatra, V., Han, M., Essa, I.: Efficient hierarchical graph-based video segmentation. In: Proceedings of IEEE Conference on Computer Vision and Pattern Recognition (2010)

12. Xu, C., Xiong, C., Corso, J.J.: Streaming hierarchical video segmentation. In: Fitzgibbon, A., Lazebnik, S., Perona, P., Sato, Y., Schmid, C. (eds.) ECCV 2012. LNCS, vol. 7577, pp. 626–639. Springer, Heidelberg (2012). doi:10.1007/978-3-642-33783-3_45

13. Liu, X., Tao, D., Song, M., Ruan, Y., Chen, C., Bu, J.: Weakly supervised multiclass video segmentation. In: Proceedings of IEEE Conference on Computer Vision and Pattern Recognition (2014)

14. Hartmann, G., Grundmann, M., Hoffman, J., Tsai, D., Kwatra, V., Madani, O., Vijayanarasimhan, S., Essa, I., Rehg, J., Sukthankar, R.: Weakly supervised learning of object segmentations from web-scale video. In: Proceedings of the 12th European Conference on Computer Vision Workshop (2012)

15. Chen, A.Y., Corso, J.J.: Propagating multi-class pixel labels throughout video frames. In: Proceedings of Western New York Image Processing Workshop (2010)

16. Lee, Y.J., Kim, J., Grauman, K.: Key-segments for video object segmentation. In: Proceedings of IEEE International Conference on Computer Vision (2011)

17. Li, F., Kim, T., Humayun, A., Tsai, D., Rehg, J.M.: Video segmentation by tracking many figure-ground segments. In: Proceedings of IEEE International Conference on Computer Vision (2013)

18. Papazoglou, A., Ferrari, V.: Fast object segmentation in unconstrained video. In: Proceedings of IEEE International Conference on Computer Vision (2013)

19. Jain, S.D., Grauman, K.: Supervoxel-consistent foreground propagation in video. In: Fleet, D., Pajdla, T., Schiele, B., Tuytelaars, T. (eds.) ECCV 2014. LNCS, vol. 8692, pp. 656–671. Springer, Cham (2014). doi:10.1007/978-3-319-10593-2_43

20. Wen, L., Du, D., Lei, Z., Li, S.Z., Yang, M.H.: Jots: joint online tracking and segmentation. In: Proceedings of IEEE Conference on Computer Vision and Pattern Recognition (2015)

21. Nagaraja, N.S., Schmidt, F., Brox, T.: Video segmentation with just a few strokes. In: Proceedings of IEEE International Conference on Computer Vision (2015)

22. Tsai, Y.H., Yang, M.H., Black, M.J.: Video segmentation via object flow. In: Proceedings of IEEE Conference on Computer Vision and Pattern Recognition (2016)
23. Brox, T., Malik, J.: Object segmentation by long term analysis of point trajectories. In: Daniilidis, K., Maragos, P., Paragios, N. (eds.) ECCV 2010. LNCS, vol. 6315, pp. 282–295. Springer, Heidelberg (2010). doi:10.1007/978-3-642-15555-0_21
24. Tsai, Y.-H., Zhong, G., Yang, M.-H.: Semantic co-segmentation in videos. In: Leibe, B., Matas, J., Sebe, N., Welling, M. (eds.) ECCV 2016. LNCS, vol. 9908, pp. 760–775. Springer, Cham (2016). doi:10.1007/978-3-319-46493-0_46
25. Rubio, J.C., Serrat, J., López, A.: Video co-segmentation. In: Lee, K.M., Matsushita, Y., Rehg, J.M., Hu, Z. (eds.) ACCV 2012. LNCS, vol. 7725, pp. 13–24. Springer, Heidelberg (2013). doi:10.1007/978-3-642-37444-9_2
26. Chiu, W.C., Fritz, M.: Multi-class video co-segmentation with a generative multi-video model. In: Proceedings of IEEE Conference on Computer Vision and Pattern Recognition (2013)
27. Fu, H., Xu, D., Zhang, B., Lin, S.: Object-based multiple foreground video co-segmentation. In: Proceedings of IEEE Conference on Computer Vision and Pattern Recognition (2014)
28. Guo, J., Cheong, L.-F., Tan, R.T., Zhou, S.Z.: Consistent foreground co-segmentation. In: Cremers, D., Reid, I., Saito, H., Yang, M.-H. (eds.) ACCV 2014. LNCS, vol. 9006, pp. 241–257. Springer, Cham (2015). doi:10.1007/978-3-319-16817-3_16
29. Zhang, D., Javed, O., Shah, M.: Video object co-segmentation by regulated maximum weight cliques. In: Fleet, D., Pajdla, T., Schiele, B., Tuytelaars, T. (eds.) ECCV 2014. LNCS, vol. 8695, pp. 551–566. Springer, Cham (2014). doi:10.1007/978-3-319-10584-0_36
30. Socher, R., Lin, C.C., Manning, C., Ng, A.Y.: Parsing natural scenes and natural language with recursive neural networks. In: Proceedings of the 28th International Conference on Machine Learning (2011)
31. Munoz, D., Bagnell, J.A., Hebert, M.: Stacked hierarchical labeling. In: Daniilidis, K., Maragos, P., Paragios, N. (eds.) ECCV 2010. LNCS, vol. 6316, pp. 57–70. Springer, Heidelberg (2010). doi:10.1007/978-3-642-15567-3_5
32. Yang, J., Price, B., Cohen, S., Yang, M.H.: Context driven scene parsing with attention to rare classes. In: Proceedings of IEEE Conference on Computer Vision and Pattern Recognition (2014)
33. Xu, J., Schwing, A.G., Urtasun, R.: Tell me what you see and I will show you where it is. In: Proceedings of IEEE Conference on Computer Vision and Pattern Recognition (2014)
34. Chen, X., Jain, A., Davis, L.S.: Object co-labeling in multiple images. In: Proceedings of IEEE Winter Conference on Applications of Computer Vision (2014)
35. Galvão, R.D.: Uncapacitated facility location problems: contributions. Pesquisa Operacional 24, 7–38 (2004)
36. Lazic, N., Givoni, I., Frey, B., Aarabi, P.: Floss: Facility location for subspace segmentation. In: Proceedings of IEEE International Conference on Computer Vision (2009)
37. Zhu, F., Jiang, Z., Shao, L.: Submodular object recognition. In: Proceedings of IEEE Conference on Computer Vision and Pattern Recognition (2014)
38. Leskovec, J., Krause, A., Guestrin, C., Faloutsos, C., VanBriesen, J., Glance, N.: Cost-effective outbreak detection in networks. In: Proceedings of the 13th ACM SIGKDD International Conference on Knowledge Discovery and Data Mining (2007)

39. Yang, F., Jiang, Z., Davis, L.S.: Submodular reranking with multiple feature modalities for image retrieval. In: Cremers, D., Reid, I., Saito, H., Yang, M.-H. (eds.) ACCV 2014. LNCS, vol. 9003, pp. 19–34. Springer, Cham (2015). doi:10. 1007/978-3-319-16865-4_2
40. Siva, P., Russell, C., Xiang, T.: In defence of negative mining for annotating weakly labelled data. In: Fitzgibbon, A., Lazebnik, S., Perona, P., Sato, Y., Schmid, C. (eds.) ECCV 2012. LNCS, vol. 7574, pp. 594–608. Springer, Heidelberg (2012). doi:10.1007/978-3-642-33712-3_43
41. Vezhnevets, A., Ferrari, V., Buhmann, J.M.: Weakly supervised semantic segmentation with a multi-image model. In: Proceedings of IEEE International Conference on Computer Vision (2011)
42. Tsai, Y.H., Shen, X., Lin, Z., Sunkavalli, K., Yang, M.H.: Sky is not the limit: Semantic-aware sky replacement. ACM Trans. Graph. (Proc. ACM SIGGRAPH) (2016)

Supervoxel-Based Segmentation of 3D Volumetric Images

Chengliang Yang[✉], Manu Sethi, Anand Rangarajan, and Sanjay Ranka

Department of Computer and Information Science and Engineering,
University of Florida, Gainesville, FL 32611, USA
ximen14@ufl.edu, {msethi,anand,ranka}@cise.ufl.edu

Abstract. While computer vision has made noticeable advances in the state of the art for 2D image segmentation, the same cannot be said for 3D volumetric datasets. In this work, we present a scalable approach to volumetric segmentation. The methodology, driven by supervoxel extraction, combines local and global gradient-based features together to first produce a low level supervoxel graph. Subsequently, an agglomerative approach is used to group supervoxel structures into a segmentation hierarchy with explicitly imposed containment of lower level supervoxels in higher level supervoxels. Comparisons are conducted against state of the art 3D segmentation algorithms. The considered applications are 3D spatial and 2D spatiotemporal segmentation scenarios.

1 Introduction

The advent of big data and ease of access to computational resources has led to increasing interest in directly analyzing 3D data instead of merely relying on their 2D image projections. Consequently, 3D supervoxels which are a natural extension to 2D superpixels, play a key role as can be seen in many computer vision applications spanning spatiotemporal video and 3D volumetric datasets. For example, high dimensional data in the form of lightfields, RGBD videos and regular video data, particle-laden turbulent flow data comprising dust storms and snow avalanches and finally 3D medical images like MRIs are avenues for further development and application of 3D supervoxel estimation methods.

The popular ultrametric contour map (UCM) framework [1] has established itself as the state of the art in 2D superpixel segmentation. However, a direct extension of this framework has not been seen in the case of 3D supervoxels. The current popular frameworks in 3D segmentation start by constructing regions based on some kind of agglomerative clustering (graph based or otherwise) of single pixel data. However, the 2D counterparts of these methods on which they are founded are not as accurate as UCM. This is because the UCM algorithm does not begin by grouping pixels into regions. Instead it first computes a high quality boundary map which is subsequently utilized toward obtaining closed regions.

This work is partially supported by AFOSR under FA9550-15-1-0047.

S.-H. Lai et al. (Eds.): ACCV 2016, Part I, LNCS 10111, pp. 37–53, 2017.
DOI: 10.1007/978-3-319-54181-5_3

The two step process in UCM is carried out by combining boundary organization with pixel clustering which leads to high quality segmentation results in 2D.
Therefore, it follows that a natural extension of this two step process in the case
of 3D should be a harbinger for success.

UCM derives its power from a combination of local and global cues which
complement each other in order to detect highly accurate boundaries. However,
all the other methods, with the exception of normalized cuts [2] (which directly
obtains regions) are inherently local and do not incorporate global image information. In sharp contrast, UCM includes global image information by estimating
eigenfunction scalar fields of graph Laplacians formed from local image features.
Despite this inherent advantage, the high computational cost of UCM is a bottleneck in the development of its 3D analog. Recent work in [3] has addressed this
issue in 2D by providing an efficient GPU-based implementation but the huge
number of voxels involved in the case of spatiotemporal volumes remains an
issue. The work in [4] provides an efficient CPU-only implementation by leveraging the structure of the underlying problem and provided a reduced order
normalized cuts approach to solve the eigenvector problem. Consequently, this
opens up a plausible route to 3D as a reduced order approach effectively solves
the same globalization problem but at a fraction of the cost. Therefore, the main
work of this paper is threefold: First, we design filters for 3D volumes (either
space-time video or volumetric data) which provide local cues at multiple scales
and at different orientations akin to the approach in 2D UCM. Second, we introduce a new method called as *oriented intervening contour cue* which extends
the idea of the intervening contour cue in [5] for constructing the graph affinity
matrix. Third, we solve the reduced order eigenvector problem by leveraging
ideas from the approach in [4]. After this globalization step, the local and global
fields are merged to obtain surface boundary fields. Subsequent application of
a watershed transform yields supervoxel tessellations (represented as relatively
uniform polyhedra that tessellate the region). The next step in this paper is to
build a hierarchy of supervoxels. While 2D UCM merges regions based on their
boundary strengths in its oriented watershed transform approach, we did not
find this approach to work well in 3D. Instead, to obtain the supervoxel hierarchy, we follow the approach of [6] which performs a graph-based merging of
regions using the method of internal variation [7].

In summary, our approach extends the popular and highly accurate *gPb-*
UCM framework to 3D resulting in a highly scalable framework based on reduced
order normalized cuts. To the best of our knowledge, there does not exist a
surface detection method in 3D which uses graph Laplacian-based globalization for estimating supervoxels. Further, our results indicate that owing to the
deployment of a high quality boundary detector as the initial step, our method
maintains a distinction between the foreground and background regions by not
causing unnecessary oversegmentation of the background at the lower levels of
hierarchy.

Road map: The next section describes the related work on 2D and 3D segmentation which influenced our work. Section 3 describes how we extract supervoxels

by extending the 2D gPb-UCM framework. Section 4 evaluates the proposed app-roach and other state of the art supervoxel methods on two different types of 3D volumetric datasets both qualitatively and quantitatively. Section 5 concludes by summarizing our contributions and also discusses the scope for future work. Throughout the paper, we use the term pixel and voxel interchangeably when referring to the basic "atom" of 2D/3D images.

2   Related Work

Normalized Cuts and gPb-UCM: Normalized cuts [2] gained immense pop-ularity in 2D image segmentation by treating the image as a graph and com-puting hard partitions. The gPb-UCM framework [1] leveraged this approach in a soft manner to introduce globalization into their contour detection process. This led to a drastic reduction in the oversegmentation arising out of gradual texture or brightness changes in the previous approaches. Being a computation-ally expensive method, the last decade saw the emergence of several techniques to speed up the underlying spectral decomposition process in [1]. These tech-niques range from various optimization techniques like multilevel solvers [8,9], to systems implementations exploiting GPU parallelism [3], and to approximate methods like reduced order normalized cuts [4,10].

3D Volumetric Image Segmentation: Segmentation techniques applied to 3D volumetric images can be found in the literature of medical imaging (usually MRI) and 2D+time video sequence segmentation. In 3D medical image segmen-tation, unsupervised techniques like region growing [11] have been well studied. In [12], normalized cuts were applied to MRIs but gained little attention. Recent literature mostly focuses on supervised techniques [13,14]. In video segmenta-tion, there are two streams of works. On one hand, [15–18] are frame based that rely on segmenting each frame into superpixels in the first place. Therefore they are not applicable to general 3D volumes. On the other hand, a variety of other methods treat the video sequences as spatiotemporal volumes and try to segment them into supervoxels. Those methods are mostly extensions of popu-lar 2D image segmentation approaches, including graph cuts [19,20], SLIC [21], mean shift [22], graph-based methods [6] and normalized cuts [9,23]. Besides these, temporal superpixels [24–26] are another set of effective approaches that extract high quality spatiotemporal volumes. A comprehensive review of super-voxel methods can be found in [27,28]. All of this development in the area of 3D supervoxels has led to a general consensus in the video segmentation community that supervoxels have favorable properties which can be leveraged later in the pipeline. These characteristics include: (i) supervoxel boundaries should stick to the meaningful image boundaries; (ii) regions within one supervoxel should be homogeneous while inter-supervoxel differences should be substantially large; (iii) it's important that supervoxels have regular topologies; (iv) a hierarchy of supervoxels is favorable as different applications have different supervoxel gran-ularity preferences.

3 Supervoxel Extraction

Our work is a natural extension of the state of the art 2D superpixel method, the *gPb*-Ultrametric Contour Map framework (*gPb*-UCM) [1], henceforth referred to as 3D-UCM. The 2D *gPb*-UCM framework consists of three major parts: image gradient features detection, globalization and agglomeration. Analogously, the workflow in the 3D-UCM framework presented here is volume gradient features detection, globalization and supervoxel agglomeration. However, the voxel cardinality of 3D volumetric images far exceeds their 2D counterparts. Therefore, certain computational considerations force us to adopt reduced order eigensystem solvers. These approximations will become clear as we proceed. The upside is that the 3D-UCM algorithm becomes scalable to handle sizable datasets.

3.1 Volume Gradient Features Detection

We first require an edge detector to help quantify the presence of boundary-surfaces. Most gradient-based edge detectors in 2D [29,30] can be extended to 3D for this purpose. We based our 3D edge detector on the *mPb* detector proposed in [1,31] which has been empirically shown to have superior performance in 2D.

The building block of the 3D *mPb* gradient detector is an oriented gradient operator $G(x, y, z, \theta, \varphi, r)$ that is described in detail in Fig. 1. To be more specific, in a 3D volumetric intensity image, we place a sphere centered at each pixel to denote its neighborhood. An equatorial plane specified by its normal vector $t(\theta, \varphi)$ splits the sphere into two half spheres. We compute the intensity histograms for both half spheres as **g** and **h**. Then we define the gradient magnitude in the direction $t(\theta, \varphi)$ as the χ^2 distance between g and h:

$$\chi^2(\mathbf{g}, \mathbf{h}) = \frac{1}{2} \sum_i \frac{(g(i) - h(i))^2}{g(i) + h(i)}. \tag{1}$$

In order to capture gradient information at multiple scales, this gradient value is calculated at different radius values r of the neighborhood sphere. Gradients obtained from different scales are then linearly combined together using

$$G(x, y, z, \theta, \varphi) = \sum_r \alpha_r G(x, y, z, \theta, \varphi, r) \tag{2}$$

where α_r weighs the gradient contribution at different scales. For multi-channel 3D images like video sequences, $G(x, y, z, \theta, \varphi)$ are calculated separately from different channels and summed up using equal weights. Finally, the measure of boundary strength at (x, y, z) is computed as the maximum response among various directions $t(\theta, \varphi)$:

$$mPb(x, y, z) = \max_{\theta, \varphi} G(x, y, z, \theta, \varphi). \tag{3}$$

In our experiments, θ and φ take values in $\{0, \frac{\pi}{4}, \frac{\pi}{2}, \frac{3\pi}{4}\}$ and $\{-\frac{\pi}{4}, 0, \frac{\pi}{4}\}$ respectively and in one special case, $\varphi = \frac{\pi}{2}$. Therefore we compute local gradients in

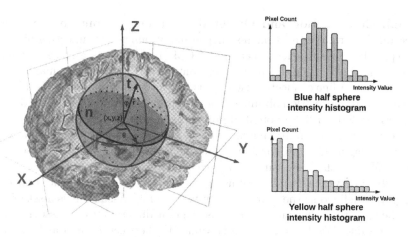

Fig. 1. The oriented gradient operator $G(x, y, z, \theta, \varphi, r)$: At location (x, y, z), the local neighborhood is defined by a sphere with radius r. An equatorial plane n (shaded green) along with its normal vector t splits the sphere into two half spheres. The one above n is shaded yellow and the one below is shaded blue. We histogram the intensity values of voxels that fall in the yellow and blue half spheres respectively. Finally we calculate the χ^2 distance between the yellow and blue histograms as the local gradient magnitude in the direction t of scale r (Color figure online).

13 different directions. Neighborhood values of 2, 4 and 6 voxels were used for r. Equal weights α_r were used to combine gradients from different scales. Also, as is standard, we always apply an isotropic Gaussian smoothing filter with $\sigma = 3$ voxels before any gradient operation.

3.2   Globalization

The core aspect of the gPb-UCM algorithms is spectral clustering. It globalizes the local cues obtained from the gradient features detection phase and specifically focuses on the most salient boundaries in the image by analyzing the eigenvectors derived from the normalized cuts problem [2]. However, this approach depends on solving a sparse eigensystem at the scale of the number of pixels in the image. Thus as the size of the image grows large, the globalization step becomes the computational bottleneck of the whole process. This problem is even more severe in the 3D setting because the voxel cardinality far exceeds the pixel cardinality of our 2D counterparts. An efficient approach was proposed in [4] to reduce the size of the eigensystem while maintaining the quality of the eigenvectors used in globalization. We generalize this method to 3D so that our approach becomes scalable to handle sizable datasets.

In the following, we describe the globalization steps: (i) graph construction and oriented intervening contour cue, (ii) reduced order normalized cuts and eigenvector computation, (iii) scale-space gradient computation on the eigenvector image and (iv) the combination of local and global gradient information.

Graph Construction and Oriented Intervening Contour Cue: In 2D, spectral clustering begins from a sparse graph obtained by connecting pixels that are spatially close to each other. gPb-UCM [1] constructs a sparse symmetric affinity matrix W using the intervening contour cue [5] that is the maximal value of mPb along a line connecting two pixels. However, this approach doesn't utilize all the useful information obtained from the previous gradient features detection step. Figure 2 describes a potential problem and how we deal with it. To improve the accuracy of the affinity matrix, we take the direction vector of the maximum gradient magnitude into consideration when calculating the pixel-wise affinity value. We call this new variant as *oriented intervening contour cue*. For any spatially close voxels i and j, we use $\overline{ij}$ to denote the line segment connecting i and j. $\boldsymbol{d}$ is defined as the unit direction vector of $\overline{ij}$. Assume P is a set of voxels that lie close to $\overline{ij}$. For any $p \in P$, $\boldsymbol{n}$ is the unit direction vector associated with its mPb value. We define the affinity value W_{ij} between i and j as follows:

$$W_{ij} = \exp(-\max_p\{mPb(p)|\langle \boldsymbol{d}, \boldsymbol{n}\rangle|\}/\rho) \qquad (4)$$

where $\langle\rangle$ is the inner product operator of the vector space and ρ is a scaling constant. In our experiments, P contains the voxels that are at most 1 voxel away from $\overline{ij}$. ρ is set to 0.1. In the affinity matrix W, each voxel is connected to voxels that fall in the 5×5 cube centered at that voxel. So the graph defined by W is very sparse.

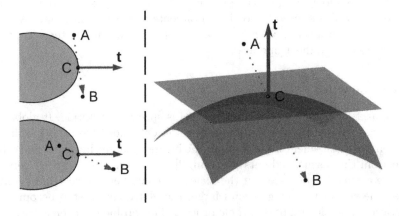

Fig. 2. Left: Suppose we want to calculate the affinity value between voxels A and B. C is the voxel with maximal mPb value that lies on the line segment $\overline{AB}$. In the upper left case, A and B belong to one region. In the lower left case, A and B are in two different regions. But the intervening contour cue of UCM [1] gives the same affinity value in both cases, which is not very satisfactory. Obviously it would be better if we consider the direction $\boldsymbol{t}$ of C's mPb. **Right**: In our *oriented intervening contour cue* approach, when calculating affinity values, we always take the product of $mPb(C)$ with the absolute value of the inner product $\langle \boldsymbol{t}, \boldsymbol{n}\rangle$, where $\boldsymbol{n}$ is the unit direction vector of line segment $\overline{AB}$. If A and B are on different sides of a boundary surface, $|\langle \boldsymbol{t}, \boldsymbol{n}\rangle|$ will be large, leading to small affinity value and vice-versa.

Reduced Order Normalized Cuts and Eigenvector Computation: At this point, the standard 2D gPb-UCM solves for the generalized eigenvectors of the sparse eigensystem

$$(D - W)v = \lambda Dv \tag{5}$$

where D is a diagonal matrix defined by $D_{ii} = \Sigma_j W_{ij}$. However, solving this eigenvalue problem is very computationally intensive. It becomes the bottleneck, both in time and memory efficiency, of the normalized cuts segmentation algorithms. To overcome this, an efficient and highly parallel GPU implementation was provided in [3]. However, this approach requires us to use GPU-based hardware and software suites—an unnecessary restriction. A clever alternative in [4,10] builds the graph on superpixels instead of pixels to reduce the size of the eigensystem. We chose to generalize [4]'s approach to 3D as (i) the superpixel solution is more scalable than the GPU solution in terms of memory requirements, (ii) specialized GPU co-processors are not commonly available in many computing platforms like smart phones and wearable devices, and (iii) the approach in [10] is specifically designed for superpixels in each frame in video segmentation, thus not easily generalizable. Finally, the approach in [4] constructs a reduced order normalized cuts system which is easier to solve. The reduced order eigensystem is denoted by

$$(L^T(D - W)L)x = \lambda' L^T DLx \tag{6}$$

where $L \in \mathbb{R}^{m \times n}, x \in \mathbb{R}^m$ and $Lx = v$. The purpose of L is to assign each pixel to a superpixel/supervoxel. In our approach, the supervoxels are generated by a watershed transform on the mPb image obtained from the volume gradient features detection step. Obviously the number of supervoxels m is much smaller than the number of voxels n in the whole 3D volumetric image. In practice, there are usually two to three orders reduction in the size of the eigensystem (from millions voxels to a few thousands supervoxels). Therefore it is much more efficient to solve Eq. (6) than Eq. (5).

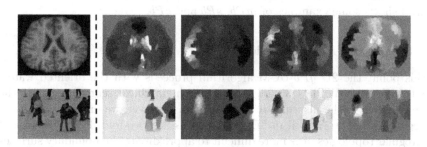

Fig. 3. Upper Left: One slice of a brain MRI from the IBSR dataset [32]. **Lower Left**: One frame of a video sequence from the BuffaloXiph dataset [33]. **Right**: The corresponding slices/frames of the first 4 eigenvectors.

Scale Space Gradient Computation on the Eigenvector Image: We solve for the generalized eigenvectors $\{x_0, x_1, \ldots, x_n\}$ of the system in (6) corresponding to the smallest eigenvalues $\{\lambda'_0, \lambda'_1, \ldots, \lambda'_n\}$. As stated in [4], λ_i in (5) will equal to λ'_i and Lx_i will match v_i modulo an irrelevant scale factor, where v_i are the eigenvectors of the original eigensystem (5). Similar to the 2D scenario [1], eigenvectors v_i carry surface information. Figure 3 shows several example eigenvectors obtained from two types of 3D volumetric datasets. In both cases, the eigenvectors distinguish salient aspects of the original image. Based on this observation, we apply the gradient operator mPb defined in (3.1) to the eigenvector images. The outcome of this procedure is denoted as 'sPb' because it represents the 'spectral' component of the boundary detector, following the convention established in [1]:

$$sPb(x, y, z) = \sum_{i=1}^{K} \frac{1}{\sqrt{\lambda_i}} mPb_{v_i}(x, y, z). \tag{7}$$

Note that this weighted summation starts from $i = 1$ because λ_0 always equals 0 and v_0 is a vanilla image. The weighting by $1/\sqrt{\lambda_i}$ is inspired by the mass-spring system in mechanics [1,34]. In our experiments, we use 16 eigenvectors, i.e. $K = 16$.

The Combination of Local and Global Gradient Information: The last step is to combine local cues mPb and global cues sPb. mPb tries to capture variations in every corner while sPb aims to obtain salient boundary surfaces. By linearly combining them together, we get a 'globalized' boundary detector gPb:

$$gPb(x, y, z) = \omega mPb(x, y, z) + (1 - \omega)sPb(x, y, z). \tag{8}$$

In practice, we use equal weights for mPb and sPb. After obtaining the gPb values, we apply a post-processing step of non-maximum suppression [29] to get thinned boundary surfaces when the resulting edges from mPb are too thick. Figure 4 shows some examples of mPb, sPb and gPb.

3.3   Supervoxel Agglomeration

At this point, the 2D gPb-UCM algorithm proceeds with the oriented watershed transform (OWT) [1,35,36] to create a hierarchical segmentation of the image resulting in the ultrametric contour map. However, we find that the same strategy does not work well in 3D. The reasons are two-fold. First, because of the irregular topologies, it is more difficult to approximate the boundary surfaces with square or triangular meshes in 3D than to approximate the boundary curves with line segments in 2D. Second, in the merging process, following OWT, only the information of the pixels on the boundaries are used when the boundaries between superpixels are greedily removed. This is not a robust design especially considering that in 3D the boundary surfaces are frequently fragmented.

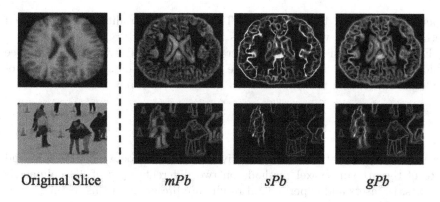

Original Slice   mPb   sPb   gPb

Fig. 4. *sPb* augments the strength of the most salient boundaries in *gPb*.

Due to the above considerations, we turn to the popular graph based image and video segmentation methods [6,7] to create the segmentation hierarchy. We first apply a watershed transform to the *gPb* strengths obtained from the previous step to get an oversegmentation. Next we iteratively merge the adjacent segments starting from this oversegmentation. The output of this procedure is a segmentation hierarchy represented by a tree-structure whose lower level segments are always contained in higher level segments. As in [6], the merge rules run on a graph. The nodes of the graph are regions. First, for any two adjacent regions R_i and R_j, we assign an edge e_{ij} to connect them on the graph. The weight of e_{ij} is set to the χ^2 distance between *Lab* space or intensity value histograms of R_i and R_j with 20 bins used. Also, for any region R, a quantity named the relaxed internal variation RInt(R) is defined:

$$\mathrm{RInt}(R) := \mathrm{Int}(R) + \frac{\tau}{|R|} \tag{9}$$

where Int(R) is defined as the maximum edge weight of its minimum spanning tree (MST). For the lowest level regions, i.e. the regions of oversegmentation obtained from the watershed transform, Int(R) is set to 0. $|R|$ is the voxel cardinality of region R. τ is a parameter to trigger the merging process and control the preferred granularity of the regions. In each iteration of merging, all the edges are traversed in ascending order. For any edge e_{ij}, we merge incident regions R_i and R_j if the weight of e_{ij} is less than the minimum of the relaxed internal variation of the two regions. Thus the merging condition is written as

$$\mathrm{weight}(e_{ij}) < \min\{\mathrm{RInt}(R_i), \mathrm{RInt}(R_j)\}. \tag{10}$$

In practice, we increase the granularity parameter τ by a factor of 1.1 in each iteration. This agglomeration process iteratively progresses until no edge meets the merging criteria. The advantage of graph based methods is that they make use of the information in all voxels in the merged regions. Furthermore, as shown in the experiments below, we see that it overcomes the weakness of fragmented supervoxels of graph based methods. This is because traditional graph based methods are built on voxel-level graphs.

Finally, we obtain a supervoxel hierarchy represented by a bottom-up tree structure. This is the final output of the 3D-UCM algorithm. The granularity of the segmentation is a user driven choice guided by the application.

4   Evaluation

We perform quantitative and qualitative comparisons between 3D-UCM and state of the art supervoxel methods on two different types of 3D volumetric datasets. Datasets and experimental results are presented in this section.

4.1   Experimental Setup

Datasets: The most typical use cases of 3D segmentation are medical images like MRIs and video sequences. We use the publicly available Internet Brain Segmentation Repository (IBSR) [32] for our medical imaging application. It contains 18 volumetric brain MRIs with their white matter, gray matter and cerebrospinal fluid labeled by human experts. These represent cortical and subcortical structures of interest in neuroanatomy. For video segmentation applications, we use the dataset BuffaloXiph introduced by [33]. The 8 video sequences in the dataset have 69–85 frames and are labeled with semantic pixels. This allows us to examine whether the algorithms have the same perception as humans in the case of 2D+time segmentation.

Methods: A comprehensive comparison of current supervoxel and video segmentation methods is available in [27,28]. Building on their approach, in our experiments, we will compare 3D-UCM to two state of the art supervoxel methods: (i) hierarchical graph based (GBH) [6] and (ii) segmentation by weighted aggregation (SWA) [8,9,37]. GBH is the standard graph based method for video segmentation and SWA is a multilevel normalized cuts solver that also generates a hierarchy of segmentations.

4.2   Qualitative Comparisons

We present some example slices from both the IBSR and BuffaloXiph datasets in Fig. 5. Even a cursory examination shows us that all the three methods differ markedly in the kind of segmentations they obtain. However, it is hard to say which method performs best when compared to the ground truth. As can be noticed, GBH has a fragmentation problem in both the IBSR and BuffaloXiph datasets. A large number of small fragments of irregular shapes are visible in its results. On the other hand, SWA has a clean segmentation in IBSR but suffers from fragmentation in video sequences. In contrast, 3D-UCM has the most regular segmentation. This difference is clearer if we look at the whole hierarchy.

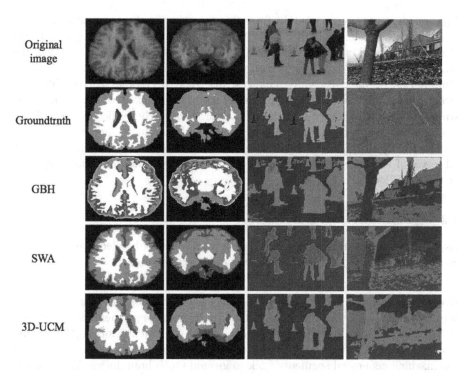

Fig. 5. The level of segmentation hierarchy is chosen as similar to ground truth granularity. **Left**: IBSR dataset results. The white, gray and dark gray regions are white matter, gray matter and cerebro-spinal fluid (CSF) respectively. **Right**: BuffaloXiph dataset results.

Fig. 6. The hierarchy of segmentation of all three methods of the same skating frame in Fig. 5. The leftmost column is the finest segmentation in each hierarchy.

Figure 6 shows the segmentation from fine to coarse of the same skating frame as in Fig. 5. Obviously, GBH and SWA generate unnecessary oversegmentations of the background at finer levels. This is because the first building blocks of the GBH and SWA hierarchies are single voxels. This makes them sensitive to local illumination or intensity value changes. 3D-UCM overcomes this problem by integrating global cues to obtain an initial set of supervoxels which leverage the strength of a high quality boundary detector. Hence, the basic building blocks of our hierarchy are these initial supervoxels and not the elementary voxels. It suffices to say that 3D-UCM generates meaningful and compact supervoxels at all levels of the hierarchy while GBH and SWA have a fragmentation problem at the lower levels.

4.3   Quantitative Measures

As both the IBSR and BuffaloXiph datasets are densely labeled, we are able to compare the supervoxels methods on a variety of measures.

Boundary Quality: The precision-recall curve is the most recommended measure and has found widespread use in comparing image segmentation methods [1,31]. This was introduced into the world of video segmentation in [16]. We use it as the boundary quality measure in our benchmarks. It measures how well the machine generated segments stick to ground truth boundaries. More importantly, it shows the tradeoff between the positive predictive value (precision) and true positive rate (recall). For a set of machine generated boundary pixels

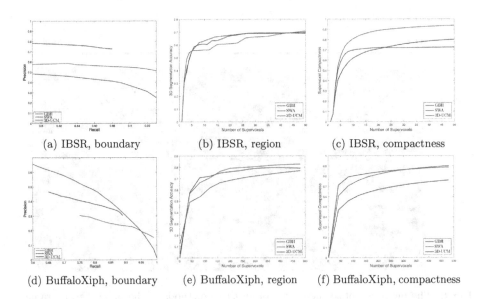

(a) IBSR, boundary   (b) IBSR, region   (c) IBSR, compactness

(d) BuffaloXiph, boundary   (e) BuffaloXiph, region   (f) BuffaloXiph, compactness

Fig. 7. Quantitative measures.

S_b and human labeled boundary pixels G_b, precision and recall are defined as follows:

$$\text{precision} = \frac{|S_b \cap G_b|}{|S_b|}, \quad \text{recall} = \frac{|S_b \cap G_b|}{|G_b|}. \tag{11}$$

We show the precision-recall curves on IBSR and BuffaloXiph datasets in Fig. 7a and d respectively. 3D-UCM performs the best on IBSR and is the second best on BuffaloXiph. GBH does not perform well on BuffaloXiph while SWA is worse on IBSR. One limitation of 3D-UCM is that there is an upper limit of its boundary recall because it is based on supervoxels. GBH and SWA can have arbitrarily fine segmentations. Thus they can achieve a recall rate arbitrarily close to 1, though the precision is usually low in these situations.

Region Quality: Measures based on overlaps of regions such as Dice's coefficients are widely used in evaluating region covering performances of voxel-wise segmentation approaches [38,39]. We use the 3D segmentation accuracy introduced in [27,28] to measure the average fraction of ground truth segments that is correctly covered by the machine generated supervoxels. Given that $G_v = \{g_1, g_2, \ldots, g_m\}$ are ground truth volumes, $S_v = \{s_1, s_2, \ldots, s_n\}$ are supervoxels generated by the algorithms and V represents the whole volume, the 3D segmentation accuracy is defined as

$$\text{3D segmentation accuracy} = \frac{1}{m} \sum_{i=1}^{m} \frac{\sum_{j=1}^{n} |s_j \cap g_i| \times \mathbf{1}(|s_j \cap g_i| \geq |s_j \cap \overline{g}_i|)}{|g_i|} \tag{12}$$

where $\overline{g}_i = V \setminus g_i$. We plot the 3D segmentation accuracy against the number of supervoxels in Figs. 7b and e. GBH and SWA again perform differently in IBSR and BuffaloXiph datasets. But 3D-UCM consistently showed the best performance, especially when the number of supervoxels is low.

Supervoxel Compactness: Compact supervoxels of regular shapes are always favored because they benefit further higher level tasks in computer vision. The compactness of superpixels generated by a variety of image segmentation algorithms in 2D was investigated in [40]. It uses a measure inspired by the isoperimetric quotient to measure compactness. We use another quantity defined similarly to specific surface area in material science and biology. In essence, specific surface area and isoperimetric quotient both try to quantify the total surface area per unit mass or volume. Formally, given that $S_v = \{s_1, s_2, \ldots, s_n\}$ are supervoxels generated by algorithms, the specific surface area to measure supervoxel compactness is defined as

$$\text{specific surface area} = \frac{1}{n} \sum_{i=1}^{n} \frac{\text{Surface}(s_i)}{\text{Volume}(s_i)} \tag{13}$$

where Surface() and Volume() count voxels on the surfaces and in the supervoxels respectively. Lower values of specific surface area imply more compact supervoxels. The compactness comparisons on IBSR and BuffaloXiph are shown in Figs. 7c and f. We see that 3D-UCM always generates the most compact supervoxels except at small supervoxel granularity on IBSR. GBH does not perform well in this measure because of its fragmentation problem, which is consistent with our qualitative observations.

Our quantitative measures cover boundary quality, region quality and supervoxel compactness that are the most important aspects of supervoxel qualities. 3D-UCM always performs the best or the second best in all measures on both IBSR and BuffaloXiph datasets. In contrast, GBH and SWA fail on some measures. In conclusion, we have empirically shown that 3D-UCM is a very competitive general purpose supervoxel method on 3D volumetric datasets with respect to the proposed measures albeit on a few datasets.

5   Discussion

In this paper, we presented the 3D-UCM supervoxel framework, an extension of the most successful 2D image segmentation technique, gPb-UCM, to 3D. Experimental results show that our approach outperforms the current state of the art in most benchmark measures on two different types of 3D volumetric datasets. Furthermore, we deployed a reduced order normalized cuts technique to overcome a computational bottleneck of the traditional gPb-UCM framework. This immediately allows our method to scale up to large datasets. When we jointly consider supervoxel quality and computational efficiency, we believe that 3D-UCM can become the standard bearer for massive 3D volumetric datasets. We expect applications of 3D-UCM in a wide range of vision tasks, including video semantic understanding, video object tracking and labeling in high-resolution medical imaging, etc.

Since this is a new and fresh approach to 3D segmentation, 3D-UCM still has several limitations from our perspective. First, because it is a general purpose technique, the parameters of 3D-UCM have not been tuned using supervised learning for specific applications. In immediate future work, we plan to follow the metric learning framework as in [1] to deliver higher performance. Second, since it is derived from the modular framework of gPb-UCM, 3D-UCM has numerous alternative algorithmic paths that can be considered. In image and video segmentation, a better boundary detector was proposed in [18], with different graph structures deployed in [17,41,42] followed by graph partitioning alternatives in [43,44]. A careful study of these alternative options may result in improved versions of 3D-UCM going forward. Finally, 3D-UCM at the moment does not incorporate prior knowledge for segmentation [45,46]. Prior information like object shape and optical flow motion cues could greatly improve segmentation performance. These represent interesting opportunities for future work.

References

1. Arbelaez, P., Maire, M., Fowlkes, C., Malik, J.: Contour detection and hierarchical image segmentation. IEEE Trans. Pattern Anal. Mach. Intell. **33**, 898–916 (2011)
2. Shi, J., Malik, J.: Normalized cuts and image segmentation. IEEE Trans. Pattern Anal. Mach. Intell. **22**, 888–905 (2000)
3. Catanzaro, B., Su, B.Y., Sundaram, N., Lee, Y., Murphy, M., Keutzer, K.: Efficient, high-quality image contour detection. In: 2009 IEEE 12th International Conference on Computer Vision, pp. 2381–2388. IEEE (2009)
4. Taylor, C.: Towards fast and accurate segmentation. In: Proceedings of the IEEE Conference on Computer Vision and Pattern Recognition, pp. 1916–1922 (2013)
5. Fowlkes, C., Martin, D., Malik, J.: Learning affinity functions for image segmentation: combining patch-based and gradient-based approaches. In: Proceedings of the 2003 IEEE Computer Society Conference on Computer Vision and Pattern Recognition, vol. 2, pp. II-54. IEEE (2003)
6. Grundmann, M., Kwatra, V., Han, M., Essa, I.: Efficient hierarchical graph-based video segmentation. In: 2010 IEEE Conference on Computer Vision and Pattern Recognition (CVPR), pp. 2141–2148. IEEE (2010)
7. Felzenszwalb, P.F., Huttenlocher, D.P.: Efficient graph-based image segmentation. Int. J. Comput. Vision **59**, 167–181 (2004)
8. Sharon, E., Brandt, A., Basri, R.: Fast multiscale image segmentation. In: Proceedings of the IEEE Conference on Computer Vision and Pattern Recognition, vol. 1, pp. 70–77. IEEE (2000)
9. Sharon, E., Galun, M., Sharon, D., Basri, R., Brandt, A.: Hierarchy and adaptivity in segmenting visual scenes. Nature **442**, 810–813 (2006)
10. Galasso, F., Keuper, M., Brox, T., Schiele, B.: Spectral graph reduction for efficient image and streaming video segmentation. In: Proceedings of the IEEE Conference on Computer Vision and Pattern Recognition, pp. 49–56 (2014)
11. Pohle, R., Toennies, K.D.: Segmentation of medical images using adaptive region growing. In: International Society for Optics and Photonics on Medical Imaging 2001, pp. 1337–1346 (2001)
12. Carballido-Gamio, J., Belongie, S.J., Majumdar, S.: Normalized cuts in 3-D for spinal MRI segmentation. IEEE Trans. Med. Imaging **23**, 36–44 (2004)
13. Fischl, B., Salat, D.H., Busa, E., Albert, M., Dieterich, M., Haselgrove, C., Van Der Kouwe, A., Killiany, R., Kennedy, D., Klaveness, S., et al.: Whole brain segmentation: automated labeling of neuroanatomical structures in the human brain. Neuron **33**, 341–355 (2002)
14. Deng, Y., Rangarajan, A., Vemuri, B.C.: Supervised learning for brain MR segmentation via fusion of partially labeled multiple atlases. In: IEEE International Symposium on Biomedical Imaging (ISBI) (2016)
15. Galasso, F., Cipolla, R., Schiele, B.: Video segmentation with superpixels. In: Lee, K.M., Matsushita, Y., Rehg, J.M., Hu, Z. (eds.) ACCV 2012. LNCS, vol. 7724, pp. 760–774. Springer, Heidelberg (2013). doi:10.1007/978-3-642-37331-2_57
16. Galasso, F., Nagaraja, N.S., Cardenas, T.J., Brox, T., Schiele, B.: A unified video segmentation benchmark: Annotation, metrics and analysis. In: IEEE International Conference on Computer Vision (2013)
17. Khoreva, A., Galasso, F., Hein, M., Schiele, B.: Classifier based graph construction for video segmentation. In: 2015 IEEE Conference on Computer Vision and Pattern Recognition (CVPR), pp. 951–960. IEEE (2015)

18. Khoreva, A., Benenson, R., Galasso, F., Hein, M., Schiele, B.: Improved image boundaries for better video segmentation. arXiv preprint arXiv:1605.03718 (2016)
19. Boykov, Y., Veksler, O., Zabih, R.: Fast approximate energy minimization via graph cuts. IEEE Trans. Pattern Anal. Mach. Intell. **23**, 1222–1239 (2001)
20. Boykov, Y.Y., Jolly, M.P.: Interactive graph cuts for optimal boundary & region segmentation of objects in N-D images. In: Proceedings of thr Eighth IEEE International Conference on Computer Vision, ICCV 2001, vol. 1, pp. 105–112. IEEE (2001)
21. Achanta, R., Shaji, A., Smith, K., Lucchi, A., Fua, P., Susstrunk, S.: SLIC superpixels compared to state-of-the-art superpixel methods. IEEE Trans. Pattern Anal. Mach. Intell. **34**, 2274–2282 (2012)
22. Paris, S., Durand, F.: A topological approach to hierarchical segmentation using mean shift. In: IEEE Conference on Computer Vision and Pattern Recognition, CVPR 2007, pp. 1–8. IEEE (2007)
23. Fowlkes, C., Belongie, S., Malik, J.: Efficient spatiotemporal grouping using the Nystrom method. In: Proceedings of the 2001 IEEE Computer Society Conference on Computer Vision and Pattern Recognition, CVPR 2001, vol. 1, pp. I–231. IEEE (2001)
24. Chang, J., Wei, D., Fisher, J.: A video representation using temporal superpixels. In: Proceedings of the IEEE Conference on Computer Vision and Pattern Recognition, pp. 2051–2058 (2013)
25. Reso, M., Jachalsky, J., Rosenhahn, B., Ostermann, J.: Temporally consistent superpixels. In: Proceedings of the IEEE International Conference on Computer Vision, pp. 385–392 (2013)
26. Bergh, M., Roig, G., Boix, X., Manen, S., Gool, L.: Online video seeds for temporal window objectness. In: Proceedings of the IEEE International Conference on Computer Vision, pp. 377–384 (2013)
27. Xu, C., Corso, J.J.: Evaluation of super-voxel methods for early video processing. In: 2012 IEEE Conference on Computer Vision and Pattern Recognition (CVPR), pp. 1202–1209. IEEE (2012)
28. Xu, C., Corso, J.J.: LIBSVX: a supervoxel library and benchmark for early video processing. Int. J. Comput. Vis., 1–19 (2016)
29. Canny, J.: A computational approach to edge detection. IEEE Trans. Pattern Anal. Mach. Intell. **8**(6), 679–698 (1986)
30. Meer, P., Georgescu, B.: Edge detection with embedded confidence. IEEE Trans. Pattern Anal. Mach. Intell. **23**, 1351–1365 (2001)
31. Martin, D.R., Fowlkes, C.C., Malik, J.: Learning to detect natural image boundaries using local brightness, color, and texture cues. IEEE Trans. Pattern Anal. Mach. Intell. **26**, 530–549 (2004)
32. Worth, A.: The Internet Brain Segmentation Repository (IBSR) (2016). http://www.nitrc.org/projects/ibsr
33. Chen, A.Y., Corso, J.J.: Propagating multi-class pixel labels throughout video frames. In: 2010 Western New York Image Processing Workshop (WNYIPW), pp. 14–17. IEEE (2010)
34. Belongie, S., Malik, J.: Finding boundaries in natural images: a new method using point descriptors and area completion. In: Burkhardt, H., Neumann, B. (eds.) ECCV 1998. LNCS, vol. 1406, pp. 751–766. Springer, Heidelberg (1998). doi:10.1007/BFb0055702
35. Arbelaez, P., Maire, M., Fowlkes, C., Malik, J.: From contours to regions: an empirical evaluation. In: IEEE Conference on Computer Vision and Pattern Recognition, CVPR 2009, pp. 2294–2301. IEEE (2009)

36. Najman, L., Schmitt, M.: Geodesic saliency of watershed contours and hierarchical segmentation. IEEE Trans. Pattern Anal. Mach. Intell. **18**, 1163–1173 (1996)
37. Corso, J.J., Sharon, E., Dube, S., El-Saden, S., Sinha, U., Yuille, A.: Efficient multilevel brain tumor segmentation with integrated Bayesian model classification. IEEE Trans. Med. Imaging **27**, 629–640 (2008)
38. Menze, B.H., Jakab, A., Bauer, S., Kalpathy-Cramer, J., Farahani, K., Kirby, J., Burren, Y., Porz, N., Slotboom, J., Wiest, R., et al.: The multimodal brain tumor image segmentation benchmark (BRATS). IEEE Trans. Med. Imaging **34**, 1993–2024 (2015)
39. Malisiewicz, T., Efros, A.A.: Improving spatial support for objects via multiple segmentations. In: Proceedings of the British Machine Vision Conference (BMVC), pp. 1–10 (2007)
40. Schick, A., Fischer, M., Stiefelhagen, R.: Measuring and evaluating the compactness of superpixels. In: 2012 21st International Conference on Pattern Recognition (ICPR), pp. 930–934. IEEE (2012)
41. Ren, X., Malik, J.: Learning a classification model for segmentation. In: Proceedings of the Ninth IEEE International Conference on Computer Vision, pp. 10–17. IEEE (2003)
42. Briggman, K., Denk, W., Seung, S., Helmstaedter, M.N., Turaga, S.C.: Maximin affinity learning of image segmentation. In: Advances in Neural Information Processing Systems, pp. 1865–1873 (2009)
43. Brox, T., Malik, J.: Object segmentation by long term analysis of point trajectories. In: Daniilidis, K., Maragos, P., Paragios, N. (eds.) ECCV 2010. LNCS, vol. 6315, pp. 282–295. Springer, Heidelberg (2010). doi:10.1007/978-3-642-15555-0_21
44. Palou, G., Salembier, P.: Hierarchical video representation with trajectory binary partition tree. In: Proceedings of the IEEE Conference on Computer Vision and Pattern Recognition, pp. 2099–2106 (2013)
45. Vu, N., Manjunath, B.: Shape prior segmentation of multiple objects with graph cuts. In: IEEE Conference on Computer Vision and Pattern Recognition, CVPR 2008, pp. 1–8. IEEE (2008)
46. Chen, F., Yu, H., Hu, R., Zeng, X.: Deep learning shape priors for object segmentation. In: Proceedings of the IEEE Conference on Computer Vision and Pattern Recognition, pp. 1870–1877 (2013)

Message Passing on the Two-Layer Network for Geometric Model Fitting

Xing Wang, Guobao Xiao, Yan Yan, and Hanzi Wang[(✉)]

Fujian Key Laboratory of Sensing and Computing for Smart City,
School of Information Science and Technology, Xiamen University, Xiamen, China
xingwang_xmu@163.com, x-gb@163.com, {yanyan,hanzi.wang}@xmu.edu.cn

Abstract. In this paper, we propose a novel model fitting method to recover multiple geometric structures from data corrupted by noises and outliers. Instead of analyzing each model hypothesis or each data point separately, the proposed method combines both the consensus information in all model hypotheses and the preference information in all data points into a two-layer network, in which the vertices in the first layer represent the data points and the vertices in the second layer represent the model hypotheses. Based on this formulation, the clusters in the second layer of the network, corresponding to the true structures, are detected by using an effective Two-Stage Message Passing (TSMP) algorithm. TSMP can not only accurately detect multiple structures in data without specifying the number of structures, but also handle data even with a large number of outliers. Experimental results on both synthetic data and real images further demonstrate the superiority of the proposed method over several state-of-the-art fitting methods.

1 Introduction

The task of recovering geometric structures from data, i.e., model fitting, plays an important role in many applications of computer vision, such as face clustering [1,2], homography/fundamental matrix estimation [3,4], motion segmentation [5,6], etc. In practice, geometric model fitting is a challenging problem because real-world data usually contain multiple structures, and are often contaminated by severe noises, gross outliers, and pseudo-outliers [7]. For the purpose of being robust to noises and outliers, many fitting methods (e.g., RANSAC [8] and its variants [9–12]) adopt the "hypothesize-and-verify" framework, where a number of model hypotheses are firstly generated from randomly sampled subsets of data, and then evaluated based on a given quality measure. In order to deal with multi-structure data, some methods (e.g., [13,14]) use the "fit-and-remove" procedure, which sequentially fits one structure and removes the corresponding inliers from data. However, the inaccurate estimation of the first several structures may seriously affect the estimation of the remaining structures.

Recently, a new category of fitting methods has revealed that the problem of geometric model fitting can be formulated as the "representation-and-clustering" framework (e.g., [3,5,7,15–17]). Furthermore, these methods can be divided

© Springer International Publishing AG 2017
S.-H. Lai et al. (Eds.): ACCV 2016, Part I, LNCS 10111, pp. 54–69, 2017.
DOI: 10.1007/978-3-319-54181-5_4

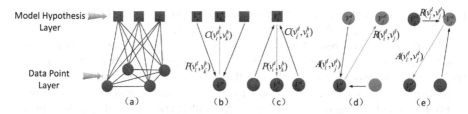

Fig. 1. Illustration of the main stages of the proposed model fitting method: (a) Two-layer network modelling in which each vertex of the data point layer represents a data point and each vertex of the model hypothesis layer represents a model hypothesis. (b) and (c) The first stage of the proposed algorithm where the "consensu" $C(v_i^d, v_k^h)$ and the "preference" $P(v_i^d, v_k^h)$ are exchanged between the vertices of the two layers. (d) and (e) The second stage of the proposed algorithm where the "responsibility" $R(v_i^d, v_j^d)$ and the "availability" $A(v_i^d, v_j^d)$ are exchanged among the vertices of the pruned data point layer.

into the consensus-based methods and the preference-based methods. For the consensus-based methods (e.g., AKSWH [7], MSH [16]), each model hypothesis corresponds to a consensus set which is defined as the set of corresponding inliers belonging to the model hypothesis. For the preference-based methods (e.g., KF [5], J-linkage [15], T-linkage [3]), each data point is represented by a set of preferred model hypotheses, to which this data point is an inlier. After the representation step, the clustering step is applied to segment model hypotheses or data points.

However, the above clustering-based fitting methods only consider either the consensus information or the preference information, which shows some limitations. Specifically, the consensus-based methods need to distinguish the true structures from the redundant structures (i.e., the structures of the same model with slightly different parameters). The preference-based methods have difficulties in dealing with intersecting structures. In addition, some preference-based methods require the posteriori information during the fitting steps. For example, J-linkage requires a user-specified threshold to filter out insignificant clusters so as to determine the number of structures.

In this paper, we propose a novel and effective message passing algorithm on the two-layer network to fit and segment multi-structure data. Instead of only considering either the consensus information or the preference information, the proposed method (called as TSMP) combines both of them into a two-layer network, which is consisted of a data point layer and a model hypothesis layer. Based on the network, a two-stage message passing algorithm is proposed. In the first stage, messages are exchanged between the vertices in the data point layer and the vertices in the model hypothesis layer to prune the data point layer. In the second stage, the affinity propagation clustering approach [18] is applied to detect clusters in the pruned data point layer (corresponding to the structures in the data) based on a novel similarity measure. Finally, the number and the parameters of structures are estimated according to the detected clusters. An overview of the proposed method is shown in Fig. 1.

Compared with current state-of-the-art clustering based model fitting methods, the proposed method (TSMP) shows three significant advantages: Firstly, the constructed two-layer network can effectively describe the complex relationships between data points and model hypotheses, and it can be easily implemented. Secondly, TSMP can not only fit multi-structure data with a large number of outliers (more than 90%), but also effectively handle unbalanced data (i.e., the numbers of inliers belonging to different structures are significant unbalanced). Thirdly, TSMP does not require a user-specified threshold to determine the number of structures in data, since it uses an effective message passing algorithm on the two-layer network to automatically estimate the number of structures.

Note that, FLOSS [6] also uses a message passing algorithm to deal with model fitting problems. However, the differences between FLOSS and the proposed method are significant. (1) FLOSS selects an optimal subset of facilities (i.e., structures) and assigns each customer (i.e., a data point) to one facility by minimizing a cost function composed by two terms: the facility cost and the distance between customers to their assigned facilities. The effectiveness of FLOSS is mainly depended on choosing a good trade-off between the two terms. However, the proposed method attempts to directly estimate the number of structures without relying on the trade-off. (2) FLOSS is intrinsically designed for outlier-free data. To cope with the data contaminated by outliers, FLOSS requires to add an outlier model into the optimization. Moreover, the performance of FLOSS deteriorates rapidly as the outlier ratio surpasses 35% [19]. In contrast, the proposed method contains an effective outlier removal procedure, and it can handle data even with a large number of outliers (more than 90%). Therefore, the proposed method is much more effective than FLOSS.

2   Two-Layer Network Modeling

In this paper, the geometric model fitting problem is formulated as a message passing problem on a two-layer network, in which the vertices in the first layer represent the data points and the vertices in the second layer represent the model hypotheses. In Sect. 2.1, we introduce the modeling of the two-layer network, and express the relationships between data points and model hypotheses by using the two-layer network. The weighting score is introduced to weight the edge in Sect. 2.2.

2.1   The Two-Layer Network

A two-layer network $\mathbf{NW} = \{V^d, V^h, E^{dd}, E^{dh}, W^{dd}, W^{dh}\}$ consists of vertices V^d in the first layer (i.e., the data point layer), vertices V^h in the second layer (i.e., the model hypothesis layer), edges E^{dd} and E^{dh}, weights W^{dd} and W^{dh}. Each edge in E^{dh} connects a vertex in V^d and a vertex in V^h, and it is measured by a weighting score in W^{dh}. Each edge in E^{dd} connects a pair of vertices in V^d, and it is weighted by a score in W^{dd}. Note that, there is no edge between the vertices in the model hypothesis layer, since there is no message exchanged between these vertices.

To construct a two-layer network, we first sample M p-subsets from the N data points $\mathbf{X} = \{x_i\}_{i=1}^N$, where p is the minimum number of data points required to estimate the parameters of a structure (e.g., 2 for line fitting and 3 for circle fitting). Then, we generate M model hypotheses $\Theta = \{\theta_k\}_{k=1}^M$ using the sampled p-subsets, and estimate the inlier noise scale of each hypothesis. In this paper, we use IKOSE [7] to estimate the inlier noise scale due to its good performance. After that, each data point is represented as a vertex of the first layer and each model hypothesis is represented as a vertex of the second layer. Then, we connect each pair of vertices in the data point layer, and connect each vertex in the data point layer to each vertex in the model hypothesis layer. Therefore, the complex relationships between model hypotheses and data points are effectively described by the two-layer network.

2.2   Weighting Score

Each edge in the two-layer network **NW** is associated with a weighting score, while the weighting scores W^{dh} and W^{dd} measure different kinds of information. A weighting score $W^{dh}(i, k)$, corresponding to an edge in E^{dh}, measures the affinity between a vertex $v_i^d \in V^d$ (i.e., a data point x_i) and a vertex $v_k^h \in V^h$ (i.e., a model hypothesis θ_k), and it is defined as follows:

$$W^{dh}(i, k) = exp(-d(x_i, \theta_k)/\sigma_k), \qquad (1)$$

where $d(x_i, \theta_k)$ is the residual from the model hypothesis θ_k to the data point x_i; σ_k is the inlier noise scale of θ_k. While the weighting scores in W^{dd}, corresponding to the edges in E^{dd}, represent the similarities between the pairs of vertices in V^d, and the values of W^{dd} are computed as described in Sect. 3.3.

Ideally, in E^{dh}, an edge connecting a vertex in V^d (corresponding to a data point belonging to the inliers of a true structure) and a vertex in V^h (corresponding to a model hypothesis belonging to the true structure) should be assigned with a high weighting score. Accordingly, in E^{dd}, an edge derived from a pair of vertices in V^d (corresponding to a pair of data points belonging to the inliers of a true structure) should also be assigned with a high weighting score. Recall that any two vertices in the data point layer connected by an edge with a high weighting score are similar. Thus, the vertices in the data point layer, corresponding to the inliers of a true structure, should form a cluster in this layer. In this manner, we can directly detect clusters on the two-layer network for model fitting.

3   The Two-Stage Message Passing Algorithm

Based on the constructed two-layer network, the task of fitting structures in data is formulated as the problem of detecting clusters in the data point layer.

To detect clusters, we propose to utilize the affinity propagation (AP) [18] approach to cluster the vertices in the data point layer (described in Sect. 3.4). AP is an effective message passing based approach which automatically determines the number of clusters. However, AP cannot deal with a large number of bad vertices in the data point layer, corresponding to the outliers in data which are usuall encountered in model fitting. In order to cope with the problem involved in AP, we further propose an effective two-stage message passing (TSMP) algorithm. At the first stage of TSMP, messages are exchanged between the vertices of the data point layer and the vertices of the model hypothesis layer (as described in Sect. 3.1). Based on the result of the first stage, the data point layer can be pruned by removing the bad vertices and the corresponding edges (described in Sect. 3.2). At the second stage of TSMP, AP is applied to detect clusters in the pruned data point layer (described in Sect. 3.4). In addition, we develop an effective similarity measure for pairs of vertices in the pruned data point layer (described in Sect. 3.3) to enhance the performance of AP.

3.1   Message Passing Between the Two Layers

For the purpose of pruning the data point layer, we present an algorithm that iteratively exchanges messages along the edges between the data point layer and the model hypothesis layer to distinguish good vertices (corresponding to inliers) from bad vertices (corresponding to outliers). Note that good vertices should obtain more messages than bad vertices during the message passing process (see Sect. 3.2).

There are two types of messages exchanged between vertices in the data point layer and vertices in the model hypothesis layer. The first type of messages, called "consensus", are sent from the vertices in the data point layer to the vertices in the model hypothesis layer. A consensus $C(v_i^d, v_k^h)$ reflects how appropriate a vertex v_i^d is to be a good vertex for a vertex v_k^h, taking into account other vertices in V^h for the vertex v_i^d (see Fig. 1(b)). The second type of messages, called "preference", are sent from the vertices in the model hypothesis layer to the vertices in the data point layer. A preference $P(v_i^d, v_k^h)$ reflects the accumulated evidence for how well-suited the vertex v_i^d is to be good vertex for the vertex v_k^h, taking into account the supports from the vertices in V^d for the vertex v_k^h (see Fig. 1(c)).

At the beginning, all the preferences are initialized to $\frac{1}{M}$, i.e., $P(v_i^d, v_k^h) = \frac{1}{M}$. Then a consensus $C(v_i^d, v_k^h)$, sent from the vertex v_i^d to the vertex v_k^h, is calculated as follows:

$$C(v_i^d, v_k^h) = \sum_{k'=1}^{M} P(v_i^d, v_{k'}^h) \cdot W^{dh}(i, k). \tag{2}$$

At the first iteration, since all the preferences are equal to $\frac{1}{M}$, $C(v_i^d, v_k^h)$ is set to the affinity between the vertex v_i^d in the data point layer and the vertex v_k^h in the model hypothesis layer. After several iterations, when some vertices in the data point layer are weakly supported by some vertices in the model hypothesis layer, their preferences will be assigned with small values than the others according to the update rule introduced below (see Eq. (3)). These preferences with small values will weaken the support of the corresponding vertices in the data point layer to the vertices in the model hypothesis layer.

The above consensus update makes each vertex in the model hypothesis layer gather supports from the vertices in the data point layer. On the other hand, the following preference update makes each vertex in the data point layer gather supports from the vertices in the model hypothesis layer:

$$P(v_i^d, v_k^h) = \sum_{i'=1}^{N} C(v_{i'}^d, v_k^h) \cdot W^{dh}(i, k).$$ (3)

The preference $P(v_i^d, v_k^h)$ is set to be the sum of the product of the consensuses of the vertex v_k^h gathering from the other vertices in the data point layer and the affinity (derived from the vertex v_i^d and the vertex v_k^h).

The above procedure can be terminated within a limited number of iterations. Specifically, in our experiments, after only two or three iterations, the distinction of the updated preferences between the vertices in the data point layer is often evident. Once the above procedure is terminated, the preferences can be used to remove bad vertices in the data point layer.

3.2   Pruning the Data Point Layer

To prune the data point layer, each vertex v_i^d in the data point layer is assigned a weighting score based on the corresponding preferences, and the weighting score is calculated as follows:

$$\pi(v_i^d) = \sum_{k=1}^{M} P(v_i^d, v_k^h).$$ (4)

The weighting score $\pi(v_i^d)$ is set to be the sum of the preferences that the vertex v_i^d obtained from all vertices in the model hypothesis layer. Since the vertices corresponding to inliers should obtain more preferences than the vertices corresponding to outliers during the above message passing procedure (see Fig. 2(a) for an illustration), we can select good vertices (with high weighting scores) while discarding bad vertices (with low weighting scores) by a threshold ψ.

As shown in Fig. 2(b), we observe that the histogram of weighting scores of vertices in the data point layer has two distinct modes (i.e., good vertices and bad vertices). Based on this observation, we can use a threshold to distinguish good vertices from bad vertices. Instead of manually setting a threshold value, we use a Gaussian Mixture Model (GMM) based approach as [5] to automatically estimate the threshold value.

Specifically, given the vertices V^d and the corresponding weighting scores π, we fit a 1D GMM with two components:

$$F(\pi) = \sum_{c=1,2} \lambda_c \mathcal{N}\{\pi|\mu_c, \sigma_c\},$$ (5)

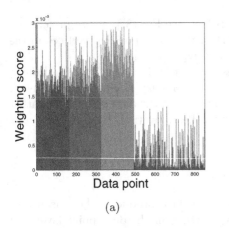
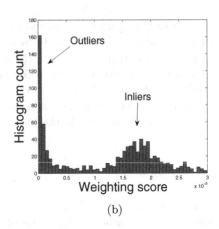

(a)   (b)

Fig. 2. An example illustrating the weighting scores of the vertices in the data point layer and the corresponding histogram. (a) The weighting scores of the vertices in the data point layer (ordered by cluster) corresponding to the data points in the "3 Lines" data (see in Fig. 3). (b) The histogram of the weighting scores of the data points in the "3 Lines" data.

where $\mathcal{N}$ is a Gaussian distribution; μ_c and σ_c are the mean and stand deviation, respectively; λ_c is the weight of the c-th component. This can be solved by using the EM algorithm [20]. Then the threshold value ψ can be obtained as the average between the two means [5], i.e.,

$$\psi = (\mu_1 + \mu_2)/2. \tag{6}$$

After obtaining the threshold value, we can remove the cluster of vertices with lower weighting scores and the corresponding edges, by which we obtain a pruned data point layer. Therefore, we can directly utilize AP to detect clusters in the pruned data point layer.

3.3 Similarity Measure

To detect the clusters in the pruned data point layer by using the AP approach, we develop an effective similarity measure for pairs of the vertices.

Let $\tilde{V}^d = \{\tilde{v}_i^d\}_{i=1}^{N'}$ be the vertices in the pruned data point layer, where N' $(N' \ll N)$ is the number of the remaining vertices. Denote the preferences vector of vertex $\tilde{v}_i^d$ by $\mathbf{P}(\tilde{v}_i^d) = [P(\tilde{v}_i^d, v_1^h), \ldots, P(\tilde{v}_i^d, v_M^h)]$. The Tanimoto-like similarity between two vertices $\tilde{v}_i^d$ and $\tilde{v}_j^d$ based on the corresponding preferences vector is given by [21]

$$S(\tilde{v}_i^d, \tilde{v}_j^d) = \frac{< \mathbf{P}(\tilde{v}_i^d), \mathbf{P}(\tilde{v}_j^d) >}{\left\| \mathbf{P}(\tilde{v}_i^d) \right\| + \left\| \mathbf{P}(\tilde{v}_j^d) \right\| - < \mathbf{P}(\tilde{v}_i^d), \mathbf{P}(\tilde{v}_j^d) >} - 1, \tag{7}$$

where the notation $< \cdot, \cdot >$ indicates the standard inner product and $\|\cdot\|$ indicates the corresponding induced norm. The similarity measure ranges in $[-1, 0]$.

Algorithm 1. The two-stage message passing method for geometric model fitting

---

Input: Data points $\mathbf{X}$, and the K value for IKOSE.
Output: The structures in data and the corresponding inliers.
1: Generate a set of model hypotheses Θ.
2: Estimate the inlier noise scale of each model hypothesis by IKOSE.
3: Construct a two-layer network (described in Sect. 2.1).
4: Calculate the weighting scores of all the edges between the data point layer and the model hypothesis layer by Eq. (1).
5: Exchange messages between the data point layer and the model hypothesis layer (described in Sect. 3.1).
6: Prune the data point layer (described in Sect. 3.2).
7: Calculate the similarities between each pair of vertices in the pruned data point layer by Eq. (7).
8: Exchange messages among the pruned data point layer to detect clusters (described in Sect. 3.4).
9: Estimate the parameters of the structure derived from the vertices of each cluster.

Recall that a preference $P(v_i^d, v_k^h)$, sent from a vertex v_k^h to a vertex v_i^d, contains all the information that vertex v_k^h gathering from all the vertices in the data point layer as in Eq. (2). Thus, each element $P(\tilde{v}_i^d, v_k^h)$ in the preference vector $\mathbf{P}(\tilde{v}_i^d)$ not only measures the affinity between the vertex $\tilde{v}_i^d$ and the vertex v_k^h, but also considers the relationships between the vertex $\tilde{v}_i^d$ and the other vertices in the data point layer. Therefore, the proposed similarity measure can characterize more complex relationships among the vertices in the data point layer.

Besides the similarity measure for each pair of vertices, AP also requires a number $S(\tilde{v}_i^d, \tilde{v}_i^d)$ for each vertex. The vertex v_i^d with a larger value of $S(\tilde{v}_i^d, \tilde{v}_i^d)$ is more likely to be chosen as the cluster center. In this paper, $S(\tilde{v}_i^d, \tilde{v}_i^d)$ for all the vertices is set to be a common value, which is equal to the minimum value of the similarities for pairs of vertices in the pruned data point layer.

3.4   Detecting Clusters

We utilize the AP approach [18] to detect clusters in the pruned data point layer, where each cluster of vertices defines a structure in data. AP finds the clusters by finding a set of center points based on a message passing algorithm. Each center point represents a cluster, and each other data point belongs to one of the center points.

Messages are exchanged between the vertices in the pruned data point layer until a set of center vertices and the corresponding clusters gradually converges. Similar to message passing between the two layers as described in Sect. 3.1, there are two kinds of messages exchanged among the vertices of the pruned data point layer. The first kind of message, i.e., the responsibility $R(\tilde{v}_i^d, \tilde{v}_j^d)$, sent from the vertex $\tilde{v}_i^d$ to the candidate center vertex $\tilde{v}_j^d$, describes how well-suited the vertex $\tilde{v}_j^d$ is to serve as the center vertex of the vertex $\tilde{v}_i^d$. A responsibility between a vertex $\tilde{v}_i^d$ and another vertex $\tilde{v}_j^d$ is calculated as follows [18]:

$$R(\tilde{v}_i^d, \tilde{v}_j^d) = S(\tilde{v}_i^d, \tilde{v}_j^d) - \max_{j' \ s.t. \ j' \neq j} \{A(\tilde{v}_i^d, \tilde{v}_{j'}^d) + S(\tilde{v}_i^d, \tilde{v}_{j'}^d)\}, \tag{8}$$

where $S(\tilde{v}_i^d, \tilde{v}_j^d)$ is the similarity between the vertex $\tilde{v}_i^d$ and the vertex $\tilde{v}_j^d$ as described before. The second kind of message, i.e., the availability $A(\tilde{v}_i^d, \tilde{v}_j^d)$, sent from the candidate center vertex $\tilde{v}_j^d$ to the vertex $\tilde{v}_i^d$, reflects the probability of the vertex $\tilde{v}_i^d$ choosing $\tilde{v}_j^d$ as its center vertex. An availability between a vertex $\tilde{v}_i^d$ and another vertex $\tilde{v}_j^d$ is updated as follows [18]:

$$A(\tilde{v}_i^d, \tilde{v}_j^d) = \min\{0, R(\tilde{v}_j^d, \tilde{v}_j^d) + \sum_{i' \ s.t. \ i' \notin \{i,j\}} \max\{0, R(\tilde{v}_{i'}^d, \tilde{v}_j^d)\}\}. \tag{9}$$

After obtaining the set of center vertices and the corresponding clusters, we use the vertices of each cluster to estimate a structure.

4   The Complete Method

With all the ingredients described in the previous sections, we now present the complete fitting method in this section. We summarize the proposed Two-Stage Message Passing (TSMP) method for geometric model fitting in Algorithm 1.

The proposed TSMP fits structures by using an effective two-stage message passing algorithm to detect clusters based on a two-layer network. TSMP can not only automatically estimate the number of structures in data, but also handle data with a large number of outliers. The computational complexity of TSMP mainly consists of the first stage for exchanging messages between the data point layer and the model hypothesis layer, and the second stage for exchanging messages among the pruned data point layer. For the first stage, only a few number of iterations is required to distinguish good vertices from bad vertices (as described in Sect. 3.1). For the second stage, the number of good vertices N' is much smaller than the total number of vertices N in the data point layer. Therefore, the proposed method is quite efficient (more results are presented in Sect. 5).

5   Experiments

In this section, we test the performance of the proposed model fitting method (i.e., TSMP) on synthetic and real data. To evaluate the effectiveness of the proposed method, we compare TSMP against four state-of-the-art fitting methods, i.e., KF [5], J-linkage [15], AKSWH [7] and T-linkage [3]. We choose these representative methods because they are clustering based methods as the proposed method. The parameters of all the methods have been tuned for the best performance.

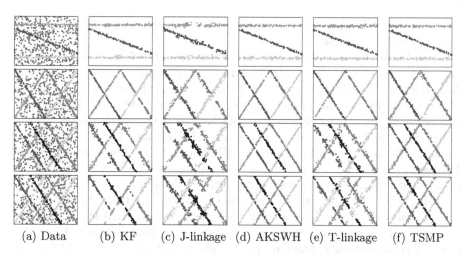

(a) Data   (b) KF   (c) J-linkage   (d) AKSWH   (e) T-linkage   (f) TSMP

Fig. 3. Examples for 2D line fitting. 1st to 4th rows respectively fit three, four, five and six lines. (a) The input data. (b) to (f) The fitting results obtained by KF, J-linkage, AKSWH, T-linkage and TSMP, respectively.

Table 1. The fitting errors (in percentage) for 2D line fitting on the four data (the best results are boldfaced)

	KF		J-linkage		AKSWH		T-linkage		TSMP	
	Avg	Med	Avg	Med	Avg	Med	Avg	Med	Avg	Med
3 Lines	5.98	5.31	7.44	7.51	2.25	2.31	4.87	4.85	**1.83**	**1.88**
4 Lines	7.01	6.56	9.89	9.84	5.40	5.17	7.67	7.54	**2.70**	**2.72**
5 Lines	17.20	17.30	21.11	21.13	7.44	7.45	17.33	17.31	**3.45**	**3.50**
6 Lines	28.59	28.50	27.15	27.08	8.01	8.00	19.27	19.05	**4.13**	**4.09**

In each experiments, the same set of model hypotheses is generated for all the fitting methods by using the proximity sampling [15]. More specifically, we generate 5,000 model hypotheses for line fitting (Sect. 5.1) and circle fitting (Sect. 5.2), 10,000 model hypotheses for homography based segmentation (Sect. 5.3) and 20,000 model hypotheses for two-view based motion segmentation (Sect. 5.4). We repeat each experiment 50 times and report the segmentation errors as [3]. All experiments are ran on a windows machine equipped with a 3.6 GHz Intel Core i7 processor and 32 GB RAM.

5.1 Line Fitting

We evaluate the performance of the five methods on line fitting using four challenging synthetic 2D line data. The corresponding outlier ratios of the four data are respectively 87.5% for the "3 Lines data", 88.8% for the "4 Lines data",

90.0% for the "5 Lines data" and 90.9% for the "6 Lines data". The inlier noise scales of the four data are set to 1.0.

From Fig. 3 and Table 1, we can see that: (1) For the "3 Lines data" and "4 Lines data", all five fitting methods correctly estimate the number of lines in data. However, TSMP achieves the best results among the five fitting methods. (2) For the "5 Lines data" and "6 Lines data", the five methods succeed in estimating the number of the lines in data. But KF, J-linkage and T-linkage can not effectively segment data points. In contrast, AKSWH and TSMP correctly fit the lines with lower fitting errors, while TSMP achieves the lowest fitting errors. This is because: Firstly, the outlier ratios of the two data are too high for KF, J-linkage and T-linkage to capture the complex relationship between data points; Secondly, there exist several intersections in the data, and the three methods cannot deal with the data points near the intersections. AKSWH obtains relatively lower fitting errors among the four competing methods, since it clusters model hypotheses rather than data points, which makes it not very sensitive to data distribution.

5.2   Circle Fitting

In this subsection, we evaluate the performance of the five methods on circle fitting using four challenging synthetic data (see Fig. 4). The corresponding outlier ratios of the four data are respectively 85.0% for the "4 Circles data", 86.9% for the "5 Circles data", 88.4% for the "6 Circles data" and 89.6% for the "7 Circles data". The inlier noise scales of the four data are set to 0.7.

From Fig. 4 and Table 2 we can see that: (1) For the "4 Circles data", the five fitting methods correctly fit all the four circles, where TSMP achieves the lowest average fitting error. Both AKSWH and TSMP achieve the lowest median fitting errors. (2) For the "5 Circles data" and "6 Circles data", all the five fitting methods succeed in estimating the number of circles in data. However, TSMP achieves the best performance among the five fitting methods. KF, J-linkage and T-linkage wrongly discard many inliers while preserving some outliers. (3) For the "7 Circles data", TSMP correctly fits the seven circles and achieves the best performance among the five fitting methods. KF wrongly estimates the number of circles in data. This is because KF is sensitive to the user specified weighting

Table 2. The fitting errors (in percentage) for 2D circle fitting on four data (the best results are boldfaced)

	KF		J-linkage		AKSWH		T-linkage		TSMP	
	Avg	Med	Avg	Med	Avg	Med	Avg	Med	Avg	Med
4 Circles	6.98	6.60	7.88	7.64	1.65	**1.40**	1.52	1.48	**1.38**	**1.40**
5 Circles	16.26	15.70	24.82	24.75	2.82	2.83	15.15	15.11	**2.57**	**2.61**
6 Circles	13.70	17.92	17.74	17.51	6.25	6.29	11.62	11.70	**3.13**	**3.08**
7 Circles	35.99	35.59	36.13	36.05	5.02	4.34	22.14	22.06	**3.68**	**3.79**

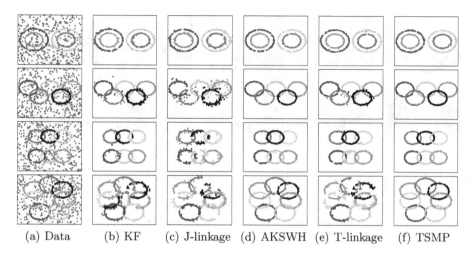

(a) Data (b) KF (c) J-linkage (d) AKSWH (e) T-linkage (f) TSMP

Fig. 4. Examples for 2D circle fitting. 1st to 4th rows respectively fit four, five, six and seven circles. (a) The input data. (b) to (f) The fitting results obtained by KF, J-linkage, AKSWH, T-linkage and TSMP, respectively.

ratio between the fitting error and the model complexity. J-linkage, AKSWH and T-linkage also succeed in estimating the number of circles in data. However, J-linkage and T-linkage obtain higher fitting errors than AKSWH and TSMP. Since there exist several intersections in the data, J-linkage and T-linkage cannot effectively deal with the data points near the intersection.

5.3 Homography Based Segmentation

We test the performance of the five fitting methods for the task of homography based segmentation using the five real image pairs from the AdelaideRMF dataset [22][1].

As shown in Fig. 5 and Table 3, TSMP correctly estimates the numbers of planes, achieving the lowest average and the lowest median fitting errors in 4 out of 5 data. AKSWH and T-linkage achieve relatively low fitting errors. However, KF and J-linkage obtain worse results in most cases. We note that KF may wrongly preserve many outliers when the data is unbalanced, and J-linkage is sensitive to the cutoff threshold. In contrast, due to the two-layer network framework, TSMP is not sensitive to the data distribution.

5.4 Two-View Based Motion Segmentation

We also test the performance of the five fitting methods for the task of two-view based motion segmentation by using five real image pairs from the AdelaideRMF dataset [22].

---

[1] http://cs.adelaide.edu.au/~hwong/doku.php?id=data.

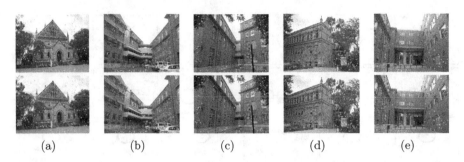

(a)   (b)   (c)   (d)   (e)

Fig. 5. Examples of Homography based segmentation on five image pairs, namely (a) Elderhalla, (b) Nese, (c) Sene, (d) Oldclassicswing and (e) Neem. The first and second rows are the ground truth and the segmentation results obtained by the proposed method respectively. The results obtained by other 4 methods are not shown due to the space limit.

Table 3. The fitting errors (in percentage) for homography based segmentation on five data (the best results are boldfaced)

	KF		J-linkage		AKSWH		T-linkage		TSMP	
	Avg	Med	Avg	Med	Avg	Med	Avg	Med	Avg	Med
Elderhalla	3.93	3.93	6.96	7.52	1.27	1.15	2.85	2.57	**0.98**	**0.93**
Nese	12.99	12.45	14.57	14.48	1.57	1.57	1.30	1.18	**0**	**0**
Sene	13.52	13.06	16.40	16.40	1.60	1.64	1.28	1.20	**0.64**	**0.60**
Oldclassicswing	11.08	11.05	17.90	17.90	3.69	3.66	**3.54**	**3.17**	3.64	3.54
Neem	15.15	14.26	17.47	17.71	6.71	6.69	4.56	4.52	**3.18**	**3.05**

From Fig. 6 and Table 4, we can see that the results of KF and J-linkage are not good in most cases. The reason is that 8 data points are required to generate a model hypothesis in this task, and a large number of model hypotheses should be generated to cover all the motions in data. Thus, a large proportion of bad model hypotheses is generated, which results in the inaccurate similarity measure between data points in KF and J-linkage. The performance of AKSWH is relatively better than both KF and J-linkage. However, during the procedure of selecting significant hypotheses in AKSWH, some good model hypotheses may be wrongly removed, which leads to a high fitting error. TSMP obtains both the lowest average and median fitting errors in 4 out of 5 data. For the "Cubebreadtoychips" data, TSMP also achieves a competitive result compared with the best result obtained by T-linkage.

| (a) | (b) | (c) | (d) | (e) |

Fig. 6. Examples of two-view based motion segmentation on five image pairs, namely (a) Cubetoy, (b) Camebiscuit, (c) Cubechips, (d) Breadtoycar and (e) Cubebreadtoy-chips. The first and second rows are the ground truth and the segmentation results obtained by the proposed method, respectively.

Table 4. The fitting errors (in percentage) for two-view based motion segmentation on five data (the best results are boldfaced)

	KF		J-linkage		AKSWH		T-linkage		TSMP	
	Avg	Med	Avg	Med	Avg	Med	Avg	Med	Avg	Med
Cubetoy	6.31	6.35	12.83	12.71	5.32	5.50	3.14	3.05	**1.49**	**1.20**
Cubechips	10.56	10.92	12.11	12.06	4.73	4.68	5.52	5.17	**0.67**	**0.70**
Gamebiscuit	10.18	10.40	14.85	12.96	7.28	7.51	7.32	7.41	**3.14**	**3.05**
Breadtoycar	16.20	15.00	17.83	17.66	9.06	9.01	4.33	4.21	**3.13**	**3.11**
Cubebreadtoychips	16.73	15.50	30.73	30.09	15.66	15.08	**3.05**	**3.11**	5.20	5.05

6 Conclusion

This paper formulates geometric model fitting as a message passing problem on a two-layer network which consists of a data point layer and a model hypothesis layer. Based on the formulated two-layer network, a two-stage message passing algorithm is proposed to detect clusters (corresponding to structures in data) by exchanging messages among the two-layer network. The proposed method (TSMP) simultaneously estimates the number of structures and the corresponding parameters of each structure without specifying the number of structures in data. Furthermore, TSMP can not only handle data with a large proportion of outliers, but also alleviate sensitivity to unbalanced data. Experimental results on both synthetic data and real images demonstrate the proposed method outperforms several state-of-the-art fitting methods.

Acknowledgment. This work was supported by the National Natural Science Foundation of China under Grants U1605252, 61472334 and 61571379.

References

1. Mittal, S., Anand, S., Meer, P.: Generalized projection-based M-estimator. IEEE Trans. Pattern Anal. Mach. Intell. **34**, 2351–2364 (2012)
2. Purkait, P., Chin, T.J., Ackermann, H., Suter, D.: Clustering with hypergraphs: the case for large hyperedges. In: Proceedings of the European Conference on Computer Vision, pp. 672–687 (2014)
3. Magri, L., Fusiello, A.: T-linkage: a continuous relaxation of J-linkage for multi-model fitting. In: Proceedings of the IEEE Conference on Computer Vision and Pattern Recognition, pp. 3954–3961 (2014)
4. Raguram, R., Chum, O., Pollefeys, M., Matas, J., Frahm, J.: USAC: a universal framework for random sample consensus. IEEE Trans. Pattern Anal. Mach. Intell. **35**, 2022–2038 (2013)
5. Chin, T.J., Wang, H., Suter, D.: Robust fitting of multiple structures: the statistical learning approach. In: Proceedings of the IEEE Conference on Computer Vision and Pattern Recognition, pp. 413–420 (2009)
6. Lazic, N., Givoni, I., Frey, B., Aarabi, P.: FLoSS: facility location for subspace segmentation. In: Proceedings of the IEEE Conference on International Conference on Computer Vision, pp. 825–832 (2009)
7. Wang, H., Chin, T.J., Suter, D.: Simultaneously fitting and segmenting multiple-structure data with outliers. IEEE Trans. Pattern Anal. Mach. Intell. **34**, 1177–1192 (2012)
8. Fischler, M.A., Bolles, R.C.: Random sample consensus: a paradigm for model fitting with applications to image analysis and automated cartography. Commun. ACM **24**, 381–395 (1981)
9. Torr, P.H., Zisserman, A.: Mlesac: a new robust estimator with application to estimating image geometry. Comput. Vis. Image Underst. **78**, 138–156 (2000)
10. Chum, O., Matas, J., Kittler, J.: Locally optimized RANSAC. In: Michaelis, B., Krell, G. (eds.) DAGM 2003. LNCS, vol. 2781, pp. 236–243. Springer, Heidelberg (2003). doi:10.1007/978-3-540-45243-0_31
11. Chum, O., Matas, J.: Matching with PROSAC-progressive sample consensus. In: Proceedings of the IEEE Conference on Computer Vision and Pattern Recognition, pp. 220–226 (2005)
12. Frahm, J.M., Pollefeys, M.: RANSAC for (quasi-) degenerate data (QDEGSAC). In: Proceedings of the IEEE Conference on Computer Vision and Pattern Recognition, pp. 453–460 (2006)
13. Vincent, E., Laganiere, R.: Detecting planar homographies in an image pair. In: Proceedings of the International Symposium on Image and Signal Processing and Analysis, pp. 182–187 (2001)
14. Kanazawa, Y., Kawakami, H.: Detection of planar regions with uncalibrated stereo using distributions of feature points. In: Proceedings of the British Machine Vision Conference, pp. 1–10 (2004)
15. Toldo, R., Fusiello, A.: Robust multiple structures estimation with J-linkage. In: Proceedings of the European Conference on Computer Vision, pp. 537–547 (2008)
16. Wang, H., Xiao, G., Yan, Y., Suter, D.: Mode-seeking on hypergraphs for robust geometric model fitting. In: Proceedings of the IEEE International Conference on Computer Vision, pp. 2902–2910 (2015)
17. Magri, L., Fusiello, A.: Robust multiple model fitting with preference analysis and low-rank approximation. In: Proceedings of the British Machine Vision Conference, pp. 1–12 (2015)

18. Frey, B.J., Dueck, D.: Clustering by passing messages between data points. Science **315**, 972–976 (2007)
19. Yu, J., Chin, T.J., Suter, D.: A global optimization approach to robust multi-model fitting. In: Proceedings of the IEEE Conference on Computer Vision and Pattern Recognition, pp. 2041–2048 (2011)
20. Dempster, A.P., Laird, N.M., Rubin, D.B.: Maximum likelihood from incomplete data via the EM algorithm. J. Roy. Stat. Soc. **39**, 1–38 (1977)
21. Tanimoto, T.T.: Elementary mathematical theory of classification and prediction. Internal IBM Technical Report (1957)
22. Wong, H.S., Chin, T.J., Yu, J., Suter, D.: Dynamic and hierarchical multi-structure geometric model fitting. In: Proceedings of the IEEE International Conference on Computer Vision, pp. 1044–1051 (2011)

Deep Supervised Hashing with Triplet Labels

Xiaofang Wang$^{(\boxtimes)}$, Yi Shi, and Kris M. Kitani

Carnegie Mellon University, Pittsburgh, PA 15213, USA
{xiaofan2,ys1}@andrew.cmu.edu, kkitani@cs.cmu.edu

Abstract. Hashing is one of the most popular and powerful approximate nearest neighbor search techniques for large-scale image retrieval. Most traditional hashing methods first represent images as off-the-shelf visual features and then produce hashing codes in a separate stage. However, off-the-shelf visual features may not be optimally compatible with the hash code learning procedure, which may result in sub-optimal hash codes. Recently, deep hashing methods have been proposed to simultaneously learn image features and hash codes using deep neural networks and have shown superior performance over traditional hashing methods. Most deep hashing methods are given supervised information in the form of pairwise labels or triplet labels. The current state-of-the-art deep hashing method DPSH [1], which is based on pairwise labels, performs image feature learning and hash code learning simultaneously by maximizing the likelihood of pairwise similarities. Inspired by DPSH [1], we propose a triplet label based deep hashing method which aims to maximize the likelihood of the given triplet labels. Experimental results show that our method outperforms all the baselines on CIFAR-10 and NUS-WIDE datasets, including the state-of-the-art method DPSH [1] and all the previous triplet label based deep hashing methods.

1 Introduction

With the rapid growth of image data on the Internet, much attention has been devoted to approximate nearest neighbor (ANN) search. Hashing is one of the most popular and powerful techniques for ANN search due to its computational and storage efficiencies. Hashing aims to map high dimensional image features into compact hash codes or binary codes so that the Hamming distance between hash codes approximates the Euclidean distance between image features.

Many hashing methods have been proposed and they can be categorized into data-independent and data-dependent methods. Compared with the data-dependent methods, data-independent methods need longer codes to achieve satisfactory performance [2]. Data-dependent methods can be further categorized into unsupervised and supervised methods. Compared to unsupervised methods, supervised methods usually can achieve competitive performance with fewer bits due to the help of supervised information, which is advantageous for search speed and storage efficiency [3].

Most existing hashing methods first represent images as off-the-shelf visual features such as GIST [4], SIFT [5] and hash code learning procedure is independent of the features of images. However, off-the-shelf visual features may not

© Springer International Publishing AG 2017
S.-H. Lai et al. (Eds.): ACCV 2016, Part I, LNCS 10111, pp. 70–84, 2017.
DOI: 10.1007/978-3-319-54181-5_5

be optimally compatible with hash code learning procedure. In other words, the similarity between two images may not be optimally preserved by the visual features and thus the learned hash codes are sub-optimal [3]. Therefore, those hashing methods may not be able to achieve satisfactory performance in practice. To address the drawbacks of hashing methods that rely on off-the-shelf visual features, feature learning based deep hashing methods [1,3,6,7] have been proposed to simultaneously learn image feature and hash codes with deep neural networks and have demonstrated superior performance over traditional hashing methods. Most proposed deep hashing methods fall into the category of supervised hashing methods. Supervised information is given in the form of pairwise labels or triplet labels, a special case of ranking labels.

DPSH [1] is the current state-of-the-art deep hashing method, which is supervised by pairwise labels. Similar to LFH [7], DPSH aims to maximize the likelihood of the pairwise similarities, which is modeled as a function of the Hamming distance between the corresponding data points.

We argue that triplet labels inherently contain richer information than pairwise labels. Each triplet label can be naturally decomposed into two pairwise labels. Whereas, a triplet label can be constructed from two pairwise labels only when the same query image is used in a positive pairwise label and a negative pairwise label simultaneously. A triplet label ensures that in the learned hash code space, the query image is close to the positive image and far from the negative image simultaneously. However, a pairwise label can only ensure that one constraint is observed. Triplet labels explicitly provide a notion of relative similarities between images while pairwise labels can only encode that implicitly.

Therefore, we propose a triplet label based deep hashing method, which performs image feature learning and hash code learning simultaneously by maximizing the likelihood of the given triplet labels. As shown in Fig. 1, the proposed model has three key components: (1) image feature learning component: a deep neural network to learn visual features from images, (2) hash code learning component: one fully connected layer to learn hash codes from image features and (3) loss function component: a loss function to measure how well the given triplet labels are satisfied by the learned hash codes by computing the likelihood of the given triplet labels. Extensive experiments on standard benchmark datasets such as CIFAR-10 and NUS-WIDE show that our proposed deep hashing method outperforms all the baselines, including the state-of-the-art method DPSH [1] and all the previous triplet label based deep hashing methods.

Contributions: (1) We propose a novel triplet label based deep hashing method to simultaneously perform image feature and hash code learning in an end-to-end manner; (2) We present a novel formulation of the likelihood of the given triplet labels to evaluate the quality of learned hash codes; (3) We provide ablative analysis of our loss function to help understand how each term contributes to performance; (4) We obtain state-of-the-art performance on benchmark datasets.

2 Related Work

Hashing methods can be categorized into data-independent and data-dependent methods, based on whether they are independent of training data. Representative data-independent methods include Locality Sensitive Hashing (LSH) [8] and Shift-Invariant Kernels Hashing (SIKH) [9]. Data-independent methods generally need longer hash codes for satisfactory performance, compared to data-dependent methods. Data-dependent methods can be further divided into unsupervised and supervised methods, based on whether supervised information is provided or not.

Unsupervised hashing methods only utilize the training data points to learn hash functions, without using any supervised information. Notable examples for unsupervised hashing methods include Spectral Hashing (SH) [10], Binary Reconstructive Embedding (BRE) [11], Iterative Quantization (ITQ) [2], Isotropic Hashing (IsoHash) [12], graph-based hashing methods [13–15] and two deep hashing methods: Semantic Hashing [16] and the hashing method proposed in [17].

Supervised hashing methods leverage labeled data to learn hash codes. Typically, the supervised information are provided in one of three forms: point-wise labels, pairwise labels or ranking labels [1]. Representative point-wise label based methods include CCA-ITQ [2] and the deep hashing method proposed in [18]. Representative pairwise label based hashing methods include Minimal Loss Hashing (MLH) [19], Supervised Hashing with Kernels (KSH) [20], Latent Factor Hashing (LFH) [21], Fash Supervised Hashing (FASTH) [22] and two deep hashing methods: CNNH [23] and DPSH [1]. Representative ranking label based hashing methods include Ranking-based Supervised Hashing (RSH) [24], Column Generation Hashing (CGHASH) [25] and the deep hashing methods proposed in [3,6,7,17]. One special case of ranking labels is triplet labels.

Most existing supervised hashing methods represent images as off-the-shelf visual features and perform hash code learning independent of visual features, including some deep hashing methods [16,17]. However, off-the-shelf features may not be optimally compatible with hash code learning procedure and thus results in sub-optimal hash codes.

CNNH [23], supervised by triplet labels, is the first proposed deep hashing method without using off-the-shelf features. However, CNNH cannot learn image features and hash codes simultaneously and still has limitations. This has been verified by the authors of CNNH themselves in a follow-up work [3]. Other ranking label or triplet label based deep hashing methods include Network in Network Hashing (NINH) [3], Deep Semantic Ranking based Hashing (DSRH) [6], Deep Regularized Similarity Comparison Hashing (DRSCH) [7] and Deep Similarity Comparison Hashing (DSCH) [7]. While these methods can simultaneously perform image feature learning and hash code learning given supervision of triplet labels, we present a novel formulation of the likelihood of the given triplet labels to evaluate the quality of learned hash codes.

Deep Pairwise-Supervised Hashing (DPSH) [1] is the first proposed deep hashing method to simultaneously perform image feature learning and hash code learning with pairwise labels and achieves highest performance compared to

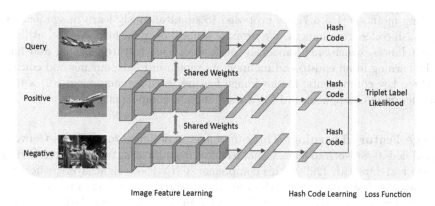

Fig. 1. Overview of the proposed end-to-end deep hashing method. (Color figure online)

other deep hashing methods. Our method is supervised by triplet labels because triplet labels inherently contain richer information than pairwise labels.

3 Approach

We first give the formal definition of our problem and then introduce our proposed end-to-end method which simultaneously performs image feature learning and hash code learning from triplet labels.

3.1 Problem Definition

Given N training images $\mathcal{I} = \{I_1, \ldots, I_N\}$ and M triplet labels $\mathcal{T} = \{(q_1, p_1, n_1), \ldots, (q_M, p_M, n_M)\}$ where the triplet of image indices (q_m, p_m, n_m) denotes that the query image of index $q_m \in \{1 \ldots N\}$ is more similar to the positive image of index $p_m \in \{1 \ldots N\}$ than to the negative image of index $n_m \in \{1 \ldots N\}$. One possible way of generating triplet labels is by selecting two images from the same semantic category (I_{q_m} and I_{p_m}) and selecting the negative (I_{n_m}) from a different semantic category.

Our goal is to learn a hash code $\mathbf{b}_n$ for each image I_n, where $\mathbf{b} \in \{+1, -1\}^L$ and L is the length of hash codes. The hash codes $\mathcal{B} = \{\mathbf{b}_n\}_{n=1}^N$ should satisfy all the triplet labels $\mathcal{T}$ as much as possible in the Hamming space. More specifically, $\mathrm{dist}_H(\mathbf{b}_{q_m}, \mathbf{b}_{p_m})$ should be smaller than $\mathrm{dist}_H(\mathbf{b}_{q_m}, \mathbf{b}_{n_m})$ as much as possible, where $\mathrm{dist}_H(\cdot, \cdot)$ denotes the Hamming distance between two binary codes. Generally speaking, we aim to learn a hash function $h(\cdot)$ to map images to hash codes. We can write $h(\cdot)$ as $[h_1(\cdot), \ldots, h_L(\cdot)]$ and for image I_n, its hash code can be denoted as $\mathbf{b}_n = h(I_n) = [h_1(I_n), \ldots, h_L(I_n)]$.

3.2 Learning the Hash Function

Most previous hashing methods rely on off-the-shelf visual features, which may not be optimally compatible with the hash code learning procedure. Thus deep hashing methods [1,3,6,7] are proposed to simultaneously learn image features and hash codes from images. We propose a novel deep hashing method utilizing triplet labels, which simultaneously performs image feature learning and hash code learning in an end-to-end manner. As shown in Fig. 1, our method consists of three key components: (1) an image feature learning component, (2) a hash code learning component and (3) a loss function component.

Image Feature Learning. This component is designed to employ a Convolutional neural network to learn visual features from images. We adopt the CNN-F network architecture [26] for this component. CNN-F has eight layers, where the last layer is designed to learn the probability distribution over category labels. So only the first 7 layers of CNN-F are used in this component. Other networks like AlexNet [27], residual network [28] can also be used in this component.

Hash Code Learning. This component is designed to learn hash codes of images. We use one fully connected layer for this component and we want this layer to output hash codes of images. In particular, the number of neurons of this layer equals the length of targeted hash codes. Multiple fully connected layers or other architectures like the divide-and-encode module proposed by [3] can also be applied here. We do not focus on this in this work and leave this for future study.

Loss Function. This component measures how well the given triple labels are satisfied by the learned hash codes by computing the likelihood of the given triplet labels. Inspired by the likelihood of the pairwise similarities proposed in LFH [21], we present our formulation of the likelihood for a given triplet label. We call this the *triplet label likelihood* throughout the text.

Let Θ_{ij} denote half of the inner product between two hash codes $\mathbf{b}_i, \mathbf{b}_j \in \{+1, -1\}^L$:

$$\Theta_{ij} = \frac{1}{2}\mathbf{b}_i^T\mathbf{b}_j \tag{1}$$

Then the triplet label likelihood is formulated as:

$$p(\mathcal{T} \mid \mathcal{B}) = \prod_{m=1}^{M} p((q_m, p_m, n_m) \mid \mathcal{B}) \tag{2}$$

with

$$p((q_m, p_m, n_m) \mid \mathcal{B}) = \sigma(\Theta_{q_m p_m} - \Theta_{q_m n_m} - \alpha) \tag{3}$$

where $\sigma(x)$ is the sigmoid function $\sigma(x) = \frac{1}{1+e^{-x}}$, α is the margin, a positive hyper-parameter and $\mathcal{B}$ is the set of all hash codes.

We now show how maximizing the triplet label likelihood matches the goal to preserve the relative similarity between the query image, positive image and negative image. We first prove the following relationship between the Hamming distance between two binary codes and their inner product:

$$\text{dist}_H(\mathbf{b}_i, \mathbf{b}_j) = \frac{1}{2}(L - 2\Theta_{ij}) \tag{4}$$

According to Eq. 4, we can have

$$\text{dist}_H(\mathbf{b}_{q_m}, \mathbf{b}_{p_m}) - \text{dist}_H(\mathbf{b}_{q_m}, \mathbf{b}_{n_m}) = -(\Theta_{q_m p_m} - \Theta_{q_m n_m}) \tag{5}$$

According to Eq. 3, we know that the larger $p((q_m, p_m, n_m) \mid \mathcal{B})$ is, the larger $(\Theta_{q_m p_m} - \Theta_{q_m n_m} - \alpha)$ will be. Since α is a constant number here, the larger $(\Theta_{q_m p_m} - \Theta_{q_m n_m} - \alpha)$ is, the smaller $(\text{dist}_H(\mathbf{b}_{q_m}, \mathbf{b}_{p_m}) - \text{dist}_H(\mathbf{b}_{q_m}, \mathbf{b}_{n_m}))$ will be. Thus, by maximizing the triplet label likelihood $p(\mathcal{T} \mid \mathcal{B})$, we can enforce the Hamming distance between the query and the positive image to be smaller than that between the query and the negative image. The margin α here can regularize the distance gap between $\text{dist}_H(\mathbf{b}_{q_m}, \mathbf{b}_{p_m})$ and $\text{dist}_H(\mathbf{b}_{q_m}, \mathbf{b}_{n_m})$. The margin α can also help speed up training our model as explained later in this Section and verified in our experiments in Sect. 4.4.

Now we define our loss function as the negative log triplet label likelihood as follows,

$$\begin{aligned} L &= -\log p(\mathcal{T} \mid \mathcal{B}) \\ &= -\sum_{m=1}^{M} \log p((q_m, p_m, n_m) \mid \mathcal{B}) \end{aligned} \tag{6}$$

Plug Eq. 3 into the above equation, we can drive that

$$L = -\sum_{m=1}^{M} (\Theta_{q_m p_m} - \Theta_{q_m n_m} - \alpha - \log(1 + e^{\Theta_{q_m p_m} - \Theta_{q_m n_m} - \alpha})) \tag{7}$$

Minimizing the loss defined in (7) is an intractable discrete optimization problem [1]. One can choose to relax $\{\mathbf{b}_n\}$ from discrete to continuous, i.e., relax $\{\mathbf{b}_n\}$ to $\{\mathbf{u}_n\}$, where $\mathbf{u}_n \in \mathbb{R}^{L \times 1}$ and then minimize the loss. This is the strategy employed by LFH [21], but this may be harmful to the performance due to the relaxation error [29]. In the context of hashing, the relaxation error is actually the quantization error. Although optimal real vectors $\{\mathbf{u}_n\}$ are learned, we still need to quantize them to binary codes $\{\mathbf{b}_n\}$. This process induces quantization error. To reduce it, we propose to also consider the quantization error when solving the relaxed problem.

Concretely, we relax binary codes $\{\mathbf{b}_n\}$ to real vectors $\{\mathbf{u}_n\}$ and re-define Θ_{ij} as

$$\Theta_{ij} = \frac{1}{2}\mathbf{u}_i^T \mathbf{u}_j \tag{8}$$

and our loss function becomes

$$
\begin{aligned}
L = &- \sum_{m=1}^{M} (\Theta_{q_m p_m} - \Theta_{q_m n_m} - \alpha - \log(1 + e^{\Theta_{q_m p_m} - \Theta_{q_m n_m} - \alpha})) \\
&+ \lambda \sum_{n=1}^{N} ||\mathbf{b}_n - \mathbf{u}_n||_2^2
\end{aligned}
\tag{9}
$$

where λ is a hyper-parameter to balance the negative log triplet likelihood and the quantization error and $\mathbf{b}_n = sgn(\mathbf{u}_n)$, where $sgn(\cdot)$ is the sign function and $sgn(\mathbf{u}_n^{(k)})$ equals to 1 if $\mathbf{u}_n^{(k)} > 0$ and -1, otherwise. The quantization error term $\sum_{n=1}^{N} ||\mathbf{b}_n - \mathbf{u}_n||_2^2$ is also adopted as a regularization term in DPSH [1]. However, they do not interpret it as quantization error.

Model Learning. The three key components of our method can be integrated into a Siamese-triplet network as shown in Fig. 1, which takes triplets of images as input and output hash codes of images. The network consists of three sub-networks with exactly the same architecture and shared weights. In our experiments, the sub-network is a fully connected layer on top of the first seven layers of CNN-F [26]. We train our network by minimizing the loss function:

$$
\begin{aligned}
L(\theta) = &- \sum_{m=1}^{M} (\Theta_{q_m p_m} - \Theta_{q_m n_m} - \alpha - \log(1 + e^{\Theta_{q_m p_m} - \Theta_{q_m n_m} - \alpha})) \\
&+ \lambda \sum_{n=1}^{N} ||\mathbf{b}_n - \mathbf{u}_n||_2^2
\end{aligned}
\tag{10}
$$

where θ denotes all the parameters of the sub-network, $\mathbf{u}_n$ is the output of the sub-network with n^{th} training image and $\mathbf{b}_n = sgn(\mathbf{u}_n)$. We can see that L is differentiable with respect to $\mathbf{u}_n$. Thus the back-propagation algorithm can be applied here to minimize the loss function.

Once training is completed, we can apply our model to generate hash codes for new images. For a new image I, we pass it into the trained sub-network and take the output of the last layer $\mathbf{u}$. Then the hash code $\mathbf{b}$ of image I is $\mathbf{b} = sgn(\mathbf{u})$.

Impact of Margin α. We argue that a positive margin α can help speed up the training process. We now analyze this theoretically by looking at the derivative of the loss function. In particular, for n^{th} image, we compute the derivative of the loss function L with respect to $\mathbf{u}_n$ as follows:

$$
\begin{aligned}
\frac{\partial L}{\partial \mathbf{u}_n} = & -\frac{1}{2} \sum_{m:(n,p_m,n_m)\in\mathcal{T}} (1 - \sigma(\Theta_{q_m p_m} - \Theta_{q_m n_m} - \alpha))(\mathbf{u}_{p_m} - \mathbf{u}_{n_m}) \\
& -\frac{1}{2} \sum_{m:(q_m,n,n_m)\in\mathcal{T}} (1 - \sigma(\Theta_{q_m p_m} - \Theta_{q_m n_m} - \alpha))\mathbf{u}_{q_m} \\
& +\frac{1}{2} \sum_{m:(q_m,p_m,n)\in\mathcal{T}} (1 - \sigma(\Theta_{q_m p_m} - \Theta_{q_m n_m} - \alpha))\mathbf{u}_{q_m} \\
& + 2\lambda(\mathbf{u}_n - \mathbf{b}_n)
\end{aligned}
\tag{11}
$$

where $\sigma(x)$ is the sigmoid function $\sigma(x) = \frac{1}{1+e^{-x}}$ and $\mathcal{T}$ is the set of triplet labels. In the derivative shown above, we observe this term $(1 - \sigma(\Theta_{q_m p_m} - \Theta_{q_m n_m} - \alpha))$. We know that $\sigma(x)$ saturates very quickly, *i.e.*, being very close to 1, as x increases. If $\alpha = 0$, when $(\Theta_{q_m p_m} - \Theta_{q_m n_m})$ becomes positive, the term $(1 - \sigma(\Theta_{q_m p_m} - \Theta_{q_m n_m} - \alpha))$ will be very close to 0. This will make the magnitude of the derivative very small and further make the model hard to train. A positive margin α adds a negative offset on $(\Theta_{q_m p_m} - \Theta_{q_m n_m})$ and can prevent $(1 - \sigma(\Theta_{q_m p_m} - \Theta_{q_m n_m} - \alpha))$ from being very small. Further this makes the model easier to train and helps speed up the training process. We give experiment results to verify this in Sect. 4.4.

4   Experiment

4.1   Datasets and Evaluation Protocol

We conduct experiments on two widely used benchmark datasets, CIFAR-10 [30] and NUS-WIDE [31]. The CIFAR-10 [30] dataset contains 60,000 color images of size 32×32, which can be divided into 10 categories and 6,000 images for each category. Each image is only associated with one category. The NUS-WIDE [31] dataset contains nearly 27,000 color images from the web. Different from CIFAR-10, NUS-WIDE is a multi-label dataset. Each image is annotated with one or multiple labels in 81 semantic concepts. Following the setting in [1,3,7,23], we only consider images annotated with 21 most frequent labels. For each of the 21 labels, at least 5,000 images are annotated with the label. In addition, NUS-WIDE provides links to images for downloading and some links are now invalid. This causes some differences between the image set used by previous work and our work. In total, we use 161,463 images from the NUS-WIDE dataset.

We employ mean average precision (MAP) to evaluate the performance of our method and baselines similar to most previous work [1,3,7,23]. Two images in CIFAR-10 are considered similar if they belong to the same category. Two images in NUS-WIDE are considered similar if they share at least one label.

4.2   Baselines and Setting

Following [1], we consider the following baselines:

1. Traditional unsupervised hashing methods using hand-crafted features, including SH [10], ITQ [2].
2. Traditional supervised hashing methods using hand-crafted features, including SPLH [32], KSH [20], FastH [22], LFH [21] and SDH [33].
3. The above traditional hashing methods using features extracted by CNN-F network [26] pre-trained on ImageNet.
4. Pairwise label based deep hashing methods: CNNH [23] and DPSH [1].
5. Triplet label based deep hashing methods: NINH [3], DSRH [6], DSCH [7] and DRSCH [7].

When using hand-crafted features, we use a 512-dimensional GIST descriptor [4] to represent CIFAR-10 images. For NUS-WIDE images, we represent them by a 1134-dimensional feature vector, which is the concatenation of a 64-D color histogram, a 144-D color correlogram, a 73-D edge direction histogram, a 128-D wavelet texture, a 225-D block-wise color moments and a 500-D BoW representation based on SIFT descriptors.

Following [1,6], we initialize the first seven layers of our network with the CNN-F network [26] pre-trained on ImageNet. In addition, the hyper-parameter α is set to half of the length of hash codes, *e.g.*, 16 for 32-bit hash codes and the hyper-parameter λ is set to 100 unless otherwise stated.

We compare our method to most baselines under the following experimental setting. Following [1,3,23], in CIFAR-10, 100 images per category, *i.e.*, in total 1,000 images, are randomly sampled as query images. The remaining images are used as database images. For unsupervised hashing methods, all the database images are used as training images. For supervised hashing methods, 500 database images per category, *i.e.*, in total 5,000 images, are randomly sampled as training images. In NUS-WIDE, 100 images per label, *i.e.*, in total 2,100 images, are randomly sampled as query images. Likewise, the remaining images are used as database images. For unsupervised hashing methods, all the database images are used as training images. For supervised hashing methods, 500 database images per label, *i.e.*, in total 10,500 images, are randomly sampled as training images. Since NUS-WIDE contains a huge number of images, when computing MAP for NUS-WIDE, only the top 5,000 returned neighbors are considered.

We also compare our method to DSRH [6], DSCH [7], DRSCH [7] and DPSH [1] under a different experimental setting. In CIFAR-10, 1,000 images per category, *i.e.*, in total 10,000 images, are randomly sampled as query images. The remaining images are used as database images and all the database images are used as training images. In NUS-WIDE, 100 images per label, *i.e.*, in total 2,100 images, are randomly sampled as query images. The remaining images are used as database images and still, all the database images are used as training images. Under this setting, when computing MAP for NUS-WIDE, we only consider the top 50,000 returned neighbors.

4.3   Performance Evaluation

Comparison to Traditional Hashing Methods using Hand-crafted Features. As shown in Table 1, we can see that on both datasets, our method

Table 1. Mean Average Precision (MAP) under the first experimental setting. The MAP for NUS-WIDE is computed based on the top 5,000 returned neighbors. The best performance is shown in boldface. DPSH* denotes the performance we obtain by running the code provided by the authors of DPSH in our experiments.

Method	CIFAR-10				Method	NUS-WIDE			
	12 bits	24 bits	32 bits	48 bits		12 bits	24 bits	32 bits	48 bits
Ours	0.710	**0.750**	**0.765**	**0.774**	Ours	**0.773**	**0.808**	**0.812**	**0.824**
DPSH	**0.713**	0.727	0.744	0.757	DPSH*	0.752	0.790	0.794	0.812
NINH	0.552	0.566	0.558	0.581	NINH	0.674	0.697	0.713	0.715
CNNH	0.439	0.511	0.509	0.522	CNNH	0.611	0.618	0.625	0.608
FastH	0.305	0.349	0.369	0.384	FastH	0.621	0.650	0.665	0.687
SDH	0.285	0.329	0.341	0.356	SDH	0.568	0.600	0.608	0.637
KSH	0.303	0.337	0.346	0.356	KSH	0.556	0.572	0.581	0.588
LFH	0.176	0.231	0.211	0.253	LFH	0.571	0.568	0.568	0.585
SPLH	0.171	0.173	0.178	0.184	SPLH	0.568	0.589	0.597	0.601
ITQ	0.162	0.169	0.172	0.175	ITQ	0.452	0.468	0.472	0.477
SH	0.127	0.128	0.126	0.129	SH	0.454	0.406	0.405	0.400

outperforms previous hashing methods using hand-crafted features significantly. In Table 1, the results of NINH, CNNH, KSH and ITQ are from [3,23] and the results of other methods except our method are from [1]. This is reasonable as we use the same experimental setting and evaluation protocol.

Comparison to Traditional Hashing Methods using Deep Features. When we train our model, we initialize the first 7 layers of our network with CNN-F network [26] pre-trained on ImageNet. Thus one may argue that the boost in performance comes from that network instead of our method. To further validate our method, we compare our method to traditional hashing methods using deep features extracted by CNN-F network. As shown in Table 2, we can see that our method can significantly outperform traditional methods on CIFAR-10 and obtain comparable performance with the best performing traditional methods on NUS-WIDE. The results in Table 2 are copied from [1], which is reasonable as we used the same experimental setting and evaluation protocol.

Comparison to Deep Hashing Methods. Now we compare our method to other deep hashing methods. In particular, we compare our method to CNNH, NINH and DPSH under the first experimental setting in Table 1 and DSRH, DSCH, DRSCH and DPSH under the second experimental setting in Table 3. The results of DSRH, DSCH and DRSCH are directly from [7]. We can see that our method significantly outperforms all previous triplet label based deep hashing methods, including NINH, DSRH, DSCH and DRSCH.

Table 2. Mean Average Precision (MAP) under the first experimental setting. The MAP for NUS-WIDE is computed based on the top 5,000 returned neighbors. The best performance is shown in boldface.

Method	CIFAR-10				NUS-WIDE			
	12 bits	24 bits	32 bits	48 bits	12 bits	24 bits	32 bits	48 bits
Ours	**0.710**	**0.750**	**0.765**	**0.774**	0.773	**0.808**	0.812	0.824
FastH + CNN	0.553	0.607	0.619	0.636	0.779	0.807	**0.816**	**0.825**
SDH + CNN	0.478	0.557	0.584	0.592	**0.780**	0.804	0.815	0.824
KSH + CNN	0.488	0.539	0.548	0.563	0.768	0.786	0.79	0.799
LFH + CNN	0.208	0.242	0.266	0.339	0.695	0.734	0.739	0.759
SPLH + CNN	0.299	0.33	0.335	0.33	0.753	0.775	0.783	0.786
ITQ + CNN	0.237	0.246	0.255	0.261	0.719	0.739	0.747	0.756
SH + CNN	0.183	0.164	0.161	0.161	0.621	0.616	0.615	0.612

Table 3. Mean Average Precision (MAP) under the second experimental setting. The MAP for NUS-WIDE is computed based on the top 50,000 returned neighbors. The best performance is shown in boldface. DPSH* denotes the performance we obtain by running the code provided by the authors of DPSH in our experiments.

Method	CIFAR-10				NUS-WIDE			
	16 bits	24 bits	32 bits	48 bits	16 bits	24 bits	32 bits	48 bits
Ours	**0.915**	**0.923**	**0.925**	**0.926**	**0.756**	**0.776**	**0.785**	**0.799**
DPSH	0.763	0.781	0.795	0.807	0.715	0.722	0.736	0.741
DRSCH	0.615	0.622	0.629	0.631	0.618	0.622	0.623	0.628
DSCH	0.609	0.613	0.617	0.62	0.592	0.597	0.611	0.609
DSRH	0.608	0.611	0.617	0.618	0.609	0.618	0.621	0.631
DPSH*	0.903	0.885	0.915	0.911	N/A			

We now compare our method to the current state-of-the-art method DPSH under the first experimental setting. As shown in Table 1, our method outperforms DPSH by about 2% on both CIFAR-10 and NUS-WIDE datasets. Note that on NUS-WIDE, we are comparing to DPSH* instead of DPSH. DPSH represents the performance reported in [1] and DPSH* represents the performance we obtain by running the code of DPSH provided by the authors of [1] on NUS-WIDE. We re-run their code on NUS-WIDE because the NUS-WIDE dataset does not provide the original images to download instead of the links to image, which results in some differences between the images used by them [1] and us.

As shown in Table 3, our method outperforms DPSH by more than 10% on CIFAR-10 and about 5% on NUS-WIDE under the second experimental setting. We also re-run the code of DPSH on CIFAR-10 under the same setting and

Fig. 2. Retrieval examples. Left: Query images. Middle: Top images retrieved by hash codes learnt by our method. Right: Top images retrieved by hash codes learnt by DPSH [1].

we can obtain much higher the performance then what was reported in [1].[1] The performance we obtain is denoted by DPSH* in Table 3 and we can see our method still outperforms DPSH* by about 1%. We also show some retrieval examples on NUS-WIDE with hash codes learned by our method and DPSH respectively in Fig. 2.

4.4 Ablation Studies

Impact of the hyper-parameter α. Figure 3a shows the effect of the margin α at 32 bits on CIFAR-10 dataset. We can see that within the same number of training epochs, we can obtain better performance with a larger margin. This verifies our previous analysis in Sect. 3.

Impact of the hyper-parameter λ. Figure 3b shows the effect of the hyper-parameter λ in Eq. 9 at 12 bits and 32 bits on CIFAR-10 dataset. As one can see, for both 12 bits and 32 bits, there is a significant performance drop in terms of MAP when λ becomes very small (*e.g.*, 0.1) or very large (*e.g.*, 1000). This is reasonable since λ is designed to balance the negative log triplet likelihood and the quantization error. Setting λ to small or to large will lead to inbalance between these two terms.

Impact of the number of training images. We also study the impact of the number of training images on the performance. Figure 3c shows the performance

---

[1] We communicated with the authors of DPSH. The main difference between our experiments and their experiments is the step size and decay factor for learning rate change. They say that with our parameters, they can also get better results than what is reported in their paper [1].

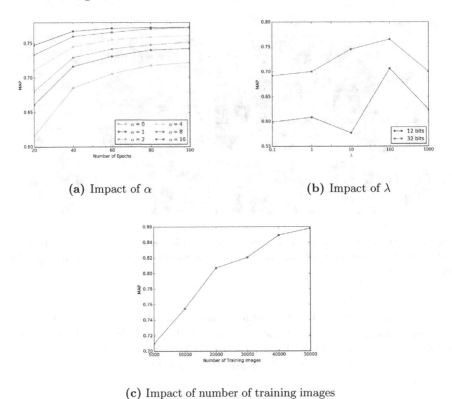

(a) Impact of α (b) Impact of λ

(c) Impact of number of training images

Fig. 3. Ablation studies.

of our method at 12 bits on CIFAR-10 using different number of training images. We can see that more training images will incur noticeable improvements.

5   Conclusion

In this paper, we have proposed a novel deep hashing method to simultaneously learn image features and hash codes given the supervision of triplet labels. Our method learns high quality hash codes by maximizing the likelihood of given triplet labels under learned hash codes. Extensive experiments on standard benchmark datasets show that our method outperforms all the baselines, including the state-of-the-art method DPSH [1] and all the previous triplet label based deep hashing methods.

Acknowledgement. This work was sponsored by DARPA under agreement number FA8750-14-2-0244. The U.S. Government is authorized to reproduce and distribute reprints for Governmental purposes notwithstanding any copyright notation thereon. The views and conclusions contained herein are those of the authors and should not be interpreted as necessarily representing the official policies or endorsements, either expressed or implied, of DARPA or the U.S. Government.

References

1. Li, W.J., Wang, S., Kang, W.C.: Feature learning based deep supervised hashing with pairwise labels. arXiv preprint arXiv:1511.03855 (2015)
2. Gong, Y., Lazebnik, S.: Iterative quantization: a procrustean approach to learning binary codes. In: 2011 IEEE Conference on Computer Vision and Pattern Recognition (CVPR), 817–824. IEEE (2011)
3. Lai, H., Pan, Y., Liu, Y., Yan, S.: Simultaneous feature learning and hash coding with deep neural networks. In: Proceedings of the IEEE Conference on Computer Vision and Pattern Recognition, pp. 3270–3278 (2015)
4. Oliva, A., Torralba, A.: Modeling the shape of the scene: a holistic representation of the spatial envelope. Int. J. Comput. Vision 42, 145–175 (2001)
5. Lowe, D.G.: Distinctive image features from scale-invariant keypoints. Int. J. Comput. Vision 60, 91–110 (2004)
6. Zhao, F., Huang, Y., Wang, L., Tan, T.: Deep semantic ranking based hashing for multi-label image retrieval. In: Proceedings of the IEEE Conference on Computer Vision and Pattern Recognition, pp. 1556–1564 (2015)
7. Zhang, R., Lin, L., Zhang, R., Zuo, W., Zhang, L.: Bit-scalable deep hashing with regularized similarity learning for image retrieval and person re-identification. IEEE Trans. Image Process. 24, 4766–4779 (2015)
8. Andoni, A., Indyk, P.: Near-optimal hashing algorithms for approximate nearest neighbor in high dimensions. In: 47th Annual IEEE Symposium on Foundations of Computer Science, FOCS 2006, pp. 459–468. IEEE (2006)
9. Raginsky, M., Lazebnik, S.: Locality-sensitive binary codes from shift-invariant kernels. In: Advances in Neural Information Processing Systems, pp. 1509–1517 (2009)
10. Weiss, Y., Torralba, A., Fergus, R.: Spectral hashing. In: Advances in Neural Information Processing Systems, pp. 1753–1760 (2009)
11. Kulis, B., Darrell, T.: Learning to hash with binary reconstructive embeddings. In: Advances in Neural Information Processing Systems, pp. 1042–1050 (2009)
12. Kong, W., Li, W.J.: Isotropic hashing. In: Advances in Neural Information Processing Systems, pp. 1646–1654 (2012)
13. Liu, W., Wang, J., Kumar, S., Chang, S.F.: Hashing with graphs. In: Proceedings of the 28th International Conference on Machine Learning (ICML 2011), pp. 1–8 (2011)
14. Liu, W., Mu, C., Kumar, S., Chang, S.F.: Discrete graph hashing. In: Advances in Neural Information Processing Systems, pp. 3419–3427 (2014)
15. Jiang, Q.Y., Li, W.J.: Scalable graph hashing with feature transformation. In: Proceedings of the International Joint Conference on Artificial Intelligence (2015)
16. Salakhutdinov, R., Hinton, G.: Semantic hashing. Int. J. Approximate Reasoning 50, 969–978 (2009)
17. Erin Liong, V., Lu, J., Wang, G., Moulin, P., Zhou, J.: Deep hashing for compact binary codes learning. In: Proceedings of the IEEE Conference on Computer Vision and Pattern Recognition, pp. 2475–2483 (2015)
18. Lin, K., Yang, H.F., Hsiao, J.H., Chen, C.S.: Deep learning of binary hash codes for fast image retrieval. In: Proceedings of the IEEE Conference on Computer Vision and Pattern Recognition Workshops, pp. 27–35 (2015)
19. Norouzi, M., Fleet, D.J.: Minimal loss hashing for compact binary codes. In: ICML, vol. 1, p. 2 (2011)

20. Liu, W., Wang, J., Ji, R., Jiang, Y.G., Chang, S.F.: Supervised hashing with kernels. In: 2012 IEEE Conference on Computer Vision and Pattern Recognition (CVPR), pp. 2074–2081. IEEE (2012)

21. Zhang, P., Zhang, W., Li, W.J., Guo, M.: Supervised hashing with latent factor models. In: Proceedings of the 37th International ACM SIGIR Conference on Research & Development in Information Retrieval, pp. 173–182. ACM (2014)

22. Lin, G., Shen, C., Shi, Q., Hengel, A., Suter, D.: Fast supervised hashing with decision trees for high-dimensional data. In: Proceedings of the IEEE Conference on Computer Vision and Pattern Recognition, pp. 1963–1970 (2014)

23. Xia, R., Pan, Y., Lai, H., Liu, C., Yan, S.: Supervised hashing for image retrieval via image representation learning. In: AAAI, vol. 1, p. 2 (2014)

24. Wang, J., Liu, W., Sun, A., Jiang, Y.G.: Learning hash codes with listwise supervision. In: Proceedings of the IEEE International Conference on Computer Vision, pp. 3032–3039 (2013)

25. Li, X., Lin, G., Shen, C., Van den Hengel, A., Dick, A.: Learning hash functions using column generation. In: Proceedings of The 30th International Conference on Machine Learning, pp. 142–150(2013)

26. Chatfield, K., Simonyan, K., Vedaldi, A., Zisserman, A.: Return of the devil in the details: delving deep into convolutional nets. arXiv preprint arXiv:1405.3531 (2014)

27. Krizhevsky, A., Sutskever, I., Hinton, G.E.: ImageNet classification with deep convolutional neural networks. In: Advances in Neural Information Processing Systems, pp. 1097–1105 (2012)

28. He, K., Zhang, X., Ren, S., Sun, J.: Deep residual learning for image recognition. arXiv preprint arXiv:1512.03385 (2015)

29. Kang, W.C., Li, W.J., Zhou, Z.H.: Column sampling based discrete supervised hashing (2016)

30. Krizhevsky, A., Hinton, G.: Learning multiple layers of features from tiny images (2009)

31. Chua, T.S., Tang, J., Hong, R., Li, H., Luo, Z., Zheng, Y.: NUS-WIDE: a real-world web image database from National University of Singapore. In: Proceedings of the ACM International Conference on Image and Video Retrieval, p. 48. ACM (2009)

32. Wang, J., Kumar, S., Chang, S.F.: Sequential projection learning for hashing with compact codes. In: Proceedings of the 27th International Conference on Machine Learning (ICML 2010), pp. 1127–1134 (2010)

33. Shen, F., Shen, C., Liu, W., Tao Shen, H.: Supervised discrete hashing. In: Proceedings of the IEEE Conference on Computer Vision and Pattern Recognition, pp. 37–45 (2015)

Boosting Zero-Shot Image Classification via Pairwise Relationship Learning

Hanhui Li[1,2], Hefeng Wu[3,1]($\boxtimes$), Shujin Lin[1], Liang Lin[2], Xiaonan Luo[1,4], and Ebroul Izquierdo[5]

[1] National Engineering Research Center of Digital Life,
Sun Yat-sen University, Guangzhou 510006, China
lihanhui@mail2.sysu.edu.cn, wuhefeng@gmail.com,
{linshjin,lnslxn}@mail.sysu.edu.cn
[2] School of Data and Computer Science,
Sun Yat-sen University, Guangzhou 510006, China
linliang@ieee.org
[3] School of Informatics, Guangdong University of Foreign Studies,
Guangzhou 510006, China
[4] Beijing Key Laboratory of Multimedia and Intelligent Software Technology,
College of Metropolitan Transportation, Beijing University of Technology,
Beijing 100124, China
[5] Queen Mary, University of London, London, UK
ebroul.izquierdo@qmul.ac.uk

Abstract. Zero-shot image classification (ZSIC) is one of the emerging challenges in the communities of computer vision, artificial intelligence and machine learning. In this paper, we propose to exploit the pairwise relationships between test instances to increase the performance of conventional methods, e.g. direct attribute prediction (DAP), for the ZSIC problem. To infer pairwise relationships between test instances, we introduce two different methods, a binary classification based method and a metric learning based method. Based on the inferred relationships, we construct a similarity graph to represent test instances, and then employ an adaptive graph anchors voting method to refine the results of DAP iteratively: In each iteration, we partition the similarity graph with the normalized spectral clustering method, and determine the class label of each cluster via the voting of graph anchors. Extensive experiments validate the effectiveness of our method: with the properly learned pairwise relationships, we successfully boost the mean class accuracy of DAP on two standard benchmarks for the ZSIC problem, **Animal with Attribute** and **aPascal-aYahoo**, from 57.46% to 84.43% and 26.59% to 70.09%, respectively. Besides, experimental results on the **SUN Attribute** also suggest our method can obtain considerable performance improvement for the large-scale ZSIC problem.

1 Introduction

Recently, an emerging problem called zero-shot image classification (ZSIC) [1], has attracted the attention of the communities of computer vision, artificial intelligence

© Springer International Publishing AG 2017
S.-H. Lai et al. (Eds.): ACCV 2016, Part I, LNCS 10111, pp. 85–99, 2017.
DOI: 10.1007/978-3-319-54181-5_6

and machine learning. Distinguished from traditional classification problems, in the training phase of zero-shot classification, examples of the test classes are not available. Due to this constraint, most conventional classification methods become infeasible.

To the best of our knowledge, most methods tackle the ZSIC problem by introducing some mid-level information which can be shared among the training classes and test classes, e.g. attributes [1,2]. Undoubtedly, these methods provide us with practical solutions to the ZSIC problem. However, the performance of these methods is still obviously lower than that of conventional classification methods, since attributes are the only information they can exploit.

In this paper, we propose to explore a high-level concept, the pairwise relationships between test images, to help to better tackle the ZSIC problem. More specifically, by pairwise relationship, we refer to the relationship that *whether two images in a pair belong to the same class*. Employing pairwise relationships in ZSIC is natural: as in the example presented in Fig. 1, given the images of several classes of animals that we have not seen before, we still can divide these images into different classes. Therefore, if this ability of identifying pairwise relationships can be learned, we may be able to use it to correct the mistaken predictions of current methods for the ZSIC problem.

Fig. 1. Demonstration of the intuition of this paper. Human has the ability of identifying images belonging to the same class, even without knowing their names (class labels). This ability should be helpful in zero-shot image classification. In this figure, images belonging to the same class are labeled with the same color. (Color figure online)

Based on such motivation, we focus on two problems in this paper: First, can a model learned on a disjointed training set be used to describe the pairwise relationships between test images? Second, can these pairwise relationships be used to help to solve the zero-shot image classification problem?

To address the aforementioned problems, we propose a unified pairwise relationship aid framework to boost the performance of current zero-shot learning methods. Our framework consists of two major procedures: First, in the training phase, a pairwise relationship predictor is learned on the disjoint training set. We present two kinds of methods, a binary classification based method and a metric learning based method, to infer the pairwise relationships between images. Second, in the test phase, we construct a similarity graph based on the predicted pairwise similarities of test images. With the similarity graph, we employ normalized spectral clustering to divide test images into several semantic groups. We assume that images in the same group share the same class label. Furthermore, we introduce an adaptive graph anchors voting method to classify the semantic groups. We select the most representative test images as anchors based on the predictions of existing ZSIC methods, and use the majority voting of the selected anchors to decide the class label of each semantic groups. Since we do not have any training images of the test classes, the selected anchors may be unreliable. Therefore, a heuristic iterative strategy is proposed to perform anchors voting adaptively.

Our solution is inspired by the recent success in semi-supervised learning [3–10], which aims at boosting supervised learning with unlabeled data. The two assumptions in semi-supervised learning, the *smooth assumption* that similar images should share the same class label, and the *cluster assumption* that images tend to form discrete clusters, are naturally satisfied in the ZSIC problem. However, there is an essential difference between our goal and semi-supervised clustering: the former does not have any training example of the test classes. This difference increases the difficulty of utilizing pairwise relationships as well.

The contributions of this paper are three-fold: Firstly, we explore a novel pairwise relationship aid framework to help to handle the zero-shot image classification problem; Secondly, we propose two different methods, a binary classification based method and a metric learning based method, to capture the pairwise relationships between images; Last but not least, an adaptive graph anchor voting method is introduced to refine the predicted results of zero-shot learning methods.

2 Related Work

The first practical solution of zero-shot image classification is proposed in [1], where attributes are introduced as an intermediate layer to bridge the gap between low-level image features and classes. Since then, extensive attribute-based methods and semantic embedding based methods are proposed to tackle the ZSIC problem [11–22].

Akata et al. [11] proposed to embed classes into the attribute space to transform the classification problem into a label embedding problem. Romera-Paredes and Torr [12] proposed to represent the attribute learning and zero-shot learning into a joint optimization problem, and provided a simple solution to it. In [13], the unreliability of attributes is taken into consideration during the process of training random forest for the ZSIC problem. Socher et al. [23] replaced attributes with semantic word vectors, and they proposed to combine the prediction of seen and unseen classes into a Bayesian framework and considered detecting unseen classes as an outlier detection problem. A similar idea was presented in [14], where they considered unseen classes as the absorbing states in a Markov chain process. Fu et al. [24] proposed a transductive multi-view embedding framework to include low-level feature space, semantic word space and attribute representation into a common space. The most related previous research to our work is [20], which tries to model the pairwise relationships between the unseen and seen classes via sparse coding. What makes our method differ from [20] is that our method directly predicts the pairwise relationships between unseen classes.

Except for image classification, zero-shot learning also appears in applications like activity recognition [21,25,26] and event detection [27,28]. Due to the cost of collecting training data in conventional classification methods, zero-shot learning has become more and more popular in the related research areas.

Exploiting pairwise relationships in supervised learning or semi-supervised learning is considerable because training instances are available [4,6–10,20]. However, no prior study has explored whether it is suitable to do so in zero-shot learning. Therefore, this situation makes our method differ from conventional methods with pairwise relationships.

3   Zero-Shot Image Classification

The zero-shot image classification problem can be formulated as follows: Let X be an arbitrary feature space of images, $Y = \{y_1, ..., y_T\}$ be a set of T classes, $Z = \{z_1, ..., z_L\}$ be a set of L classes and $Y \cap Z = \emptyset$. Given the training instances and their corresponding class labels, our task is to learn a function $f : X \to Z$.

As mentioned above, most practical solutions for this problem build on the attribute-based representation. Here we introduce Direct Attribute Prediction (DAP) [29], which is a straightforward solution and will be used as a building block of our framework. DAP introduces M binary attributes to represent both Y and Z, e.g., a training class y can be represented as $\mathbf{a}^y = (a_1^y, ...a_M^y)$, which satisfies $a_m^y = 1, m \in 1 ... M$ if class y has the m-th attribute and is 0 otherwise. Then the posterior probability of a given instance $\mathbf{x}$ belonging to test class z is calculated via the Bayes' rule:

$$p(z|\mathbf{x}) = \sum_{\mathbf{a}\in\{0,1\}^M} p(z|\mathbf{a})p(\mathbf{a}|\mathbf{x}) = \frac{p(z)}{p(\mathbf{a}^z)} \prod_{m=1}^{M} p(a_m^z|\mathbf{x}), \tag{1}$$

where $p(z)$ is the prior probability of test class z and $p(\mathbf{a}^z)$ is the attribute prior probability; $p(a_m^z|\mathbf{x})$ denotes the post probability of the presence of the m-th attribute that accords with its assignment in class z, given the instance $\mathbf{x}$. Since $p(z)$ and $p(\mathbf{a}^z)$ are constants, DAP just needs to determine the value of $p(a_m^z|\mathbf{x})$, which can be achieved by learning an attribute classifier on the training set.

4   The PRA Framework

In this section we present the Pairwise Relationship Aid (PRA) framework for zero-shot image classification in details. We first overview the proposed PRA framework in Sect. 4.1; then we introduce two different methods, a binary classification based method and a metric learning based method, to infer relationships in image pairs in Sect. 4.2; the learned pairwise relationships are used to construct a similarity graph in Sect. 4.3; Finally, in Sect. 4.4, an adaptive graph anchors voting method is proposed, which exploits pairwise relationships to refine classification results.

4.1   Overview of the PRA Framework

The intuition behind the PRA framework is that pairwise relationships between test images can help to correct prediction errors. The diagram of the PRA framework is demonstrated in Fig. 2: Given the test images, we first employ a deep Convolutional Neural Network (CNN) to extract the corresponding image features. Secondly, we obtain the coarse classification results by DAP. Thirdly, we construct a graph based representation for the test images, with the similarity matrix calculated via a binary relationship classifier or a Mahalanobis distance function learned on the training images. Lastly, we refine the coarse classification results iteratively: in each iteration, we perform spectral clustering to segment the graph of test instances into L connected components, each connected component represents a test class and then its label is determined by the adaptively selected graph anchors in the component. Note that any attribute-based method for the ZSIC problem can be employed as the base classifier in our method, and we choose DAP because of its simplicity.

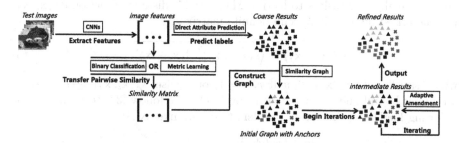

Fig. 2. Diagram of the proposed PRA framework. Given the test images, we first extract their features via a pre-trained CNN, and apply DAP to get coarse classification results. We then predict the pairwise relationships between test images and use them to construct a similarity graph to represent test images. At last, based on the similarity graph, we refine the coarse results iteratively via an adaptive graph anchors voting method.

4.2 Pairwise Relationship Learning

In this paper, we present two learning methods to infer the pairwise relationship between a pair of images: a binary classification based method and a metric learning based method. Each of the two methods represents a typical strategy for pairwise relationship learning. The inferred pairwise relationships between all test images will be used to construct a similarity graph in the following section.

Binary classification based method (BCBM): BCBM considers pairwise relationship learning as a binary classification problem since, given a pair of images, they either belong to the same class or belong to different classes. Specifically, let $r_{ij} = 1$ denote image i and image j belong to the same class, otherwise $r_{ij} = -1$. We minimize the following hinge loss function to learn a binary relationship classifier f_b:

$$\min_{f_b} \sum_{(i,j) \in S} \max(0, 1 - f_b(\mathbf{x}_i, \mathbf{x}_j) r_{ij}), \tag{2}$$

where S denotes an image pair set constructed by randomly sampling on the training set of the zero-shot classification problem. The choice of f_b is plenty. We choose neural networks because of their ability of nonlinear mapping. Our implementation is a two-layer fully connected neural network which takes the concatenation of a pair of image features as its input: the first layer compresses the input pairwise image features, while the second layer predicts the value of r. We adopt the Rectified Linear Units [30] as the activation function. We employ the standard mini-batch gradient descent to minimize Eq. (2). Implementation details of our network is presented in the experiment section.

Metric learning based method (MLBM): MLBM assumes that distances between images in the same class are small while those from different classes are large. Formally, our task is to learn a Mahalanobis distance function $d_A(\mathbf{x}_i, \mathbf{x}_j)$ parameterized by a positive semi-definite matrix A:

$$d_A(\mathbf{x}_i, \mathbf{x}_j) = \sqrt{(\mathbf{x}_i - \mathbf{x}_j)^T A(\mathbf{x}_i - \mathbf{x}_j)}, \tag{3}$$

which satisfies $d_A(\mathbf{x}_i, \mathbf{x}_j) \leq u$ if $r_{ij} = 1$, otherwise $d_A(\mathbf{x}_i, \mathbf{x}_j) \geq l$. u and l are upper bound and lower bound respectively. We follow [31] to minimize the following LogDet optimization problem to obtain matrix A:

$$\begin{aligned} \min_{A \succeq 0} \quad & D_{ld}(A, A_0), \\ s.t. \quad & d_A(\mathbf{x}_i, \mathbf{x}_j) \leq u, if \quad r_{ij} = 1, \\ & d_A(\mathbf{x}_i, \mathbf{x}_j) \geq l, \quad otherwise, \end{aligned} \tag{4}$$

where $D_{ld}(A, A_0)$ denotes the LogDet divergence between A and A_0. A_0 is the initial matrix and we set it as the matrix that parameterizes the Euclidean distance. u and l are set as the minimum and the maximum Euclidean distances calculated on the training image set. We can solve Eq. (4) by performing Bregman projection iteratively. Interested readers can refer to [31] for more details.

Both BCBM and MLBM have their own advantages and disadvantages. Compared with MLBM, pairwise relationships learned by BCBM are more straightforward, and we can directly tell whether a pair of images belong to the same class by looking at the output of BCBM. However, MLBM can capture more complex relationships, because MLBM considers not only the distances between different classes, but also the distances in the same class, while BCBM opts for finding a decision boundary.

4.3 Similarity Graph Construction

To fully utilize the pairwise relationships between test images, we propose to construct a similarity graph to represent test images and their relationships. Let $G = (V, E)$ be an undirected weighted graph, with vertex set V and edge set E. Each vertex represents a test image and we assume $|V| = n$. An edge between two vertexes indicates the corresponding images belonging to the same class (or indicates they are similar), while its corresponding weight can be considered as the "confidence".

Corresponding to the aforementioned two pairwise relationship learning methods, we propose two ways to incorporate pairwise relationships into the construction of similarity graphs. As to BCBM, we employ the ε-neighborhood graph, which means vertexes are connected if the confidence between them is larger than ε. We define the weight w_{ij} of the edge between two vertex i and j as $w_{ij} = \frac{f_b(\mathbf{x}_i, \mathbf{x}_j) + f_b(\mathbf{x}_j, \mathbf{x}_i)}{2}$ to ensure $w_{ij} = w_{ji}$, and set $\varepsilon = 0$ to connect images belonging to the same class. As to MLBM, we first calculate the distances between all test images via the learned Mahalanobis distance function, and normalize the distances to $[0, 1]$. Then we adopt the fully connected graph, with weights defined as $w_{ij} = 1 - d_A^*(\mathbf{x}_i, \mathbf{x}_j)$, where $d_A^*(\mathbf{x}_i, \mathbf{x}_j)$ denotes the normalized distance.

As in many spectral clustering methods [32], once we obtain the similarity graph representation of test instances, detecting clusters in the graph can be done trivially if we project the test instances into the space spanned by several eigenvectors of a graph Laplacian. We choose the symmetric normalized graph Laplacian [33] to do so because it can generate balanced clusters. In the following sections, we assume test instances are presented in the new presentation.

4.4 Adaptive Graph Anchors Voting

In this section, we propose an adaptive graph anchors voting method to classify clusters. The idea of "anchors" is first introduced in [34], which refers to a subset of vertexes used for representing the whole graph. In our case, anchors are used to decide the class labels of the rest instances.

Our adaptive graph anchors voting method proceeds in an iterative scheme: at the begin of each iteration, we select the instances with the top-K highest probabilities (calculated by DAP) of belonging to each test class as the graph anchors, because the predicted labels of these instances are more likely to be correct. Note that we allow duplicate anchors so that an anchor can vote for multiple classes. In this way, we obtain L groups of anchors and each group contains K anchors. We consider each group as a center and cluster each test instance by assigning it to its nearest center. Let $\mathbf{x}_{c_{ij}}$ denote the j-th anchor point in the i-th group, then the distance between a test instance $\mathbf{x}$ and a center c_i is calculated as:

$$d(\mathbf{x}, c_i) = \sum_{j=1}^{K} d_E(\mathbf{x}, \mathbf{x}_{c_{ij}}), \tag{5}$$

where d_E denotes the Euclidean distance function. Once every instance is clustered, we determine the class label of each cluster by the majority voting of the anchors in it.

After each iteration, we increase the value of K by a fixed interval K_{iter} to include more anchors for voting. This is because the more correctly predicted anchors, the more likely their voting results are correct. However, including wrong anchors would achieve the opposite effect. Therefore, we propose a heuristic strategy to prevent including too many "wrong" anchors. Let P_t denote the set of anchors selected in the t-th iteration, and $f_t(p_i)$ denote the predicted class label of the anchor point p_i. Note that we do not update the initial predicted probabilities of images belonging to each class, therefore we have $P_{t-1} \subseteq P_t$. We define the stability of the anchor set in iteration t, $f_s(P_t)$, as follows:

$$f_s(P_t) = \frac{\sum_{p_i \in P_{t-1}} [f_{t-1}(p_i) = f_t(p_i)]}{|P_{t-1}|}, t \geq 2, \tag{6}$$

where $[\cdot]$ denotes the Iverson bracket operator ($[O] = 1$ if the statement O is true, otherwise $[O] = 0$) and $|P_{t-1}|$ denotes the cardinality of P_{t-1}. We let $f_s(P_1) = 0$. Equation (6) actually calculates the proportion of anchors whose class labels remaining unchanged after an iteration. Since anchors changing their class labels are not supportive enough for voting, we consider $f_s(P_t) - f_s(P_{t-1}) < 0$ as the stop criterion of our adaptive voting process.

5   Experiments

In this section, we conduct a series of experiments to validate the effectiveness of the PRA framework.

Datasets: Our PRA framework is tested on two standard datasets for the ZSIC problem: **Animal with Attribute**[1] (AWA) and **aPascal-aYahoo**[2] (aP). AWA [29] contains 85 attributes, 40 training classes and 10 test classes (6,180 test images). aP [2] contains 64 attributes, 20 training classes and 12 test classes (2,644 test images). Besides, we also use the **Sun Attribute** dataset[3] [35] for a

---

[1] http://attributes.kyb.tuebingen.mpg.de/.
[2] http://vision.cs.uiuc.edu/attributes/.
[3] http://cs.brown.edu/~gen/sunattributes.html.

large-scale test. Training classes and test classes are disjoint in all datasets. We evaluate the performance of each classifier by calculating its confusion matrix, and use the mean of the confusion matrix (mean class accuracy) as our representing score.

Implementation Details: We use the outputs of the "fc7" layer of the pre-trained vgg-19 neural network [36] as our image features (4096-D). We utilize the LIBLINEAR library [37] with l_2-norm regularized logistic regression setting to implement DAP [29]. The neural network for binary pairwise relationship classification is trained for 20 epochs, each epoch consists of 5,000 mini-batches and each mini-batch consists of 8 randomly sampled image pairs (an even mixed of positive and negative pairs). The learning rate of the network is set as 20 logarithmically equally spaced values between 10^{-2} and 10^{-4}. The number of nodes in the first layer of the network is 4096. We implement MLBM via the code provided in [31] with its default parameters. The initial number of anchors in each class K is set to 10 and the interval K_{iter} is set to 5.

Results: In Table 1, we compare the accuracy of the PRA framework with several state-of-the-art methods. It is obvious that, on both datasets, the proposed PRA framework outperforms all the state-of-the-art methods significantly. Compared with DAP, which is used to obtain the initial prediction in our framework, PRA with MLBM increases the performance of DAP by 43.03% on AWA and 50.99% on aP, respectively. These results demonstrate the considerable potential of exploiting pairwise relationships between test instances in the zero-shot learning problem.

Table 1. Accuracy of the proposed PRA framework with two pairwise relationship learning methods on AWA and aP. Results of six state-of-the-art methods are presented for comparison.

	Feature type	AWA	aP
DAP [29]	Hand-crafted	41.4	19.1
IAP [29]	Hand-crafted	42.2	16.9
UA [13]	Hand-crafted	43.01	26.02
ESZSL [12]	Hand-crafted	49.03	15.11
LRL [38]	Hand-crafted	40.05	24.71
AMP [14]	CNN based	66.00	-
SSE-ReLU [15]	CNN based	76.33	46.23
SMS [19]	-	78.47 (CNN)	39.03 (hand-crafted)
RKT [20]	CNN based	82.43	-
JLSE [22]	CNN based	80.46	50.35
PRA-BCBM	CNN based	82.73	53.87
PRA-MLBM	CNN based	84.43	70.09

Table 2. Accuracy of the proposed PRA framework with different image features. Components with hand-crafted features are denoted as "hfeat" while those with features extracted by CNNs are denoted as "CNN". "Pairwise Acc" denotes the accuracy of pairwise relationship prediction.

	Multi class Acc		Pairwise Acc	
	AWA	aP	AWA	aP
DAP (hfeat)	28.07	17.29	-	-
DAP (CNN)	57.46	26.59	-	-
DAP (hfeat) + BCBM (hfeat)	13.68	24.32	44.04	55.63
DAP (hfeat) + MLBM (hfeat)	27.00	23.84	40.49	57.34
DAP (CNN) + BCBM (hfeat)	15.77	23.01	44.04	55.63
DAP (CNN) + MLBM (hfeat)	30.32	26.93	40.49	57.34
DAP (hfeat) + BCBM (CNN)	69.35	51.16	88.68	84.01
DAP (hfeat) + MLBM (CNN)	51.36	55.03	93.69	94.63

To further analyze the performance of the PRA framework, we calculate the confusion matrices of DAP, PRA with BCBM, and PRA with MLBM on both datasets, as presented in Fig. 3. From these confusion matrices, we can observe that a distinguished advantage of PRA is that, it can discover potential classes which cannot be predicted by DAP, e.g., "giant panda" on AWA. We owe this advantage to the selecting strategy of graph anchors.

However, we must point out that most conventional methods are based on hand-crafted features (except SSE-ReLU [15] which employed vgg-19 as well), therefore our performance gain might owe to the CNN features, instead of the PRA framework itself. Besides, how the accuracy of initial predictions and pairwise relationship learning affect the final results remains unclear. To figure out these problems, we conduct an experiment in which components of PRA with different types of features are tested. We also record the accuracy of the two pairwise relationship learning methods. We consider the accuracy of BCBM as the proportion of image pairs it predicts correctly in all image pairs. As to MLBM, since it is hard to tell whether a distance function is correct or wrong, we roughly estimate the accuracy of MLBM via the following way: for each test image, we combine it with its nearest neighbor to construct an image pair. If the images in a pair belong to the same class, then we consider this pair as a correct prediction. In this way, we calculate the accuracy of MLBM as the proportion of correct predictions in all pairs. Results of this experiment are presented in Table 2.

From our experimental results, we can obtain two important conclusions. Firstly, a precise pairwise relationship model is vital for PRA. This conclusion can be validated from two aspects: on the one hand, results of DAP (hfeat) + BCBM (CNN) and DAP (hfeat) + MLBM (CNN) show that, pairwise relationship models with high accuracy can increase the performance of a weak model significantly; on the other hand, even with a strong classifier (DAP (CNN)), an

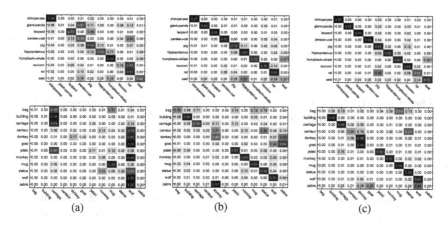

(a)   (b)   (c)

Fig. 3. Confusion matrices of (a) Direct Attribute Prediction, (b) PRA with BCBM and (c) PRA with MLBM on AWA (top) and aP (bottom), respectively.

unreliable pairwise relationship predictor (BCBM with hand-crafted features on AWA) will degrade the final performance of PRA. Secondly, for the pairwise relationship learning problem, employing features extracted by CNNs is more effective. In our opinion, pairwise relationship that whether two images belong to the same class is a high-level information, which cannot be captured easily via hand-crafted features. Hand-crafted features are prone to clustering objects with similar appearances into the same class, while CNNs can successfully learn some high-level information, as indicated in [39]. However, general results in Table 2 show that, no matter what kind of features we are using, properly learned pairwise relationships can help to tackle the zero-shot image classification problem.

Another issue we are curious about is how many test instances we need to obtain a good result. Therefore, we randomly split test instances into $N_s \in [1, \ldots, C]$ equal-sized subsets for 10 times and calculate the mean accuracy of PRA on them. To ensure each group have enough anchors for voting, we set $C = 100$

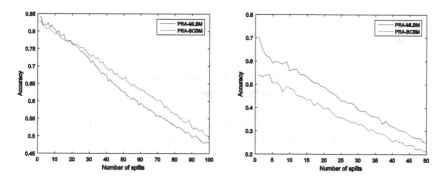

Fig. 4. Accuracy of PRA with different splits on AWA (left) and aP (right).

and $C = 50$ for AWA and aP, respectively. We report the mean accuracy versus the number of splits N_s in Fig. 4. Our experimental results demonstrate that the performance of PRA is negative related to the number of splits. This property does accord with the scheme of the PRA framework because, with more splits, the average chance of selecting properly predicted anchors becomes smaller.

The Sun Attribute dataset contains 717 categories of images and 101 attributes, thus it is suitable for evaluating the performance of the proposed method facing with massive categories. We perform random train/test split for 5 times, and report the mean accuracy in Fig. 5. In our experiments, we found that with the increasing number of unseen classes, the performance of DAP decreases dramatically (from 37.70% with 50 classes to 9.62% with 500 classes). Such performance degradation is caused by the insufficiency of attributes for distinguishing similar classes. However, the proposed method still increases the performance of the baseline significantly, e.g. MLBM obtains a performance gain of 14.74% with 50 classes, and 5.33% with 500 classes. Note that the expected accuracy of random guess with 500 classes is 0.2%, thus the effect of the proposed method is validated.

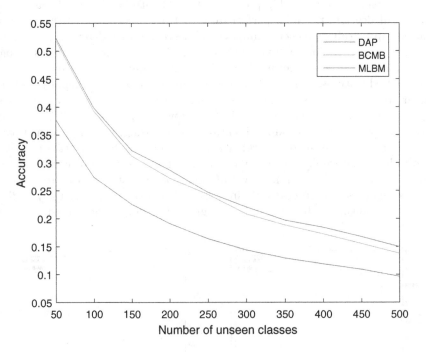

Fig. 5. Accuracy of PRA on the SUN attribute dataset. This figure demonstrates the effectiveness of the proposed method for the large-scale ZSIC problem.

6   Conclusions

In this paper, we have proposed a pairwise relationship aid framework to help to tackle the zero-shot image classification problem. We explore two different ways, a binary classification based method and a metric learning based method, to learn a pairwise relationship predictor on the training set and use it to infer the pairwise relationships between test images. Then we construct a similarity graph based on the pairwise relationships and perform adaptive anchors voting to refine the results of direct attribute prediction. Experimental results on three datasets show that we can enjoy a considerable performance gain with properly learned pairwise relationships in zero-shot image classification problem.

Acknowledgement. This research is supported by National Natural Science Foundation of China (61320106008, 61232011, 61402120, 61572531, 61622214), Educational Commission of Guangdong Province (2013CXZDB001), and Natural Science Foundation of Guangdong Province (2014A030310348). The corresponding author is Hefeng Wu.

References

1. Lampert, C.H., Nickisch, H., Harmeling, S.: Learning to detect unseen object classes by between-class attribute transfer. In: Proceedings of IEEE Conference on Computer Vision and Pattern Recognition (CVPR), pp. 951–958 (2009)
2. Farhadi, A., Endres, I., Hoiem, D., Forsyth, D.A.: Describing objects by their attributes. In: Proceedings of IEEE Conference on Computer Vision and Pattern Recognition (CVPR), pp. 1778–1785(2009)
3. Chapelle, O., Schölkopf, B., Zien, A., et al.: Semi-supervised Learning. MIT Press, Cambridge (2006)
4. Hu, J., Lu, J., Tan, Y.: Discriminative deep metric learning for face verification in the wild. In: Proceedings of IEEE Conference on Computer Vision and Pattern Recognition (CVPR), pp. 1875–1882(2014)
5. Li, H., Li, D., Luo, X.: BAP: bimodal attribute prediction for zero-shot image categorization. In: Proceedings of the ACM International Conference on Multimedia, pp. 1013–1016 (2014)
6. Maggini, M., Melacci, S., Sarti, L.: Learning from pairwise constraints by similarity neural networks. Neural Netw. **26**, 141–158 (2012)
7. Li, Z., Liu, J., Tang, X.: Pairwise constraint propagation by semidefinite programming for semi-supervised classification. In: Proceedings of the Twenty-Fifth International Conference on Machine Learning (ICML), pp. 576–583 (2008)
8. Baghshah, M.S., Shouraki, S.B.: Semi-supervised metric learning using pairwise constraints. In: Proceedings of the 21st International Joint Conference on Artificial Intelligence (IJCAI), pp. 1217–1222 (2009)
9. Zhu, G., Yan, S., Ma, Y.: Image tag refinement towards low-rank, content-tag prior and error sparsity. In: Proceedings of the 18th ACM International Conference on Multimedia, pp. 461–470 (2010)
10. Hong, S., Choi, J., Feyereisl, J., Han, B., Davis, L.S.: Joint image clustering and labeling by matrix factorization. IEEE Trans. Pattern Anal. Mach. Intell. **38**, 1411–1424 (2016)

11. Akata, Z., Perronnin, F., Harchaoui, Z., Schmid, C.: Label-embedding for attribute-based classification. In: Proceedings of IEEE Conference on Computer Vision and Pattern Recognition (CVPR), pp. 819–826 (2013)
12. Romera-Paredes, B., Torr, P.H.S.: An embarrassingly simple approach to zero-shot learning. In: Proceedings of the 32nd International Conference on Machine Learning (ICML), pp. 2152–2161 (2015)
13. Jayaraman, D., Grauman, K.: Zero-shot recognition with unreliable attributes. In: Proceedings of Advances in Neural Information Processing Systems (NIPS), pp. 3464–3472 (2014)
14. Fu, Z., Xiang, T.A., Kodirov, E., Gong, S.: Zero-shot object recognition by semantic manifold distance. In: Proceedings of IEEE Conference on Computer Vision and Pattern Recognition (CVPR), pp. 2635–2644 (2015)
15. Zhang, Z., Saligrama, V.: Zero-shot learning via semantic similarity embedding. In: Proceedings of IEEE International Conference on Computer Vision (ICCV), pp. 4166–4174 (2015)
16. Mensink, T., Gavves, E., Snoek, C.G.M.: COSTA: co-occurrence statistics for zero-shot classification. In: Proceedings of IEEE Conference on Computer Vision and Pattern Recognition (CVPR), pp. 2441–2448 (2014)
17. Elhoseiny, M., Saleh, B., Elgammal, A.M.: Write a classifier: zero-shot learning using purely textual descriptions. In: Proceedings of IEEE International Conference on Computer Vision (ICCV), pp. 2584–2591 (2013)
18. Da, Q., Yu, Y., Zhou, Z.: Learning with augmented class by exploiting unlabeled data. In: Proceedings of the Twenty-Eighth AAAI Conference on Artificial Intelligence, pp. 1760–1766 (2014)
19. Guo, Y., Ding, G., Jin, X., Wang, J.: Transductive zero-shot recognition via shared model space learning. In: Proceedings of the Thirtieth AAAI Conference on Artificial Intelligence, pp. 3434–3500 (2016)
20. Wang, D., Li, Y., Lin, Y., Zhuang, Y.: Relational knowledge transfer for zero-shot learning. In: Proceedings of the Thirtieth AAAI Conference on Artificial Intelligence, pp. 2145–2151 (2016)
21. Gan, C., Lin, M., Yang, Y., de Melo, G., Hauptmann, A.G.: Concepts not alone: exploring pairwise relationships for zero-shot video activity recognition. In: Proceedings of the Thirtieth AAAI Conference on Artificial Intelligence, pp. 3487–3493 (2016)
22. Zhang, Z., Saligrama, V.: Zero-shot learning via joint latent similarity embedding. In: Proceedings of IEEE Conference on Computer Vision and Pattern Recognition (CVPR), pp. 6034–6042 (2016)
23. Socher, R., Ganjoo, M., Manning, C.D., Ng, A.Y.: Zero-shot learning through cross-modal transfer. In: Proceedings of Advances in Neural Information Processing Systems (NIPS), pp. 935–943 (2013)
24. Fu, Y., Hospedales, T.M., Xiang, T., Gong, S.: Transductive multi-view zero-shot learning. IEEE Trans. Pattern Anal. Mach. Intell. **37**, 2332–2345 (2015)
25. Guadarrama, S., Krishnamoorthy, N., Malkarnenkar, G., Venugopalan, S., Mooney, R.J., Darrell, T., Saenko, K.: YouTube2Text: Recognizing and describing arbitrary activities using semantic hierarchies and zero-shot recognition. In: Proceedings of IEEE International Conference on Computer Vision (ICCV), pp. 2712–2719 (2013)
26. Cheng, H., Griss, M.L., Davis, P., Li, J., You, D.: Towards zero-shot learning for human activity recognition using semantic attribute sequence model. In: Proceedings of the ACM International Joint Conference on Pervasive and Ubiquitous Computing, pp. 355–358 (2013)

27. Chang, X., Yang, Y., Hauptmann, A.G., Xing, E.P., Yu, Y.: Semantic concept discovery for large-scale zero-shot event detection. In: Proceedings of the Twenty-Fourth International Joint Conference on Artificial Intelligence (IJCAI), Buenos Aires, Argentina, pp. 2234–2240 (2015)
28. Wu, S., Bondugula, S., Luisier, F., Zhuang, X., Natarajan, P.: Zero-shot event detection using multi-modal fusion of weakly supervised concepts. In: Proceedings of IEEE Conference on Computer Vision and Pattern Recognition (CVPR), Columbus, OH, USA, pp. 2665–2672, 23–28 June 2014
29. Lampert, C.H., Nickisch, H., Harmeling, S.: Attribute-based classification for zero-shot visual object categorization. IEEE Trans. Pattern Anal. Mach. Intell. **36**, 453–465 (2014)
30. Krizhevsky, A., Sutskever, I., Hinton, G.E.: Imagenet classification with deep convolutional neural networks. In: Proceedings of Advances in Neural Information Processing Systems (NIPS), Lake Tahoe, Nevada, United States, pp. 1106–1114, 3–6 December 2012
31. Davis, J.V., Kulis, B., Jain, P., Sra, S., Dhillon, I.S.: Information-theoretic metric learning. In: Proceedings of the Twenty-Fourth International Conference on Machine Learning (ICML), Corvallis, Oregon, USA, pp. 209–216, 20–24 June 2007
32. von Luxburg, U.: A tutorial on spectral clustering. Stat. Comput. **17**, 395–416 (2007)
33. Ng, A.Y., Jordan, M.I., Weiss, Y.: On spectral clustering: analysis and an algorithm. In: Proceedings of Advances in Neural Information Processing Systems (NIPS), Vancouver, British Columbia, Canada, pp. 849–856, 3–8 December 2001
34. Liu, W., He, J., Chang, S.: Large graph construction for scalable semi-supervised learning. In: Proceedings of the 27th International Conference on Machine Learning (ICML), Haifa, Israel, pp. 679–686, 21–24 June 2010
35. Patterson, G., Xu, C., Su, H., Hays, J.: The SUN attribute database: beyond categories for deeper scene understanding. Int. J. Comput. Vision **108**, 59–81 (2014)
36. Simonyan, K., Zisserman, A.: Very deep convolutional networks for large-scale image recognition. In: Proceedings of International Conference on Learning Representations (ICLR) (2015)
37. Fan, R., Chang, K., Hsieh, C., Wang, X., Lin, C.: LIBLINEAR: a library for large linear classification. J. Mach. Learn. Res. **9**, 1871–1874 (2008)
38. Li, X., Guo, Y., Schuurmans, D.: Semi-supervised zero-shot classification with label representation learning. In: Proceedings of IEEE Conference on Computer Vision and Pattern Recognition (CVPR), pp. 4211–4219 (2015)
39. Escorcia, V., Niebles, J.C., Ghanem, B.: On the relationship between visual attributes and convolutional networks. In: Proceedings of IEEE Conference on Computer Vision and Pattern Recognition (CVPR), pp. 1256–1264 (2015)

Segmentation and Semantic Segmentation

Hierarchical Supervoxel Graph for Interactive Video Object Representation and Segmentation

Xiang Fu[1], Changhu Wang[2]([✉]), and C.-C. Jay Kuo[1]

[1] University of Southern California, Los Angeles, CA, USA
[2] Microsoft Research, Beijing, China
chw@microsoft.com

Abstract. In this paper, we study the problem of how to represent and segment objects in a video. To handle the motion and variations of the internal regions of objects, we present an interactive hierarchical supervoxel representation for video object segmentation. First, a hierarchical supervoxel graph with various granularities is built based on local clustering and region merging to represent the video, in which both color histogram and motion information are leveraged in the feature space, and visual saliency is also taken into account as merging guidance to build the graph. Then, a supervoxel selection algorithm is introduced to choose supervoxels with diverse granularities to represent the object(s) labeled by the user. Finally, based on above representations, an interactive video object segmentation framework is proposed to handle complex and diverse scenes with large motion and occlusions. The experimental results show the effectiveness of the proposed algorithms in supervoxel graph construction and video object segmentation.

1 Introduction

Interactive video object segmentation plays a significant role in region of interest extraction in diverse video-related applications, such as movie post-production, object boundary tracking, pose estimation, content based information retrieval, security/surveillance, and video summarization. Decades ago, we were impressed by the girl in red in the famous black-and-white movie "Schindler's List", as shown in Fig. 1. If the target object in the movie is able to be automatically segmented and tracked, it becomes possible to further edit the extracted object or put it to other backgrounds. It can also help further analyze the motion, pose, or activity of that object. However, how to effectively represent and segment moving objects from a video remains a big challenge, due to the diversity of scenes with temporal incoherence, motion blur, occlusion, etc.

In the last decade, several classic interactive video segmentation systems were developed. In 2005, Li et al.[1] extended the traditional graph cut algorithm [2] to a 3D version for videos, but they only considered the color information and neglected other information cues. Wang et al.[3] designed an interactive hierarchical version of Mean Shift [4] and Graph Cut [2] scheme for video object segmentation. In spite of smoothness preservation across both space and time,

© Springer International Publishing AG 2017
S.-H. Lai et al. (Eds.): ACCV 2016, Part I, LNCS 10111, pp. 103–120, 2017.
DOI: 10.1007/978-3-319-54181-5_7

Fig. 1. The girl in red in the famous black-and-white movie "Schindler's List" (best view in color). If the target object in the movie is able to be automatically segmented and tracked, it becomes possible to further edit and analyze the extracted object (Color figure online).

its user interface is not natural for common users and thus limits its applications. In 2009, Price *et al.*[5] applied a graph-cut optimization framework and propagated the selection forward frame by frame using many cues. Bai *et al.*[6] proposed the Video SnapCut system based on localized classifiers, which was further transferred to Adobe After Effect CS5 in 2010 as a state-of-the-art video object segmentation system. However, after processing over twenty frames, this system will have large errors, and cannot extract objects with large motion and occlusions. In addition, most of existing methods on video segmentation were restricted to object-level segmentation, which neglect abundant information inside the regions of moving objects.

To make a complete representation for each patch of a moving object in the video, we base the video object segmentation framework on the supervoxel representation of a video. Supervoxel is a 3D extension of superpixel in image representation, and describes the motion status of a patch along the time axes. Typical superpixel and supervoxel algorithms were summarized in [7,8], including Normalized Cut [9], Mean Shift [4], graph-based [10], Quick Shift [11], TurboPixels [12], and CCP [13], etc. However, most of existing supervoxel algorithms only use one-level supervoxels to represent a video at a time for further analysis, which cannot support arbitrary object representation interactively labeled by a user at once. As shown in Fig. 2, if the supervoxel granularity is small (low-level), the object will consist of too many trivial regions which is hard to model or track; while if the granularity is large (high-level), the supervoxels cannot cover small or narrow parts of objects such as hands. Thus, a hierarchical supervoxel representation, which connects different levels of supervoxels, is needed.

Recently Grundmann *et al.* [14] extended the image segmentation approach [10] (GB) and presented a hierarchical supervoxel structure with different granularities (GBH) for video representation. However, it is intrinsically a local approach focusing on the color variance of local regions, and thus suffers from the problems of occluded objects and scattered regions with uncontrollable shape and size for low-level supervoxels (Fig. 7). Some latest work on supervoxel such as real-time superpixels [15], trajectory binary partition tree [16], temporal superpixels [17], temporally consistent superpixels [18], and flattening supervoxel hierarchies [19] also showed improvements on supervoxel generation. But they did

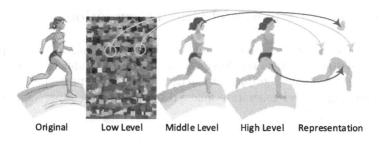

Original   Low Level   Middle Level   High Level   Representation

Fig. 2. Example of hierarchical superovxel graph and its representation of the face, the hands and the legs of the runner. For display purpose, we ignore the time axes. The hierarchical supervoxel graph is shown in the middle three columns. One color indicates the same supervoxel in the same level. For middle and high levels, we use color averaging instead of random color to denote each supervoxel. Based on the hierarchical supervoxel graph, different parts of the runner can be represented by supervoxels at different levels, as shown in the rightmost column (Color figure online).

not leverage both motion and saliency features, and didn't further study how to leverage supervoxel graph in video object representation and segmentation.

In this paper, we propose a novel hierarchical supervoxel graph construction approach in a bottom-up manner to represent a video. First, the bottom-level supervoxels are obtained by local clustering using color and motion cues. Then, higher-level supervoxels are composed of lower-level ones based on a region merging algorithm by leveraging color, motion, and visual saliency of objects. The proposed graph has potential to overcome some challenges suffered in existing work such as occlusion, motion blurring, and inaccurate merging (Figs. 7 and 9).

Given the constructed supervoxel graph, we propose a hierarchical supervoxel selection algorithm to represent a target object using supervoxels with different granularities. The target object can be labeled using the efficient Paint Selection tool [20], and thus provides an easy-to-use interface. The supervoxel selection algorithm tries to use the least number of supervoxels to fit the labeled regions. As shown in Fig. 2, the hands and the legs can be represented and segmented at the same time even if they obviously belong to different levels.

Based on the proposed video and object representations, an interactive video object segmentation framework is proposed. Extensive experiments show the effectiveness of the proposed algorithms compared with state-of-the-art approaches in supervoxel graph construction and video object segmentation.

The rest of the paper is organized as follows. The supervoxel graph construction algorithm for videos and interactive supervoxel representation algorithm for target objects are introduced in Sects. 2 and 3, followed by the video object segmentation framework in Sect. 4. In Sects. 5 and 6 we show the experimental results and conclude the paper.

2   Hierarchical Supervoxel Graph Generation

In this section, we present how to build the hierarchical supervoxel graph for video representation. First, the hierarchical supervoxel graph is briefly introduced. Then the detailed graph construction algorithm is presented, including bottom-level supervoxel generation followed by higher-level supervoxel generation.

2.1   Hierarchical Supervoxel Graph

As a 3D extension of superpixel for video representation [21], supervoxel tries to capture internal motions and variations of the video, and simplify it into disjoint 3D regions in some level. Different algorithm and parameter configurations will result in supervoxels with different granularities. Smaller granularity means lower level and finer representation, while bigger granularity corresponds to higher level and coarser representation. As shown in Fig. 2, low-level supervoxels can illustrate the detail parts like hair, neck, and hands; while high-level ones can illustrate the large parts like arms, clothes, legs, or even the entire human body.

We try to build a tree structured supervoxels with different granularities, so that each (3D) region can have multiple representations, e.g. a small number of big-granularity supervoxels, or a large number of small-granularity supervoxels, all combinations of different granuaries. This is the foundation of interactive video object representation to be introduced in Sect. 3.

In mathematics, hierarchical supervoxel graph can be described as follows. For a given video V, we consider it as a spatio-temporal lattice $\Omega = \Pi \times T$, where Π denotes the 2D pixel lattice for each frame and T is the time space. Each supervoxel can be represented as s_r^l, where level $l \in \{1, 2, \cdots, L\}$, and region $r \in \{1, 2, \cdots, R(l)\}$. L is the largest available level of the hierarchical supervoxel graph, the connected voxels with the same label r at the level l form a region r (not necessarily in the same frame), and $R(l)$ is the largest region id at level l. The hierarchical supervoxel graph results in L levels of individual supervoxels $S = \{S^1, S^2, \cdots, S^L\}$, where each level S^l is a set of supervoxels $\{s_1^l, s_2^l, \cdots, s_{R(l)}^l\}$, having $s_r^l \subset \Omega$, $\cup_r s_r^l = \Omega$, and $s_i^l \cap s_j^l = \emptyset$ for every (i, j) pair. In some frames, some region id r might be missing due to occlusion or large motion, which means region id does not necessarily include all the regions from 1 to $R(l)$.

We first introduce how to generate the bottom-level supervoxels S^1, followed by the construction of other levels based on the bottom level.

2.2   Bottom-Level Supervoxels

We first extend the SLIC algorithm [7] to video analysis with some modifications, and then proposed a modified version for robust supervoxel generation.

Extension of SLIC to Videos. SLIC [7] tries to produce uniform superpixels for the image by local K-means clustering. It was claimed to yield state-of-the-art adherence to image boundaries and outperform existing methods when used for segmentation with high efficiency.

In this work, we extend SLIC to the video scenario in a better manner. Each video shot is considered as a spatio-temporal 3D cube and each pixel is one node in the graph. Local clustering is applied for the whole video space to build the bottom-level supervoxels with large number of supervoxels. Starting from initial uniform-distributed seeds as the centroids of the supervoxels, each pixel in the space-time neighborhood of the supervoxel seed will be reassigned to the closest supervoxel in the space-time neighborhood. The distance d_{ij} between the pixel i and the supervoxel center j, shown in Eq. 1, is measured by CIELAB color distance d_c and space-time distance d_s in [7].

$$
\begin{aligned}
d_{ij} &= \sqrt{d_c^2 + \gamma d_s^2}, \\
d_c^2 &= (l_i - l_j)^2 + (a_i - a_j)^2 + (b_i - b_j)^2, \\
d_s^2 &= (x_i - x_j)^2 + (y_i - y_j)^2 + (t_i - t_j)^2,
\end{aligned}
\tag{1}
$$

where γ is the compact factor to balance between color distance and space-time distance. The supervoxels will be iteratively updated until no update occurs. This convergent local-clustering solution will result in the bottom-level supervoxels.

Modified SLIC (MSLIC). To better encode the motion and temporal information when conducting local clustering in videos, we modify the SLIC algorithm in two ways: encode motion feature, and add a spatio-temporal factor when measuring the distance between each pixel and the supervoxel center.

First, we add motion vectors to the feature space for pixel reassignment. Since the motion orientation for each pixel in the same bottom-level supervoxel is supposed to be similar, besides the pixel's color in CIELAB color space $[l, a, b]^T$ and the position $[x, y, t]^T$, pixel-based motion vector $[v_x, v_y]^T$ is also applied to measure the distance between the pixel and the supervoxel, as shown in Eq. 2. We obtain the motion vectors from advanced optical flow [22].

$$
\begin{aligned}
\hat{d}_{ij} &= \sqrt{d_c^2 + \gamma d_s^2 + \delta d_v^2}, \\
d_v^2 &= (v_{xi} - v_{xj})^2 + (v_{yi} - v_{yj})^2,
\end{aligned}
\tag{2}
$$

where δ is the balance factor between motion and the other two spaces: color and space-time.

The second modification is about spatio-temporal distance measurement. Since the spatial domain and temporal domain have different measurements, it is not reasonable to use the same weight as SLIC to leverage the two distances. We design a spatio-temporal factor ST to balance the spatial and temporal distances. The spatio-temporal distance between the pixel i and the supervoxel center j is:

$$
\hat{d}_s^2 = (x_i - x_j)^2 + (y_i - y_j)^2 + ST \cdot (t_i - t_j)^2.
\tag{3}
$$

In fact, the spatio-temporal factor ST depends on the motion of the video, including the movement from both camera and objects. If the video tends to be static, ST could be set large; otherwise, it could be relatively small.

2.3 Higher-Level Supervoxels

Based on the bottom-level supervoxels, appearance and motion are leveraged to merge neighbor supervoxels from lower level to higher level. All the bottom-level supervoxels are considered as nodes in the graph, and the spatio-temporal neighbor supervoxels are connected by weighted edges. The weight of each edge D_{ij} is determined by the distances of CIELAB color histograms D_c and average motion directions D_v between supervoxels i and j, as follows:

$$D_{ij} = \sqrt{D_c^2 + \alpha D_v^2},$$
$$D_c^2 = \|L_i - L_j\|^2 + \|A_i - A_j\|^2 + \|B_i - B_j\|^2, \tag{4}$$
$$D_v^2 = (v_{xi} - v_{xj})^2 + (v_{yi} - v_{yj})^2,$$

where α is a parameter to balance color and motion space. (L_i, A_i, B_i) is the quantified color histogram of supervoxel i, weighted by the reciprocal of its size.

Figure 3 shows the comparisons between the output of SLIC (using average color without motion as feature space) and our method (using color histogram with motion as feature space). For SLIC, some internal regions of the fish are merged to the background, because the average color of this part is quite close to that of the background. Similarly, the underpant is also merged unless the motion cue is leveraged. But in the proposed method, both cases work quite well, owning to the use of color histogram (the fish case) and the motion cue (the underpant case).

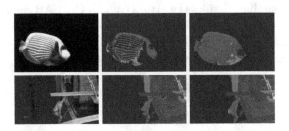

Fig. 3. Supervoxel comparison between SLIC (using color averaging without motion as feature space, middle column) and the proposed MSLIC (using color histogram with motion as feature space, right column). We can see that, for SLIC, some internal regions of the fish are merged to the background, because the average color of this part is quite close to that of the background. Similarly, the underpant is also merged unless the motion is leveraged. But in the proposed method, both cases work quite well, owning to the leverage of color histogram (the fish case) and the motion cue (the underpant case) (Color figure online).

In addition, we also consider region-based saliency maps as guidance for supervoxel merging, as shown in Fig. 4. Based on the observation, supervoxels inside or outside the salient regions have higher priorities to merge together compared with the boundary ones. Thus, the derivative of the saliency map is combined to Eq. 4 to calculate the edge weight, as shown in Eq. 5.

$$\hat{D}_{ij} = \sqrt{D_c^2 + \alpha D_v^2 + \beta D_{\nabla Grad}^2}$$
$$D_{\nabla Grad}^2 = (g_i - g_j)^2, \tag{5}$$

where β is a factor to balance gradient of saliency and the other two spaces: color and motion. g_i is the gradient averaging of saliency for supervoxel i. We adopt context-based saliency and shape prior (CBS) [23], which integrates bottom-up salient stimuli and object-level shape prior, leading to clear boundaries.

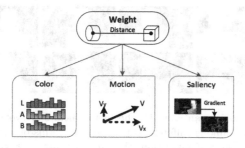

Fig. 4. Illustration of the weights between two supervoxels. When merging supervoxels from lower level to higher level, the weight between two neighbor supervoxels is measured by the combination of distances in CIELAB color histogram, motion direction, and gradient of saliency map (Color figure online).

We merge two supervoxels with least weighted edge one by one, until the number of the remaining supervoxels arrives the expected value (a predefined fixed number, in this paper, we take 64, 128, 256, and so on). In this tree structure, each low-level supervoxel is only covered by one high-level supervoxel, but each high-level supervoxel might correspond to multiple low-level supervoxels. Figure 5 shows an example of a video shot (row 1). The hierarchical supervoxel graph for level #30, #41, and #44 are illustrated in Fig. 5 (row 2–4). We use the same color to denote one supervoxel in different frames at the same level. For each frame, supervoxels are merged from lower level to higher level. At the same level, supervoxels are tracked along the time. This process is illustrated by the graph in Fig. 5 (row 5). In this example, the user labeled man (red sketches) can be represented by different supervoxels in different levels (#41 and #44 in row 7).

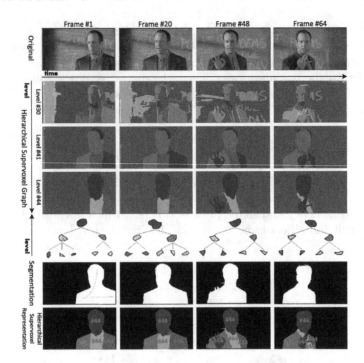

Fig. 5. For the video of a man (row 1), the hierarchical supervoxel graph (row 2–4) shows the connectivity among different supervoxels in terms of two dimensions: time and level. For each frame, disjoint regions that are spatially approximate and coherent in appearance with similar motions will be merged into larger regions; for each level, the regions with the same color or ids indicate the same supervoxel. This graph can be abstracted in row 5. When a user labels the target object in the first frame (row 6 col 1, red sketches are used for display purpose), hierarchical supervoxel representation for this man is calculated, where each supervoxel can be from different levels in the graph. The corresponding segmentation results are shown in row 6, which is represented by different supervoxels in different levels (row 7) (Color figure online).

3   Interactive Hierarchical Supervoxel Representation

In this section, we introduce how to represent (an) arbitrary user-labeled object(s) in one frame using a compact set of supervoxels from different levels in the hierarchical supervoxel graph.

There are two constraints we need to follow. On the one hand, the union set of the resulting supervoxels on that frame should accurately match the labeled regions. On the other hand, the number of supervoxels should be as few as possible. This will avoid the trivial supervoxels in lower regions, and make further modeling and tracking easier.

Algorithm 1. Hierarchical Supervoxel Selection

Require: Hierarchical supervoxel graph s_r^l, where $l \in \{1, 2, \cdots, L\}$, $r \in \{1, 2, \cdots, R(l)\}$; Ground truth segmentation for the key frame Seg;
Ensure: Supervoxel list SV_List;
 1: Set $SV\_List = NULL$
 2: Set $Reg = Seg$
 3: **for** each level from high to low $l = L, L - 1, ..., 1$ **do**
 4:   **for** each region $r = 1, 2, ..., R(l)$ **do**
 5:     **if** $\frac{|s_r^l \cap Reg|}{|s_r^l|} \geq \min e + (\max e - \min e)\frac{l}{L}$ **then**
 6:       $SV\_List = SV\_List + (l, r)$
 7:       $Reg = Reg - s_r^l$
 8:       **if** $|Reg| < \varepsilon |Seg|$ **then**
 9:         Break out all the loops
10:       **end if**
11:     **end if**
12:   **end for**
13: **end for**
14: **return** SV_List;

Let's denote the labeled regions on the frame as Seg. Then the optimization formula is given by (see notations in Sect. 2.1):

$$\underset{(l,r)}{\arg \min} E = \left(1 - \frac{\left| \bigcup\limits_{i=1}^{|R|} s_{r(i)}^l \cap Seg \right|}{\left| \bigcup\limits_{i=1}^{|R|} s_{r(i)}^l \cup Seg \right|} \right)^2 + \mu |R|^2, \tag{6}$$

where μ is a weight to balance the matching term and the number of supervoxels. Notice that here we only consider the regions on the labeled frame. There are $|R|$ supervoxels $s_{r(i)}^l$ that are not overlapped with each other. When $s_r^m = s_r^n$ for $m > n$, we select the larger level m. Thus a list of (l, r) non-overlapped supervoxels will be obtained through Eq. 6.

To speed up the object model construction, we propose a greedy or a relaxed version of the optimal solution, called hierarchical supervoxel selection algorithm, which is developed to minimize the error between the represented supervoxel list and the labeled regions for the key frame given the smallest number of supervoxels. We start from scanning all the supervoxels s_r^l from the highest level to see whether there is any supervoxel can be covered by the labeled region Seg. If yes, this region of the supervoxel on this frame will be subtracted from the frame until there is almost no remaining area any more. Since the size of the supervoxels becomes smaller from higher level to lower level, we can tolerate more covering errors between s_r^l and the remaining segmentation at lower levels than higher levels. After this stage, the obtained supervoxel list can represent the objects on the labeled frame, where each supervoxel can be from different levels of the hierarchical supervoxel graph.

The hierarchical supervoxel selection algorithm is summarized in Algorithm 1, where $\min e$ and $\max e$ are the acceptance error rates at the lowest and highest levels respectively. ε is a very small number to guarantee that the majority of the object can be covered by the given supervoxels.

This hierarchical supervoxel representation is capable of characterizing any object(s) labeled by users even if they are not in the same level (like in Fig. 2), or at different locations. The supervoxel list can be easily propagated to subsequent frames. The objects along the time can be represented by the supervoxel list sequences, which makes the real-time object segmentation available.

4   Video Object Segmentation Framework

A block diagram of the proposed video object segmentation framework is shown in Fig. 6. It contains three stages: offline hierarchical supervoxel graph generation, interactive object representation for the key frame, and representation propagation and refinement.

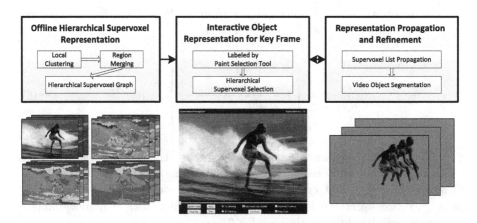

Fig. 6. The video object segmentation framework contains three stages. The offline supervoxel representation stage builds hierarchical supervoxel graph based on local clustering and region merging. In the interactive object representation stage, the users label interested objects via Paint Selection tool [20], which is further represented by a compact set of supervoxels obtained by the proposed hierarchical supervoxel selection algorithm. Finally, the representation propagation and refinement stage will reduce segmentation errors to achieve final video object segmentation.

4.1   Hierarchical Supervoxel Graph Generation

For an input video, the hierarchical supervoxel graph is firstly constructed to represent the video using the algorithms introduced in Sect. 2. Some tiny refinement on supervoxel representation results are also conducted to meet the basic hypothesis of hierarchical supervoxel graphs, such as supervoxel connectivity enforcement. Boundaries of all the supervoxels are further smoothed by Gaussian filter to avoid roughness. This stage is conducted offline, and is prepared for real-time interactive object segmentation.

4.2 Interactive Object Representation

Our goal is to enable users to label (an) arbitrary object(s) in a video shot in an easy way, then the system can automatically segment and track the labeled object in the shot. Thus, an easy-to-use label tool is necessary for users to label objects. We embedded the Paint Selection [20] tool in our system, which is an interactive image segmentation tool to convert sparse scribbles into full segmentation. It shows the instant feedback to users as they drag the mouse/pen, which is quite efficient to extract the objects.

Based on the labeled objects, the proposed hierarchical supervoxel selection algorithm in Sect. 3 is applied to model and represent the objects by a set of supervoxels.

4.3 Representation Propagation and Refinement

We assume that the input video shot has no shot cut and the labeled objects nearly appear all the time without large appearance change. Thus, all the object regions along the time can be covered by the supervoxels which are part of the supervoxel set in the labeled frame. To achieve the segmentations of the labeled objects, we propagate the supervoxel set from the labeled frame to the following frames using the same supervoxels in that level, followed by some post-processing techniques to refine the boundary of salient objects, such as segmentation label connectivity enforcement and Gaussian filtering. These operations help avoid multihole objects and bumpy boundaries. Eventually we could achieve the video object segmentation results for all the frames in this shot.

If the representation propagation does not work well for some subsequent frames, user can continue to label that frame for the segmentation of next frames. The disadvantage is that this will cost more human efforts. However, in this paper, all the segmentation results are just based on one labeled frame.

5   Experimental Results

In this section, we first describe the experimental data set and the generation of ground truth labels. Then, we compare the proposed algorithms with state-of-the-art methods in the two aspects, i.e. hierarchical supervoxel graph construction, and interactive video object segmentation.

5.1   Dataset

Three public video datasets were used in our experiments. One is the dataset used in Video SnapCut [6], which is a state-of-the-art video object segmentation algorithm, and has successfully applied to Adobe After Effect CS5. Video Snap-Cut dataset has seven available video shots with around 100 frames for each shot at a resolution from 320×240 to 720×480. To make a larger dataset for a more objective evaluation, SegTrack database [24], and UCF Sports Action Data Set [25] are also leveraged. SegTrack database has five available video shots ranging in length from 21 to 70 frames for each shot at a resolution from 320×240 to 414×352. UCF Sports Action Data Set consists of a set of actions collected from various sports, which contains 40 video sequences at a resolution of 720×480.

Except SegTrack, the other two datasets do not provide the ground-truth segmentations. To quantitatively evaluate the segmentation performance, we develop an interactive interface to generate the ground truth frame by frame for the entire database using Paint Selection tool [20].

The three datasets were combined together as our experimental dataset for diversity, including videos with diverse style, length, and movement speed, one fifth of which were randomly selected as the training set for parameter tuning, and the other videos compose the testing set. In bottom-level supervoxel generation, we set different spatial factors γ for different levels but the same motion factor δ. We first tuned γ to achieve the best visual results, and then δ. In higher-level generation, we first tuned the other motion factor α and then the saliency factor β. In supervoxel selection, we tried different combinations of $\min e$, $\max e$, and ϵ, and chose the best one in the training set. After getting all parameters, we evaluated on the testing set. Since we have already included videos with diverse style, length, and movement in our dataset, the parameters are shown to be not sensitive to different videos.

5.2   Performance of Hierarchical Supervoxel Graph

Two state-of-the-art algorithms were compared with the proposed hierarchical supervoxel graph generation algorithm. One is the hierarchical graph-based approach (GBH) [14], and the other is the SLIC algorithm [7]. We compare these methods at different granularities of the graph, i.e. with different numbers of supervoxels as 64, 128, 256, 512, and 1024 respectively.

We follow the supervoxel segmentation evaluation measurement discussed in [8], including 3D Under-segmentation Error (3D UE), 3D Boundary Recall (3D BR), 3D Segmentation Accuracy (3D ACCU), and Explained Variation. Except 3D UE, the other three are the larger, the better. The evaluation measures of compared algorithms with increasing number of supervoxels are shown in Fig. 8. We can see that, for all the measurements the proposed methods consistently outperform the compared methods. Specifically, GBH is worse than the proposed algorithm in all granularities, while the performance of SLIC drops quickly when the granularity of supervoxels becomes larger (the number of supervoxels is smaller).

Fig. 7. Examples of hierarchical supervoxel graphs produced by various methods: GBH, SLIC and the proposed method. From left to right: the original frames, GBH with 64 supervoxels, a closeup of GBH (64), SLIC with 64 supervoxels, a closeup of SLIC (64), the proposed method with 64 supervoxels, and a closeup of our method (64). For the "girl" video, the right arm is occluded for several frames. Our method can successfully detect it after its reappearance, and group it to the same supervoxel before its disappearance because both arms are in the same supervoxel in our design, while other methods failed to detect it or consider it as a new part. Similarly in the "referee" video, the reappeared right hand can only be recaptured by our method.

Figure 7 shows the supervoxel graphs produced by the three algorithms for two video shots. We can see that, for the "girl" video with a large motion, the right arm is occluded in several frames. When this arm reappears in the frame, GBH and SLIC consider it as a new part or fail to detect it. In contrast, our method can detect that arm before its disappearance and after its reappearance, and then group them into one supervoxel. Similarly, in the "referee" video, the reappeared right hand can only be recaptured by our method.

5.3 Performance of Video Object Segmentation

We compare our solution in video object segmentation with the state-of-the-art solution Video SnapCut [6], which has been transferred to Adobe After Effects CS5[1] (AAE). Object accuracy A_O and boundary accuracy A_B [26] were used to measure the segmentation performance frame by frame. A_O is given by

---

[1] We directly use Adobe After Effects CS5 for video object segmentation in the experiments.

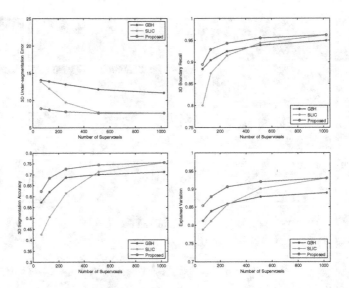

Fig. 8. Comparison of hierarchical supervoxel construction. (a) 3D Under-segmentation Error, (b) 3D Boundary Recall, (c) 3D Segmentation Accuracy, and (d) Explained Variation versus the number of supervoxels.

$$A_O = \frac{|G_O \cap S_O|}{|G_O \cup S_O|}, \tag{7}$$

where G_O is the set of all the pixels inside the ground-truth object, and S_O is the set of all the pixels in the object segmented by algorithms. A_B is given by:

$$A_B = \frac{\sum_x \min \left(\tilde{G}_B(x), \tilde{S}_B(x) \right)}{\sum_x \max \left(\tilde{G}_B(x), \tilde{S}_B(x) \right)}, \tag{8}$$

where

$$\tilde{G}_B(x) = \exp \left(-\frac{\|x - \hat{x}\|^2}{2\sigma^2} \right),$$

$$\tilde{S}_B(x) = \exp \left(-\frac{\|x - \hat{x}\|^2}{2\sigma^2} \right), \tag{9}$$

$$\hat{x} = \arg \min_{y \in G_B \, or \, S_B} \|x - y\|.$$

Here x is the location of each pixel for the segmentation map. G_B is the set of border pixels for the ground-truth object. $\tilde{G}_B$ and $\tilde{S}_B$ are the fuzzy set for the border of the ground-truth object and that of the object segmented by algorithms respectively. We set the bandwidth parameter σ as 2.

Table 1. Comparison on video object segmentation

Algorithm	Object accuracy	Boundary accuracy
AAE	80.00%	53.06%
Our method	88.07%	63.26%

In the cause of fairness, we manually segment the main object in the first frame of each video shot exactly the same, and check the segmentation performance of subsequent frames that are automatically generated. The more accurate the interactive video object segmentation is, the less human effort it will take. The results in Table 1 show that our segmentation method can produce 8%+ performance improvements over AAE in average object accuracy and average boundary accuracy.

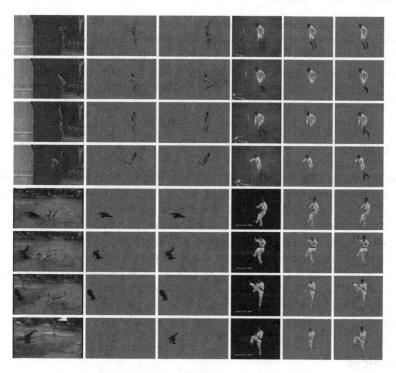

Fig. 9. Visual comparison of video segmentation produced by Adobe After Effects (AAE) and our method. Every four rows and three columns represent a video example. From left to right: original frames, AAE, and our method. The first row of each example is labeled frame. These cases are challenging because of large motion, motion blur or ambiguous colors. We can see that the proposed solution outperforms AAE in above cases (Color figure online).

Figure 9 is the visual comparison of four video segmentation cases of the two methods. For each case, the first row is the labeled frame, and other rows are testing frames. These four videos have large motions. For the "diver" video, when the motion is large enough, AAE cannot capture the arms and disconnect the body and the legs in some frames. For the "soccer player" video, the moving and blurring leg are ignored by AAE. For the "monkey" video, AAE cannot capture the monkey with large motions. For the "kicker" video, the occluded leg is ignored or disconnected from the body for some frames. In contrast, our method is able to deal with these challenging cases.

It is worth noting that our method is capable of segmenting disconnected objects at the same time. However, AAE can only extract one connected object along the time and could only propagate the segmentation for twenty frames to avoid large errors.

5.4 Time Cost Analysis

The system contains offline graph generation, and online interactive object segmentation. The online part is fast, around 0.01 second/frame on average with the resolution 320×240 using ten threads on a common machine, and achieved real-time response. The offline part is slower, about 0.8 second/frame under the same setting, one tenth slower than SLIC. Additional time cost was mainly from optical flow and saliency detection.

6 Conclusions

In this paper, we proposed a hierarchical supervoxel graph construction algorithm for video representation, as well as an interactive supervoxel representation to model target objects. Based on these representations, a video object segmentation framework was developed, which has potential to handle challenging scenarios such as large motion, motion blur or ambiguous colors in videos. We have compared the proposed algorithms with state-of-the-art methods in hierarchical supervoxel graph construction and video object segmentation. Although the concept is fairly simple, the proposed algorithms perform well with high efficiency for online representation, which valuably provides a new angle to deal with the problem of interactive video object segmentation.

References

1. Li, Y., Sun, J., Shum, H.Y.: Video object cut and paste. ACM Trans. Graph. (TOG) **24**, 595–600 (2005)
2. Yuri, B., Marie-Pierre, J.: Interactive graph cuts for optimal boundaryand region segmentation of objects in N-D images. In: 2001 IEEE International Conference on Computer Vision (ICCV), vol. 112 (2001)
3. Wang, J., Bhat, P., Colburn, R.A., Agrawala, M., Cohen, M.F.: Interactive video cutout. ACM Trans. Grap. (TOG) **24**, 585–594 (2005)

4. Comaniciu, D., Meer, P.: Mean shift: a robust approach toward feature space analysis. IEEE Trans. Pattern Anal. Mach. Intell. **24**, 603–619 (2002)
5. Price, B.L., Morse, B.S., Cohen, S.: Livecut: Learning-based interactive video segmentation by evaluation of multiple propagated cues. In: 2009 IEEE International Conference on Computer Vision (ICCV), pp. 779–786. IEEE (2009)
6. Bai, X., Wang, J., Simons, D., Sapiro, G.: Video snapcut: robust video object cutout using localized classifiers. ACM Trans. Graph. (TOG) **28**, 70 (2009)
7. Achanta, R., Shaji, A., Smith, K., Lucchi, A., Fua, P., Susstrunk, S.: Slic superpixels compared to state-of-the-art superpixel methods. IEEE Trans. Pattern Anal. Mach. Intell. **34**, 2274–2282 (2012)
8. Xu, C., Corso, J.J.: Evaluation of super-voxel methods for early video processing. In: 2012 IEEE Conference on Computer Vision and Pattern Recognition (CVPR), pp. 1202–1209. IEEE (2012)
9. Shi, J., Malik, J.: Normalized cuts and image segmentation. IEEE Trans. Pattern Anal. Mach. Intell. **22**, 888–905 (2000)
10. Felzenszwalb, P.F., Huttenlocher, D.P.: Efficient graph-based image segmentation. Int. J. Comput. Vision **59**, 167–181 (2004)
11. Vedaldi, A., Soatto, S.: Quick shift and kernel methods for mode seeking. In: Forsyth, D., Torr, P., Zisserman, A. (eds.) ECCV 2008. LNCS, vol. 5305, pp. 705–718. Springer, Heidelberg (2008). doi:10.1007/978-3-540-88693-8_52
12. Levinshtein, A., Stere, A., Kutulakos, K.N., Fleet, D.J., Dickinson, S.J., Siddiqi, K.: Turbopixels: fast superpixels using geometric flows. IEEE Trans. Pattern Anal. Mach. Intell. **31**, 2290–2297 (2009)
13. Fu, X., Wang, C.Y., Chen, C., Wang, C., Kuo, C.C.J.: Robust image segmentation using contour-guided color palettes. In: 2015 IEEE International Conference on Computer Vision (ICCV), pp. 1618–1625. IEEE (2015)
14. Grundmann, M., Kwatra, V., Han, M., Essa, I.: Efficient hierarchical graph-based video segmentation. In: 2010 IEEE Conference on Computer Vision and Pattern Recognition (CVPR), pp. 2141–2148. IEEE (2010)
15. Van den Bergh, M., Van Gool, L.: Real-time stereo and flow-based video segmentation with superpixels. In: 2012 IEEE Workshop on Applications of Computer Vision (WACV), pp. 89–96. IEEE (2012)
16. Palou, G., Salembier, P.: Hierarchical video representation with trajectory binary partition tree. In: 2013 IEEE Conference on Computer Vision and Pattern Recognition (CVPR), pp. 2099–2106. IEEE (2013)
17. Chang, J., Wei, D., Fisher, J.W.: A video representation using temporal superpixels. In: 2013 IEEE Conference on Computer Vision and Pattern Recognition (CVPR), pp. 2051–2058. IEEE (2013)
18. Reso, M., Jachalsky, J., Rosenhahn, B., Ostermann, J.: Temporally consistent superpixels. In: 2013 IEEE International Conference on Computer Vision (ICCV), pp. 385–392. IEEE (2013)
19. Xu, C., Whitt, S., Corso, J.: Flattening supervoxel hierarchies by the uniform entropy slice. In: 2013 IEEE International Conference on Computer Vision (ICCV), pp. 2240–2247 (2013)
20. Liu, J., Sun, J., Shum, H.Y.: Paint selection. ACM Trans. Graph. (ToG) **28**, 69 (2009)
21. Ren, X., Malik, J.: Learning a classification model for segmentation. In: 2003 IEEE International Conference on Computer Vision (ICCV), pp. 10–17. IEEE (2003)
22. Liu, C.: Beyond pixels: exploring new representations and applications for motion analysis. PhD thesis. Citeseer (2009)

23. Jiang, H., Wang, J., Yuan, Z., Liu, T., Zheng, N., Li, S.: Automatic salient object segmentation based on context and shape prior. In: Proceedings of the British Machine Vision Conference, vol. 6, p. 9 (2011)
24. Tsai, D., Flagg, M., Nakazawa, A., Rehg, J.M.: Motion coherent tracking using multi-label MRF optimization. Int. J. Comput. Vis. **100**, 190–202 (2012)
25. Rodriguez, M.D., Ahmed, J., Shah, M.: Action mach a spatio-temporal maximum average correlation height filter for action recognition. In: 2008 IEEE Conference on Computer Vision and Pattern Recognition (CVPR), pp. 1–8. IEEE (2008)
26. McGuinness, K., O'connor, N.E.: A comparative evaluation of interactive segmentation algorithms. Pattern Recogn. **43**, 434–444 (2010)

Learning to Generate Object Segment Proposals with Multi-modal Cues

Haoyang Zhang[1,2(✉)], Xuming He[1,2], and Fatih Porikli[1,2]

[1] The Australian National University, Canberra, Australia
{haoyang.zhang,xuming.he,fatih.porikli}@anu.edu.au
[2] Data61, CSIRO, Canberra, Australia
haoyang.zhang@data61.csiro.au

Abstract. This paper presents a learning-based object segmentation proposal generation method for stereo images. Unlike existing methods which mostly rely on low-level appearance cue and handcrafted similarity functions to group segments, our method makes use of learned deep features and designed geometric features to represent a region, as well as a learned similarity network to guide the grouping process. Given an initial segmentation hierarchy, we sequentially merge adjacent regions in each level based on their affinity measured by the similarity network. This merging process generates new segmentation hierarchies, which are then used to produce a pool of regional proposals by taking region singletons, pairs, triplets and 4-tuples from them. In addition, we learn a ranking network that predicts the objectness score of each regional proposal and diversify the ranking based on Maximum Marginal Relevance measures. Experiments on the Cityscapes dataset show that our approach performs significantly better than the baseline and the current state-of-the-art.

1 Introduction

Object proposal generation, which aims to produce a set of high-quality object candidates in an image, has become a core component in modern object detection [1–3] and segmentation pipelines. By focusing on a relatively small set of object-like regions, it enables us to use better object representations and improves significantly the accuracy of target vision tasks. While most work in object proposal generation focus on generating bounding boxes for object detection [4–7], object segments or region proposals play an important role in semantic segmentation and object segmentation [8,9].

Compared to bounding box proposals, generating object segment candidates is more challenging due to inaccuracies in bottom-up segmentation processes. Early work incorporate boundary consistency and smoothness priors through superpixel grouping [5,9] or MRF-based segmentation [8,10,11]. They rely on handcrafted image features to group pixels into region proposals. More recent approaches use deep ConvNets to learn the feature representation and directly predict class-agnostic object masks [12,13]. However, such end-to-end learning of a deep network makes it difficult to incorporate additional input data from

© Springer International Publishing AG 2017
S.-H. Lai et al. (Eds.): ACCV 2016, Part I, LNCS 10111, pp. 121–136, 2017.
DOI: 10.1007/978-3-319-54181-5_8

other sensor modalities, such as depth cues [14,15]. It may require retraining of the full system using a large dataset with instance-level annotations, which can be expensive and time-consuming.

In this work, we consider the problem of generating object segmentation proposals with stereo image inputs. To efficiently incorporate the depth cues computed from the stereo, we take an alternative deep learning approach, and learn an iterative merging process for generating a diverse set of high-quality region proposals. Unlike the previous global approaches, we mainly focus on learning a representation for object-driven perceptual grouping, which is an easier problem due to its local nature and potential to be modeled by a simpler network. More importantly, it enables us to design a late fusion strategy to incorporate the noisy depth cues into grouping without retraining the full deep network pipeline.

Specifically, our method consists of two stages. We start from an initial segmentation hierarchy of the left image and sequentially merge neighboring regions in each level of the hierarchy based on affinity scores predicted by a learned similarity network. This merging process generates new hierarchies of image segments, which is used to produce a pool of regional proposals by taking single, pair, triple and 4-tuple neighboring segments from the hierarchies. We then learn a ranking network to predict the objectness score of each region proposal. Our similarity and ranking network use a combination of learned deep features for appearance and designed geometric features for depth cue. While the similarity network predicts how likely two regions belong to the same object instance or the same background class, the ranking network estimates the overlap ratio with respect to the ground-truth for each candidate region.

We evaluate our algorithm on the Cityscapes dataset [16] with comparisons to Selective Search baseline and several stat-of-the-art methods, including Multiscale Combinatorial Grouping (MCG) and Geodesic Object Proposals (GOP). Our results show that we achieves significant improvement over these methods. The main contributions of our work are three folds: first, we propose a deep learning approach to the multi-modal object segmentation proposal generation; second, we design an alternative method to produce region proposals with a learned merging network and ranking network; and finally, our method achieves superior performance to the strong baselines on the challenging Cityscapes dataset.

2 Related Work

Generating high-quality object proposals plays an important role in the recent advance of object detection [1,2], and has drawn much attention in computer vision literature [17]. Many early work use handcrafted features to score bounding box hypotheses [4,6,7] and generate a set of object candidates mainly for detection task. More recent work in object detection literature [3], however, begin to learn the proposal generation as an integrated component of detection networks and do not rely on a preprocessing step for bounding box generation any more.

In contrast to bounding box generation, much less progress has been made in object segmentation proposal generation. One strategy is to formulate the problem as a series of foreground segmentation tasks [8,10,11,18]. By solving multiple graph-cut problems with diverse seeds, they can generate a large set of region proposals. In particular, Lee et al. [19] learns a parametric energy function to combine handcrafted mid-level cues in order to generate a diverse set of regional proposals. An alternative approach is to group superpixels into a hierarchical segmentation and seek semantic meaningful region proposals [5,9,20]. The MCG [9] generates multiscale UCMs and takes singletons, pairs, triplets and 4-tuples in their hierarchical segmentations as object proposals. [21] integrates both global foreground and local grouping strategies. Yanulevskaya et al. [22] also learns a grouping method for proposal generation, but their method uses manually designed appearance features and predefined similarity metrics. By contrast, our method directly learns a similarity network to group regions.

Deep learning based methods have been applied to object proposal generation, including bounding boxes [3,23,24] and segmentation [12,13]. Deep segmentation proposal methods usually take an end-to-end framework and use a global network to produce final candidate masks. In contrast, we adopt a learning-to-merge strategy and learn a simpler network structure. Such strategy has been successfully applied to semantic segmentation [25]. We note that our similarity network does not compute a distance metric between two regions [26]. Instead, it is an affinity score for grouping.

Few work have addressed the multi-modal object proposal generation task. Chen et al. [15] use depth information in an energy minimization framework to generate object proposals in a form of 3D bounding boxes. Bleyer et al. [14] design an iterative labeling strategy to segment object proposals from stereo images, which is computationally expensive.

Most of existing object segmentation proposal methods use ranking to improve the quality of the candidate pool. For example, [8,9] both learn a Random Forest with a set of low-level features to rank the object proposals. By contrast, we learn a neural network to rank object proposals.

We build our work on top of several existing techniques. Our model uses FCN network [27] and Hypercolumn feature [28] trained on PASCAL-Context dataset [29]. We extract deep features using "feature masking" technique introduced in [30].

3   Our Method

We aim to generate a set of object segmentation proposals and their objectness scores from a pair of stereo images. To this end, we design a segmentation proposal generation pipeline that learns to fuse multi-modal cues and to merge oversegmentation into object candidates. Figure 1 illustrates an overview of our approach. We first estimate a dense depth map of the scene using the stereo images, and build a segmentation hierarchy of the left image. Given the initial segmentation hierarchy, we represent all the regions in the hierarchy using convolutional and depth features. We then train a neural network to predict the affinity of neighboring regions, and rebuild multiple segmentation hierarchies by

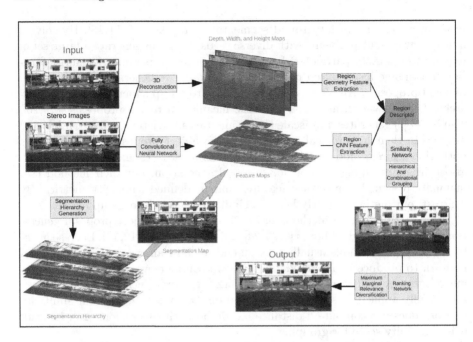

Fig. 1. Overview: Our system takes as input a pair of stereo images. We first generate a segmentation hierarchy, compute the convolutional feature maps and reconstruct the 3D scene. Then, we extract descriptors for regions in the segmentation hierarchy. Next, we iteratively merge adjacent regions based on their affinity score predicted by a similarity network to generate object proposals. Finally, we rank these object proposals through a ranking network and diversify the ranking.

incrementally merging adjacent regions from all the levels based on the learned similarity. From the new segmentation hierarchies, we extract region singletons, pairs, triplets and 4-tuples as object segmentation proposals. Finally, we rank these object proposals through a learned ranking network and diversify the ranking based on Maximum Marginal Relevance measures [8]. We now describe each stage of our pipeline in detail.

3.1   Initial Segmentation Hierarchy Generation

The first step of our method constructs an initial segmentation hierarchy of the left image. To generate the segmentation hierarchy, we use the Structured Edge Detection [31] on the left image to obtain an edge map for its efficiency and accuracy. An Ultrametric Contour Map (UCM) [9] is generated based on the estimated edge probability map. Then we threshold the UCM at five different levels to create the segmentation hierarchy. The thresholds are chosen such that the numbers of regions from the base level to the top level are roughly 1024, 768, 512, 384 and 256, respectively. For every region, we also record its child regions in the hierarchy, which enables efficient propagation of region descriptors from the base level to higher levels in the hierarchy.

3.2 Multi-modal Region Representation

For each region in the segmentation hierarchy, we extract two types of features to capture its appearance and 3D geometric properties. We take an efficient bottom-up approach to compute the region features at all the hierarchy levels. We only need to calculate those features explicitly for the base level regions and use max-pooling or weighted average-pooling to obtain features of higher level regions recursively.

Appearance Features. We extract a set of rich deep features to encode the appearance of a region. We first feed the left image into a Fully Convolutional Network (FCN) [27] to generate multiple layers of feature maps for the entire image. We choose the FCN-8s model trained on PASCAL-Context dataset [29] for the scene labeling task due to its superior performance and diverse set of 59 semantic classes, including *sky, ground, grass, building, road, person, bicycle and car etc.*. The feature map outputs from *pool1, pool2, pool3, pool4, pool5 and fc7 layers* are used as our representation, inspired by the "Hypercolumns" concept proposed by Hariharan et al. [28].

Given the feature maps, we compute the appearance features of a region by masking and max-pooling. As the feature maps of different layers are not of the same size, the deep features of a region cannot be directly masked out from these maps. A straightforward way to solve this problem is to upsample the feature maps to the same size as the image [28]. However, due to high dimensionality and varying sizes of the feature maps, e.g. the output from *fc7 layer* has 4096 dimensions and a very small size (17×33 in our case), such upsampling is very time-consuming and memory-costly.

To tackle this issue, we adopt the convolutional feature masking technique proposed by Dai et al. [30]. Specifically, we first compute the receptive field for every neuron activation in each layer according to the receptive field geometry [32]. Then we project each neuron activation onto the image plane, which is located at the center of its receptive field. We define a "domain of influence" for a neuron on the image plane, which has the same center as its receptive field and a smaller width (or height) that equals to the distance between neighboring receptive field centers. For example, the neuron at location $(1, 1)$ in *pool5 layer* may have a square "domain of influence" with its center at $(16, 16)$ and its side length as 32 in the image domain. If over 50% of the "domain of influence" of a neuron is covered by a region mask, we label this neuron activation as active for this region and it will be included in the calculation of region feature. By this labeling process, we project the base level region masks in the image plane onto the feature maps and then we do max-pooling in the projected masks on the feature maps to extract regions' deep features. Figure 2 (left) shows an example of our feature computation process. This generates a 5568-dimensional feature to

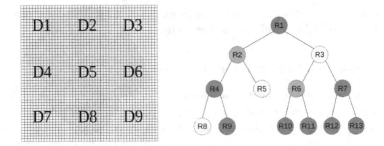

Fig. 2. Left: Illustration of "domain of influence" and feature masking. D1–D9 red rectangles are the domains of influence of activations A1–A9 in *pool5 layer*. The yellow mask is a region and only A2, A4, A5, A6 and A8 are activated by this region, as over half of their domains of influence are overlapped by this region. **Right:** Illustration of combinatorial grouping. Singletons:R1–R13. Pairs:(R2, R6). Triplets:(R3, R5, R8). (Color figure online)

encode the region's appearance. Note that when computing the deep features of regions in higher levels of the hierarchy, we only need to do max-pooling among their child region features.

3D Geometric Features. To encode geometric properties of a region, we extract two sets of 3D geometric features. We first estimate the dense depth map using the method [33] and convert it into a point cloud representation in the camera coordinate system according to the provided camera parameters.

Given the point cloud and a base level region mask, we segment out the subset of the point cloud using the mask. The subset is used to compute two sets of features to describe the region's geometric properties. Denoting the position of a 3D point as (x, y, z), we first compute the center of the region as one set of features, including *mean x, mean y, and mean z*. Another set of features describe the spatial distribution of the point cloud, consisting of three histograms, one for each dimension of the point cloud. Specifically, for the width, height and depth dimension, we evenly divide the spatial ranges $[-50\,\text{m}, 40\,\text{m}], [-40\,\text{m}, 3\,\text{m}]$ and $[-1\,\text{m}, -100\,\text{m}]$ into 256 *bins*, 128 *bins* and 256 *bins* in the log space, respectively. These spatial ranges are obtained from the statistics of the point clouds in the training set. The spatial histograms are computed based on these bins and then normalised by their L_1 norm. The two sets of features are concatenated to form a 643-dimensional feature G_{r_i} to encode the region's 3D geometric properties. The geometric features of higher levels can be efficiently computed through the hierarchy by weighted average-pooling of child region features as follows,

$$G_{r_{parent}} = \frac{\sum_{r_i \in children} G_{r_i} \times area(r_i)}{\sum_{r_i \in children} area(r_i)}. \tag{1}$$

3.3 Similarity Network

Given the segmentation hierarchy, we want to learn a merging process that generates a high-quality object candidate set from the initial oversegmentation. To achieve this, we design and train a neural network to compute the affinity between two adjacent regions and use the network to merge region pairs recursively. Unlike the manually designed similarity scores used in [5, 22], our network enables us to learn a more effective merging criterion in the multi-modal space.

Network Architecture. Our similarity network takes a concatenation of feature descriptors from two adjacent regions as input and consists of three fully-connected layers. Each layer has 512 neurons and uses RELU as activation function except the last layer. We add the dropout layer after the first two layers to prevent overfitting. The output is the affinity score between two input regions in the range of $[0, 1]$ and indicates how likely two regions belong to the same object. We use the MatConvNet [32] to implement our networks in this work.

Network Training. We obtain the training examples from the initial segmentation hierarchy. As we are learning a similarity network for object proposal generation, we expect that the network is able to output a high similarity score for two regions from the same object instance or the same background class, and to output a low score for two regions from different object instances. This network can be viewed as re-weighing the boundary strength between regions in the original UCM.

We formulate the network training as a binary classification problem. Each positive example is a pair of neighbouring regions overlapped with the same object instance and both regions have an overlap score larger than a threshold $t_{p1} = 0.7$. Here we define the overlap score as the intersection of a region and an object instance divided by the area of that region. We also take pairs overlapped with the same background class. In particular, we hope that regions belonging to the same background class around the object instance can merge together so that they do not interfere with the grouping of those regions from this object instance. To balance the positive examples from the object instance and from the background classes, we keep the proportion between them roughly at $1 : 2$.

For negative examples, we first take pairs of neighbouring regions in which one has an overlap with the object instance higher than $t_{p1} = 0.7$ while the other overlaps with the same object instance less than $t_n = 0.6$. Similar to the positive examples, we also include adjacent background region pairs which satisfy the same overlapping condition. We keep the proportion of negative examples from the object instance and from the background classes at about $1 : 1$.

To mimic the process of grouping at test time, we scan regions from all levels of the segmentation hierarchy and obtain about 4,120,000 positive and 3,370,000 negative training examples. As there are two ways to concatenate features from two adjacent regions, we use both orders in the network training and the total number of training samples is doubled.

We train the similarity network to minimize the *log loss* using stochastic gradient descent with a batch size of 2,000 examples, momentum of 0.9, weight decay of 0.0005 and for 15 epochs. The learning rate we use for each epoch changes from 0.001 *to* 0.000001 evenly in the log space.

3.4   Hierarchical and Combinatorial Grouping

Given the region features in every level of the segmentation hierarchy and a learned similarity network, we generate a set of object proposals by a hierarchical and combinatorial grouping process.

Hierarchical Grouping. We start from the initial regions in a single level of the segmentation hierarchy and re-group them by applying the similarity network. Specifically, we first compute the affinities between all adjacent regions via forwarding the feature descriptor of neighbouring regions through the similarity network. Then two most similar regions are merged into a new region and the descriptor for this new region is computed. This can be easily done by max-pooling (for appearance feature) or weighted average-pooling (for geometric feature) as described in Sect. 3.2. Next the affinities between this new region and its neighbours inherited from its child regions are updated using the similarity network. This merging process is repeated until the whole image becomes a single region. We apply this hierarchical grouping procedure to all five levels of the initial segmentation hierarchy, and take all single regions (region singletons) in the five new segmentation hierarchies as our initial set of object proposals.

Combinatorial Grouping. Selecting the region singletons in the segmentation hierarchies only, however, is insufficient to generate a high quality pool of object proposals. We follow a combinatorial grouping procedure similar to [9] to generate a larger object proposal set. In particular, we empirically select 10,000 region pairs, 10,000 region triplets and 5,000 region 4-tuples from every newly generated segmentation hierarchy to expand our object proposal pool, which performs well in our experiments. Figure 2 (right) shows an example of region singletons, pair and triplet in the segmentation hierarchy. We perform Non-Maximal Suppression (NMS) afterwards, which significantly reduces the number of candidates, since those region pairs, triplets and 4-tuples from the same segmentation hierarchy are heavily overlapped. The final pool of object proposals contains less than 10,000 proposals per image on average.

3.5   Ranking Network

In the final step, we want to estimate the quality of each object proposal, or its objectness score. This allows us to obtain good trade-off between the number and the quality of object proposals under different settings. We achieve this by training a ranking neural network to predict the IoU of each object proposal with the matched ground truth as in [8].

Network Architecture. Our ranking network is a regression network, which has a similar architecture to the similarity network except the input and output layer. It also consists of three fully-connected layers and each layer has 512 neurons. The input is the feature descriptor of a single object proposal, which can be computed efficiently as follows. Proposals defined by region singletons have their descriptors precomputed during the merging process. For those proposals formed by region pair, triplet or 4-tuple, their descriptors can be computed using the same max-pooling or average-pooling method described before. The output layer of the network is a linear layer that predicts the IoU between the input proposal and the corresponding ground truth. We use the mean squared loss during network training. In the training stage, we also add a dropout layer after the first two layers to prevent overfitting.

Network Training. We build the training dataset by choosing four types of training examples. The first type includes all the ground truths and the corresponding target IoUs are 1.0. The remaining training examples come from the object proposals generated on the training set. We split these object proposals into three categories according to their IoU to the ground truth: $IoU >= 0.5$, $0 < IoU < 0.5$ and $IoU = 0$. For the first category, we take all proposals in this group as training examples and denote its size as N. As to the latter two categories, we randomly select $3N$ and $3N$ examples from their pools respectively, which balances the training dataset. Finally, we obtain about 5,000,000 training examples in total.

We train the ranking network using stochastic gradient descent with a batch size of $2,000$ examples, momentum of 0.9, weight decay of 0.0005 and for 10 epochs. The learning rate we use for each epoch changes from 0.01 *to* 0.00001 evenly in the log space.

Diversifying the Ranking. After assigning every proposal a ranking score, we diversify the ranking to reduce redundancy. Following [8], we achieve this based on Maximum Marginal Relevance measure, which is used to remove redundant object proposals. We apply the same re-ranking procedure as in [8] to lower the rank of the segment proposals that heavily overlap with higher-ranked proposals.

4 Experiments

In this section, we evaluate our multi-modal object proposal generation approach on the publicly available Cityscapes dataset [16]. To the best of our knowledge, Cityscapes dataset is the only public dataset with stereo images and object instance segmentation ground truth, which are required by our method for quantitative evaluation.

Dataset. Cityscapes [16] is a newly released large-scale dataset for semantic urban scene understanding. It is comprised of a large diverse set of stereo video sequences recorded on streets from 50 different cities. 5,000 of these images have high quality instance-level annotations for humans and vehicles and they are split into separate training (2,975 images), validation (500 images) and test (1,525 images) sets. This dataset is very challenging as it is biased towards busy and cluttered scenes where many, often highly occluded, objects occur at various scales. Figure 3 shows some examples.

In our experiments, we further split the training set into two subsets: one for training (2,614 images) and the other for validation (361 images taken at Tubingen, Ulm and Zurich). We use their validation set (500 images) to evaluate the approaches, as the ground truth of the test set is withheld and their evaluation server does not provide results on proposal generation. The original image size is 1024×2048, which is too large to feed into the GPU memory when forwarding the image through the FCN-8s. So we downscale the original image by a factor of 4 into 512×1024. The dataset only provides instance-level annotations for humans (*person and rider*) and vehicles (*car, truck, bus, bicycle, motorbicycle, caravan and trailer*), which are considered as object proposal ground-truth in our experiments.

Fig. 3. Illustration of the Cityscapes dataset.**Top:** RGB images. **Bottom:** instance-level ground truth.

Evaluation Measures. We employ the recall vs. number of proposals with a fixed IoU threshold and the average recall (AR) as the evaluation metrics. As discussed in Hosang et al.'s work [17], AR has been shown to have a strong correlation with the final detection performance. In our experiments, we compute the AR between IoU 0.5 to 1 and report AR vs. number of proposals.

Baseline and State-of-the-Art. As we focus on object segmentation proposals generation, we mainly compare our approach (Ours-Depth-Seg) against two widely-used top-performing segmentation proposal generation methods: MCG [9] and SelectiveSearch [5], as well as our approach without geometric feature (Ours-NoDetph-Seg). In addition, we compare to the more recent 'Geodesic Object Proposals' (GOP) method [34] (GOP(200,15) and GOP(140,4)), which has publicly available code.

We use the default parameters in MCG to generate the proposals. For Selective Search, we adopt the parameters used in RCNN [1], and keep the segmentation proposals instead of bounding boxes. The "Quality" version of Selective search (SeSe-Quality-60k) uses four different initial segmentations, five color spaces and four similarity functions to diversify object proposals and over 60,000 proposals are generated per image on average. To make a fair comparison, we randomly select 10,000 proposals (SeSe-Quality-10k) from the SeSe-Quality-60k and evaluate their quality. We repeat this for 5 times and take the average results as their performance. The "Fast" version (SeSe-fast) uses only two different initial segmentations, two color spaces and two similarity functions for diversification and about 12,000 proposals on average are generated per image.

Furthermore, in order to demonstrate that our method can also generate high-quality bounding box proposals, we conduct experiments to compare with the EdgeBoxes [7]. We use the tightest boxes enclosing our segmentation proposals as the output to evaluate our method.

Segmentation Results. Figure 4 shows the recall rate when varying the number of object proposals under different IoU thresholds. We can see that our approach constantly and significantly outperforms MCG, SeSe-Quality-10k, SeSe-Fast and GOP. The recall of our approach attains 44.8% at about 5,000 proposals

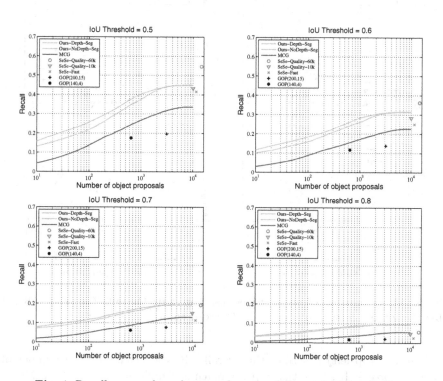

Fig. 4. Recall vs. number of proposals under different IoU thresholds.

when IoU threshold is 0.5, while MCG attains 33.6%, SeSe-Fast 41.7%, and SeSe-Quality-10k 44.0%. The performance of both versions of GOP is much lower than the above methods. With 1,000 proposals and IoU threshold as 0.5, the recall of our approach is above 40.0% while MCG just gets 26.8%. When the IoU threshold increases, we can see that the performance of Selective Search drops much faster than Ours and MCG, and particularly when the IoU threshold equals to 0.7, our method has a similar recall as SeSe-Quality-60k. This indicates that the quality of our proposals is better than Selective Search.

On the other hand, the performance of our approach using geometry information is always better than that without geometry information. Surprisingly, the upper bound of recall is not boosted by geometry information. This might be due to the noisy depth cues computed from the stereo images and that the geometric feature we manually designed is relatively weak. However, the ranking of proposals indeed benefits from the additional geometry information, as geometry information like the 3D height of a region is a good indicator of the objectness in street scenes.

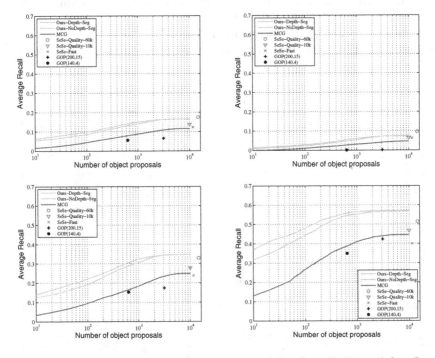

Fig. 5. AR vs. number of object proposals. **Top left:** Overall. **Top right:** Small objects. **Bottom left:** Medium objects. **Bottom right:** Large objects.

Figure 5 (top left) describes the overall AR when changing the number of object proposals. It shows that our approach is significantly better than MCG, SeSe-Fast, SeSe-Quality-10k and GOP, but slightly worse than SeSe-Quality-60k

Table 1. AR at different number of proposals (100, 1,000, 5,000 and total number of proposals (N)), overall AUC (AR averaged across all proposal counts) and also AUC at different scales (small, medium and large objects denoted by superscripts S, M and L).

Method	AR@100	AR@1000	AR@5000	AR@N	AUC	AUC^S	AUC^M	AUC^L
SeSe-Fast	-	-	-	0.122	0.106	0.052	0.206	0.358
SeSe-Quality-10k	-	-	-	0.137	0.108	0.047	0.221	0.402
SeSe-Quality-60k	-	-	-	**0.174**	0.145	**0.077**	0.278	0.451
MCG	0.045	0.087	0.113	0.115	0.107	0.041	0.229	0.432
GOP(200,15)	0.032	0.059	0.065	0.065	0.063	0.001	0.169	0.406
GOP(140,4)	0.032	0.056	0.056	0.056	0.055	0.001	0.151	0.344
Ours-noDepth-Seg	0.086	0.140	0.165	0.166	0.159	0.069	0.335	0.559
Ours-Depth-Seg	**0.099**	**0.150**	**0.165**	0.166	**0.160**	0.070	**0.337**	**0.566**

which uses much more proposals. With 1,000 proposals, our method achieves an AR of 15.0%, while MCG just 8.7% and this number is consistent with the performance of instance segmentation task reported by Cordts et al. [16] who use MCG object proposals in their experiments.

Following [12], we also report the AR vs. the number of object proposals at different object scales, as the size of object in Cityscapes dataset varies in a quite wide range. We split the ground truth into three sets according to object pixel area a: small ($a < 32^2$), medium ($32^2 \leq a \leq 96^2$) and large ($a \geq 96^2$).

Fig. 6. Qualitative examples of our object proposals. **Left:** RGB images. **Middle:** Ground truth. **Right:** Our best proposals.

Figure 5 describes the performance at each scale. All methods perform poorly on small objects (top right), which leads to the low overall AR. By contrast, when it comes to the categories of medium (Bottom left) and large (Bottom right) object, the AR by all approaches has a considerable increase and our method performs substantially better than MCG, SeSe-Fast, SeSe-Quality-10k and GOP, and also slightly better than SeSe-Quality-60k.

More detailed quantitative results are shown in Table 1, which reports the AR at selected proposal numbers and the averaged overall AR across all proposal numbers (AUC), as well as AUC at different object scales. Finally, examples of generated proposals with the highest IoU to the ground truth on selected images are shown in Fig. 6.

Bounding Boxes Results. We also compare our method against bounding box proposal generation method, the EdgeBoxes [7], using the metric of AR. Figure 7 shows that our approach generate much better bounding box proposals than the EdgeBoxes. With 1,000 proposals, our approach gets an AR of 27.3%, which is over 2.5× higher than the EdgeBoxes' (10.5%). The upper bound of our method (31.2%) is also much higher than the EdgeBoxes's (25.0%).

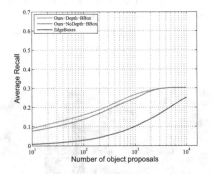

Fig. 7. AR vs. number of object proposals for Bounding Box proposals.

5   Conclusion

In this paper, we propose a learning-based object segmentation proposal generation method for stereo images, which exploits both deep features and the depth cue. We extract features from convolutional feature maps and geometry maps to describe a region. We learn a similarity network to estimate the affinity between two adjacent regions, sequentially merge regions from a segmentation hierarchy based on the affinity to generate object proposals and learn a ranking network to predict the objectness of a proposal. Experiments on the Cityscapes dataset show that our approach achieves much better average recall than the state-of-the-art and depth cue can improve the ranking of proposals.

Acknowledgment. Data61 is part of the Commonwealth Scientific and Industrial Research Organisation (CSIRO) which is the federal government agency for scientific research in Australia. The Tesla K40 used for this research was donated by the NVIDIA Corporation.

References

1. Girshick, R., Donahue, J., Darrell, T., Malik, J.: Region-based convolutional networks for accurate object detection and segmentation. IEEE Trans. Pattern Anal. Mach. Intell. **38**(1), 142–158 (2015)
2. Girshick, R.B.: Fast R-CNN. CoRR abs/1504.08083 (2015)
3. Ren, S., He, K., Girshick, R., Sun, J.: Faster R-CNN: Towards real-time object detection with region proposal networks. In: Advances in Neural Information Processing Systems, pp. 91–99 (2015)
4. Alexe, B., Deselaers, T., Ferrari, V.: What is an object? In: 2010 IEEE Conference on Computer Vision and Pattern Recognition (CVPR), pp. 73–80. IEEE (2010)
5. Uijlings, J.R., van de Sande, K.E., Gevers, T., Smeulders, A.W.: Selective search for object recognition. Intl. J. Comput. Vis. **104**, 154–171 (2013)
6. Cheng, M.M., Zhang, Z., Lin, W.Y., Torr, P.: Bing: Binarized normed gradients for objectness estimation at 300fps. In: IEEE CVPR (2014)
7. Zitnick, C.L., Dollár, P.: Edge boxes: locating object proposals from edges. In: Fleet, D., Pajdla, T., Schiele, B., Tuytelaars, T. (eds.) ECCV 2014. LNCS, vol. 8693, pp. 391–405. Springer, Heidelberg (2014). doi:10.1007/978-3-319-10602-1_26
8. Carreira, J., Sminchisescu, C.: CPMC: automatic object segmentation using constrained parametric min-cuts. IEEE Trans. Pattern Anal. Mach. Intell. **34**, 1312–1328 (2012)
9. Pont-Tuset, J., Arbeláez, P., Barron, J., Marques, F., Malik, J.: Multiscale combinatorial grouping for image segmentation and object proposal generation (2015). arXiv:1503.00848
10. Endres, I., Hoiem, D.: Category independent object proposals. In: Daniilidis, K., Maragos, P., Paragios, N. (eds.) ECCV 2010. LNCS, vol. 6315, pp. 575–588. Springer, Heidelberg (2010). doi:10.1007/978-3-642-15555-0_42
11. Krähenbühl, P., Koltun, V.: Learning to propose objects. In: CVPR (2015)
12. Pinheiro, P.O., Collobert, R., Dollar, P.: Learning to segment object candidates. In: Advances in Neural Information Processing Systems, pp. 1981–1989 (2015)
13. Dai, J., He, K., Sun, J.: Instance-aware semantic segmentation via multi-task network cascades, arXiv preprint (2015). arXiv:1512.04412
14. Bleyer, M., Rhemann, C., Rother, C.: Extracting 3D scene-consistent object proposals and depth from stereo images. In: Fitzgibbon, A., Lazebnik, S., Perona, P., Sato, Y., Schmid, C. (eds.) ECCV 2012. LNCS, vol. 7576, pp. 467–481. Springer, Heidelberg (2012). doi:10.1007/978-3-642-33715-4_34
15. Chen, X., Kundu, K., Zhu, Y., Berneshawi, A., Ma, H., Fidler, S., Urtasun, R.: 3d object proposals for accurate object class detection. In: NIPS (2015)
16. Cordts, M., Omran, M., Ramos, S., Rehfeld, T., Enzweiler, M., Benenson, R., Franke, U., Roth, S., Schiele, B.: The cityscapes dataset for semantic urban scene understanding, arXiv preprint (2016). arXiv:1604.01685
17. Hosang, J., Benenson, R., Dollár, P., Schiele, B.: What makes for effective detection proposals? arXiv preprint (2015). arXiv:1502.05082

18. Humayun, A., Li, F., Rehg, J.M.: The middle child problem: revisiting parametric min-cut and seeds for object proposals. In: Proceedings of the IEEE International Conference on Computer Vision, pp. 1600–1608 (2015)
19. Lee, T., Fidler, S., Dickinson, S.: Learning to combine mid-level cues for object proposal generation. In: Proceedings of the IEEE International Conference on Computer Vision, pp. 1680–1688 (2015)
20. Wang, C., Zhao, L., Liang, S., Zhang, L., Jia, J., Wei, Y.: Object proposal by multi-branch hierarchical segmentation. In: Proceedings of the IEEE Conference on Computer Vision and Pattern Recognition, pp. 3873–3881 (2015)
21. Rantalankila, P., Kannala, J., Rahtu, E.: Generating object segmentation proposals using global and local search. In: Proceedings of the IEEE Conference on Computer Vision and Pattern Recognition, pp. 2417–2424 (2014)
22. Yanulevskaya, V., Uijlings, J., Sebe, N.: Learning to group objects. In: Proceedings of the IEEE Conference on Computer Vision and Pattern Recognition, pp. 3134–3141 (2014)
23. Kuo, W., Hariharan, B., Malik, J.: Deepbox: Learning objectness with convolutional networks, arXiv preprint (2015). arXiv:1505.02146
24. Ghodrati, A., Diba, A., Pedersoli, M., Tuytelaars, T., Van Gool, L.: Deepproposal: hunting objects by cascading deep convolutional layers. In: Proceedings of the IEEE International Conference on Computer Vision, pp. 2578–2586 (2015)
25. Sharma, A., Tuzel, O., Liu, M.Y.: Recursive context propagation network for semantic scene labeling. In: Advances in Neural Information Processing Systems, pp. 2447–2455 (2014)
26. Zagoruyko, S., Komodakis, N.: Learning to compare image patches via convolutional neural networks. In: Proceedings of the IEEE Conference on Computer Vision and Pattern Recognition, pp. 4353–4361 (2015)
27. Long, J., Shelhamer, E., Darrell, T.: Fully convolutional networks for semantic segmentation. In: Proceedings of the IEEE Conference on Computer Vision and Pattern Recognition, pp. 3431–3440 (2015)
28. Hariharan, B., Arbeláez, P., Girshick, R., Malik, J.: Hypercolumns for object segmentation and fine-grained localization. In: Proceedings of the IEEE Conference on Computer Vision and Pattern Recognition, pp. 447–456 (2015)
29. Mottaghi, R., Chen, X., Liu, X., Cho, N.G., Lee, S.W., Fidler, S., Urtasun, R., Yuille, A.: The role of context for object detection and semantic segmentation in the wild. In: IEEE Conference on Computer Vision and Pattern Recognition (CVPR) (2014)
30. Dai, J., He, K., Sun, J.: Convolutional feature masking for joint object and stuff segmentation. In: Proceedings of the IEEE Conference on Computer Vision and Pattern Recognition, pp. 3992–4000 (2015)
31. Dollár, P., Zitnick, C.L.: Fast edge detection using structured forests. IEEE Trans. Pattern Anal. Mach. Intell. 37, 1558–1570 (2015)
32. Vedaldi, A., Lenc, K.: Matconvnet: convolutional neural networks for matlab. In: Proceedings of the 23rd Annual ACM Conference on Multimedia Conference, pp. 689–692. ACM (2015)
33. Yamaguchi, K., McAllester, D., Urtasun, R.: Efficient joint segmentation, occlusion labeling, stereo and flow estimation. In: Fleet, D., Pajdla, T., Schiele, B., Tuytelaars, T. (eds.) ECCV 2014. LNCS, vol. 8693, pp. 756–771. Springer, Heidelberg (2014). doi:10.1007/978-3-319-10602-1_49
34. Krähenbühl, P., Koltun, V.: Geodesic object proposals. In: Fleet, D., Pajdla, T., Schiele, B., Tuytelaars, T. (eds.) ECCV 2014. LNCS, vol. 8693, pp. 725–739. Springer, Heidelberg (2014). doi:10.1007/978-3-319-10602-1_47

Saliency Detection via Diversity-Induced Multi-view Matrix Decomposition

Xiaoli Sun[1(✉)], Zhixiang He[1], Xiujun Zhang[2], Wenbin Zou[3], and George Baciu[4]

[1] College of Mathematics and Statistics, Shenzhen University, Shenzhen, China
xlsun@szu.edu.cn
[2] School of Electronic and Communication Engineering,
Shenzhen Polytechnic, Shenzhen, China
[3] College of Information Engineering, Shenzhen University,
Shenzhen, China
[4] Department of Computing, The Hong Kong Polytechnic University,
Hung Hom, Hong Kong

Abstract. In this paper, a diversity-induced multi-view matrix decomposition model (DMMD) for salient object detection is proposed. In order to make the background cleaner, Schatten-p norm with an appropriate value of p in $(0,1]$ is used to constrain the background part. A group sparsity induced norm is imposed on the foreground (salient part) to describe potential spatial relationships of patches. And most importantly, a diversity-induced multi-view regularization based Hilbert-Schmidt Independence Criterion (HSIC), is employed to explore the complementary information of different features. The independence between the multiple features will be enhanced. The optimization problem can be solved through an augmented Lagrange multipliers method. Finally, high-level priors are merged to boom the salient regions detection. Experiments on the widely used MSRA-5000 dataset show that the DMMD model outperforms other state-of-the-art methods.

1 Introduction

Saliency is an important representation of human vision. It reflects that the attention of the human eyes to a certain region of an image is concentrated or not, and captures the most attractive regions. Saliency detection plays a significant role in computer vision. It is widely applied in the area of image processing, such as image resizing, image recognition, video coding and so on. So far, there are two typical categories for saliency detection: independent goals without prior knowledge, and specific goals with prior knowledge. These two types of methods are also named bottom-up method and top-down method. Typically, the bottom-up approaches usually focus on low-level features, such as color, texture, brightness, orientation, spatial position and so on. The saliency maps are generated by these features, and each feature is computed respectively. During this procedure, each feature is normalized and integrated in a certain way to

© Springer International Publishing AG 2017
S.-H. Lai et al. (Eds.): ACCV 2016, Part I, LNCS 10111, pp. 137–151, 2017.
DOI: 10.1007/978-3-319-54181-5_9

generate a final saliency map. However, the top-down methods are usually influenced by high level knowledge such as location prior, semantic prior and color prior, which are generally based on human perception. Recently, a fashion trend is that bottom-up and top-down are combined to work together. A sequence of models for salient object detection have been motivated by the traditional low-rank matrix recovery (LRMR) theory [13]. Some researchers also incorporate high level knowledge into low-level features. For instance, Shen et al. [6] propose a unified low-rank model (ULR) with a learning linear transformation of the feature space based on LRMR model to incorporate general low-level features with high-level knowledge. Lang et al. propose a multi-task sparsity pursuit model (LRR) [11] to integrate multiple types of features for saliency detection collaboratively. It is derived from low-rank representation [10]. Peng et al. propose a low-rank and structured sparse matrix decomposition model, which explores the underlying structure of image patches [15] and enlarges the distance between background and foreground by a Laplacian regularization [24]. The Laplacian regularization built on the assumption of the local invariance [27] is also incorporated into the proposed DMMD model. Besides, Zou et al. propose a segmentation driven low-rank matrix recovery model (SLR) [12] that a bottom-up segmentation prior guides the matrix recovery, and a more smooth saliency map may be obtained by post-smoothing module.

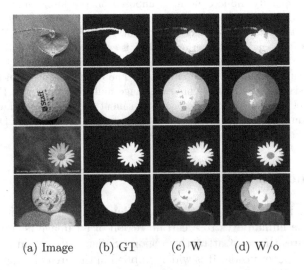

(a) Image   (b) GT   (c) W   (d) W/o

Fig. 1. Integrating the diversity term further improves the performance of the detection. First column is the input images. Second column is the groundtruth. The saliency maps without and with diversity term are shown in third column and the last column respectively. Saliency maps in the third column is more complete and more similar to the groundtruth than the last column. ('W' denotes that images detect with diversity term; 'W/o' denotes that images detect without diversity term.)

These methods are based on a common assumption that background blocks are highly correlated, and usually belong to a low-dimensional feature subspace while the salient objects are sparse. In other words, the feature matrix may be decomposed into a low-rank matrix and a sparse matrix. Although promising results may be obtained by these previous models, there are still some problems:

- If the background of image is complex, the salient map may be scattered.
- If the background and the foreground are in a similar appearance, the saliency maps of previous methods may be incomplete compared to the ground truth or be mixed with the background.
- The salient values of salient objects generated by previous approaches are not consistent.

To address these problems, a novel diversity-induced multi-view matrix decomposition model (DMMD) is proposed for the visual saliency detection. It is an extension of the classical LRMR theory. The main contributions of the proposed method are summarized as follows:

- To suppress the background, a Schatten-p norm (or S_p-norm, for short) is introduced to constrain the background, which may force the background in a lower rank space.
- We model the image as a graph by clustering. A group sparsity induced norm is introduced to encode potential relationships of patches in feature space. The relative patches of image will share similar representations.
- We propose a diversity-induced multi-view term to enhance the salient objects by exploring the complementary information of each pair of features, which is based on Hilbert-Schmidt Independence Criterion (HSIC). The diversity-induced multi-view term effectively enhances the salient objects that have complementary information cross multiple features. Some examples are shown in Fig. 1.

The rest of this paper is organized as follows. Section 2 discusses the related work. Section 3 describes the proposed DMMD model and the optimization algorithm. Section 4 performs the salient object detection. The experiment results are shown in Sect. 5. Finally, the conclusion is given in Sect. 6.

2   Related Work

Up to date, a series of methods have been proposed for salient object detection. Based on the LRMR theory, a brief discussion on LRR [11] and structured sparsity matrix decomposition (SMD) [24] is as follows.

Firstly, LRR approach assumes that the background patches are highly correlated, and are represented by a low-rank coefficient matrix multiplied by the feature matrix itself. Salient object is usually a small and special area. Collaborative multiple features and high-level priors are integrated into multitask sparsity pursuit model. It can be written as

$$\min_{\substack{Z_1,...,Z_K \\ E_1,...,E_K}} \sum_{i=1}^{i=K} \|Z_i\|_* + \lambda\|E\|_{2,1} \quad st. \quad X_i = X_i Z_i + E_i, i = 1, ..., K. \tag{1}$$

where $E = [E_1; ...; E_K]$ is the salient object, K is the number of features. X_i is the feature matrix, and Z_i is the coefficient matrix corresponding to i-th feature. Ω is a given prior vector. $\|\cdot\|_*$ is the nuclear norm (the sum of the singular values) which is a convex approximation to the rank function. $\|\cdot\|_{2,1}$ forces columns of E sparse (the sum of ℓ_2 norm of the columns of a matrix). In LRR [11], it combines multiple types of feature for saliency detection collaboratively. The jointly sparse matrix E may be not accurate, since it ignores the spatial relations and the feature affinities of image patches. In DMMD model, they are all taken into account and a diversity regularization HSIC is added to explore the complementarity of different views.

Secondly, structured sparsity matrix decomposition (SMD) [24] model is written as follows

$$\min_{L,S} \quad \|L\|_* + \alpha\Omega(S) + \beta\Theta(L,S) \text{ s.t. } \quad F = L + S. \tag{2}$$

where the nuclear norm $\|\cdot\|_*$ is a convex relaxation of the matrix rank function. α and β are the balance parameters. $\Omega(\cdot)$ denotes the tree-structured sparsity regularized term

$$\Omega(S) = \sum_{i}^{d} \sum_{j}^{n_i} v_j^i \|S_{G_i}\|_p^p. \tag{3}$$

It captures the potential structure of the image and enforces patches from the same object to have similar saliency values. $\Theta(\cdot,\cdot)$ is the Laplacian regularized term

$$\Theta(L,S) = Tr(SMS^T). \tag{4}$$

where M is a Laplacian matrix, which is introduced in [24]. It is based on the local invariance assumption: if two adjacent image patches are similar with respect to their features, their representations should be close to each other in the subspace, and vise versa. Tr denotes the matrix trace. It enlarges the distance between the background L and the salient object S.

In SMD, it replaces $\|S\|_1$ norm by a tree-structured sparsity-inducing norm to guarantee the extraction of salient targets. In other side, a Laplacian regularization under the assumption of the local invariance [27] is added on SMD. And we also append it to the proposed DMMD model for all views respectively, since it commendably preserves the local structure of the image features and enlarges the difference between L and S. Comparing with SMD [24], DMMD method has two major contributions. First, a S_p-norm is introduced to instead of $\|\cdot\|_*$, and the value of p resides in $(0, 1]$. It is a stronger constraint of low-rankness on the background. And the background is better suppressed with an appropriate value of p. Second, a diversity-induced multi-view term is employed to explore the underlying complementary information of different features.

3 The Proposed Model

3.1 Problem Formulation

Similar to [6], a given image is over-segmented into N non-overlapping patches $P = \{P_1, P_2, \cdots, P_N\}$. Here, c-th low-level feature vector of patch P_i denotes as $F_{ci} \in \mathbb{R}^{D_c}$ (D_c is the feature dimension of c-th feature), forming a feature matrix $F_c = [F_{c1}, F_{c2}, ..., F_{cN}] \in \mathbb{R}^{D_c \times N}$, then multiple feature matrix $F = [F_1; F_2; ...; F_c] \in \mathbb{R}^{D \times N}$ is constructed by vertically concatenating $F_1, F_2, ..., F_c$ along columns. Complementarity of the multiple features is explored thoroughly by minimizing the HSIC norm of sparse matrix of c-th feature S_c. The problem is to build a model to decompose the feature matrix of an image into a low-rank part L (non-salient regions) and a sparse part S (salient regions). DMMD model is denoted by the following formula:

$$\min_{L,S} \sum_{c=1}^{C} [\Upsilon(L_c) + \lambda_1 \Lambda(S_c) + \lambda_2 \Theta(L_c, S_c) + \lambda_3 \Phi(S_c; S_1, \cdots, S_C)]$$

$$s.t. \quad F_c = L_c + S_c, \quad c = 1, \cdots, C.$$

(5)

where $\Upsilon(\cdot)$ is a low-rank constraint on background. $\Lambda(\cdot)$ is a group sparsity norm. $\Theta(\cdot) = Tr(S_c M_c S_c^T)$ is a term to enlarge the distance between the L_c and S_c, which makes the foreground easier separated from the image background. Here, M_c is a Laplacian matrix corresponding to c-th feature. $\Phi(\cdot)$ is a diversity regularization term to explore the underlying complementarity information between image of multiple features. C is the number of the features. F_c is the feature matrix corresponding to the c-th view. Similarly, L_c and S_c denote the low-rank part and the salient object part respectively, which is corresponding to the c-th view. Besides, $\lambda_1, \lambda_2, \lambda_3$ are positive parameters. We have to get the solution $L = [L_1; L_2; ...; L_C]$, $S = [S_1; S_2; \cdots; S_C]$.

Low-Rank Term for Background. In [24], the authors have demonstrated the rationality of the low-rank constraint on the image background. Instead of nuclear norm, we use $\| \cdot \|_{S_p}$ norm to constrain the low-rank part. Since it is a stronger constraint on low-rank. $\| \cdot \|_{S_p}$ norm is widely applied and researched in image processing. In addition, the works [20–22,31] show that S_p-norm will result in a lower solution than S_1-norm. By exploiting S_p-norm to research the problem of sparse representation, a group sparsity induced norm which is to pursuit group sparsity and the underlying relationship among patches of S is taken into account.

Group Sparsity for Salient Region. A valid segmentation result bears some of the potential information of the image. So, it is imposed on the salient part S as a constraint, which can be encoded by a group sparsity norm. Specifically, the group sparsity norm is defined as follows

$$\Lambda(S_c) = \sum_{i=1}^{i=n} v_i \|S_{cG_i}\|_F,$$

(6)

where $v_i > 0$ is a prior weight for the node G_i, S_{cG_i} is a sub-matrix of S_c corresponding to i-th node G_i in graph cut. And $S_{cG_i} \in \mathbb{R}^{D_c \times |G_i|}$ ($|\cdot|$ is the cardinality of a set), n is the number of nodes. $\|\cdot\|_F$ is the Frobenius norm (F-norm). In fact, $\Lambda(\cdot)$ is a weighted group sparsity norm over a graph.

On one side, it forces the patches within the same group to have more similar representations, so that they have similar saliency values. On the other side, it stresses patches from different groups to have different representations.

Diversity-Induced Multi-view Regularization. Generally, there are many different features of an image. A single feature may not completely characterize an image, since each feature of an image only focuses on one aspect and may contain some information that other feature do not contain. Therefore, exploration of complementary information of different features is significant. Some theoretical researches [25, 26, 30] indicate that the independence of different views can be used as a descriptor of complementarity of different views. Then, we explore the underlying complementary information and apply it to salient object detection. The sparse matrix of different features will be more diverse. The saliency map is more accurate than the previous methods.

In this section, HSIC is employed to measure the dependence of features. Fortunately, HSIC transforms into the trace of product of the data matrix in [23]. It makes our question to be solved more efficiently. Here, we use the inner product kernel for HSIC, $K_c = S_c^T S_c$, and the Hilbert-Schmit norm is as follows

$$HSIC(S_c, S_{\tilde{c}}) = Tr(S_c K_{\tilde{c}} S_c^T). \tag{7}$$

Then, the diversity regularization corresponding to c-th view is

$$\begin{aligned} \Phi(S_c; S_1, \cdots, S_C) &= \sum_{\tilde{c}=1, \tilde{c} \neq c}^{C} HSIC(S_c, S_{\tilde{c}}) = \sum_{\tilde{c}=1, \tilde{c} \neq c}^{C} Tr(S_c K_{\tilde{c}} S_c^T) \\ &= Tr(S_c K^c S_c^T), \end{aligned} \tag{8}$$

with

$$K^c = \sum_{\tilde{c}=1, \tilde{c} \neq c}^{C} H K_{\tilde{c}} H, \quad H = I - \frac{1}{n} 1. \tag{9}$$

where I is identity matrix, and 1 is a full one matrix.

3.2 Optimization Procedure

The problem in Eq. 5 is challenging, since S_p norm is a bit tricky. We use the alternating direction method (ADM), a variant of augment Lagrange multipliers (ALM) method [14], to extract the group sparsity matrix. The complete algorithm is shown in Algorithm 1.

Algorithm 1. DMMD Algorithm

0: Input: Feature matrix F, graph $G = \{G_i\}_1^n$, the weight of node G_i is v_i, parameters $p, \lambda_1, \lambda_2, \lambda_3$.

1: Initialize: L and S is initialized by the NRMD [32] model.

2: For c = 1 : C

3: And set $Y^0 = 0$, $W^0 = 0$, $\mu = 0.1$, $\mu_{max} = 10^6$, $\rho = 1.1$, $k = 0$

4: While not converged do

5: $L_c^{k+1} = argmin\mathcal{L}(L_c, S_c^k, T_c^k, Y_c^k, W_c^k, \mu^k)$

6: $T_c^{k+1} = argmin\mathcal{L}(L_c^{k+1}, S_c^k, T_c, Y_c^k, W_c^k, \mu^k)$

7: $S_c^{k+1} = argmin\mathcal{L}(L_c^{k+1}, S_c, T_c^{k+1}, Y_c^k, W_c^k, \mu^k)$

8: $Y_c^{k+1} = Y_c^k + \mu^k(F_c - L_c^{k+1} - S_c^{k+1})$

9: $W_c^{k+1} = W_c^k + \mu^k(F_c - L_c^{k+1} - S_c^{k+1})$

10: $\mu^{k+1} = min(\rho\mu^k, \mu_{max})$

11: $K^c = \Sigma_{\tilde{c}=1}^c HS_{\tilde{c}}^{{k+1}^T}S_{\tilde{c}}^{k+1}H + \Sigma_{\tilde{c}=c+2}^C HS_{\tilde{c}}^{k^T}S_{\tilde{c}}^k H$

12: $k = k + 1$

13: Converge condition: $F_c - L_c - S_c \to 0$ and $S_c - T_c \to 0$

14: End While

15: End For

16: $L^k = [L_1^k; \cdots; L_C^k];$

17: $S^k = [S_1^k; \cdots; S_C^k];$

18: Output: The optimal solution L and S

First, problem in Eq. 5 is equivalent to the following equation:

$$\min_{L,S} \quad \sum_{c=1}^{C}[\|L_c\|_{S_p}^p + \lambda_1 \sum_i v_i\|S_{cG_i}\|_F + \lambda_2 Tr(T_c M_c T_c^T) + \lambda_3 Tr(T_c K^c T_c^T)]$$

$$s.t. \quad F_c = L_c + S_c, \quad S_c = T_c, \quad c = 1, \cdots, C. \tag{10}$$

where $\| \cdot \|_{S_p}^p$ is a Schatten norm to enforce the matrix low-rank. $\Lambda(S)$ is a group sparsity induced norm on S to preserve the relationship between patches and groups in S. This object function is separable by introducing T, where $T = [T_1, T_2, \cdots, T_C]$ and T_c is a submatrix of matrix T. Denote the augment Lagrangian function by

$$\mathcal{L}(L, S, T, Y, W, \mu) = \sum_{c=1}^{C}[\|L_c\|_{S_p}^p + \lambda_1 \sum_i^d v_i\|S_{cG_i}\|_F + \lambda_2 Tr(T_c M_c T_c^T)$$

$$+ \lambda_3 Tr(T_c K^c T_c^T) + \langle Y_c, F_c - L_c - S_c\rangle + \langle W_c, S_c - T_c\rangle \tag{11}$$

$$+ \frac{\mu}{2}(\|F_c - L_c - S_c\|_F^2 + \|S_c - T_c\|_F^2)].$$

where Y_c and W_c are the Lagrange multipliers. M_c is a Laplacian matrix corresponding to c-th feature, and $\mu > 0$ is a penalty factor. The solution of Eq. 10 is equivalent to minimize the augment Lagrangian function $\mathcal{L}$. The optimal solution L_c and S_c can be obtained by alternating direction method (ADM). Next, pay attention to update the variables in each iteration.

Updating L_c. First, fixing S_c and Y_c, seeking L_c to minimize the function $\mathcal{L}$, and solving the following equation:

$$
\begin{aligned}
\mathbf{L}_c^{k+1} &= \underset{\mathbf{L}_c}{argmin}\ \mathcal{L}(\mathbf{L}_c, \mathbf{S}_c^k, \mathbf{T}_c^k, \mathbf{Y}_c^k, \mathbf{W}_c^k, \mu^k) \\
&= \underset{\mathbf{L}_c}{argmin}\ \|\mathbf{L}_c\|_{\mathbf{S}_{1/2}} + \langle \mathbf{Y}_c^k, \mathbf{F}_c - \mathbf{L}_c - \mathbf{S}_c^k \rangle + \frac{\mu^k}{2}\|\mathbf{F}_c - \mathbf{L}_c - \mathbf{S}_c^k\|_F^2 \quad (12) \\
&= \underset{\mathbf{L}_c}{argmin}\ \lambda\|\mathbf{L}_c\|_{\mathbf{S}_{1/2}} + \frac{1}{2}\|\mathbf{L}_c - \mathbf{M}_{L_c}\|_F^2
\end{aligned}
$$

where $\lambda = \frac{1}{\mu^k}$, $M_{L_c} = (F_c - S_c + \frac{Y_c^k}{\mu^k})$. The solution of Eq. 12 can be written as $L_c^{k+1} = \mathbf{U}T_p(\Sigma; \lambda)\mathbf{V}^T$, where $(\mathbf{U}, \Sigma, \mathbf{V}^T) = \mathcal{SVD}(M_{L_c})$.

$$
T_p(\sigma; \lambda) = \begin{cases} 0 & , \sigma \le \tau_p(\lambda) ; \\ sign(\sigma)S_p(\sigma; \lambda) & , \sigma > \tau_p(\lambda) . \end{cases} \quad (13)
$$

$$
\tau_p(\lambda) = (2\lambda(1-p))^{\frac{1}{2-p}} + \lambda p(2\lambda(1-p))^{\frac{p-1}{2-p}} . \quad (14)
$$

where Σ is the singular value matrix of M_{L_c}. $T_p(\cdot; \cdot)$ in Eq. 13 is the generalized soft-thresholding operator (GST) [1]. $S_p(\sigma; \lambda)$ can be solved by an iterative algorithm [1], which is based on the *Theorems* 1 and *Theorems* 2 in [1].

Updating T_c. When fixing other variables, T_c^{k+1} is equivalent to Eq. 15.

$$
\begin{aligned}
\mathbf{T}_c^{k+1} &= \underset{\mathbf{T}_c}{argmin}\ \mathcal{L}(\mathbf{L}_c^{k+1}, \mathbf{S}_c^k, \mathbf{T}_c, \mathbf{Y}_c^k, \mathbf{W}_c^k, \mu^k) \\
&= \underset{\mathbf{T}_c}{argmin}\ \lambda_2 Tr(\mathbf{T}_c \mathbf{M}_c \mathbf{T}_c^T) + \lambda_3 Tr(\mathbf{T}_c \mathbf{K}^c \mathbf{T}_c^T) + \langle \mathbf{W}_c^k, \mathbf{S}_c^k - \mathbf{T}_c \rangle + \frac{\mu}{2}\|\mathbf{S}_c^k - \mathbf{T}_c\|_F^2 \\
&= \underset{\mathbf{T}_c}{argmin}\ Tr[\mathbf{T}_c(\lambda_2 \mathbf{M}_c + \lambda_3 \mathbf{K}^c)\mathbf{T}_c^T] + \langle \mathbf{W}_c^k, \mathbf{S}_c^k - \mathbf{T}_c \rangle + \frac{\mu^k}{2}\|\mathbf{S}_c^k - \mathbf{T}_c\|_F^2
\end{aligned}
$$
$$(15)$$

Equation 15 is a smooth convex program. Taking derivative of the objective function with respect to T_c and setting it to zero, we have

$$
T_c^{k+1} = (W_c^k + \mu S_c^k)[2(\lambda_2 M_c + \lambda_3 K^c) + \mu I]^{-1}. \quad (16)
$$

Updating S_c. Fixing others and minimizing the augment Lagrangian function, we have

$$
\begin{aligned}
\mathbf{S}_c^{k+1} &= \underset{\mathbf{S}_c}{argmin}\ \mathcal{L}(\mathbf{L}_c^{k+1}, \mathbf{S}_c, \mathbf{T}_c^{k+1}, \mathbf{Y}_c^k, \mathbf{W}_c^k, \mu^k) \\
&= \underset{\mathbf{S}_c}{argmin}\ \lambda_1 \sum_i^n v_i \|\mathbf{S}_{c\mathbf{G}_i}\|_F \\
&\quad + \langle \mathbf{Y}_c^k, \mathbf{F}_c - \mathbf{L}_c^{k+1} - \mathbf{S}_c \rangle + \langle \mathbf{W}_c^k, \mathbf{S}_c - \mathbf{T}_c^{k+1} \rangle \\
&\quad + \frac{\mu^k}{2}(\|\mathbf{F}_c - \mathbf{L}_c^{k+1} - \mathbf{S}_c\|_F^2 + \|\mathbf{S}_c - \mathbf{T}_c^{k+1}\|_F^2) \\
&= \underset{\mathbf{S}_c}{argmin}\ \eta \sum_i^n v_i \|\mathbf{S}_{c\mathbf{G}_i}\|_F + \frac{1}{2}\|\mathbf{S}_c - \mathbf{X}_{\mathbf{S}_c}\|_F^2
\end{aligned}
$$
$$(17)$$

where $\eta = \frac{\lambda_1}{2\mu^k}$, $X_{S_c} = \frac{1}{2}(F_c - L_c^{k+1} + T_c^{k+1} + \frac{Y_c^k - W_c^k}{\mu^k})$. Optimization of above problem can be solved by the proximal operator, which is equal to compute the orthogonal projection of the matrix onto the ball of the dual norm $\|\cdot\|_F$. Details are shown in Algorithm 2.

Algorithm 2. *Group Sparsity Algorithm*

0: Input: The set of points $\{G_i\}$ in segmentation graph G, weight v_i, the matrix M_{L_c}, parameter $\lambda_1 = 1.1$
1: Initialize: Set $\eta = \frac{\lambda_1}{2\mu^k}$, $S_c = X_{S_c}$
2: For $i = 1$ to n do
3:

$$S_{cG_i} = \begin{cases} 0 & , \; \|S_{cG_i}\|_F \leq \eta v_i; \\ \frac{\|S_{cG_i}\|_F - \eta v_i}{\|S_{cG_i}\|_F} S_{cG_i} & , \; \|S_{cG_i}\|_F > \eta v_i. \end{cases} \tag{18}$$

4: $i = i + 1$
5: End For
6: Output: S_c

4   DMMD for Salient Object Detection

This section performs the salient object detection by DMMD model. There are two main parts, the first part elaborates on low-level saliency detection, and the second part integrates the high-level prior information. The framework of the proposed DMMD model is shown in Fig. 2.

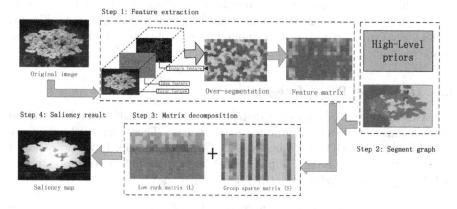

Fig. 2. The framework of the DMMD model for saliency detection.

4.1   Low-Level Salient Object Detection

Feature Extraction. According to [6], by extracting color, steerable pyramids [16] and Gabor filters [17] three different low-level features of an image, we first get a feature representation. Then, the image is segmented into N patches $P = \{P_1, P_2, \cdots, P_N\}$ by simple linear iterative clustering (SLIC) [18] method. The feature vector of patch P_i corresponding to c-th feature denotes as $f_{ci} \in \mathbb{R}^{D_c}$, forming a c-th feature matrix $F_c = [f_{c1}, f_{c2}, ..., f_{cN}] \in \mathbb{R}^{D_c \times N}$. Then it multiples feature matrix, denoted as $F = [F_1; F_2; ...; F_C] \in \mathbb{R}^{D \times N}$. Here $D = \sum D_c$ is 53.

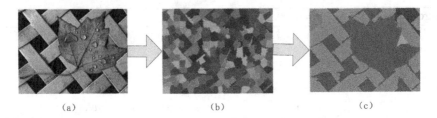

(a)   (b)   (c)

Fig. 3. Graph cut. (a): the input image. (b): the original graph G^0. (c): the graph after cutting G.

Graph Cut. In this step, a graph of an image will be obtained. The segment graph is first defined. It is based on a graph-based image segmentation algorithm [19] which determines whether patches reside in a same group according to their degrees of relationships. Each node of the graph includes a set of indices (corresponding to the patches in this work). More specifically, a segment graph with a original graph G^0 (see Fig. 3(b)) is obtained by the over-segmentation. Let G_i be the i-th node and $G = \{G_1, G_2, \cdots, G_n\}$, where n is the number of nodes in the segment graph and inversely proportional respect to the segmentation threshold T. At the same time, the nodes in the graph have to satisfy the following two conditions: (1) the indexes of nodes do not overlap, i.e. for any $1 \leq i, j \leq n$, and $i \neq j$, $G_i \cap G_j = \emptyset$. (2) G_i is a node in the graph, then $\bigcup_i G_i = G$. Finally, a rough segmentation result of an input image is obtained. Figure 3 shows an simple example.

4.2 High-Level Priors Integration

Following Shen et al. [6], Gaussian distribution of color, location and background priors [9] are integrated to produce a prior map. Then high-level priors are embedded into DMMD model as a weight. For each superpixel P_i, prior value of π_i means the possibility that superpixel P_i is a salient object corresponding to high-level prior knowledge. Specially, it is defined as

$$v_i = 1 - \max_j(\{\pi_j : P_j \in G_i\}). \tag{19}$$

This means that a large punishment is imposed on the nodes which are in small prior values, and vice versa.

Saliency Values. In this step, the saliency map obtained from the corresponding saliency value of each patch will be generated. According to the group sparsity matrix S, a saliency value function $Sal(\cdot)$ respect to S_i in patch P_i is defined as follows:

$$Sal(P_i) = \|S_i\|_\infty. \tag{20}$$

where S_i is the i-th column of S. $\| \cdot \|_\infty$ means the maximum absolute value. A larger $Sal(P_i)$ represents a higher possibility that P_i is a salient region.

After all patches are computed, an appearance-based smooth [29] process is implemented. The final saliency map is generated.

5   Experiment

5.1   Parameter Settings

The rank of the background may not be inconsistent for different images. An image may get a perfect result for a certain p, but anther one is not desired. For easily understanding this work, we only consider the low-rank term and the group sparse term to choose a validate value of p. As shown in Fig. 4, the background pixels in saliency detection map with $p = 0.5$ are thoroughly suppressed, but the rest tend to be more scattered. After comprehensive consideration, the parameter p is set to be 0.5. Segmentation threshold T is set to be 2000 for segmenting and generating a graph. In DMMD model, the bandwidth parameter is set to be $\delta = 0.05$. And we empirically set the model balanced parameters λ_1, λ_2 and λ_3 to 1.1, 0.35 and 2 respectively. In addition, to demonstrate the robustness and the contribution of DMMD model, all experimental parameters are uniformly fixed.

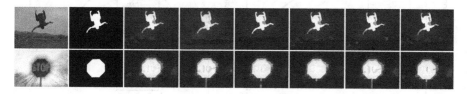

(a) Image   (b) GT   (c) p=0.1   (d) p=0.3   (e) p=0.5   (f) p=0.7   (g) p=0.9   (h) p=1

Fig. 4. The resulting saliency maps in different values ($p = 0.1, 0.3, 0.5, 0.7, 0.9, 1$). The background pixels in the generated saliency map with $p = 0.5$ are better suppressed and more similar to the groundtruth (GT), while others tend to be more scattered.

5.2   Evaluation Metrics

Two metrics are taken into account, including the precision-recall (PR) curve and the F-measure curve. In PR space, *precision* is on the y-axis and *recall* is on the x-axis. Precision measures that the possibility of positive samples that are truly positive. Recall measures that the possibility of samples categorized as positive that are correctly labels. The goal is to be in the right-up hand corner. In F-measure space, the *threshold* is on the x-axis and the F-measure is on the y-axis. The threshold value belongs to $[0, 255]$. F-measure is a combined measure that evaluates the precision and recall tradeoff (weighted harmonic mean). In the experiment, we use the balanced $F_{0.3}$ measure ($\beta^2 = 0.3$). Denote it by

$$F = \frac{(\beta^2 + 1)\, PR}{\beta^2 P + R}. \qquad (21)$$

where P and R are the averaged value of precision and recall in the test database individually.

5.3   Comparisons to the State-of-the-art

The proposed salient detect model is evaluated on the widely used MSRA-5000 dataset and compared with 9 state-of-the-art results, including BL [28], RBD [9], HS [2], GBMR [3], SF [4], CB [5], RC [7], ULR [6]. These methods were mostly proposed recently. Among these approaches, the saliency optimization from BL [28] archives the best performance on the MSRA-5000 dataset. Figure 5 shows the precision-recall curves and F-measure curves using these methods and DMMD. The precision-recall curves of these approaches and DMMD on the MSRA-5000 dataset are shown in Fig. 5(a). It can be seen that the precision-recall curve of the proposed DMMD is higher than RBD [9] and achieves the state-of-the-art performance.

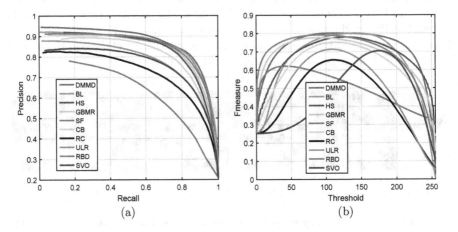

(a)   (b)

Fig. 5. Comparison on dataset MSRA-5000. (a): PR Curves of different methods. (b): F-measure curves of different methods.

On the other hand, the F-measure curve is also compared. The results of the F-measure curves on the MSRA-5000 dataset are shown in Fig. 5(b). It can be seen that the F-measure curves of the proposed DMMD has achieved the highest one among these methods, and it is more parallel to the recall-axis and more stable. This demonstrates the performance of the proposed approach is superior to that of other methods. For easily understanding, some results are shown in Fig. 6.

5.4   Experiment Analysis

For further understanding of our model, in Table 1, we display the comparison results of MAE (mean absolute error) on the MSRA-B dataset. MAE is defined as the mean absolute difference between the saliency maps S and the ground truths G [4]. As shown in Table 1, our model has achieved the smallest MAE value than

Table 1. Results on the MSRA-B dataset in term of MAE.

Metric	BL	RBD	HS	GB	SF	ULR	RC	CB	Test1[a]	Test2[b]	Our
MAE	0.171	0.131	0.162	0.128	0.179	0.222	0.263	0.184	0.111	0.124	0.095

[a]Test1 is the results of our model without diversity term.
[b]Test2 is the results of our model without diversity term and without S_p-norm term.

the other methods. Especially, we have tested the results of our model without diversity term or without S_p-norm term, as shown in the columns of Test1 and Test2. When our model is without diversity term, the MAE performance has declined from 0.095 to 0.111. When our model is not only without diversity term, but also without S_p-norm term, the MAE performance has further declined from 0.111 to 0.124. These comparisons could validate the effectiveness of the diversity term and the S_p-norm term.

A directly visual comparison with other methods is shown in Fig. 1. We can see that there is a big difference between the saliency objects generated by the DMMD model without diversity term and saliency objects detected by the proposed DMMD with diversity term. DMMD performs on the images with different scenes, even the salient objects are similar to the background in appearance.

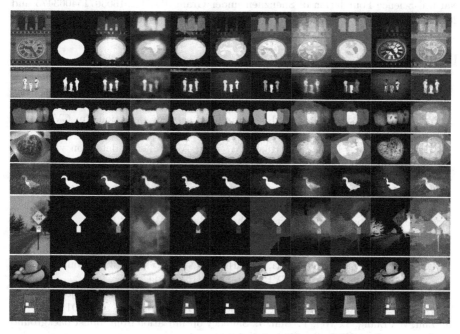

(a) Img (b) Gt (c) Our (d) Bl (e) Rb- (f) Gb (g) Hs (h) Ulr (i) Cb (j) Sf (k) Rcd

Fig. 6. The examples of saliency detection by the proposed method and other methods.

In addition, DMMD makes the saliency objects highlighted and completed which is hard to achieve by previous approaches. This due to the diversity term.

6   Conclusions

In this paper, we presented a novel approach, the diversity-induced multi-view matrix decomposition model (DMMD), for salient object detection. We have improved two terms, background and foreground, and added a regularized term into the model. A S_p-norm ($0 < p < 1$) has been proposed to suppress the background. A group sparsity induced norm has been proposed to encode the potential relations between patches of the image. And a diversity-induced multi-view regularization based Hilbert-Schmidt Independence Criterion (HSIC) has been employed to explore the complementary information of different feature and enhanced the independence between the multiple features. Additionally, high-level priors have also been embedded into the model to improve the saliency detection. Experiments on the widely used MSRA-5000 dataset show that the DMMD model outperforms the other models.

Acknowledgement. This work is supported in part by the National Natural Science Funds of China (Grant Nos. 61472257, 61402290, 61401287) and in part by the Natural Science Foundation of Shenzhen under Grant JCYJ 20160307154003475 and 2016050617251253.

References

1. Zuo, W., Meng, D., Zhang, L., Feng, X., Zhang, D.: A generalized iterated shrinkage algorithm for non-convex sparse coding. In: ICCV, pp. 217–224 (2013)
2. Yan, Q., Xu, L., Shi, J., Jia, J.: Hierarchical saliency detection. In: CVPR, pp. 1155–1162 (2013)
3. Yang, C., Zhang, L., Lu, H., Ruan, X., Yang, M.: Saliency detection via graph-based manifold ranking. In: CVPR, pp. 3166–3173 (2013)
4. Perazzi, F., Krähenbühl, P., Pritch, Y., Hornung, A.: Saliency filters: Contrast based filtering for salient region detection. In: CVPR, pp. 733–740 (2012)
5. Jiang, H., Wang, J., Yuan, Z., Liu, T., Zheng, N., Li, S.: Automatic salient object segmentation based on context and shape prior. In: BMVC (2011)
6. Shen, X., Wu, Y.: A unified approach to salient object detection via low rank matrix recovery. In: CVPR, pp. 853–860 (2012)
7. Cheng, M.-M., Zhang, G., Mitra, N.J., Huang, X., Hu, S.-M.: Global contrast based salient region detection. In: CVPR, pp. 409–416 (2011)
8. Chang, K.-Y., Liu, T.-L., Chen, H.-T., Lai, S.-H.: Fusing generic objectness and visual saliency for salient object detection. In: ICCV, pp. 914–921 (2011)
9. Zhu, W., Liang, S., Wei, Y., Sun, J.: Saliency optimization from robust background detection. In: CVPR, pp. 2814–2821 (2014)
10. Liu, G., Lin, Z., Yu, Y.: Robust subspace segmentation by low-rank representation. In: ICML, pp. 663–670 (2010)
11. Lang, C., Liu, G., Yu, J., Yan, S.: Saliency detection by multitask sparsity pursuit. IEEE Trans. Image Process. **21**, 1327–1338 (2012)

12. Zou, W., Kpalma, K., Liu, Z., Ronsin, J.: Segmentation driven low-rank matrix recovery for saliency detection. In: BMVC (2013)

13. Candès, E., Li, X., Ma, Y., Wright, J.: Robust principal component analysis? J. ACM **58**, 11–20 (2011)

14. Lin, Z., Chen, M.-M., Ma, Y.: The augmented lagrange multiplier method for exact recovery of corrupted low-rank matrices. arXiv preprint arxiv:1009.5055 (2010)

15. Peng, H., Li, B., Ji, R., Hu, W., Xiong, W., Lang, C.: Salient object detection via low-rank and structured sparse matrix decomposition. In: AAAI, pp. 796–802 (2013)

16. Simoncelli, E., Freeman, W.: The steerable pyramid: A flexible architecture for multi-scale derivative computation. In: ICIP (1995)

17. Feichtinger, H.G., Strohmer, T.: Gabor Analysis and Algorithms: Theory and Applications. Birkhäuser, Basel (2012)

18. Achanta, R., Shaji, A., Smith, K., Lucchi, A., Fua, P., Susstrunk, S.: SLIC superpixels compared to state-of-the-art superpixel methods. IEEE Trans. Pattern Anal. Mach. Intell. **12**, 2274–2282 (2012)

19. Felzenszwalb, P., Huttenlocher, D.: Efficient graph-based image segmentation. Int. J. Comput. Vis. **59**, 167–181 (2004)

20. Cetin, M., Karl, W.C.: Feature-enhanced synthetic aperture radar image formation based on nonquadratic regularization. IEEE Trans. Image Process. **10**, 623–631 (2001)

21. Chartrand, R.: Exact reconstruction of sparse signals via nonconvex minimization. IEEE Sig. Process. Lett. **14**, 707–710 (2007)

22. Chartrand, R., Staneva, V.: Restricted isometry properties and nonconvex compressive sensing. Inverse Prob. **24**, 657–682 (2008)

23. Cao, X., Zhang, C., Fu, H., Liu, S., Zhang, H.: Diversity-induced multi-view subspace clustering. In: CVPR, pp. 586–594 (2015)

24. Peng, H., Li, B., Ling, H., Hu, W., Xiong, W., Maybank, S.J.: Salient object detection via structured matrix decomposition. IEEE Trans. Pattern Anal. Mach. Intell. (2016). In Press

25. Chaudhuri, K., Kakade, S.M., Livescu, K., Sridharan, K.: Multi-view clustering via canonical correlation analysis. In: ICML, pp. 129–136 (2009)

26. Tang, W., Lu, Z., Dhillon, I.S.: Clustering with multiple graphs. In: ICDM, pp. 1016–1021 (2009)

27. Cai, D., He, X., Han, J., Huang, T.S.: Graph regularized nonnegative matrix factorization for data representation. IEEE Trans. Pattern Anal. Mach. Intell. **33**, 1548–1560 (2011)

28. Tong, N., Lu, H., Ruan, X., Yang, M.-H.: Salient object detection via bootstrap learning. In: CVPR, pp. 1884–1892 (2015)

29. Li, X., Lu, H., Zhang, L., Ruan, X., Yang, M.-H.: Saliency detection via dense and sparse reconstruction. In: ICCV, pp. 2976–2983 (2013)

30. Zhang, X., Sun, X., Xu, C., Baciu, G.: Multiple feature distinctions based saliency flow model. Pattern Recogn. **54**, 190–205 (2016)

31. Zhang, X., Xu, C., Sun, X., Baciu, G.: Schatten-q regularizer constrained low rank subspace clustering model. Neurocomputing **182**, 36–47 (2016)

32. He, Z., Sun, X., Zhang, X., Xu, C.: Saliency detection via nonconvex regularization based matrix decomposition. In: International Conference on Computational Intelligence and Security, pp. 243–247 (2015)

Parallel Accelerated Matting Method
Based on Local Learning

Xiaoqiang Li$^{(\boxtimes)}$ and Qing Cui

School of Computer Engineering and Science, Shanghai University, Shanghai, China
xqli@shu.edu.cn

Abstract. To pursue effective and fast matting method is of great importance in digital image editing. This paper proposes a scheme to accelerate learning based digital matting and implement it on modern GPU in parallel, which involves learning stage and solving stage. Firstly, we present GPU-based method to accelerate the pixel-wise learning stage. Then, trimap skeleton based algorithm is proposed to divide the image into blocks and process blocks in parallel to speed up the solving stage. Experimental results demonstrated that the proposed scheme achieves a maximal 12+ speedup over previous serial methods without degrading segmentation precision.

1 Introduction

Digital image matting is the fundamental technique in fields such as film editing and image processing, whose goal is to estimate fractional opacity of foreground layer from an image and recover the colors of the foreground/background layers respectively. Typically, given an image I, assumes that I is formed through linearly blending foreground image F and background image B with coefficients α:

$$I = \alpha F + (1 - \alpha)B \tag{1}$$

α ranges between 0 and 1. Image matting refers to the process of image foreground and background segmentation in form of calculating opacity mask α.

Learning based matting (LBM) [1] is a precise and easy-to-use algorithm for solving matting problems. It can handle difficult problems such as translucent foreground and complicated border which make it perfect candidate for current widely used grab cut serious. LBM algorithm consist of two stages: learning stage and solving stage. The first stage solve a serious of small dense matrix equation. The second stage constructs a huge sparse matrix equation and then solve it. However, both of the two steps require heavy computation which makes LBM unsuitable for real-time interactive matting scene.

Until now, various GPU-based approach [2–7] have been applied to accelerate matting-related algorithms. Acceleration methods deal with two stage individually. Zhu [2] noticed that the calculation of linear model for each pixel is relatively independent in the learning stage. They leverage the parallel computing power of GPU to implement the algorithm of Grow Cut and the running speed is improved

© Springer International Publishing AG 2017
S.-H. Lai et al. (Eds.): ACCV 2016, Part I, LNCS 10111, pp. 152–162, 2017.
DOI: 10.1007/978-3-319-54181-5_10

significantly. He et al. [7] use CUDA-based Conjugate Gradient (CG) algorithm to help solving sparse matrix expression which accelerates solving stage slightly when image is small enough. KD-tree is used to divide the image into blocks [3,7] which enables parallel computation for blocks.

In this paper, we propose trimap skeleton based (TSB) algorithm to generate image blocks in parallelized LBM. This image separation method can help to reduce a amount of computation time compared with KD-tree based method. TSB algorithm is also universal and can be easily applied to other matting algorithms [8,9]. The rest of the paper is organized as follows: Sect. 2 brings the review of local learning based matting. Section 3 describes in detail the complete process flow and our blocking strategy which help our algorithm gains more speedup than others. In Sect. 4, we present the experimental results and analyze the improvement and bottleneck of proposed method. Section 5 is the conclusion of this paper.

2   Serial Matting Model

Learning based digital matting can be carried out through two related but different approaches: local learning based approach and global learning based approach. For simplicity we only focus on local learning based approach. The local learning method is based on the assumption that each pixels alpha value can be represented by the linear combination of the neighboring pixels. For each pixel in the image, we get its extended value in RGB space $x_i = [r_i, g_i, b_i]^T$, By denoting it and all its eight neighboring pixels as $N_i = \{j_0, j_1 \ldots j_8\}$ we get matrix $X_i = [x_{j_0}, x_{j_1} \ldots x_{j_8}]^T$ containing all the pixel values in a 3*3 window. Then the coefficient for representing the center pixel using its neighboring pixels can be calculated by the following expression:

$$f_i = (X_i X_i^T + tI)^{-1} X_i x_i^T \tag{2}$$

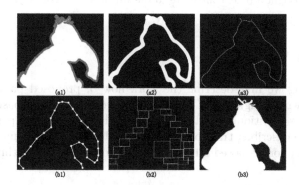

Fig. 1. Procedure of proposed method (a1) Input trimap (a2) Region to be processed (a3) Skeleton (b1) Sectionalized skeleton (b2) Image Blocks (b3) Alpha matte

Each coefficient f_i corresponds to the non-zero values in a row vector of matrix F. Matrix C is diagonal whose jth diagonal element takes the constant value c if alpha value of pixel j is already known. Input column major trimap is reshaped to get vector α^*. The matting result α of I is calculated as below:

$$\alpha = ((I - F)(I - F)^T + C)^{-1}C\alpha^* \tag{3}$$

The precision of alpha matte is closely related to the size of representation window. Size 5*5 is commonly used since it is more accurate than 3*3 and is easier to compute than 7*7. In stead of calculate f_i for 25 pixels in 5*5 window directly, we use a slide window strategy and calculate f_i for each 3*3 window. Figure 1a illustrate using top-left 3*3 window to represent the center gray pixel of 5*5 window. After calculating f_i of all the 9 sub-window, we directly add all the weight of each pixel together. The number of weight a pixel contribute to f_i is shown in Fig. 2d. The importance of a pixel is related to the distance to window center.

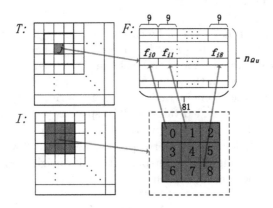

Fig. 2. Learning model for 5*5 window (a) (b) (c) Three sub-window of the big window (d) Number of weight a pixel contribute to the whole weight vector f_i

3   Parallel Work Flow

In the following we first describe the work flow of proposed parallel matting model, as shown in Fig. 3. The parallel framework obeys the basic procedure while introduces partitioning and merging operations required for parallelization. The next part describes trimap skeleton based image partitioning algorithm which is the key for proposed scheme to outperform other algorithms.

3.1   Parallel Local Learning

The learning stage calculate the laplace matrix L. The GPU process data in parallel which means it can not use dynamic memory allocation strategy.

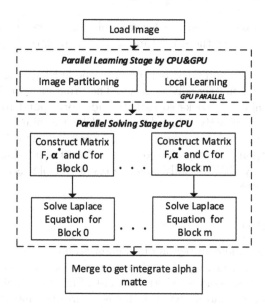

Fig. 3. Work flow for parallel matting procedure

We create a new trimap F to solve this problem. F not only recode whether this pixels need to be calculated but also the sequence number in a row-first major. If current pixel j is the ith pixel that needs to be calculated, then the corresponding value of F is i, $T_j = i$, or else $T_j = 0$. After T and i are uploaded to the device memory, GPU launch as many thread as the pixel number. Each thread correspond to a pixel. At the begging executing of each thread check its value from T to see weather it needs to be calculated. If the flag is not, it returns directly. If the flag is true, it will read the image pixel intensity of a 3*3 window and calculate f_i, then write the result to the ith row of F. Since each length of f_i is 9 and we have 9 window for a pixel, so there are 9*9 = 8 1elements in a row of F. The row number of equals to Ω_u, so the size of f is $n\Omega_u$*81;

3.2 Integrate Work Flow

In GPU-assisted learning stage, we launch as many threads as the number of image pixels. Each thread computes the linear coefficients vector f_i for a single pixel. However, only pixels marked unknown need to be processed and demand GPU memory to store the calculation result. So we set up a search table T for each pixel and write its linear coefficients vector into device memory by its offset.

More memory utilization strategies [2,3] would help to decrease the time consumption of GPU calculation. Since calculation of F is fast enough after above acceleration, solving stage become system bottleneck of our proposed scheme. Learning stage do not require extra optimization.

Image blocking is computed using CPU in parallel with GPU kernel. The detail of TSB to divided image into blocks is described in the next section.

TSB algorithm is executed on the lower resolution image to ensure it always finished earlier than or at the same time as the local learning procedure based on GPU kernel.

After both the CPU and GPU part of stage one finished, the work flow come to parallel block-wise matrix solving stage. The program then launch as many thread as the number of blocks so each threads correspond to a single block. Notice that number of active thread is limited by the number of CPU cores, usually 4 or 8. Each thread construct its own coefficient matrix F, annotation matrix α^* and diagonal matrix C according to its corresponding block position and data returned by GPU. Finally, matting results for blocks are merged to get integrate alpha matte.

3.3   Image Partition

In LBM algorithm, unknown region of different trimaps are irregular but share similarities in shape. Generally, an unknown region of a trimap for single foreground object image is a connected strip area like Fig. 1(b2). This kind of regions have distinct structural information that helps separating images into blocks and we focus on this type of trimap in our paper. We propose trimap skeleton based (TSB) algorithm to separate image into blocks which is able to generate blocks of relatively consistent size. In order to get blocks that cover all the pixels of unknown region with the least overlap, We first extract the skeleton of image and turn it into line segments. Then we detect the critical points according to its geometry position. Thirdly, the blocks are calculated using line segments and critical points. Some blocks are merged based on critical points. The blocks can be expanded to cover the unknown region outside the blocks. In this way, blocks can be safely expanded to cover the entire image.

Extraction Major Skeleton of Trimap. We use the thinning algorithm [10] to get image skeleton. The skeleton image can be regarded as a undirected graph $G = (V, E)$, where V is the vertex set and E is the edge set. The major skeleton of image can be extracted by searching for the longest routes of the graph.

There are three types of pixels in skeleton image. Two of them correspond to two kinds of vertex in G. One type of pixel $\Phi_{terminal}$ correspond to terminal vertex which connect to only one edge. The second type of pixel $\Phi_{internal}$ correspond to internal vertex which connect to at least two edges. The last type of pixels Φ_{edge} construct the edge of graph. A pixel p_0 with 8 neighboring pixels $[p1, ..., p8]$ in a clockwise sequence. We denote $A(p_0) =$ and get following definition:

$$\Phi_{terminal} = p|A(p) == 1, p \in I \tag{4}$$

$$\Phi_{edge} = p|A(p) == 2, p \in I \tag{5}$$

$$\Phi_{internal} = p|A(p) >= 3, p \in I \tag{6}$$

The vertex of is denoted as. The edge set is get through traversing the graph by breadth first search. We denote: current point, unvisited neighbor pixel,

Algorithm 1. Prepare for Blocking

---

while s not empty **do**
  $p \leftarrow pop(s)$
  while $!endOfCurve$ **do**
    set p visited
    if $p \in \Phi_{terminal}$ **then**
      $e \leftarrow e \cup p$
      $endOfCurve$=true
    end if
    if $p \in \Phi_{edge}$ **then**
      $e \leftarrow e \cup p$
      $p \leftarrow$ random select from N_p^u
    end if
    if $p \in \Phi_{internal}$ **then**
      $s \leftarrow s \cup pN_p$
      $endOfCurve$=true
    end if
    $E \leftarrow E \cup e$
    $e \leftarrow \emptyset$
  end while
end while
return B %$output\_array\_of\_blocks$%

---

a stack storing pixels. The algorithm of getting edges is as follows: get a point, (a) if is edge pixel, we put into current edge set and then visit the next unvisited pixel of. (b) if correspond to terminal vertex, current edge is terminated and we put edge into edge set, and we get a pixel from and visit it. (c) if correspond to internal points, search of current edge is terminated. We store into edge set and push neighboring pixels of into stack. Then we get a new point from s and visit it. The above algorithm is described in Algorithm 1.

Calculating Critical Nodes. The major skeleton is further simplified to its approximate line segments through algorithm [11]. We denote the set of all the sequenced line segments as Ψ and the endpoints of all segments as sequenced node set Φ. The unknown region of trimap is then represented by Ψ and Φ. In Fig. 4a, $\Phi = \{1, ..., 9, a, ..., v\}$.

Segments in Ψ can be divided into two categories: horizontal segments Ψ_h and vertical segments Ψ_v according to angle to the axis. The nodes of horizontal segments or vertical segments are denoted as Φ_h or Φ_v. The horizontal segments are separated into groups ϕ_h according to adjoining or not. Each grouped horizontal segments corresponds to a node set Φ_{gh_i} and the collection of the sets are denoted as $\phi_h = \{\Phi_{gh_0}, ..., \Phi_{gh_{n_\phi}}\}$. As shown in Fig. 4(b), the isolated horizontal line l_{78} construct first line group $\{l_{78}\}$ and $\Phi_{gh_0} = \{7, 8\}$. Since segments of l_{bc} and l_{cd} are connected by node c, we get second line group $\Phi_{gh_1} = \{b, c, d\}$. As to the total image, ϕ_h equals to $\{\{7,8\}, \{b,c,d\}, \{q,r,s\}\}$.

If a node i satisfy that its y value is greater or less than its both two neighbors, we call it local extreme node. The local extreme points are denoted as Φ_{le} and $\Phi_{le} = \{c, n, r\}$ in Fig. 4(b). If a node $i \in \Phi_{le}$ and i do not belong to any horizontal segment sets, i is the first type of critical nodes denoted as $\Phi_{critic1} \subset \Phi_{le}$. Node n is only the first type of critical node in Fig. 4(b) so $\Phi_{critic1}$ is a set that contains only one element $\{n\}$.

A horizontal node set Φ_{gh_i} contains local extreme node is recognized as the candidate critical node set. For example, horizontal node set $\Phi_{gh_0} = \{7, 8\}$ is not belong to this candidate critical node set. Both first node and last node of each candidate set connect to a horizontal segment and a vertical segment, these two nodes are denoted as $\Phi_{critic2}$ and $\Phi_{critic2}$ equals to $\{b, d, q, s\}$ in Fig. 4b. Total critical node set are union of the two sets $\Phi_{critic} = \Phi_{critic1} + \Phi_{critic2}$ which is $\{b, d, n, q, s\}$ in sample image.

Algorithm 2. get edge for graph G

---

for p in Φ_{le} **do**
  if $\exists s \in \phi_h \rightarrow p \in s$ **then**
    $\Phi_{critic2} = \Phi_{critic2} \cup \{s[first], s[last]\}$
  else
    $\Phi_{critic1} = \Phi_{critic1} \cup \{p\}$
  end if
end for
for i in Φ **do**
  if $i \in \Phi_{critic}$ **then**
    $B.pop()$
    $rect \leftarrow boundingRect(i - 1, i, i + 1)$
  else
    $rect \leftarrow boundingRect(i, i + 1)$
  end if
  $B.push(rect)$
end for
return B %output_array_of_blocks%

Image Partition with Least Block Overlap. Separating image into blocks is based on the line segments and critical points. First, blocks are generated through calculating rectangles using each line segment as diagonal. Then, traverse all the blocks and merge the ones whose diagonals are connected to critical node. Finally, if some unknown regions in trimap that still have not been covered by the blocks, nearest blocks to each region are expanded to cover them. Blocks whose corresponding line segment belongs to one line groups in ϕ_h are expanded in vertical direction while others are expanded in horizontal direction. Each merged block detect its own expanding direction where there are the least overlap among blocks. Sample of block expanding direction is annotated by blue arrows in Fig. 4(d). The procedure how to calculate the critical nodes and separate image into blocks is described in Algorithm 2.

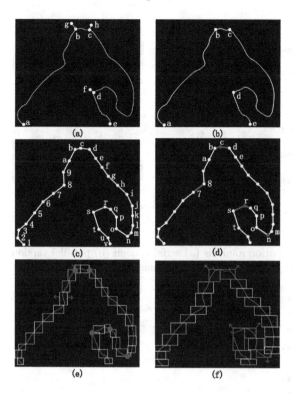

Fig. 4. Partitioning procedure (a) Skeleton of image trimap (b) Major skeleton (c) Sectionalized skeleton (d) Detected horizontal lines (red) and local extreme point (blue) (e) Initial blocking (blue) (f) Merged blocks (red). Beside the red blocks, all the other blocks are expanded either horizontally or vertically to cover the unknown region of trimap (Color figure online)

4   Experiments and Analysis

Our experiments are based on public test set provided by [12] with ground truth mattes available. We choose images from the data set that satisfy our requirements mentioned in Sect. 3. We compare proposed scheme with KD-tree based method and GPU accelerated CG algorithm. Experiments results reveals that our proposed scheme achieved higher acceleration under given hardware resources.

4.1   Two Stages Time Occupation

In the serial version, as shown in Fig. 5(a), learning stage takes up to 80 percent of the total time consumption when images are small and goes down to little less than 40% when image become large. Anyhow, the computational complexity of two stages are equally important for most images.

The parallel version of scheme reveals the same phenomenon to some extent. Learning stage takes a decreasing percent of total time when images become bigger as show in Fig. 5(b). However, the decrease is much faster than the serial version

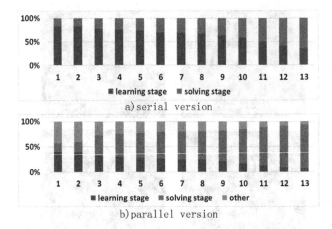

a) serial version

b) parallel version

Fig. 5. Time occupation analysis (a) serial version (b) parallel version. The horizontal axis of above figures represents different image sizes ranging from 96*68 to 2400*1692

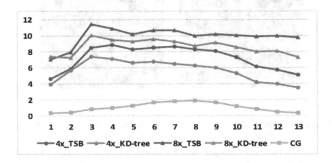

Fig. 6. Global acceleration comparison. TSB and KD-tree are both tested on 4 and 8 cores CPU. CG is tested on GPU. Horizontal axis is the same as Fig. 5.

since GPU parallel brings higher speed up for stage one than that CPU brings for stage two. Then the bottleneck of parallel version becomes the solving stage and total acceleration become more sensitive to speedup of this stage than stage one.

4.2   Acceleration Analysis

Proposed scheme outperforms the KD-tree based method under same hardware resources, as shown in Fig. 6. The maximum acceleration is achieved at image size 3 (400*282) with 8-cores CPU. Take this image size as example. Serial version takes 0.821s while TSB cost only 0.097 with 4-thread parallel and 0.072 with 8-thread parallel. The KD-tree method is slower, cost 0.111s and 0.082s with 4 and 8 cores. However, since this image is small, CG algorithm even failed to compete with serial version and cost 0.965s. Our method get highest acceleration of 12x while KD-tree approach also reach the maximum of 10x at this image size. With the resources of either 4 or 8 core CPU, our method always generates higher speedup than the KD-tree method. The set of blocks generated by KD-tree method must cover all the image area while that produced by TSB-based method do not obey this role. In other word, total workload of proposed method is smaller.

The CUDA accelerated CG algorithm is yet faster than the CPU serial version but can hardly bring 2 times of acceleration. The same result is presented in [4]. The reason can be explained through mathematic point of view. CG is an iterative algorithm. When the coefficient matrix grow big, both the number of iterations required to converge and computational complexity of each iteration increase sharply. The GPU resources are powerful but still unable to make up innate speed gap between CG and other methods that are highly optimized at both instruction and algorithm level, such as UMFPACK [13] or LDLT [14] running on the CPU.

4.3   Run-Time System

The proposed method was carried out on two PCs with different computing ability. Windows 7 64 bit operating system is installed on both of them. One PC is equipped with 4 core Intel core i5 CPU and Geforce 750Ti graphics card and the other owns CPU with 8 cores: XEON E5-1620. Programs are compiled using Visual C++ 2013 with CUDA 7.5.

5   Conclusion

This paper presents the CPU and GPU cooperated parallel work flow and TSB image partitioning algorithm to accelerate local learning based matting method. With the help of proposed approach, we make full utilization of GPU and

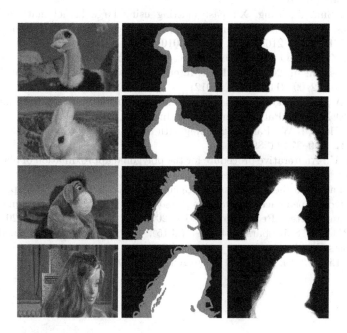

Fig. 7. Extra images. From left to right there are initial images, trimap and alpha matte

CPU power and achieved noteworthy acceleration of matting problem. Proposed scheme has significant speed advantage over CG-based method and outperform KD-tree on most images of various sizes.

Experiments result listed in Fig. 7 shows that there is no obvious differences between results of initial serial version and that of proposed method. In future work, we will conduct research on new sparse solver to directly speed up solving stage for each block.

References

1. Zheng, Y., Kambhamettu, C.: Learning based digital matting. In: 2009 IEEE 12th International Conference on Computer Vision, pp. 889–896. IEEE (2009)
2. Zhu, J., Lu, G., Zhang, D.: Growmatting: a GPU-based real-time interactive method for image matting. In: 2010 25th International Conference of Image and Vision Computing New Zealand (IVCNZ), pp. 1–8. IEEE (2010)
3. Xiao, C., Liu, M., Xiao, D., Dong, Z., Ma, K.L.: Fast closed-form matting using a hierarchical data structure. IEEE Trans. Circ. Syst. Video Technol. **24**, 49–62 (2014)
4. Sun, X., Wang, Z., Chen, G.: Parallel active contour with lattice boltzmann scheme on modern GPU. In: 2012 19th IEEE International Conference on Image Processing (ICIP), pp. 1709–1712. IEEE (2012)
5. Zhang, Q., Xiao, C.: Cloud detection of RGB color aerial photographs by progressive refinement scheme. IEEE Trans. Geosci. Remote Sens. **52**, 7264–7275 (2014)
6. Zhao, M., Fu, C.W., Cai, J., Cham, T.J.: Real-time and temporal-coherent foreground extraction with commodity RGBD camera. IEEE J. Sel. Top. Sign. Process. **9**, 449–461 (2015)
7. He, K., Sun, J., Tang, X.: Fast matting using large kernel matting laplacian matrices. In: 2010 IEEE Conference on Computer Vision and Pattern Recognition (CVPR), pp. 2165–2172. IEEE (2010)
8. Zhang, Z., Zhu, Q., Xie, Y.: Learning based alpha matting using support vector regression. In: 2012 19th IEEE International Conference on Image Processing (ICIP), pp. 2109–2112. IEEE (2012)
9. Levin, A., Lischinski, D., Weiss, Y.: A closed-form solution to natural image matting. IEEE Trans. Pattern Anal. Mach. Intell. **30**, 228–242 (2008)
10. Guo, Z., Hall, R.W.: Parallel thinning with two-subiteration algorithms. Commun. ACM **32**, 359–373 (1989)
11. Ramer, U.: An iterative procedure for the polygonal approximation of plane curves. Comput. Graph. Image Process. **1**, 244–256 (1972)
12. Rhemann, C., Rother, C., Wang, J., Gelautz, M., Kohli, P., Rott, P.: A perceptually motivated online benchmark for image matting. In: IEEE Conference on Computer Vision and Pattern Recognition, CVPR 2009, pp. 1826–1833. IEEE (2009)
13. UMFPACK: Suitesparse 4.4.6. (2015). http://faculty.cse.tamu.edu/davis/suitesparse.html
14. LDLT: Eigen 3.2.4. (2015). http://eigen.tuxfamily.org

Semi-supervised Domain Adaptation for Weakly Labeled Semantic Video Object Segmentation

Huiling Wang[1(✉)], Tapani Raiko[1], Lasse Lensu[2], Tinghuai Wang[3],
and Juha Karhunen[1]

[1] Aalto University, Espoo, Finland
huling.wang@tut.fi
[2] Lappeenranta University of Technology, Lappeenranta, Finland
[3] Nokia Technologies, Tampere, Finland

Abstract. Deep convolutional neural networks (CNNs) have been immensely successful in many high-level computer vision tasks given large labelled datasets. However, for video semantic object segmentation, a domain where labels are scarce, effectively exploiting the representation power of CNN with limited training data remains a challenge. Simply borrowing the existing pre-trained CNN image recognition model for video segmentation task can severely hurt performance. We propose a semi-supervised approach to adapting CNN image recognition model trained from labelled image data to the target domain exploiting both semantic evidence learned from CNN, and the intrinsic structures of video data. By explicitly modelling and compensating for the domain shift from the source domain to the target domain, this proposed approach underpins a robust semantic object segmentation method against the changes in appearance, shape and occlusion in natural videos. We present extensive experiments on challenging datasets that demonstrate the superior performance of our approach compared with the state-of-the-art methods.

1 Introduction

Semantically assigning each pixel in video with a known class label can be challenging for machines due to several reasons. Firstly, acquiring the prior knowledge about object appearance, shape or position is difficult. Secondly, gaining pixel-level annotation for training supervised learning algorithms is prohibitively expensive comparing with image-level labelling. Thirdly, background clutters, occlusion and object appearance variations introduce visual ambiguities that in turn induce instability in boundaries and the potential for localised under- or over-segmentation. Recent years have seen encouraging progress, particularly in terms of generic object segmentation [1–6], and the success of convolutional neural networks (CNNs) in image recognition [7–9] also sheds light on semantic video object segmentation.

Generic object segmentation methods [2,3,5,10–12] largely utilise category independent region proposal methods [13,14], to capture object-level description

© Springer International Publishing AG 2017
S.-H. Lai et al. (Eds.): ACCV 2016, Part I, LNCS 10111, pp. 163–179, 2017.
DOI: 10.1007/978-3-319-54181-5_11

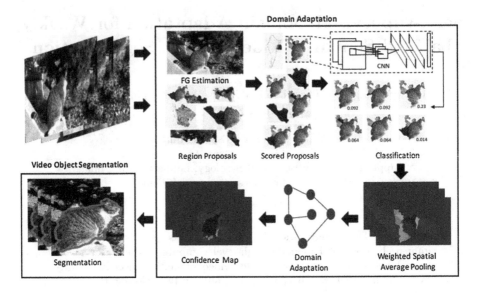

Fig. 1. Overview of our proposed method.

of the generic object in the scene incorporating motion cues. These approaches address the challenge of visual ambiguities to some extent, seeking the weak prior knowledge of what the object may look like and where it might be located. However, there are generally two major issues with these approaches. Firstly, the generic detection has very limited capability to determine the presence of an object. Secondly, such approaches are generally unable to determine and differentiate unique multiple objects, regardless of categories. These two bottlenecks limit these approaches to segmenting one single object or all foreground objects regardless classes or identifies.

Deep convolutional neural networks have been proven successful [7–9] in many high-level computer vision tasks such as image recognition and object detection. However, stretching this success to the domain of pixel-level classification or labelling, i.e., semantic segmentation, is not naturally straightforward. This is not only owing to the difficulties of collecting pixel-level annotations, but also due to the nature of large receptive fields of convolutional neural networks. Furthermore, the aforementioned challenges present in video data demand a data-driven representation of the video object in order to give a spatio-temporal coherent segmentation. This motivates us to develop a framework for adapting image recognition models (e.g., CNN) trained on static images to a video domain for the demanding task of pixel labelling. This goal is achieved by proposing a semi-supervised domain adaptation approach to forming a data-driven object representation which incorporates both the semantic evidence from pre-trained CNN image recognition model and the constraint imposed by the intrinsic structure of video data. We exploit the constraint in video data that when the same object is recurring between video frames, the spatio-temporal coherence implies the associated unlabelled data to be the same label. This data-driven object representation underpins a robust object segmentation method for weakly labelled natural videos.

The paper is structured as follows: We firstly review related work in video object segmentation (Sect. 2). Our method introduced in Sects. 3 and 4 consists of domain adaptation and segmentation respectively, as shown in Fig. 1. Evaluations and comparisons in Sect. 5 show the benefits of our method. We conclude this paper with our findings in Sect. 6.

2   Related Work

Video object segmentation has received considerable attention in recent years, with the majority of research effort categorised into three groups based on the level of supervisions: (semi-)supervised, unsupervised and weakly supervised methods.

Methods in the first category normally require an initial annotation of the first frame, which either perform spatio-temporal grouping [15, 16] or propagate the annotation to drive the segmentation in successive frames [17–20].

Unsupervised methods have been proposed as a consequence of the prohibitive cost of human-in-the-loop operations when processing ever-growing large-scale video data. Bottom-up approaches [4, 21, 22] largely utilise spatio-temporal appearance and motion constraints, while motion segmentation approaches [23, 24] perform long-term motion analysis to cluster pixels or regions in video data. Giordano et al. [25] extended [4] by introducing 'perceptual organisation' to improve segmentation. Taylor et al. [26] inferred object segmentation through long-term occlusion relations, and introduced a numerical scheme to perform partition directly on pixel grid. Wang et al. [27] exploited saliency measure using geodesic distance to build global appearance models. Several methods [2, 3, 5, 6, 11] propose to introduce a top-down notion of object by exploring recurring object-like regions from still images by measuring generic object appearance (e.g., [13]) to achieve state-of-the-art results. However, due to the limited recognition capability of generic object detection, these methods normally can only segment foreground objects regardless of semantic label.

The proliferation of user-uploaded videos which are frequently associated with semantic tags provides a vast resource for computer vision research. These semantic tags, albeit not spatially or temporally located in the video, suggest visual concepts appearing in the video. This social trend has led to an increasing interest in exploring the idea of segmenting video objects with weak supervision or labels. Hartmann et al. [28] firstly formulated the problem as learning weakly supervised classifiers for a set of independent spatio-temporal segments. Tang et al. [29] learned discriminative model by leveraging labelled positive videos and a large collection of negative examples based on distance matrix. Liu et al. [30] extended the traditional binary classification problem to multi-class and proposed nearest-neighbour-based label transfer algorithm which encourages smoothness between regions that are spatio-temporally adjacent and similar in appearance. Zhang *et al.* [31] utilised pre-trained object detector to generate a set of detections and then pruned noisy detections and regions by preserving spatio-temporal constraints.

3   Domain Adaptation

We set out our approach to first semantically discovering possible objects of interest from video. We then adapt the source domain from image recognition to the target domain, i.e., pixel or superpixel level labelling. This approach is built by additionally incorporating constraints obtained from a given similarity graph defined on unlabelled target instances.

3.1   Object Discovery

Proposal Scoring.   Unlike image classification or object detection, semantic object segmentation requires not only localising objects of interest within an image, but also assigning class label for pixels belonging to the objects. One potential challenge of using image classifier to detect objects is that any regions containing the object or even part of the object, might be "correctly" recognised, which results in a large search space to accurately localise the object. To narrow down the search of targeted objects, we adopt category-independent bottom-up object proposals.

As we are interested in producing segmentations and not just bounding boxes, we require region proposals. We consider those regions as candidate object hypotheses. The objectness score associated with each proposal from [13] indicates how likely it is for an image region contain an object of any class. However, this objectness score does *not* consider context cues, e.g. motion, object categories and temporal coherence etc., and reflects only the generic object-like properties of the region (saliency, apparent separation from background, etc.). We incorporate motion information as a context cue for video objects. There has been many previous works on estimating local motion cues and we adopt a motion boundary based approach as introduced in [4] which roughly produces a binary map indicating whether each pixel is inside the motion boundary after compensating camera motion. After acquiring the motion cues, we score each proposal r by both appearance and context,

$$s_r = \mathcal{A}(r) + \mathcal{C}(r)$$

where $\mathcal{A}(r)$ indicates region level appearance score computed using [13] and $\mathcal{C}(r)$ represents the contextual score of region r which is defined as:

$$\mathcal{C}(r) = \text{Avg}(M^t(r)) \cdot \text{Sum}(M^t(r))$$

where $\text{Avg}(M^t(r))$ and $\text{Sum}(M^t(r))$ compute the average and total amount of motion cues [4] included by proposal r on frame t respectively. Note that appearance, contextual and combined scores are normalised.

Proposal Classification. On each frame t we have a collection of region proposals scored by their appearance and contextual information. These region proposals may contain various objects present in the video. In order to identify the objects of interest specified by the video level tag, region level classification is performed. We consider proven classification architectures such as VGG-16 nets [8] which did exceptionally well in ILSVRC14. VGG-16 net uses 3×3 convolution interleaved with max pooling and 3 fully-connected layers.

In order to classify each region proposal, we firstly warp the image data in each region into a form that is compatible with the CNN (VGG-16 net requires inputs of a fixed 224×224 pixel size). Although there are many possible transformations of our arbitrary-shaped regions, we warp all pixels in a bounding box around it to the required size, regardless its original size or shape. Prior to warping, we expand the tight bounding box by a certain number of pixels (10 in our system) around the original box, which was proven effective in the task of using image classifier for object detection task [32].

After the classification, we collect the confidence of regions with respect to the specific classes associated with the video and form a set of scored regions,

$$\{\mathcal{H}_{w_1}, \ldots, \mathcal{H}_{w_K}\}$$

where

$$\mathcal{H}_{w_k} = \{(r_1, s_{r_1}, c_{r_1, w_k}), \ldots, (r_N, s_{r_N}, c_{r_N, w_k})\}$$

with s_{r_i} is the original score of proposal r_i and c_{r_i, w_k} is its confidence from CNN classification with regard to keyword or class w_k. Figure 1 shows the positive detections with confidence higher than a predefined threshold (0.01), where higher confidence does not necessarily correspond to good proposals. This is mainly due to the nature of image classification where the image frame is quite often much larger than the tight bounding box of the object. In the following discussion we drop the subscript of classes, and formulate our method with regard to one single class for the sake of clarity, albeit our method works on multiple classes.

Spatial Average Pooling. After the initial discovery, a large number of region proposals are positively detected with regard to a class label, which include overlapping regions on the same objects and spurious detections. We adopt a simple weighted spatial average pooling strategy to aggregate the region-wise score, confidence as well as their spatial extent. For each proposal r_i, we rescore it by multiplying its score and classification confidence, which is denoted by $\tilde{s}_{r_i} = s_{r_i} \cdot c_{r_i}$. We then generate score map $\mathcal{S}_{r_i}$ of the size of image frame, which is composited as the binary map of current region proposal multiplied by its score $\tilde{s}_{r_i}$. We perform an average pooling over the score maps of all the proposals to compute a confidence map,

$$C^t = \frac{\sum_{r_i \in \mathcal{R}^t} \mathcal{S}_{r_i}}{\sum_{r_i \in \mathcal{R}^t} \tilde{s}_{r_i}} \tag{1}$$

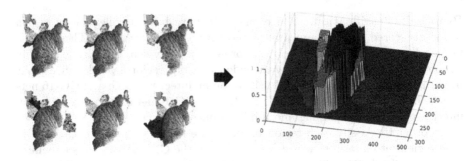

Fig. 2. An illustration of the weighted spatial average pooling strategy.

where $\sum_{r_i \in \mathcal{R}^t} \mathcal{S}_{r_i}$ performs element-wise operation and $\mathcal{R}^t$ represents the set of candidate proposals from frame t.

The resulted confidence map $\mathcal{C}^t$ aggregates not only the region-wise score but also their spatial extent. The key insight is that good proposals coincide with each other in the spatial domain and their contribution to the final confidence map are proportional to their region-wise score. An illustration of the weighted spatial average pooling is shown in Fig. 2.

3.2 Semi-supervised Domain Adaptation

To perform domain adaptation from image recognition to video object segmentation, we define a weighted space-time graph $\mathcal{G}_d = (\mathcal{V}_d, \mathcal{E}_d)$ spanning the whole video or a shot with each node corresponding to a superpixel, and each edge connecting two superpixels based on spatial and temporal adjacencies. Temporal adjacency is coarsely determined based on motion estimates, i.e., two superpixels are deemed temporally adjacent if they are connected by at least one motion vector.

We compute the affinity matrix A of the graph among spatial neighbours as

$$A_{i,j}^s = \frac{\exp(-d^c(s_i, s_j))}{d^s(s_i, s_j)} \tag{2}$$

where the functions $d^s(s_i, s_j)$ and $d^c(s_i, s_j)$ computes the spatial and color distances between spatially neighbouring superpixels s_i and s_j respectively:

$$d^c(s_i, s_j) = \frac{||c_i - c_j||^2}{2 < ||c_i - c_j||^2 >}$$

where $||c_i - c_j||^2$ is the squared Euclidean distance between two adjacent superpixels in RGB colour space, and $< \cdot >$ computes the average over all pairs i and j.

For affinities among temporal neighbours s_i^{t-1} and s_j^t, we consider both the temporal and colour distances between s_i^{t-1} and s_j^t,

$$A_{i,j}^t = \frac{\exp(-d^c(s_i, s_j))}{d^t(s_i, s_j)}$$

where

$$d^t(s_i, s_j) = \frac{1}{m_i \cdot \rho_{i,j}}, \tag{3}$$

$$m_i = \exp(-w_c \cdot \pi_i),$$

$$\rho_{i,j} = \frac{|\tilde{s}_i^{t-1} \cap s_j^t|}{|\tilde{s}_i^{t-1}|}.$$

Specifically, we define the temporal distance $d^t(s_i, s_j)$ by combining two factors, i.e., the temporal overlapping ratio $\rho_{i,j}$ and motion accuracy m_i. π_i denotes the motion coherence, and $w_c = 2.0$ is a parameter. The larger the temporal overlapping ratio is between two temporally related superpixels, the closer they are in temporal domain, subject to the accuracy of motion estimation. The temporal overlapping ratio $\rho_{i,j}$ is defined between the warped version of s_i^{t-1} following motion vectors and s_j^t, where $\tilde{s}_i^{t-1}$ is the warped region of s_i^{t-1} by optical flow to frame t, and $|\cdot|$ is the cardinality of a superpixel. The reliability of motion estimation inside s_i^{t-1} is measured by the motion coherence. A superpixel, i.e., a small portion of a moving object, normally exhibits coherent motions. We correlate the reliability of motion estimation of a superpixel with its local motion coherence. We compute quantised optical flow histograms h_i for superpixel s_i^{t-1}, and compute π_i as the information entropy of h_i. Smaller π_i indicates higher levels of motion coherence, i.e., higher motion reliability of motion estimation. An example of computed motion reliability map is shown in Fig. 3.

Fig. 3. Motion reliability map (right) computed given the optical flow between two consecutive frames (left and middle).

We follow a similar formulation with [33] to minimise an energy function $E(X)$ with respect to all superpixels confidence X ($X \in [-1, 1]$):

$$E(X) = \sum_{i,j=1}^{N} A_{ij} ||x_i d_i^{-\frac{1}{2}} - x_j d_j^{-\frac{1}{2}}||^2 + \mu \sum_{i=1}^{N} ||x_i - c_i||^2, \tag{4}$$

where μ is the regularisation parameter, and X are the desirable confidence of superpixels which are imposed by noisy confidence C in Eq. (1). We set $\mu = 0.5$. Let the node degree matrix $D = \text{diag}([d_1, \ldots, d_N])$ be defined as $d_i = \sum_{j=1}^{N} A_{ij}$, where $N = |\mathcal{V}|$. Denoting $S = D^{-1/2} A D^{-1/2}$, this energy function can be minimised iteratively as

$$X^{t+1} = \alpha S X^t + (1 - \alpha)C$$

until convergence, where α controls the relative amount of the confidence from its neighbours and its initial confidence. Specifically, the affinity matrix A of $\mathcal{G}_d$ is symmetrically normalised in S, which is necessary for the convergence of the following iteration. In each iteration, each superpixel adapts itself by receiving the confidence from its neighbours while preserving its initial confidence. The confidence is adapted symmetrically since S is symmetric. After convergence, the confidence of each unlabelled superpixel is adapted to be the class of which it has received most confidence during the iterations (Fig. 4).

We alternatively solve the optimisation problem as a linear system of equations which is more efficient. Differentiating $E(X)$ with respect to X we have

$$\nabla E(X)|_{X=X^*} = X^* - SX^* + \mu(X^* - C) = 0 \tag{5}$$

which can be transformed as

$$(I - (1 - \frac{\mu}{1+\mu})S)X^* = \frac{\mu}{1+\mu}C. \tag{6}$$

Finally we have

$$(I - (1 - \eta)S)X^* = \eta C. \tag{7}$$

where $\eta = \frac{\mu}{1+\mu}$.

The optimal solution for X can be found using the preconditioned (Incomplete Cholesky factorisation) conjugate gradient method with very fast convergence. For consistency, still let C denote the optimal semantic confidence X for the rest of this paper.

(a) Confidence maps of three consecutive frames

(b) Confidence maps after domain adaptation

Fig. 4. Proposed domain adaptation effectively adapts the noisy confidence map from image recognition to the video object segmentation domain.

4   Video Object Segmentation

We formulate video object segmentation as a superpixel-labelling problem of assigning each superpixel two classes: objects and background (not listed in the keywords). Similar to Subsect. 3.2 we define a space-time superpixel graph $\mathcal{G}_s = (\mathcal{V}_s, \mathcal{E}_s)$ by connecting frames temporally with optical flow displacement.

We define the energy function that minimises to achieve the optimal labelling:

$$E(x) = \sum_{i \in \mathcal{V}} (\psi_i^c(x_i) + \lambda_o \psi_i^o(x_i)) + \lambda_s \sum_{i \in \mathcal{V}, j \in N_i^s} \psi_{i,j}^s(x_i, x_j) + \lambda_t \sum_{i \in \mathcal{V}, j \in N_i^t} \psi_{i,j}^t(x_i, x_j)$$

$$(8)$$

where N_i^s and N_i^t are the sets of superpixels adjacent to superpixel s_i spatially and temporally in the graph respectively; λ_o, λ_s and λ_t are parameters; $\psi_i^c(x_i)$ indicates the color based unary potential and $\psi_i^o(x_i)$ is the unary potential of semantic object confidence which measures how likely the superpixel to be labelled by x_i given the semantic confidence map; $\psi_{i,j}^s(x_i, x_j)$ and $\psi_{i,j}^t(x_i, x_j)$ are spatial pairwise potential and temporal pairwise potential respectively. We set parameters $\lambda_o = 10$, $\lambda_s = 1000$ and $\lambda_t = 2000$. The definitions of these unary and pairwise terms are explained in detail next.

4.1   Unary Potentials

We define unary terms to measure how likely a superpixel is to be label as background or the object of interest according to both the appearance model and semantic object confidence map.

Colour unary potential is defined similar to [34], which evaluates the fit of a colour distribution (of a label) to the colour of a superpixel,

$$\psi_i^c(x_i) = -\log U_i^c(x_i)$$

where $U_i^c(\cdot)$ is the colour likelihood from colour model.

We train two Gaussian Mixture Models (GMMs) over the RGB values of superpixels, for objects and background respectively. These GMMs are estimated by sampling the superpixel colours according to the semantic confidence map.

Semantic unary potential is defined to evaluate how likely the superpixel to be labelled by x_i given the semantic confidence map c_i^t

$$\psi_i^o(x_i) = -\log U_i^o(x_i)$$

where $U_i^o(\cdot)$ is the semantic likelihood, i.e., for an object labelling $U_i^o = c_i^t$ and $1 - c_i^t$ otherwise.

4.2   Pairwise Potentials

We define the pairwise potentials to encourage both spatial and temporal smoothness of labelling while preserving discontinuity in the data. These terms are defined similar to the affinity matrix in Subsect. 3.2.

Superpixels in the same frame are spatially connected if they are adjacent. The spatial pairwise potential $\psi_{i,j}^s(x_i, x_j)$ penalises different labels assigned to spatially adjacent superpixels:

$$\psi_{i,j}^s(x_i, x_j) = \frac{[x_i \neq x_j]\exp(-d^c(s_i, s_j))}{d^s(s_i, s_j)}$$

where $[\cdot]$ denotes the indicator function.

The temporal pairwise potential is defined over edges where superpixels are temporally connected on consecutive frames. Superpixels s_i^{t-1} and s_j^t are deemed as temporally connected if there is at least one pixel of s_i^{t-1} which is propagated to s_j^t following the optical flow motion vectors,

$$\psi_{i,j}^t(x_i, x_j) = \frac{[x_i \neq x_j]\exp(-d^c(s_i, s_j))}{d^t(s_i, s_j)}.$$

Taking advantage of the similar definitions in computing affinity matrix in Subsect. 3.2, the pairwise potentials can be efficiently computed by reusing the affinity in Eqs. (2) and (3).

4.3   Optimisation

We adopt alpha expansion [35] to minimise Eq. (8) and the resulting label assignment gives the semantic object segmentation of the video.

4.4   Implementation

We implement our method using MATLAB and C/C++, with Caffe [36] implementation of VGG-16 net [8]. We reuse the superpixels returned from [13] which is produced by [37]. Large displacement optical flow algorithm [38] is adopted to cope with strong motion in natural videos. 5 components per GMM in RGB colour space are learned to model the colour distribution following [34]. Our domain adaptation method performs efficient learning on superpixel graph with an unoptimised MATLAB/C++ implementation, which takes around 30 s over a video shot of 100 frames. The average time on segmenting one preprocessed frame is about 3 s on a commodity desktop with a Quad-Core 4.0 GHz processor, 16 GB of RAM, and GTX 980 GPU.

We set parameters by optimising segmentation against ground truth over a sampled set of 5 videos from publicly available *Freiburg-Berkeley Motion Segmentation Dataset* dataset [39] which proved to be a versatile setting for a wide variety of videos. These parameters are fixed for the evaluation.

5   Evaluation

We evaluate our method on a large scale video dataset YouTube-Objects [40] and SegTrack [18]. YouTube-Objects consists of videos from 10 object classes with pixel-level ground truth for every 10 frames of 126 videos provided by [41]. These videos are very challenging and completely unconstrained, with objects of similar colour to the background, fast motion, non-rigid deformations, and fast camera motion. SegTrack consists of 5 videos with single or interacting objects presented in each video.

5.1 YouTube-Objects Dataset

We measure the segmentation performance using the standard *intersection-over-union* (IoU) overlap as accuracy metric. We compare our approach with 6 state-of-the-art automatic approaches on this dataset, including two motion driven segmentation [1,4], three weakly supervised approaches [29,31,40], and state-of-the-art object-proposal based approach [2]. Among the compared approaches, [1,2] reported their results by fitting a bounding box to the largest connected segment and overlapping with the ground-truth bounding box; the result of [2] on this dataset is originally reported by [4] by testing on 50 videos (5/class). The performance of [4] measured with respect to segmentation ground-truth is reported by [31]. Zhang *et al.* [31] reported results in more than 5500 frames sampled in the dataset based on the segmentation ground-truth. Wang *et al.* [27] reported the average results on 12 randomly sampled videos in terms of a different metric, i.e., per-frame pixel errors across all categories, and thus not listed here for comparison.

As shown in Table 1 and Fig. 5, our method outperforms the competing methods in 7 out of 10 classes, with gains up to 6.3%/6.6% in category/video average accuracy over the best competing method [31]. This is remarkable considering that [31] employed strongly-supervised deformable part models (DPM)

Table 1. Intersection-over-union overlap accuracies on YouTube-Objects dataset

	Brox [1]	Lee [2]	Prest [40]	Papazoglou [4]	Tang [29]	Zhang [31]	Baseline	Ours
Plane	0.539	NA	0.517	0.674	0.178	**0.758**	0.693	0.757
Bird	0.196	NA	0.175	0.625	0.198	0.608	0.590	**0.658**
Boat	0.382	NA	0.344	0.378	0.225	0.437	0.564	**0.656**
Car	0.378	NA	0.347	0.670	0.383	**0.711**	0.594	0.650
Cat	0.322	NA	0.223	0.435	0.236	0.465	0.455	**0.514**
Cow	0.218	NA	0.179	0.327	0.268	0.546	0.647	**0.714**
Dog	0.270	NA	0.135	0.489	0.237	0.555	0.495	**0.570**
Horse	0.347	NA	0.267	0.313	0.140	0.549	0.486	**0.567**
Mbike	0.454	NA	0.412	0.331	0.125	0.424	0.480	**0.560**
Train	0.375	NA	0.250	**0.434**	0.404	0.358	0.353	0.392
Cls. Avg	0.348	0.28	0.285	0.468	0.239	0.541	0.536	**0.604**
Vid. Avg	NA	NA	NA	0.432	0.228	0.526	0.523	**0.592**

as object detector while our approach only leverages image recognition model which lacks the capability of localising objects. [31] outperforms our method on *Plane* and *Car*, otherwise exhibiting varying performance across the categories — higher accuracy on more rigid objects but lower accuracy on highly flexible and deformable objects such as *Cat* and *Dog*. We owe it to that, though based on object detection, [31] prunes noisy detections and regions by enforcing spatio-temporal constraints, rather than learning an adapted data-driven representation in our approach. It is also worth remarking on the improvement in classes, e.g., *Cow*, where the existing methods normally fail or underperform due to the heavy reliance on motion information. The main challenge of the *Cow* videos is that cows very frequently stand still or move with mild motion, which the existing approaches might fail to capture whereas our proposed method excels by leveraging the recognition and representation power of deep convolutional neural network, as well as the semi-supervised domain adaptation.

(a) Aeroplane   (b) Bird   (c) Boat   (d) Car   (e) Cat

(f) Cow   (g) Dog   (h) Horse   (i) Motorbike   (j) Train

Fig. 5. Representative successful results by our approach on YouTube-Objects dataset.

Interestingly, another weakly supervised method [29] slightly outperforms our method on *Train* although all methods do not perform very well on this category due to the slow motion and missed detections on partial views of trains. This is probably owing to that [29] uses a large number of similar training videos which may capture objects in rare view. Otherwise, our method doubles or triples the accuracy of [29]. Motion driven method [4] can better distinguish rigid moving foreground objects on videos exhibiting relatively clean backgrounds, such as *Plane* and *Car*.

As ablation study, we evaluate a baseline scheme by removing the proposed domain adaptation algorithm (Sect. 3.2) from the full system. As shown in Table 1, the proposed semi-supervised domain adaptation is able to learn to successfully adapt to the target with a gain of 6.8%/6.9% in category/video average accuracies, comparing with the baseline scheme using only the semantic confidence by merging initially discovered region proposals (Sect. 3.1) for segmentation (with accuracies 0.536/0.523). This adaptation from the source domain of image recognition to the target domain of video semantic segmentation effectively compensates for the paradigm shift which is the key of our proposed method to outperform the state-of-the-art despite the use of weakly supervised image classifier.

5.2   SegTrack Dataset

We evaluate on SegTrack dataset to focus our comparison with the state-of-the-art semantic object segmentation algorithm [31] driven by object detector. We also compare with co-segmentation method [42] and the representative Figure-Ground segmentation algorithms [1–4,27,31] as baselines. To avoid confusion of segmentation results, all the compared methods only consider the primary object.

As shown in Table 2, our method outperforms the semantic segmentation [31] on *birdfall* and *monkeydog* videos, motion driven method [4] on four out of five videos, proposal ranking method [2] on four videos, proposal merging method [3] and saliency driven method [27] on two videos respectively. Clustering point tracks based method [1] results in highest error among all the methods. Co-segmentation method [42] reported the state-of-the-art results on three out of

Table 2. Quantitative segmentation results on SegTrack. Segmentation error as measured by the average number of incorrect pixels per frame.

Video (No. frames)	Ours	[1]	[4]	[3]	[2]	[42]	[31]	[27]
birdfall (30)	170	468	217	155	288	152	339	209
cheetah (29)	826	1968	890	633	905	NA	803	796
girl (21)	1647	7595	3859	1488	1785	1053	1459	1040
monkeydog (71)	304	1434	284	472	521	NA	365	562
parachute (51)	363	1113	855	220	201	189	196	207

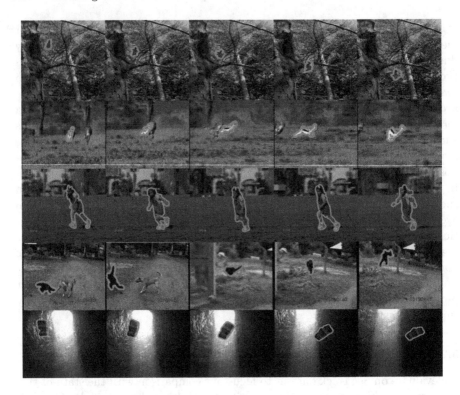

Fig. 6. Qualitative results of our method on SegTrack dataset.

five videos from SegTrack, albeit it can only segment single object as opposed to our method which can deal with objects of multiple semantic categories. Overall, our performance is about on par with the state-of-the-art semantic object segmentation method [31]. Qualitative segmentation of our approach is shown in Fig. 6.

6   Conclusion

We have proposed a semi-supervised framework to adapt CNN classifiers from image recognition domain to the target domain of semantic video object segmentation. This framework combines the recognition and representation power of CNN with the intrinsic structure of unlabelled data in the target domain to improve inference performance, imposing spatio-temporal smoothness constraints on the semantic confidence over the unlabelled video data. This proposed domain adaptation framework enables learning a data-driven representation of video objects. We demonstrated that this representation underpins a robust semantic video object segmentation method which outperforms existing methods on challenging datasets. As a future work, it would be interesting to incorporate representations learned from higher layers of CNN into the domain adaptation, which might potentially improve adaptation by propagating and combining higher level context.

References

1. Brox, T., Malik, J.: Object segmentation by long term analysis of point trajectories. In: Daniilidis, K., Maragos, P., Paragios, N. (eds.) ECCV 2010. LNCS, vol. 6315, pp. 282–295. Springer, Heidelberg (2010). doi:10.1007/978-3-642-15555-0_21
2. Lee, Y.J., Kim, J., Grauman, K.: Key-segments for video object segmentation. In: ICCV, pp. 1995–2002 (2011)
3. Zhang, D., Javed, O., Shah, M.: Video object segmentation through spatially accurate and temporally dense extraction of primary object regions. In: CVPR, pp. 628–635 (2013)
4. Papazoglou, A., Ferrari, V.: Fast object segmentation in unconstrained video. In: ICCV, pp. 1777–1784 (2013)
5. Wang, T., Wang, H.: Graph transduction learning of object proposals for video object segmentation. In: Cremers, D., Reid, I., Saito, H., Yang, M.-H. (eds.) ACCV 2014. LNCS, vol. 9006, pp. 553–568. Springer, Heidelberg (2015). doi:10.1007/978-3-319-16817-3_36
6. Wang, H., Wang, T.: Primary object discovery and segmentation in videos via graph-based transductive inference. Comput. Vis. Image Underst. **143**, 159–172 (2016)
7. Krizhevsky, A., Sutskever, I., Hinton, G.E.: Imagenet classification with deep convolutional neural networks. In: NIPS, pp. 1106–1114 (2012)
8. Simonyan, K., Zisserman, A.: Very deep convolutional networks for large-scale image recognition, arXiv preprint (2014). arXiv:1409.1556
9. Rasmus, A., Valpola, H., Honkala, M., Berglund, M., Raiko, T.: Semi-supervised learning with ladder network. In: NIPS (2015)
10. Faktor, A., Irani, M.: Video segmentation by non-local consensus voting. In: BMVC, vol. 2, p. 6 (2014)
11. Yang, J., Zhao, G., Yuan, J., Shen, X., Lin, Z., Price, B., Brandt, J.: Discovering primary objects in videos by saliency fusion and iterative appearance estimation. IEEE Trans. Circuits Syst, Video Technol (2015)
12. Perazzi, F., Wang, O., Gross, M., Sorkine-Hornung, A.: Fully connected object proposals for video segmentation. In: ICCV, pp. 3227–3234 (2015)
13. Endres, I., Hoiem, D.: Category independent object proposals. In: Daniilidis, K., Maragos, P., Paragios, N. (eds.) ECCV 2010. LNCS, vol. 6315, pp. 575–588. Springer, Heidelberg (2010). doi:10.1007/978-3-642-15555-0_42
14. Manen, S., Guillaumin, M., Gool, L.J.V.: Prime object proposals with randomized prim's algorithm. In: ICCV, pp. 2536–2543 (2013)
15. Wang, J., Xu, Y., Shum, H.Y., Cohen, M.F.: Video tooning. ACM Trans. Graph. **23**, 574–583 (2004)
16. Collomosse, J.P., Rowntree, D., Hall, P.M.: Stroke surfaces: temporally coherent artistic animations from video. IEEE Trans. Vis. Comput. Graph. **11**, 540–549 (2005)
17. Wang, T., Collomosse, J.P.: Probabilistic motion diffusion of labeling priors for coherent video segmentation. IEEE Trans. Multimed. **14**, 389–400 (2012)

18. Tsai, D., Flagg, M., Nakazawa, A., Rehg, J.M.: Motion coherent tracking using multi-label MRF optimization. Int. J. Comput. Vis. **100**, 190–202 (2012)

19. Li, F., Kim, T., Humayun, A., Tsai, D., Rehg, J.M.: Video segmentation by tracking many figure-ground segments. In: ICCV, Australia, 1–8 December 2013, pp. 2192–2199 (2013)

20. Wang, T., Han, B., Collomosse, J.P.: Touchcut: fast image and video segmentation using single-touch interaction. Comput. Vis. Image Underst. **120**, 14–30 (2014)

21. Grundmann, M., Kwatra, V., Han, M., Essa, I.A.: Efficient hierarchical graph-based video segmentation. In: CVPR, pp. 2141–2148 (2010)

22. Xu, C., Xiong, C., Corso, J.J.: Streaming hierarchical video segmentation. In: Fitzgibbon, A., Lazebnik, S., Perona, P., Sato, Y., Schmid, C. (eds.) ECCV 2012. LNCS, vol. 7577, pp. 626–639. Springer, Heidelberg (2012). doi:10.1007/978-3-642-33783-3_45

23. Wang, C., de La Gorce, M., Paragios, N.: Segmentation, ordering and multi-object tracking using graphical models. In: ICCV, pp. 747–754 (2009)

24. Sundberg, P., Brox, T., Maire, M., Arbelaez, P., Malik, J.: Occlusion boundary detection and figure/ground assignment from optical flow. In: CVPR, pp. 2233–2240 (2011)

25. Giordano, D., Murabito, F., Palazzo, S., Spampinato, C.: Superpixel-based video object segmentation using perceptual organization and location prior. In: CVPR, pp. 4814–4822 (2015)

26. Taylor, B., Karasev, V., Soatto, S.: Causal video object segmentation from persistence of occlusions. In: CVPR, pp. 4268–4276 (2015)

27. Wang, W., Shen, J., Porikli, F.: Saliency-aware geodesic video object segmentation. In: CVPR, pp. 3395–3402 (2015)

28. Hartmann, G., Grundmann, M., Hoffman, J., Tsai, D., Kwatra, V., Madani, O., Vijayanarasimhan, S., Essa, I., Rehg, J., Sukthankar, R.: Weakly supervised learning of object segmentations from web-scale video. In: Fusiello, A., Murino, V., Cucchiara, R. (eds.) ECCV 2012. LNCS, vol. 7583, pp. 198–208. Springer, Heidelberg (2012). doi:10.1007/978-3-642-33863-2_20

29. Tang, K.D., Sukthankar, R., Yagnik, J., Li, F.: Discriminative segment annotation in weakly labeled video. In: CVPR, pp. 2483–2490 (2013)

30. Liu, X., Tao, D., Song, M., Ruan, Y., Chen, C., Bu, J.: Weakly supervised multi-class video segmentation. In: CVPR, pp. 57–64 (2014)

31. Zhang, Y., Chen, X., Li, J., Wang, C., Xia, C.: Semantic object segmentation via detection in weakly labeled video. In: CVPR, pp. 3641–3649 (2015)

32. Girshick, R., Donahue, J., Darrell, T., Malik, J.: Rich feature hierarchies for accurate object detection and semantic segmentation. In: CVPR, pp. 580–587 (2014)

33. Zhou, D., Bousquet, O., Lal, T.N., Weston, J., Sch, B.: Learning with local and global consistency. In: NIPS, pp. 321–328 (2004)

34. Rother, C., Kolmogorov, V., Blake, A.: "GrabCut": interactive foreground extraction using iterated graph cuts. ACM Trans. Graph. **23**, 309–314 (2004)

35. Boykov, Y., Veksler, O., Zabih, R.: Fast approximate energy minimization via graph cuts. IEEE Trans. Pattern Anal. Mach. Intell. **23**, 1222–1239 (2001)

36. Jia, Y., Shelhamer, E., Donahue, J., Karayev, S., Long, J., Girshick, R., Guadarrama, S., Darrell, T.: Caffe: convolutional architecture for fast feature embedding. In: Proceedings of the ACM International Conference on Multimedia, pp. 675–678. ACM (2014)

37. Arbelaez, P., Maire, M., Fowlkes, C.C., Malik, J.: From contours to regions: an empirical evaluation. In: CVPR, pp. 2294–2301 (2009)

38. Brox, T., Bruhn, A., Papenberg, N., Weickert, J.: High accuracy optical flow estimation based on a theory for warping. In: Pajdla, T., Matas, J. (eds.) ECCV 2004. LNCS, vol. 3024, pp. 25–36. Springer, Heidelberg (2004). doi:10.1007/978-3-540-24673-2_3

39. Brox, T., Malik, J.: Object segmentation by long term analysis of point trajectories. In: Daniilidis, K., Maragos, P., Paragios, N. (eds.) ECCV 2010. LNCS, vol. 6315, pp. 282–295. Springer, Heidelberg (2010). doi:10.1007/978-3-642-15555-0_21

40. Prest, A., Leistner, C., Civera, J., Schmid, C., Ferrari, V.: Learning object class detectors from weakly annotated video. In: CVPR, pp. 3282–3289 (2012)

41. Jain, S.D., Grauman, K.: Supervoxel-consistent foreground propagation in video. In: Fleet, D., Pajdla, T., Schiele, B., Tuytelaars, T. (eds.) ECCV 2014. LNCS, vol. 8692, pp. 656–671. Springer, Heidelberg (2014). doi:10.1007/978-3-319-10593-2_43

42. Wang, L., Hua, G., Sukthankar, R., Xue, J., Zheng, N.: Video object discovery and co-segmentation with extremely weak supervision. In: Fleet, D., Pajdla, T., Schiele, B., Tuytelaars, T. (eds.) ECCV 2014. LNCS, vol. 8692, pp. 640–655. Springer, Heidelberg (2014). doi:10.1007/978-3-319-10593-2_42

Semantic Segmentation of Earth Observation Data Using Multimodal and Multi-scale Deep Networks

Nicolas Audebert[1,2](✉), Bertrand Le Saux[1], and Sébastien Lefèvre[2]

[1] The French Aerospace Lab, ONERA, 91761 Palaiseau, France
{nicolas.audebert,bertrand.le_saux}@onera.fr
[2] Univ. Bretagne-Sud, UMR 6074, IRISA, 56000 Vannes, France
sebastien.lefevre@irisa.fr

Abstract. This work investigates the use of deep fully convolutional neural networks (DFCNN) for pixel-wise scene labeling of Earth Observation images. Especially, we train a variant of the SegNet architecture on remote sensing data over an urban area and study different strategies for performing accurate semantic segmentation. Our contributions are the following: (1) we transfer efficiently a DFCNN from generic everyday images to remote sensing images; (2) we introduce a multi-kernel convolutional layer for fast aggregation of predictions at multiple scales; (3) we perform data fusion from heterogeneous sensors (optical and laser) using residual correction. Our framework improves state-of-the-art accuracy on the ISPRS Vaihingen 2D Semantic Labeling dataset.

1 Introduction

Over the past few years, deep learning has become ubiquitous for computer vision tasks. Convolutional Neural Networks (CNN) took over the field and are now the state-of-the-art for object classification and detection. Recently, deep networks extended their abilities to semantic segmentation, thanks to recent works designing deep networks for dense (pixel-wise) prediction, generally built around the fully convolutional principle stated by Long et al. [1]. These architectures have gained a lot of interest during the last years thanks to their ability to address semantic segmentation. Indeed, fully convolutional architectures are now considered as the state-of-the-art on most renowned benchmarks such as PASCAL VOC2012 [2] and Microsoft COCO [3]. However, those datasets focus on everyday scenes and assume a human-level point of view. In this work, we aim to process remote sensing (RS) data and more precisely Earth Observation (EO) data. EO requires to extract thematic information (e.g. land cover usage, biomass repartition, etc.) using data acquired from various airborne and/or satellite sensors (e.g. optical cameras, LiDAR). It often relies on a mapping step, that aims to automatically produce a semantic map containing various regions of interest, based on some raw data. A popular application is land cover mapping where each pixel is assigned to a thematic class, according to the type of land cover (vegetation, road, ...) or object (car, building, ...) observed at the pixel coordinates.

S.-H. Lai et al. (Eds.): ACCV 2016, Part I, LNCS 10111, pp. 180–196, 2017.
DOI: 10.1007/978-3-319-54181-5_12

As volume of EO data continuously grows (reaching the Zettabyte scale), deep networks can be trained to understand those images. However, there are several strong differences between everyday pictures and EO imagery. First, EO assumes a bird's view acquisition, thus the perspective is significantly altered w.r.t. usual computer vision datasets. Objects lie within a flat 2D plane, which makes the angle of view consistent but reduces the number of depth-related hints, such as projected shadows. Second, every pixel in RS images has a semantic meaning. This differs from most images in the PASCAL VOC2012 dataset, that are mainly comprised of a meaningless background with a few foreground objects of interest. Such a distinction is not as clear in EO data, where images may contain both semantically meaningful "stuff" (large homogeneous non quantifiable surfaces such as water bodies, roads, corn fields, ...) and "objects" (cars, houses, ...) that have different properties.

First experiments using deep learning introduced CNN for classification of EO data with a patch based approach [4]. Images were segmented using a segmentation algorithm (e.g. with superpixels) and each region was classified using a CNN. However, the unsupervised segmentation proved to be a difficult bottleneck to overcome as higher accuracy requires strong oversegmentation. This was improved thanks to CNN using dense feature maps [5]. Fully supervised learning of both segmentation and classification is a promising alternative that could drastically improve the performance of the deep models. Fully convolutional networks [1] and derived models can help solve this problem. Adapting these architectures to multimodal EO data is the main objective of this work.

In this work, we show how to perform competitive semantic segmentation of EO data. We consider a standard dataset delivered by the ISPRS [6] and rely on deep fully convolutional networks, designed for dense pixel-wise prediction. Moreover, we build on this baseline approach and present a simple trick to smooth the predictions using a multi-kernel convolutional layer that operates several parallel convolutions with different kernel sizes to aggregate predictions at multiple scale. This module does not need to be retrained from scratch and smoothes the predictions by averaging over an ensemble of models considering multiple scales, and therefore multiple spatial contexts. Finally, we present a data fusion method able to integrate auxiliary data into the model and to merge predictions using all available data. Using a dual-stream architecture, we first naively average the predictions from complementary data. Then, we introduce a residual correction network that is able to learn how to fuse the prediction maps by adding a corrective term to the average prediction.

2   Related Work

2.1   Semantic Segmentation

In computer vision, semantic segmentation consists in assigning a semantic label (i.e. a class) to each coherent region of an image. This can be achieved using pixel-wise dense prediction models that are able to classify each pixel of the image. Recently, deep learning models for semantic segmentation have started to appear. Many recent works in computer vision are actually tackling semantic segmentation with a significant success. Nearly all state-of-the-art architectures follow principles stated in [1], where semantic segmentation using Fully Convolutional Networks (FCN) has been shown to achieve impressive results on PASCAL VOC2012. The main idea consists in modifying traditional classification CNN so that the output is not a probability vector but rather a probability map. Generally, a standard CNN is used as an encoder that will extract features, followed by a decoder that will upsample feature maps to the original spatial resolution of the input image. A heat map is then obtained for each class. Following the path opened by FCN, several architectures have proven to be very effective on both PASCAL VOC2012 and Microsoft COCO. Progresses have been obtained by increasing the field-of-view of the encoder and removing pooling layers to avoid bottlenecks (DeepLab [7] and dilated convolutions [8]). Structured prediction has been investigated with integrated structured models such as Conditional Random Fields (CRF) within the deep network (CRFasRNN [9,10]). Better architectures also provided new insights (e.g. ResNet [11] based architectures [12], recurrent neural networks [13]). Leveraging analogies with convolutional autoencoders (and similarly to Stacked What-Where Autoencoders [14]), DeconvNet [15] and SegNet [16] have investigated symmetrical encoder-decoder architectures.

2.2   Scene Understanding in Earth Observation Imagery

Deep learning on EO images is a very active research field. Since the first works on road detection [17], CNN have been successfully used for classification and dense labeling of EO data. CNN-based deep features have been shown to outperform significantly traditional methods based on hand-crafted features and Support Vector Machines for land cover classification [18]. Besides, a framework using superpixels and deep features for semantic segmentation outperformed traditional methods [4] and obtained a very high accuracy in the Data Fusion Contest 2015 [19]. A generic deep learning framework for processing remote sensing data using CNN established that deep networks improve significantly the commonly used SVM baseline [20]. [21] also performed classification of EO data using ensemble of multiscale CNN, which has been improved with the introduction of FCN [22]. Indeed, fully convolutional architectures are promising as they can learn how to classify the pixels ("what") but also predict spatial structures ("where"). Therefore, on EO images, such models would be not only able to detect different types of land cover in a patch, but also to predict the shapes of the buildings, the curves of the roads, ...

3   Proposed Method

3.1   Data Preprocessing

High resolution EO images are often too large to be processed in only one pass through a CNN. For example, the average dimensions of an ISPRS tile from

Vaihingen dataset is 2493×2063 pixels, whereas most CNN are tailored for a resolution of 256×256 pixels. Given current GPU memory limitations, we split our EO images in smaller patches with a simple sliding window. It is then possible to process arbitrary large images in a linear time. In the case where consecutive patches overlap at testing time (if the stride is smaller than the patch size), we average the multiple predictions to obtain the final classification for overlapping pixels. This smoothes the predictions along the borders of each patch and removes the discontinuities that can appear.

We recall that our aim is to transpose well-known architectures from traditional computer vision to EO. We are thus using neural networks initially designed for RGB data. Therefore, the processed images will have to respect such a 3-channel format. The ISPRS dataset contains IRRG images of Vaihingen. The 3 channels (i.e. near-infrared, red and green) will thus be processed as an RGB image. Indeed, all three color channels have been acquired by the same sensor and are the consequence of the same physical phenomenon. These channels have homogeneous dynamics and meaning for our remote sensing application. The dataset also includes additional data acquired from an aerial laser sensor and consisting of a Digital Surface Model (DSM). In addition, we also use the Normalized Digital Surface Model (NDSM) from [23]. Finally, we compute the Normalized Difference Vegetation Index (NDVI) from the near-infrared and red channels. NDVI is a good indicator for vegetation and is computed as follows:

$$NDVI = \frac{IR - R}{IR + R} \ . \tag{1}$$

Let us recall that we are working in a 3-channel framework. Thus we build for each IRRG image another companion composite image using the DSM, NDSM and NDVI information. Of course, such information does not correspond to color channels and cannot be stacked as an RGB color image without caution. Nevertheless, this composite image contains relevant information that can help discriminating between several classes. In particular, the DSM includes the height information which is of first importance to distinguish a roof from a road section, or a bush from a tree. Therefore, we will explore how to process these heterogeneous channels and to combine them to improve the model prediction by fusing the predictions of two networks sharing the same topology.

3.2   Network Architecture

SegNet. There are many available architectures for semantic segmentation. We choose here the SegNet architecture [16] (cf. Fig. 1), since it provides a good balance between accuracy and computational cost. SegNet's symmetrical architecture and its use of the pooling/unpooling combination is very effective for precise relocalisation of features, which is intuitively crucial for EO data. In addition to SegNet, we have performed preliminary experiments with FCN [1] and DeepLab [7]. Results reported no significant improvement (or even no improvement at all). Thus the need to switch to more computationally

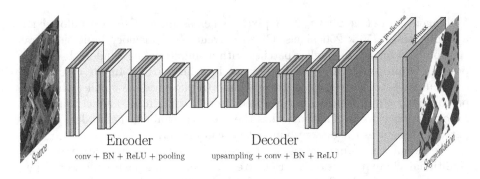

Fig. 1. Illustration of the SegNet architecture applied to EO data.

expensive architectures was not demonstrated. Note that our contributions could easily be adapted to other architectures and are not specific to SegNet.

SegNet has an encoder-decoder architecture based on the convolutional layers of VGG-16 from the Visual Geometry Group [24,25]. The encoder is a succession of convolutional layers followed by batch normalization [26] and rectified linear units. Blocks of convolution are followed by a pooling layer of stride 2. The decoder has the same number of convolutions and the same number of blocks. In place of pooling, the decoder performs upsampling using unpooling layers. This layer operates by relocating at the maximum index computed by the associated pooling layer. For example, the first pooling layer computes the mask of the maximum activations (the "$argmax$") and passes it to the last unpooling layer, that will upsample the feature map to a full resolution by placing the activations on the mask indices and zeroes everywhere else. The sparse feature maps are then densified by the consecutive convolutional layers. The encoding weights are initialized using the corresponding layers from VGG-16 and the decoding weights are initialized randomly using the strategy from [27]. We report no gain with alternative transfer functions such as ELU [28] or PReLU [27] and do not alter further the SegNet architecture. Let N be the number of pixels in a patch and k the number of classes, for a specified pixel i, let y^i denote its label and $(z_1^i, \ldots, z_k^i)$ the prediction vector; we minimize the normalized sum of the multinomial logistic loss of the softmax outputs over the whole patch:

$$loss = \frac{1}{N} \sum_{i=1}^{N} \sum_{j=1}^{k} y_j^i \log \left(\frac{\exp(z_j^i)}{\sum_{l=1}^{k} \exp(z_l^i)} \right). \tag{2}$$

As previously demonstrated in [29], visual filters learnt on generic datasets such as ImageNet can be effectively transferred on EO data. However, we suggest that remote sensing images have a common underlying spatial structure linked to the orthogonal line of view from the sky. Therefore, it is interesting to allow the filters to be optimized according to these specificities in order to leverage the

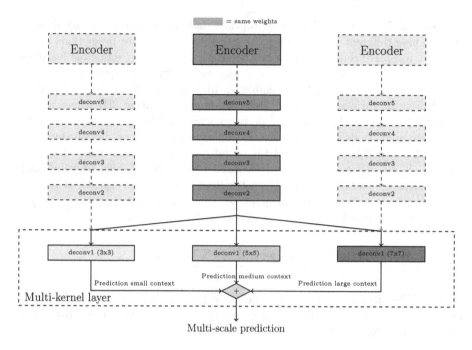

Fig. 2. Our multi-kernel convolutional layer operates at 3 multiple scales, which is equivalent to averaging an ensemble of 3 models sharing weights.

common properties of all EO images, rather than waste parameters on useless filters. To assess this hypothesis, we experiment different learning rates for the encoder (lr_e) and the decoder (lr_d). Four strategies have been experimented:

- same learning rate for both: $lr_d = lr_e$, $lr_e/lr_d = 1$,
- slightly higher learning rate for the decoder: $lr_d = 2 \times lr_e$, $lr_e/lr_d = 0.5$,
- strongly higher learning rate for the decoder: $lr_d = 10 \times lr_e$, $lr_e/lr_d = 0.1$,
- no backpropagation at all for the encoder: $lr_e = 0$, $lr_e/lr_d = 0$.

As a baseline, we also try to randomly initialize the weights of both the encoder and the decoder to train a new SegNet from scratch using the same learning rates for both parts.

Multi-kernel Convolutional Layer. Finally, we explore how to take spatial context into account. Let us recall that spatial information is crucial when dealing with EO data. Multi-scale processing has been proven effective for classification, notably in the Inception network [30], for semantic segmentation [8] and on remote sensing imagery [21]. We design here an alternative decoder whose last layer extracts information simultaneously at several spatial resolutions and aggregates the predictions. Instead of using only one kernel size of 3×3, our multi-kernel convolutional layer performs 3 parallel convolutions using kernels of size 3×3, 5×5 and 7×7 with appropriate padding to keep the image dimensions. These different kernel sizes make possible to aggregate predictions using different receptive cell sizes. This can be seen as performing ensemble learning where the models have the same topologies and weights, excepted for the last layer, as illustrated by Fig. 2. Ensemble learning with CNN has been proven to be effective in various situations, including super-resolution [31] where multiple CNN are used before the final deconvolution. By doing so, we are able to aggregate predictions at different scales, thus smoothing the predictions by combining different fields of view and taking into account different sizes of spatial context. If X_p denotes the input activations of the multi-kernel convolutional layer for the pth feature map, Z_p^s the activations after the convolution at the sth scale ($s \in \{1, \ldots, S\}$ with $S = 3$ here), Z_q' the final outputs and $W_{p,q}^s$ the qth convolutional kernel for the input map p at scale s, we have:

$$Z_q' = \frac{1}{S} \sum_{s=1}^{S} Z_p^s = \frac{1}{S} \sum_{s=1}^{S} \sum_{p} W_{p,q}^s X_p \ . \tag{3}$$

Let S denote the number of parallel convolutions (here, $S = 3$). For a given pixel at index i, if $z_k^{s,i}$ is the activation for class k and scale s, the logistic loss after the softmax in our multi-kernel variant is:

$$loss = \sum_{i=1}^{N} \sum_{j=1}^{k} y_j^i \log \left(\frac{\exp(\frac{1}{S} \sum_{s=1}^{S} z_j^{s,i})}{\sum_{l=1}^{k} \exp(\frac{1}{S} \sum_{s=1}^{S} z_l^{s,i})} \right) \ . \tag{4}$$

We can train the network using the whole multi-kernel convolutional layer at once using the standard backpropagation scheme. Alternatively, we can also train only one convolution at a time, meaning that our network can be trained at first with only one scale. Then, to extend our multi-kernel layer, we can simply drop the last layer and fine-tune a new convolutional layer with another kernel size and then add the weights to a new parallel branch. This leads to a higher flexibility compared to training all scales at once, and can be used to quickly include multi-scale predictions in other fully convolutional architectures only by fine-tuning.

This multi-kernel convolutional layer shares several concepts with the competitive multi-scale convolution [32] and the Inception module [30]. However, in our work, the parallel convolutions are used only in the last layer to perform model averaging over several scales, reducing the number of parameters to be optimized compared to performing multi-scale in every layer. Moreover, this ensures more flexibility, since the number of parallel convolutions can be simply extended by fine-tuning with a new kernel size. Compared to the multi-scale context aggregation from Yu and Koltun [8], our multi-kernel does not reduce dimensions and operates convolutions in parallel. Fast ensemble learning is then performed with a very low computational overhead. As opposed to Zhao et al. [21], we do not need to extract the patches using a pyramid, nor do we

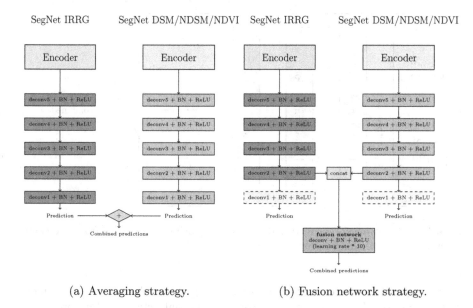

(a) Averaging strategy.   (b) Fusion network strategy.

Fig. 3. Fusion strategies of our dual-stream SegNet architecture.

need to choose the scales beforehand, as we can extend the network according to the dataset.

3.3 Heterogeneous Data Fusion with Residual Correction

Traditional 3-channel color images are only one possible type of remote sensing data. Multispectral sensors typically provide 4 to 12 bands, while hyperspectral images are made of a few hundreds of spectral bands. Besides, other data types such as DSM or radar imagery may be available. As stated in Sect. 3.1, IRRG data from the ISPRS dataset is completed by DSM, NDSM and NDVI. So we will assess if it is possible to: (1) build a second SegNet that can perform semantic segmentation using a second set of raw features, (2) combine the two networks to perform data fusion and improve the accuracy.

The naive data fusion would be to concatenate all 6 channels (IR/R/G and DSM/NDSM/NDVI) and feed a SegNet-like architecture with it. However, we were not able to improve the performance in regard to a simple IRRG architecture. Inspired by the multimodal fusion introduced in [33] for joint audio-video representation learning and the RGB-D data fusion in [34], we try a prediction-oriented fusion by merging the output activations maps. We consider here two strategies: (1) simple averaging after the softmax (Fig. 3a), (2) neural network merge (Fig. 3b). The latter uses a corrector network that can learn from both sets of activations to correct small deficiencies in the prediction and hopefully globally improve the prediction accuracy.

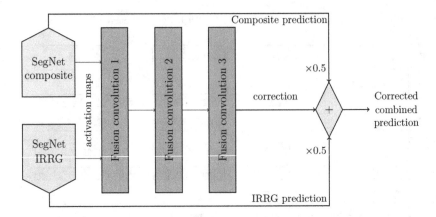

Fig. 4. Fusion network for correcting the predictions using information from two complementary SegNet using heterogeneous data.

Our original fusion network consisted in three convolutional layers which input was intermediate feature maps from the original network. More precisely, in the idea of fine-tuning by dropping the last fully connected layer before the softmax, we remove the last convolutional layer of each network and replace them by the fusion network convolutional layer, taking the concatenated intermediate feature maps in input. This allows the fusion network to have more information about raw activations, rather than just stacking the layers after the preprocessed predictions. Indeed, because of the one-hot encoding of the ground truth labels, the last layer activations tend to be sparse, therefore losing information about activations unrelated to the highest predicted class. However, this architecture does not improve significantly the accuracy compared to a simple averaging.

Building on the idea of residual deep learning [11], we propose a fusion network based on residual correction. Instead of dropping entirely the last convolutional layers from the two SegNets, we keep them to compute the average scores. Then, we use the intermediate feature maps as inputs to a 3-convolution layers "correction" network, as illustrated in Fig. 4. Using residual learning makes sense in this case, as the average score is already a good estimation of the reality. To improve the results, we aim to use the complementary channels to correct small errors in the prediction maps. In this context, residual learning can be seen as learning a corrective term for our predictive model. Let M_r denote the input of the rth stream ($r \in \{1, \ldots, R\}$ with $R = 2$ here), P_r the output probability tensor and Z_r the intermediate feature map used for the correction. The corrected prediction is:

$$P'(M_1, \ldots, M_R) = P(M_1, \ldots, M_R) + correction(Z_1, \ldots, Z_R) \tag{5}$$

where

$$P(M_1, \ldots, M_R) = \frac{1}{R} \sum_{r=1}^{R} P_r(M_r) . \tag{6}$$

Table 1. Results on the validation set with different initialization policies.

Initialization	Random	VGG-16			
Learning rate ratio $\frac{lr_e}{lr_d}$	1	1	0.5	0.1	0
Accuracy	87.0%	87.2%	**87.8%**	86.9%	86.5%

Table 2. Results on the validation set.

Type/Stride (px)	128 (no overlap)	64 (50% overlap)	32 (75% overlap)
Standard	87.8%	88.3%	88.8%
Multi-kernel	88.2%	88.6%	89.1%
Fusion (average)	88.2%	88.7%	89.1%
Fusion (correction)	88.6%	89.0%	89.5%
Multi-kernel + Average	88.5%	89.0%	89.5%
Multi-kernel + Correction	88.7%	89.3%	**89.8%**

Using residual learning should bring $\|correction\| \ll \|P\|$. This means that it should be easier for the network to learn not to add noise to predictions where its confidence is high ($\|correction\| \simeq 0$) and only modify unsure predictions. The residual correction network can be trained by fine-tuning as usual with a logistic loss after a softmax layer.

4   Experiments

4.1   Experimental Setup

To compare our method with the current state-of-the-art, we train a model using the full dataset (training and validation sets) with the same training strategy. This is the model that we tested against other methods using the ISPRS evaluation benchmark[1].

4.2   Results

Our best model achieves state-of-the art results on the ISPRS Vaihingen dataset (cf. Table 3)[2]. Figure 5 illustrates a qualitative comparison between SegNet using our multi-kernel convolutional layer and other baseline strategies on an extract of the Vaihingen testing set. The provided metrics are the global pixel-wise accuracy and the F1 score on each class:

$$F1_i = 2\,\frac{precision_i \times recall_i}{precision_i + recall_i} \text{ and } recall_i = \frac{tp_i}{C_i},\ precision_i = \frac{tp_i}{P_i}\,, \quad (7)$$

---

[1] In this benchmark, the evaluation is not performed by us or any other competing team, but directly by the benchmark organizers.

[2] http://www2.isprs.org/vaihingen-2d-semantic-labeling-contest.html.

Table 3. ISPRS 2D Semantic Labeling Challenge Vaihingen results.

Method	imp surf	building	low veg	tree	car	Accuracy
Stair Vision Library ("SVL_3") [23]	86.6%	91.0%	77.0%	85.0%	55.6%	84.8%
RF + CRF ("HUST") [35]	86.9%	92.0%	78.3%	86.9%	29.0%	85.9%
CNN ensemble ("ONE_5") [36]	87.8%	92.0%	77.8%	86.2%	50.7%	85.9%
FCN ("UZ_1")	89.2%	92.5%	81.6%	86.9%	57.3%	87.3%
FCN ("UOA") [37]	89.8%	92.1%	80.4%	88.2%	82.0%	87.6%
CNN + RF + CRF ("ADL_3") [5]	89.5%	93.2%	82.3%	88.2%	63.3%	88.0%
FCN ("DLR_2") [22]	90.3%	92.3%	82.5%	89.5%	76.3%	88.5%
FCN + RF + CRF ("DST_2")	90.5%	93.7%	83.4%	89.2%	72.6%	89.1%
Ours (multi-kernel)	**91.5%**	94.3%	82.7%	89.3%	**85.7%**	89.4%
Ours (multi-kernel + fusion)	91.0%	**94.5%**	**84.4%**	**89.9%**	77.8%	**89.8%**

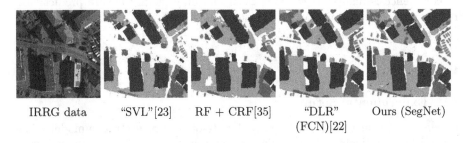

IRRG data   "SVL"[23]   RF + CRF[35]   "DLR"   Ours (SegNet)
                                                    (FCN)[22]

Fig. 5. Comparison of the generated segmentations using several methods of the ISPRS Vaihingen benchmark (patch extracted from the testing set). (white: roads, blue: buildings, cyan: low vegetation, green: trees, yellow: cars) (Color figure online)

where tp_i the number of true positives for class i, C_i the number of pixels belonging to class i, and P_i the number of pixels attributed to class i by the model. These metrics are computed using an alternative ground truth in which the borders have been eroded by a 3px radius circle.

Previous to our submission, the best results on the benchmark were obtained by combining FCN and hand-crafted features, whereas our method does not require any prior. The previous best method using only a FCN ("DLR_1") reached 88.4%, our method improving this result by 1.4%. Earlier methods using CNN for classification obtained 85.9% ("ONE_5" [36]) and 86.1% ("ADL_1" [5]). It should be noted that we outperform all these methods, including those that

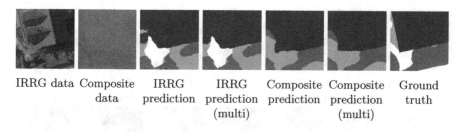

| IRRG data | Composite data | IRRG prediction | IRRG prediction (multi) | Composite prediction | Composite prediction (multi) | Ground truth |

Fig. 6. Effects of the multi-kernel convolutional layer on selected patches.

use hand-crafted features and structured models such as Conditional Random Fields, although we do not use these techniques.

4.3   Analysis

Sliding Window Overlap. Allowing an overlap when sliding the window across the tile slows significantly the segmentation process but improves accuracy, as shown in Table 2. Indeed, if we divide the stride by 2, the number of patches is multiplied by 4. However, averaging several predictions on the same region helps to correct small errors, especially around the borders of each patch, which are difficult to predict due to a lack of context. We find that a stride of 32px (75% overlap) is fast enough for most purposes and achieves a significant boost in accuracy (+1% compared to no overlap). Processing a tile takes 4 min on a Tesla K20c with a 32px stride and less than 20 s with a 128px stride. The inference time is doubled using the dual-stream fusion network.

Transfer Learning. As shown in Table 1, the model achieves highest accuracy on the validation set using a low learning rate on the encoder. This supports previous evidences hinting that fine-tuning generic filters on a specialized task performs better than training new filters form scratch. However, we suggest that a too low learning rate on the original filters impede the network from reaching an optimal bank of filters if enough data is available. Indeed, in our experiments, a very low learning rate for the encoder (0.1) achieves a lower accuracy than a moderate drop (0.5). We argue that given the size and the nature (EO data) of our dataset, it is beneficial to let the filters from VGG-16 vary as this allows the network to achieve better specialization. However, a too large learning rate brings also the risk of overfitting, as showed by our experiment. Therefore, we argue that setting a lower learning rate for the encoder part of fully convolutional architectures might act as regularizer and prevent some of the overfitting that would appear otherwise. This is similar to previous results in remote sensing [20], but also coherent with more generic observations [38].

Multi-kernel Convolutional Layer. The multi-kernel convolutional layer brings an additional boost of 0.4% to the accuracy. As illustrated in Fig. 6,

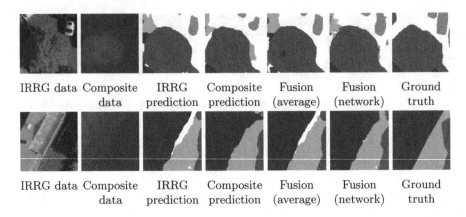

IRRG data Composite   IRRG   Composite   Fusion   Fusion   Ground
              data   prediction prediction (average) (network)   truth

IRRG data Composite   IRRG   Composite   Fusion   Fusion   Ground
              data   prediction prediction (average) (network)   truth

Fig. 7. Effects of our fusion strategies on selected patches.

it smooths the prediction by removing small artifacts isolated in large homogeneous regions. It also helps to alleviate errors by averaging predictions over several models.

This approach improves previous results on the ISPRS Vaihingen 2D labeling challenge, reaching 89.4%[3] (cf. Table 3). Improvements are significant for most classes, as this multi-kernel method obtains the best F1 score for "impervious surfaces" (+1.0%), "buildings" (+0.8%) and "cars" (+3.7%) classes. Moreover, this method is competitive on the "low vegetation" and "tree" classes. Although the cars represent only 1.2% of the whole Vaihingen dataset and therefore does not impact strongly the global accuracy, we believe this improvement to be significant, as our model is successful both on "stuff" and on objects.

Data Fusion and Residual Correction. Naive prediction fusion by averaging the maps boosts the accuracy by 0.3–0.4%. This is cumulative with the gain from the multi-kernel convolutions, which hints that the two methods are complementary. This was expected, as the latter leverages multi-scale predictions whereas the data fusion uses additional information to refine the predictions. As illustrated in Fig. 7, the fusion manages to correct errors in one model by using information from the other source. The residual correction network generates more visually appealing predictions, as it learns which network to favor for each class. For example, the IRRG data is nearly always right when predicting car pixels, therefore the correction network often keeps those. However the composite data has the advantage of the DSM to help distinguishing between low vegetation and trees. Thus, the correction network gives more weight to the predictions of the "composite SegNet" for these classes. Interestingly, if m_{avg}, m_{corr}, s_{avg} and s_{corr} denote the respective mean and standard deviation of the activations after averaging and after correction, we see that

---

[3] "ONE_6":   https://www.itc.nl/external/ISPRS_WGIII4/ISPRSIII_4_Test_results/
2D_labeling_vaih/2D_labeling_Vaih_details_ONE_6/index.html.

$m_{avg} \simeq 1.0$, $m_{corr} \simeq 0$ and $s_{avg} \simeq 5$, $s_{corr} \simeq 2$. We conclude that the network actually learnt how to apply small corrections to achieve a higher accuracy, which is in phase with both our expectations and theoretical developments [11].

This approach improves our results on the ISPRS Vaihingen 2D Labeling Challenge even further, reaching 89.8%[4] (cf. Table 3). F1 scores are significantly improved on buildings and vegetation, thanks to the discriminative power of the DSM and NDVI. However, even though the F1 score on cars is competitive, it is lower than expected. We explain this by the poor accuracy of the composite SegNet on cars, that degrades the average prediction and is only partly corrected by the network. We wish to investigate this issue further in the future.

5   Conclusion and Future Work

In this work, we investigated the use of DFCN for dense scene labeling of EO images. Especially, we showed that encoder-decoder architectures, notably Seg-Net, designed for semantic segmentation of traditional images and trained with weights from ImageNet, can easily be transposed to remote sensing data. This reinforces the idea that deep features and visual filters from generic images can be built upon for remote sensing tasks. We introduced in the network a multi-kernel convolutional layer that performs convolutions with several filter sizes to aggregate multi-scale predictions. This improves accuracy by performing model averaging with different sizes of spatial context. We investigated prediction-oriented data fusion with a dual-stream architecture. We showed that a residual correction network can successfully identify and correct small errors in the prediction obtained by the naive averaging of predictions coming from heterogeneous inputs. To demonstrate the relevance of those methods, we validated our methods on the ISPRS 2D Vaihingen semantic labeling challenge, on which we improved the state-of-the-art by 1%.

In the future, we would like to investigate if residual correction can improve performance for networks with different topologies. Moreover, we hope to study how to perform data-oriented fusion, sooner in the network, to reduce the computational overhead of using several long parallel streams. Finally, we believe that there is additional progress to be made by integrating the multi-scale nature of the data early in the network design.

Acknowledgement. The Vaihingen data set was provided by the German Society for Photogrammetry, Remote Sensing and Geoinformation (DGPF) [39]: http://www. ifp.uni-stuttgart.de/dgpf/DKEP-Allg.html.

Nicolas Audebert's work is supported by the Total-ONERA research project NAOMI. The authors acknowledge the support of the French Agence Nationale de la Recherche (ANR) under reference ANR-13-JS02-0005-01 (Asterix project).

---

[4] "ONE_7":   https://www.itc.nl/external/ISPRS_WGIII4/ISPRSIII_4_Test_results/
2D_labeling_vaih/2D_labeling_Vaih_details_ONE_7/index.html.

References

1. Long, J., Shelhamer, E., Darrell, T.: Fully convolutional networks for semantic segmentation. In: Proceedings of the IEEE Conference on Computer Vision and Pattern Recognition, pp. 3431–3440 (2015)
2. Everingham, M., Eslami, S.M.A., Gool, L.V., Williams, C.K.I., Winn, J., Zisserman, A.: The pascal visual object classes challenge: a retrospective. Int. J. Comput. Vis. **111**, 98–136 (2014)
3. Lin, T.-Y., Maire, M., Belongie, S., Hays, J., Perona, P., Ramanan, D., Dollár, P., Zitnick, C.L.: Microsoft COCO: common objects in context. In: Fleet, D., Pajdla, T., Schiele, B., Tuytelaars, T. (eds.) ECCV 2014. LNCS, vol. 8693, pp. 740–755. Springer, Heidelberg (2014). doi:10.1007/978-3-319-10602-1_48
4. Lagrange, A., Le Saux, B., Beaupere, A., Boulch, A., Chan-Hon-Tong, A., Herbin, S., Randrianarivo, H., Ferecatu, M.: Benchmarking classification of earth-observation data: from learning explicit features to convolutional networks. In: IEEE International Geosciences and Remote Sensing Symposium (IGARSS), pp. 4173–4176 (2015)
5. Paisitkriangkrai, S., Sherrah, J., Janney, P., Van Den Hengel, A.: Effective semantic pixel labelling with convolutional networks and conditional random fields. In: Proceedings of the IEEE Conference on Computer Vision and Pattern Recognition Workshops, pp. 36–43 (2015)
6. Rottensteiner, F., Sohn, G., Jung, J., Gerke, M., Baillard, C., Benitez, S., Breitkopf, U.: The ISPRS benchmark on urban object classification and 3d building reconstruction. ISPRS Ann. Photogrammetry Remote Sens. Spat. Inf. Sci. **1**, 3 (2012)
7. Chen, L.C., Papandreou, G., Kokkinos, I., Murphy, K., Yuille, A.: Semantic image segmentation with deep convolutional nets and fully connected CRFs. In: Proceedings of the International Conference on Learning Representations (2015)
8. Yu, F., Koltun, V.: Multi-scale context aggregation by dilated convolutions. In: Proceedings of the International Conference on Learning Representations (2015)
9. Zheng, S., Jayasumana, S., Romera-Paredes, B., Vineet, V., Su, Z., Du, D., Huang, C., Torr, P.H.S.: Conditional random fields as recurrent neural networks. In: Proceedings of the IEEE International Conference on Computer Vision, pp. 1529–1537 (2015)
10. Arnab, A., Jayasumana, S., Zheng, S., Torr, P.: Higher order conditional random fields in deep neural networks (2015). arXiv:1511.08119 [cs]
11. He, K., Zhang, X., Ren, S., Sun, J.: Deep residual learning for image recognition. In: Proceedings of the IEEE Conference on Computer Vision and Pattern Recognition (2016)
12. Wu, Z., Shen, C., Van Den Hengel, A.: High-performance semantic segmentation using very deep fully convolutional networks (2016). arXiv:1604.04339 [cs]
13. Yan, Z., Zhang, H., Jia, Y., Breuel, T., Yu, Y.: Combining the best of convolutional layers and recurrent layers: a hybrid network for semantic segmentation. arXiv:1603.04871 [cs] (2016)
14. Zhao, J., Mathieu, M., Goroshin, R., LeCun, Y.: Stacked what-where auto-encoders. In: Proceedings of the International Conference on Learning Representations (2015)
15. Noh, H., Hong, S., Han, B.: Learning deconvolution network for semantic segmentation. In: Proceedings of the IEEE Conference on Computer Vision and Pattern Recognition, pp. 1520–1528 (2015)

16. Badrinarayanan, V., Kendall, A., Cipolla, R.: SegNet: a deep convolutional encoder-decoder architecture for image segmentation. arXiv preprint arXiv:1511.00561 (2015)

17. Mnih, V., Hinton, G.E.: Learning to detect roads in high-resolution aerial images. In: Daniilidis, K., Maragos, P., Paragios, N. (eds.) ECCV 2010. LNCS, vol. 6316, pp. 210–223. Springer, Heidelberg (2010). doi:10.1007/978-3-642-15567-3_16

18. Penatti, O., Nogueira, K., Dos Santos, J.: Do deep features generalize from everyday objects to remote sensing and aerial scenes domains? In: Proceedings of the IEEE Conference on Computer Vision and Pattern Recognition Workshops, pp. 44–51 (2015)

19. Campos-Taberner, M., Romero-Soriano, A., Gatta, C., Camps-Valls, G., Lagrange, A., Le Saux, B., Beaupère, A., Boulch, A., Chan-Hon-Tong, A., Herbin, S., Randrianarivo, H., Ferecatu, M., Shimoni, M., Moser, G., Tuia, D.: Processing of extremely high-resolution LiDAR and RGB data: outcome of the 2015 IEEE GRSS data fusion contest part A: 2-D contest. IEEE J. Sel. Topics Appl. Earth Obs. Remote Sens. **PP**, 1–13 (2016)

20. Nogueira, K., Penatti, O.A.B., Dos Santos, J.A.: Towards better exploiting convolutional neural networks for remote sensing scene classification. arXiv:1602.01517 [cs] (2016)

21. Zhao, W., Du, S.: Learning multiscale and deep representations for classifying remotely sensed imagery. ISPRS J. Photogrammetry Remote Sens. **113**, 155–165 (2016)

22. Marmanis, D., Wegner, J.D., Galliani, S., Schindler, K., Datcu, M., Stilla, U.: Semantic segmentation of aerial images with an ensemble of CNNs. ISPRS Ann. Photogrammetry Remote Sens. Spat. Inf. Sci. **3**, 473–480 (2016)

23. Gerke, M.: Use of the stair vision library within the ISPRS 2d semantic labeling benchmark (Vaihingen). Technical report, International Institute for Geo-Information Science and Earth Observation (2015)

24. Chatfield, K., Simonyan, K., Vedaldi, A., Zisserman, A.: Return of the devil in the details: delving deep into convolutional nets. In: Proceedings of the British Machine Vision Conference, pp. 6.1–6.12. British Machine Vision Association (2014)

25. Simonyan, K., Zisserman, A.: Very deep convolutional networks for large-scale image recognition. arXiv:1409.1556 [cs] (2014)

26. Ioffe, S., Szegedy, C.: Batch normalization: accelerating deep network training by reducing internal covariate shift. In: Proceedings of the 32nd International Conference on Machine Learning, pp. 448–456 (2015)

27. He, K., Zhang, X., Ren, S., Sun, J.: Delving deep into rectifiers: surpassing human-level performance on imagenet classification. In: Proceedings of the IEEE International Conference on Computer Vision, pp. 1026–1034 (2015)

28. Clevert, D.A., Unterthiner, T., Hochreiter, S.: Fast and accurate deep network learning by exponential linear units (ELUs). In: Proceedings of the International Conference on Learning Representations (2015)

29. Marmanis, D., Datcu, M., Esch, T., Stilla, U.: Deep learning earth observation classification using imagenet pretrained networks. IEEE Geosci. Remote Sens. Lett. **13**, 105–109 (2016)

30. Szegedy, C., Liu, W., Jia, Y., Sermanet, P., Reed, S., Anguelov, D., Erhan, D., Vanhoucke, V., Rabinovich, A.: Going deeper with convolutions. In: Proceedings of the IEEE Conference on Computer Vision and Pattern Recognition, pp. 1–9 (2015)

31. Liao, R., Tao, X., Li, R., Ma, Z., Jia, J.: Video super-resolution via deep draft-ensemble learning. In: Proceedings of the IEEE International Conference on Computer Vision, pp. 531–539 (2015)
32. Liao, Z., Carneiro, G.: Competitive multi-scale convolution. arXiv:1511.05635 [cs] (2015)
33. Ngiam, J., Khosla, A., Kim, M., Nam, J., Lee, H., Ng, A.Y.: Multimodal deep learning. In: Proceedings of the 28th international conference on machine learning (ICML 2011), pp. 689–696 (2011)
34. Eitel, A., Springenberg, J.T., Spinello, L., Riedmiller, M., Burgard, W.: Multimodal deep learning for robust RGB-D object recognition. In: Proceedings of the International Conference on Intelligent Robots and Systems, pp. 681–687. IEEE (2015)
35. Quang, N.T., Thuy, N.T., Sang, D.V., Binh, H.T.T.: An efficient framework for pixel-wise building segmentation from aerial images. In: Proceedings of the Sixth International Symposium on Information and Communication Technology, p. 43. ACM (2015)
36. Boulch, A.: DAG of convolutional networks for semantic labeling. Technical report, Office national d'études et de recherchesaérospatiales (2015)
37. Lin, G., Shen, C., Van Den Hengel, A., Reid, I.: Efficient piecewise training of deep structured models for semantic segmentation. In: Proceedings of the IEEE Conference on Computer Vision and Pattern Recognition (2015)
38. Yosinski, J., Clune, J., Bengio, Y., Lipson, H.: How transferable are features in deep neural networks? In: Advances in Neural Information Processing Systems, pp. 3320–3328 (2014)
39. Cramer, M.: The DGPF test on digital aerial camera evaluation - overview and test design. Photogrammetrie - Fernerkundung - Geoinformation **2**, 73–82 (2010)

Object Boundary Guided
Semantic Segmentation

Qin Huang[✉], Chunyang Xia, Wenchao Zheng, Yuhang Song,
Hao Xu, and C.-C. Jay Kuo

University of Southern California, Los Angeles, CA, USA
qinhuang@usc.edu

Abstract. Semantic segmentation is critical to image content understanding and object localization. Recent development in fully-
convolutional neural network (FCN) has enabled accurate pixel-level
labeling. One issue in previous works is that the FCN based method
does not exploit the object boundary information to delineate segmentation details since the object boundary label is ignored in the network
training. To tackle this problem, we introduce a double branch fully
convolutional neural network, which separates the learning of the desirable semantic class labeling with mask-level object proposals guided by
relabeled boundaries. This network, called object boundary guided FCN
(OBG-FCN), is able to integrate the distinct properties of object shape
and class features elegantly in a fully convolutional way with a designed
masking architecture. We conduct experiments on the PASCAL VOC
segmentation benchmark, and show that the end-to-end trainable OBG-
FCN system offers great improvement in optimizing the target semantic
segmentation quality.

1 Introduction

The convolutional neural network (CNN) has brought a rapid progress in computer vision research and development in recent years [1–3]. Due to the availability of a large amount of image data [4–6], the performance of various CNNs
has been improved significantly. These deep learning based approaches have
been applied to high-level vision challenges such as image recognition and object
detections [1,7–9] and low-level vision problems such as semantic segmentation
[10–12]. The network learns to design tailored feature pools for a vision task
by examining deep features of discriminative properties and shallow features of
local visual patterns.

Recent developments in the fully convolutional neural network (FCN) [10]
have extended CNN's capability from image-level recognition to pixel-level decision. It allows the network to see the object location as well as the object class.
By taking the advantage of low-level pooling features and probability distributions of neighboring contents, recent studies [10–12] have further improved
segmentation accuracy on the PASCAL VOC dataset [4].

© Springer International Publishing AG 2017
S.-H. Lai et al. (Eds.): ACCV 2016, Part I, LNCS 10111, pp. 197–212, 2017.
DOI: 10.1007/978-3-319-54181-5_13

One way to refine segmentation results is to exploit the edge information [13, 14]. The ground truth labels provided by the PASCAL VOC dataset have already offered the object contour information. However, object boundaries and hard cases are marked with the same label. To avoid confusion, both object boundaries and hard cases are ignored during the loss calculation in the training stage.

In this work, we propose an end-to-end fully convolutional neural network, which takes advantage of object boundaries to guide the semantic segmentation. By relabeling the ground truth into three classes (object without class difference, object boundary and background), we first independently train an object boundary prediction FCN (OBP-FCN), which gives us an accurate prior knowledge of object localizations and shape details. This mask-level object proposal, then goes through a designed masking architecture (OBG-Mask), and is later combined with another FCN branch which specifically learn to predict the object classes, formulating the object boundary guided FCN (OBG-FCN). The finalized system is thus able to combine the strengths of two independent pre-trained FCNs and refine the output with the standard back-propagation, as illustrated in Fig. 1.

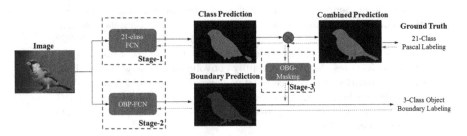

Fig. 1. The proposed fully convolutional OBG-FCN consists of three subnets: the FCN-8s, the Object Boundary Prediction FCN (OBP-FCN) and the Object Boundary Guided Mask (OBG-Mask). In the first stage, the FCN-8s is trained using the ground truth labels and used to predict object classes. In the second stage, we convert the ground truth labels into 3 categories (object without class distinction, background and object boundary) and train the OBP-FCN accordingly. In the third stage, we use the OBG-Mask subnet to pass the boundary information to the result of the FCN-8s to yield the ultimate semantic segmentation result.

We evaluate the performance of the proposed OBG-FCN method on the PASCAL VOC 2011 and 2012 semantic segmentation datasets. It offers great improvement compared with the baseline FCN model on both validation and testing sets. The experimental results demonstrate that object boundaries offer useful information in delineating object details for better semantic segmentation.

The rest of this paper is organized as follows. The related work is reviewed in Sect. 2. The label conversion and the design of the OBP-FCN are discussed in Sect. 3. The full OBG-FCN system is proposed in Sect. 4. The experimental results are presented in Sect. 5. Finally, concluding remarks and future research directions are given in Sect. 6.

2   Related Work

Being apart from the traditional segmentation task [15–17], semantic segmentation demands both pixel-wise accuracy and semantic outputs. Thus, low-level image features and high-level object knowledge have to be integrated to achieve this goal. Deep learning methods have been proposed and proven to be effective for semantic segmentation. In this section, we review several related work along this direction.

Object detection is a topic that is highly related to semantic segmentation. It has been extensively studied using CNNs, e.g., [7,8,18–20]. By predicting object bounding boxes and categories, RCNN [7], SPPnet [20] and Fast RCNN [8] can detect object regions using object proposals. Faster RCNN [9] exploits the shared convolutional features to extract object proposals, leading to a faster inference speed. Masking level proposals can also be extracted in a similar manner by sharing either convolutional features or layer outputs [21–24].

The FCN [10] allows pixel-wise regression. By leveraging the skip architecture [25] to combine the information from pooling layers, the FCN can achieve coarse segmentation with rough object boundaries. MRF/CRF-driven CNN methods have been used to train classifiers and graphical models simultaneously [11,26,27] to further improve detection accuracy and segmentation details. An end-to-end framework has been proposed in [12] to combine the conditional random field with the recursive neural network (RNN) [28] for performance enhancement to refine segmentation details.

Recent developments in instance segmentation demonstrate the advantages of multi-task learning and multi-network assembling. For example, the bounding box locations and object scores are predicted in a fully-convolutional form in [9]. Furthermore, a multi-task network cascades (MNCs) structure is proposed in [29]. This structure utilizes the result of a sub-task as a pixel-level mask to help other subtasks in the network. The network involves several subnetworks (or subnets) and considers their mutual interaction to offer a powerful solution. The network training can be simplified by adopting an independent pre-training procedure for each subnet which is then followed by a dependent learning procedure.

One way particular in multi-tasking learning is to incorporate edge/contour detection with semantic class labeling. Specifically, Bertasius et al. [14] and Chen [13] exploit features from intermediate layers of a deep network and conduct a edge detection sub-task in the similar way of [30]. In [14], Bertasius et al. improves the boundary detection with semantic segmentation, while Chen [13] designs a domain transform structure to conduct an edge-preserving filtering for segmentation.

In comparison with previous related work, we propose a multi-network system that addresses the object boundary detection and the semantic segmentation problem simultaneously. It is shown that an improved object boundary predictor can guide the object labeling task in semantic segmentation.

3   Object Boundary Prediction with OBP-FCN

The FCN in [10] is trained using the PASCAL VOC dataset for the recognition of 20 object classes, which offers good performance since it recognizes patterns of desired classes by examining both coarse-level and fine-level visual features. It generates a blob-wise result to describe the coarse shape of an object and predict its class label. Although the deconvolution layer can partially recover the lost resolution of the input in the pooling layer, its segmented result is still rough and the class label could be wrong as local features can be confusing. Edge detection is conducted in [13] with middle-level features, yet this method detects edges around and inside an object. To enhance the accuracy of segmented object boundaries, we propose a variant of the FCN, called the OBP-FCN, that offers pixel-wise object/boundary prediction in this section.

3.1   Generation of New Labels

We first process the existing labels in the PASCAL-VOC dataset and convert them into a set of new labels. Then, the new labels will be used to train the OBP-FCN for more accurate object mask prediction.

The PASCAL VOC dataset provides labels for object classes and instances as the ground truth. For each image with indexes I, N object classes are labeled in N colors, denoted by $L_c = \{l_1, l_2, \cdots, l_N\}$, where $N = 20$ for the PASCAL VOC dataset. The background area (I_b) is labeled in black, denoted by l_b, and region of the object boundary area and hard cases (I_w) are labeled in white, denoted by l_w, which are usually ignored in the penalty function calculation during the CNN training process.

To recover the accurate location information of object boundaries, we convert the existing PASCAL VOC labels into our desired 4 categories: (1) objects without class distinction (l_o), (2) object boundaries (l_{ob}), (3) background (l_b), and (4) hard cases (l_{hc}).

To begin with, we first derive the object indexes with labels $L_I \in L_c$ as object regions (I_o). We then derive the outline of each object region as the object boundary (I_{ob}). For this purpose, we compare the label of each object pixel with those of its neighbors in a 3×3 window, and label the one without uniform-class neighbor as object boundary. As a result, we can find all pixels that separate different class labeling (object classes as well as background).

Algorithm A. *Label Conversion for the PASCAL VOC dataset.*

▷Initialize all image pixels with the background label, $L_I \leftarrow l_b$
▷Assign object boundary label to extended region, $L_{I_{exb}} \leftarrow l_{ob}$
▷Assign object label to original object region, $L_{I_o} \leftarrow l_o$
▷Label the hard case regions $L_{I_{hc}} \leftarrow l_{hc}$

(a) Image   (b) Original Labels   (c) New Labels

Fig. 2. Illustration of (a) a sample input image; (b) its original labels from the PASCAL VOC dataset; and (c) its new labels with maximum width $w = 4$. (Color figure online)

Then we thicken the boundary region (I_{ob}) into the extended boundary region (I_{exb}) by dilating a pixel to four directions by w pixels. The remaining pixels with original l_w label and not within the thickened boundary region are noted as hard cases (I_{hc}).

After deriving all desirable class regions, we assign the target labels as in Algorithm A. Note that the order of the assignment is very important so as to keep the completeness of object and the accuracy of the boundary.

A sample image, its original and corresponding new labels are shown in Figs. 2(a), (b) and (c), respectively. The white color in Fig. 2(b) represents not only object boundaries, but also occluded objects in the background that are difficult to recognize. In contrast, the new label system marks both the person and the horse in red, the thickened object boundaries in green, the background in black, and the hard cases in olive (or yellow green) in Fig. 2(c). We keep the hard case region and its loss would be ignored during training as in conventional methods.

3.2 Object Boundary Prediction FCN (OBP-FCN)

With the relabeled ground truth, we then train a network that can predict object (without class distinction), boundary and background regions while ignoring the hard case region. The network structure of the proposed network, called the OBP-FCN, is shown in Fig. 3, where its first 5 layers have the same convolution, pooling and ReLU operations as in VGG while the fully connected layers 'fc-6' and 'fc-7' in VGG are replaced by two convolutional layers.

The unique characteristics of the OBP-FCN is that it considers all features of 'pool4', 'pool3' and 'pool2' so as to combine large-scale class knowledge and detail boundary information. To initialize the OBP-FCN, we begin with the VGG network [2] pre-trained on the ImageNet dataset [6]. Then, we use the new 4-category labels to train the OBP-FCN for the desired goal.

We would like to elaborate the importance of thickened object boundaries below. In the traditional FCN, the size of all kernels in VGG-16 convolution layers is 3×3. And with the help of the pooling layer, the gradually growing receptive field of each layer allows the network to see patterns on different scales. The labeled object boundary has an influence on learning local features, and it can force filters to consider its existence at deeper layers. Without the constraint

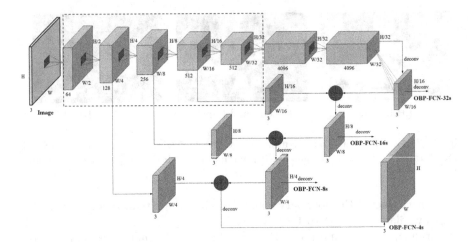

Fig. 3. The proposed OBP-FCN follows the basic structure of the FCN by combining the coarse high level information with the detailed low level information. Deconvolution is applied in upscaling. The response maps of each output are summed element-wise by following a skip scheme. All network models (i.e., OBP-FCN-4s, OBP-FCN-8s, OBP-FCN-16s and OBP-FCN-32s) and their results can be retrieved in each training step. The final detail level is OBP-FCN-4s.

of labeled object boundary, the original FCN network from [10] stops at pool-3 as the performance does not improve furthermore.

In contrast, we observe that the OBP-FCN continues to refine its object boundary detection, benefiting from features in layer pool-2. This is because the labeled maximum boundary width is two or four pixels, which can be seen on smaller scales.

4   Semantic Segmentation with OBG-FCN

In this section, we propose an enhanced semantic segmentation solution, called the object boundary guided FCN (OBG-FCN). The object shape and location information predicted by the OBP-FCN is used as a spatial mask to guide the semantic segmentation task. An overview of the OBG-FCN is given in Sect. 4.1. One important subnet of the OBG-FCN, called the OBG-Mask, is introduced in Sect. 4.2. Some implementation details are discussed in Sect. 4.3.

4.1   Overview of the OBG-FCN

The OBG-FCN system consists of three subnets; namely, FCN-8s, OBP-FCN-4s and OBG-Mask. The evolution of the filter response maps for an exemplary bird image is shown in Fig. 4. Since the output of the OBP-FCN is a 3-category map, we design a masking architecture called the OBG-Mask that passes the trained object shape and localization information to the FCN-8s that segments

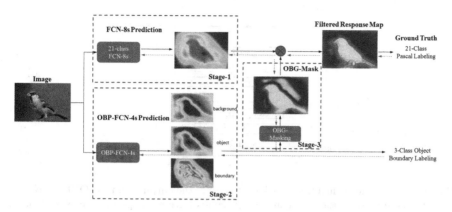

Fig. 4. Evolution of the response maps in the OBG-FCN: (1) the FCN-8s provides a coarse class label for all objects (20 classes); (2) the OBP-FCN-4s indicates the object localization without class distinction; (3) the OBG-Mask network produces an object mask, maps it to the corresponding class label, and yield a more accurate filtered score map; (4) the final output of the OBG-FCN.

21 semantic classes (namely, 20 object classes plus the background). This combination of two branches yields the final 21-class segmentation results. We show the response score maps of three subnets as well as the final output of the OBG-FCN in Fig. 4, which demonstrates the performance has improved significantly by integrating the three subnets.

Multi-task learning is popular in recent CNN-based segmentation methods, where extracted features are shared among multiple tasks. However, when we attempt to share features for the FCN-8s and the OBP-FCN, the training tends to be biased on the FCN-8s sub-branch, resulting in poor performance of the OBP-FCN. For this reason, we train the OBP-FCN separately and adopt the learned filter weights in the OBG-FCN system afterwards.

4.2 OBG-Mask

The main purpose of the OBG-Mask subnet structure is to pass the 3-category (object, boundary and background) labels obtained from the OBP-FCN to the output of the FCN-8s to yield the ultimate output of the whole OBG-FCN system. Specifically, the OBG-Mask subnet first converts the 3-category inference result into 21-class object masks. Then, the mask-level object proposals are combined with the output of FCN-8s via element-wise production.

One example of the OBG-Mask is shown in Fig. 5, which consists of three convolution and rectifier linear unit (ReLU) layers in pair plus the 4th convolution layer. The last convolution layer, conv-m4, does not include the ReLU layer. Its output is integrated with the output from FCN-8s using either element-wise multiplication or summation, which will be further discussed in the experiment section. The parameters of each conv layer are given in the bracket, indicating the number of filters, the number of input channels, the kernel height and width, respectively.

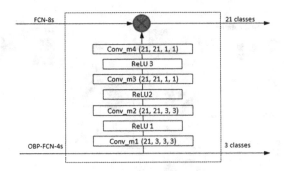

Fig. 5. An example of the OBG-Mask that accepts the output from the OBP-FCN with $w = 2$. The first two cascaded convolution layers have a compound receptive field of 5×5, which is large enough to see two adjacent boundaries. The masking architecture, then accepts the output from the FCN-8s and conducts element-wise multiplication to produce the final 21-class response score map.

The maximum boundary width w in the OBP-FCN determines the kernel size of the convolutional layers. For example, the OBG-Mask shown in Fig. 5, is designed for the case of $w = 2$. The first two cascaded convolution layers have a compound receptive field of 5×5, where two adjacent boundaries can be covered at the same time. In this way, the detected boundary can help improve object or background labeling based on the local region information and back-propagated class information. This is especially beneficial to small objects and complex regions. Although we can benefit from a larger receptive field by increasing the kernal size, this increases training complexity as well. The proposed simple structure is already sufficient to meet our needs.

4.3   Implementation Details

The full OBG-FCN system is designed to integrate the strengths of the FCN-8s and the OBP-FCN. We first discuss its training procedure. Both filter weights of FCN-8s and OBP-FCN subnets are pre-trained and their pre-trained values are used for initialization. For filter weights of the OBG-Mask subnet, we adopt a random initialization scheme.

The performance of the OBG-FCN is sensitive to its learning rate. We adopt three training rates: (1) 10^{-15} for the FCN-8s; (2) 10^{-17} for the OBP-FCN; and (3) 10^{-10} for the OBG-Mark. The training rate of 10^{-15} is commonly used for the FCN training. We adopt a lower training rate for the OBP-FCN to ensure the provided object boundary information is consistent and stable. We adopt a higher training rate for the OBG-Mask so that its weights can be adjusted more aggressively for faster converging.

We trained multiple OBP-FCNs step-by-step: OBP-FCN-32s first, OBP-FCN-16s next, OBP-FCN-8s afterwards, and OBP-FCN-4s last. The network learning rates are fixed at 10^{-10}, 10^{-13}, 10^{-14} and 10^{-15}, respectively. The momentum is set to 0.99 as we use the full image for training with a batch size of 1. For the training of OBP-FCN-32s, the weights of the first 5 convolutional layers are copied from VGG and network surgery is performed to transform the parameters of the original fully-connected form into the fully convolutional form.

The weight decay is set to 0.005 for OBP-FCN and 0.016 for OBG-FCN following the set-up of FCN and CRF-RNN, respectively. We use the standard softmax loss function, which is referred to as the log-likelihood error function in [31], and the ReLU throughout the system for non-linearity.

We implement the OBG-FCN system using the Caffe [32] library. The complete source code and trained models will be available to the public. Each training stage was conducted on a Titan-X graphic card, and the training time varies depending on the number of training images.

5 Experiments

In this section, we evaluate the performance of the proposed OBG-FCN on the PASCAL VOC dataset. We first evaluate the contributions of edge labeling on object inference with the PASCAL VOC 11 dataset. We then follow [12] to train the proposed OBG-FCN framework with both PASCAL VOC 2012 training image and an augmented PASCAL labeling [18]. The performance is evaluated on a non-overlapping subset of the PASCAL VOC 2012 validation image set. Finally, we compare the performance on both the PASCAL 2011 and 2012 test sets by submitting the proposed solution to the evaluation server.

5.1 Performance of OBP-FCN

We first evaluate the impact of object boundary labeling on the accuracy of object region prediction. As mentioned in Sect. 3.1, we can choose any desired maximum width in object boundary relabeling. We conduct experiments with three different maximum boundary widths and compare the performance on the PASCAL VOC 2011. 1112 images are used for the training of OBP-FCN, and 1111 images are used for validation. In order to compare with the original 20 object-class labels, we retrain the original FCN model with 20 object classes using the training data of the PASCAL VOC 2011. Then, we convert the segmentation result into object labels without class distinction for fair comparison.

We evaluate the performance on object region prediction by calculating the accuracy between the predicted object area and the relabeled ground truth. By following [10], four evaluation metrics are used, including pixel accuracy, mean accuracy, mean IU (Intersection over Union) and frequency weighted IU. The relabeled object boundaries are proven to be more effective in predicting objects' detail shapes than the 20-class labeling (Table 1).

We compare object shape prediction results of FCN-8s and OBP-FCN-4s with three maximum width values ($w = 0$, 2 and 4.) on a test image in Fig. 6, where the original input and the ground truth label are also provided. We see that object boundary relabeling and training does help improve object localization in providing more object details and avoiding false alarms. The case $w = 2$ gives

Table 1. Performance on object area prediction with different labeling methods.

	pixel acc.				mean acc.				mean IU				f.w. IU			
	32s	16s	8s	4s	32s	16s	8s	4s	32s	16s	8s	4s	32s	16s	8s	4s
FCN	90.4	90.7	90.9	90.3	83.6	83.9	84.0	82.6	76.8	77.4	77.7	76.2	82.4	82.8	83.1	82.0
0-pixel OBP-FCN	90.7	91.2	91.3	91.0	84.8	86.1	86.4	85.5	77.8	79.1	79.3	78.5	82.9	83.9	84.0	83.5
2-pixel OBP-FCN	**91.4**	**91.9**	**92.0**	**92.0**	**89.3**	**89.0**	**88.8**	**89.0**	**80.5**	**81.1**	**81.2**	**81.3**	**84.6**	**85.2**	**85.3**	**85.4**
4-pixel OBP-FCN	91.1	91.3	91.5	91.4	86.9	87.1	87.1	87.2	79.2	79.6	79.9	79.8	83.7	84.1	84.4	84.3

Input Image   FCN-8s   OBP-FCN-4s (w=0)   OBP-FCN-4s (w=2)   OBP-FCN-4s (w=4)   Ground Truth

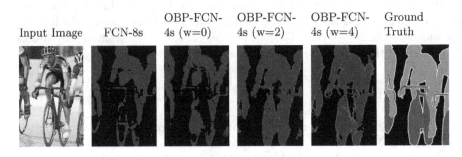

Fig. 6. Performance comparison of FCN-8s and OBP-FCN-4s with three maximum width values. (Color figure online)

the best performance. It is worthwhile to point out that we do not evaluate the performance of the object boundaries prediction since we only use boundaries in the training to help object/background or object/object segmentation. Thus, even if the predicted object boundary (in green color) may not be closed, it still contributes to the completeness of object prediction when all results are integrated in a later stage.

In addition, the results indicate that with $w = 2$, the accuracy of object inference can be further improved even on OBP-FCN-4s. In Fig. 7, we show that by gradually combining the low-level features, the object contours begin to emerge and OBP-FCN-4s gives the best result with natural and sufficient details.

OBP-FCN-32s   OBP-FCN-16s   OBP-FCN-8s   OBP-FCN-4s   New Label

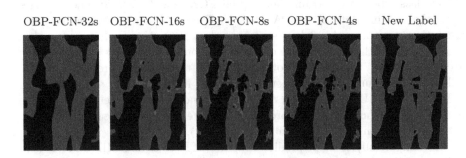

Fig. 7. Performance comparison of four OBP networks with maximum width $w = 2$: OBP-FCN-32s, OBP-FCN-16s, OBP-FCN-8s and OBP-FCN-4s.

5.2 Performance of OBG-FCN

As described in Sect. 3, we train the proposed OBP-FCN-4s with 11,685 relabeled ground truth images, including 1,464 labeled images in the PASCAL VOC 2012 trainging set and the augmented labeled images from [18]. Then, we adopt the pre-trained FCN-8s model from [10] and conduct the end-to-end training to get the final OBG-FCN. Since the labeling of augmented data is not very accurate, we use the full set of 11,685 images to train the OBP-FCN-32s only. This is done because a large amount of image data can provide rich yet coarse information of object features. Then, we train the remaining two subnets of the OBG-FCN with the 1,464 labeled images in the PASCAL VOC 2012 datset. The accurately labeled object boundaries help construct more accurate object shapes, leading to better segmentation results.

Performance on Validation Set. Since there is an overlap between the augmented labeled image set and the PASCAL VOC 12 validation set, we select a list of 346 non-overlapping images from the PASCAL VOC 12 validation set, and evaluate the performance of the proposed OBG-FCN on this subset. The results of the baseline FCN-8s, along with the proposed OBG-FCN with relabeled boundaries of 2-pixel and 4-pixel maximum widths are presented in Table 2. We see that the OBG-FCN with $w = 2$ offers the best results.

Table 2. Performance on selective PASCAL VOC 2012 validation set of OBG-FCN with different labeling pixel widths (w).

	pixel acc	mean acc	mean IU	f.w. IU
FCN-8s	90.1	74.1	61.1	82.7
OBG-FCN ($w = 2$)	**91.6**	**76.4**	**64.9**	**85.3**
OBG-FCN ($w = 4$)	91.5	75.5	64.5	84.9

Exemplary segmentation results of single-class images are shown in Fig. 8. Compared with FCN-8s, more accurate silhouettes are obtained by the OBP-FCN-4s since the relabeled boundary offers extra information to constraint the object. As a result, the OBG-FCN can improve FCN's results by either providing some lost object information (e.g. bird's wing) or correcting some false decisions (e.g. train's tail). The boundary regions are smoother and more natural.

Furthermore, we compare the segmentation results of FCN-8s and OBG-FCN for several exemplary multi-class and multi-object images in Fig. 9. We see that the proposed OBP-FCN-4s can localize some boundaries between concatenated objects (e.g. boundaries between humans in the last row). Generally speaking, the segmentation results of OBG-FCN are better than those of FCN-8s.

Input Image   FCN-8s   OBP-FCN-4s   OBG-FCN   Ground Truth

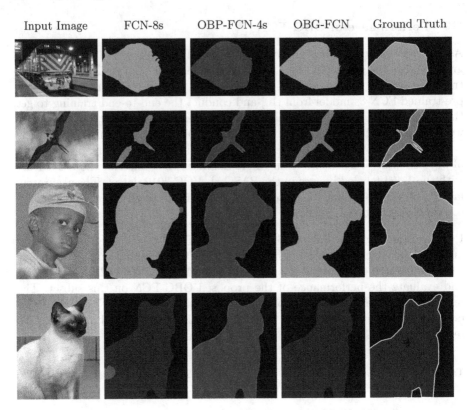

Fig. 8. Visualization of exemplary single-class segmentation results in VOC2012 validation set (from left to right and best viewed in color): input images, intermediate results of FCN-8s, intermediate results of OBP-FCN-4s, and final results of OBG-FCN, and the ground truth labels.

Comparison of Different Structure. As mentioned in Sect. 4.2, the combination of two fully convolutional branches can be either element-wise multiplication or summation. Therefore, we evaluate the performance of these two different settings to show that they bring with similar results. Furthermore, we present another set of results using only OBP-FCN-8s in stage 2 to construct the OBG-FCN system. Results in Table 3 prove that OBP-FCN-4s does provide a better benchmark result with refined object details.

Performance on PASCAL VOC Test Set. We then use the same training data set and evaluate the performance on the PASCAL VOC 2011 and 2012 test sets by submitting the results to the evaluation server. The results are given in Table 4. As shown in this table, the OBG-FCN with $w = 2$ reaches 69.5 % mean IU in VOC 2011 test and 69.1 % in VOC 2012 test, outperforming the baseline FCN by about 7 %. It has also surpassed the previous state-of-the-art methods without relying on conditional random fields.

| Input Image | FCN-8s | OBP-FCN | OBG-FCN | Ground Truth |

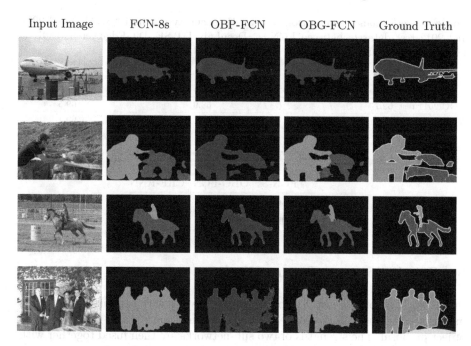

Fig. 9. Examples of multi-class and multi-object segmentation results in VOC 2012 validation set. Best viewed in color. (Color figure online)

Table 3. Comparison of mean IU performance using multiplication or summation for subnet fusion and using the OBP-FCN-4s or the OBP-FCN-8s.

OBG-Mask method	Product		Summation	
	OBP-FCN-8s	OBP-FCN-4s	OBP-FCN-8s	OBP-FCN-4s
Mean IU	64.5	64.8	64.5	64.9

Comparison of Inference Speed. We finally evaluate the inference speed of the proposed framework. As in Table 5, the OBG-FCN takes about 0.187 s/image in average on a Titan X GPU, which offers possibilities for real-time segmentation. The inference time on the Intel Core i7-5930 CPU takes about 5.6 s/image in average. Compared with FCN-8s, the proposed OBG-FCN takes about twice the inference time as we rely on two distinct FCN sub-branches. We further test on a publicly available model of CRF-RNN with 10 mean-filed iterations. And although the CRF-RNN provides the best 72.0 % mean IU in VOC12 test set, its average inference time on GPU is ten times slower than the proposed OBG-FCN.

Table 4. Performance comparison of nean IU accuracy on the PASCAL VOC 2011 and 2012 test datasets between FCN-8s, DeepLab, DT-SE, DT-EdgeNet and OBG-FCN with $w = 2$ and $w = 4$.

	FCN-8s [10]	DeepLab [11]	DT-SE [13]	OBG-FCN (w=4)	DT-EdgeNet [13]	OBG-FCN (w=2)
VOC 2011 test	62.7	/	/	68.9	/	69.5
VOC 2012 test	62.2	65.1	67.8	68.6	69.0	69.1

Table 5. Average inference time (s/image).

	FCN-8s	OBG-FCN	CRF-RNN
CPU time	2.56	5.34	5.21
GPU time	0.09	0.186	1.92

6   Conclusion and Future Work

In this work, we propose a fully-convolutional network with two distinct branches in earlier stage to specifically learn the class information and the mask-level object proposal. The strengths of two sub-networks are then fused together with a proposed OBG-Mask architecture to provide better semantic segmentation results. The method was proven to be effective in generating accurate object localizations and refined object details.

Although the proposed OBP-FCN subnet can provide a better object shape constraint and yield better semantic segmentation results, the final performance is still limited by the accuracy of the 20-class FCN subnet. It is important to find a better baseline in building the full OBG-FCN system. Also, the performance can be further improved if better labeling can be provided for more training images. Generally speaking, semantic segmentation remains to be a challenging problem for further research.

Acknowledgment. Computation for the work described in this paper was supported by the University of Southern California's Center for High-Performance Computing (hpc.usc.edu).

References

1. Krizhevsky, A., Sutskever, I., Hinton, G.E.: Imagenet classification with deep convolutional neural networks. In: Advances in Neural Information Processing Systems, pp. 1097–1105 (2012)
2. Simonyan, K., Zisserman, A.: Very deep convolutional networks for large-scale image recognition. arXiv preprint arxiv:1409.1556 (2014)
3. Szegedy, C., Liu, W., Jia, Y., Sermanet, P., Reed, S., Anguelov, D., Erhan, D., Vanhoucke, V., Rabinovich, A.: Going deeper with convolutions. In: Proceedings of the IEEE Conference on Computer Vision and Pattern Recognition, pp. 1–9 (2015)

4. Everingham, M., Van Gool, L., Williams, C.K.I., Winn, J., Zisserman, A.: The pascal visual object classes (voc) challenge. Int. J. Comput. Vis. **88**, 303–338 (2010)
5. Lin, T.-Y., Maire, M., Belongie, S., Hays, J., Perona, P., Ramanan, D., Dollár, P., Zitnick, C.L.: Microsoft COCO: common objects in context. In: Fleet, D., Pajdla, T., Schiele, B., Tuytelaars, T. (eds.) ECCV 2014. LNCS, vol. 8693, pp. 740–755. Springer, Heidelberg (2014). doi:10.1007/978-3-319-10602-1_48
6. Deng, J., Dong, W., Socher, R., Li, L.J., Li, K., Fei-Fei, L.: Imagenet: a large-scale hierarchical image database. In: IEEE Conference on Computer Vision and Pattern Recognition, CVPR 2009, pp. 248–255. IEEE (2009)
7. Girshick, R., Donahue, J., Darrell, T., Malik, J.: Rich feature hierarchies for accurate object detection and semantic segmentation. In: Proceedings of the IEEE Conference on Computer Vision and Pattern Recognition, pp. 580–587 (2014)
8. Girshick, R.: Fast R-CNN. In: Proceedings of the IEEE International Conference on Computer Vision, pp. 1440–1448 (2015)
9. Ren, S., He, K., Girshick, R., Sun, J.: Faster R-CNN: towards real-time object detection with region proposal networks. In: Advances in Neural Information Processing Systems, pp. 91–99 (2015)
10. Long, J., Shelhamer, E., Darrell, T.: Fully convolutional networks for semantic segmentation. In: Proceedings of the IEEE Conference on Computer Vision and Pattern Recognition, pp. 3431–3440 (2015)
11. Chen, L.C., Papandreou, G., Kokkinos, I., Murphy, K., Yuille, A.L.: Semantic image segmentation with deep convolutional nets and fully connected CRFs. arXiv preprint arxiv:1412.7062 (2014)
12. Zheng, S., Jayasumana, S., Romera-Paredes, B., Vineet, V., Su, Z., Du, D., Huang, C., Torr, P.H.: Conditional random fields as recurrent neural networks. In: Proceedings of the IEEE International Conference on Computer Vision, pp. 1529–1537 (2015)
13. Chen, L.C., Barron, J.T., Papandreou, G., Murphy, K., Yuille, A.L.: Semantic image segmentation with task-specific edge detection using CNNs and a discriminatively trained domain transform. (arXiv preprint arxiv:1511.03328) accepted by CVPR 2016
14. Bertasius, G., Shi, J., Torresani, L.: Semantic segmentation with boundary neural fields. arXiv preprint arxiv:1511.02674 (2015)
15. Zhang, Y.J.: A survey on evaluation methods for image segmentation. Pattern Recogn. **29**, 1335–1346 (1996)
16. Shi, J., Malik, J.: Normalized cuts and image segmentation. IEEE Trans. Pattern Anal. Mach. Intell. **22**, 888–905 (2000)
17. Felzenszwalb, P.F., Huttenlocher, D.P.: Efficient graph-based image segmentation. Int. J. Comput. Vis. **59**, 167–181 (2004)
18. Hariharan, B., Arbeláez, P., Girshick, R., Malik, J.: Simultaneous detection and segmentation. In: Fleet, D., Pajdla, T., Schiele, B., Tuytelaars, T. (eds.) ECCV 2014. LNCS, vol. 8695, pp. 297–312. Springer, Heidelberg (2014). doi:10.1007/978-3-319-10584-0_20
19. Hariharan, B., Arbeláez, P., Girshick, R., Malik, J.: Hypercolumns for object segmentation and fine-grained localization. In: Proceedings of the IEEE Conference on Computer Vision and Pattern Recognition, pp. 447–456 (2015)
20. He, K., Zhang, X., Ren, S., Sun, J.: Spatial pyramid pooling in deep convolutional networks for visual recognition. IEEE Trans. Pattern Anal. Mach. Intell. **37**, 1904–1916 (2015)

21. Arbeláez, P., Pont-Tuset, J., Barron, J., Marques, F., Malik, J.: Multiscale combinatorial grouping. In: Proceedings of the IEEE Conference on Computer Vision and Pattern Recognition, pp. 328–335 (2014)
22. Carreira, J., Sminchisescu, C.: CPMC: automatic object segmentation using constrained parametric min-cuts. IEEE Trans. Pattern Anal. Mach. Intell. **34**, 1312–1328 (2012)
23. Uijlings, J.R., van de Sande, K.E., Gevers, T., Smeulders, A.W.: Selective search for object recognition. Int. J. Comput. Vis. **104**, 154–171 (2013)
24. Dai, J., He, K., Sun, J.: Convolutional feature masking for joint object and stuff segmentation. In: Proceedings of the IEEE Conference on Computer Vision and Pattern Recognition, pp. 3992–4000 (2015)
25. Bishop, C.: Pattern Recognition and Machine Learning (2001)
26. Russell, C., Kohli, P., Torr, P.H., et al.: Associative hierarchical CRFs for object class image segmentation. In: 2009 IEEE 12th International Conference on Computer Vision, pp. 739–746. IEEE (2009)
27. Krähenbühl, P., Koltun, V.: Efficient inference in fully connected CRFs with gaussian edge potentials. arXiv preprint arxiv:1210.5644 (2012)
28. Pinheiro, P.H., Collobert, R.: Recurrent convolutional neural networks for scene parsing. arXiv preprint arxiv:1306.2795 (2013)
29. Dai, J., He, K., Li, Y., Ren, S., Sun, J.: Instance-sensitive fully convolutional networks. arXiv preprint arxiv:1603.08678 (2016)
30. Xie, S., Tu, Z.: Holistically-nested edge detection. In: Proceedings of the IEEE International Conference on Computer Vision, pp. 1395–1403 (2015)
31. Krähenbühl, P., Koltun, V.: Parameter learning and convergent inference for dense random fields. In: Proceedings of the 30th International Conference on Machine Learning (ICML 2013), pp. 513–521 (2013)
32. Jia, Y., Shelhamer, E., Donahue, J., Karayev, S., Long, J., Girshick, R., Guadarrama, S., Darrell, T.: Caffe: convolutional architecture for fast feature embedding. In: Proceedings of the ACM International Conference on Multimedia, pp. 675–678. ACM (2014)

FuseNet: Incorporating Depth into Semantic Segmentation via Fusion-Based CNN Architecture

Caner Hazirbas$^{(\boxtimes)}$, Lingni Ma, Csaba Domokos, and Daniel Cremers

Technical University of Munich, Munich, Germany
{hazirbas,lingni,domokos,cremers}@cs.tum.edu

Abstract. In this paper we address the problem of semantic labeling of indoor scenes on RGB-D data. With the availability of RGB-D cameras, it is expected that additional depth measurement will improve the accuracy. Here we investigate a solution how to incorporate complementary depth information into a semantic segmentation framework by making use of convolutional neural networks (CNNs). Recently encoder-decoder type fully convolutional CNN architectures have achieved a great success in the field of semantic segmentation. Motivated by this observation we propose an encoder-decoder type network, where the encoder part is composed of two branches of networks that simultaneously extract features from RGB and depth images and fuse depth features into the RGB feature maps as the network goes deeper. Comprehensive experimental evaluations demonstrate that the proposed fusion-based architecture achieves competitive results with the state-of-the-art methods on the challenging SUN RGB-D benchmark obtaining 76.27% global accuracy, 48.30% average class accuracy and 37.29% average intersection-over-union score.

1 Introduction

Visual scene understanding in a glance is one of the most amazing capability of the human brain. In order to model this ability, semantic segmentation aims at giving a class label for each pixel on the image according to its semantic meaning. This problem is one of the most challenging tasks in computer vision, and has received a lot of attention from the computer vision community [1–7].

Convolutional neural networks (CNNs) have recently attained a breakthrough in various classification tasks such as semantic segmentation. CNNs have been shown to be powerful visual models that yields hierarchies of features. The key success of this model mainly lies in its general modeling ability for complex visual scenes. Currently CNN-based approaches [3,4,8] provide the state-of-the-art performance in several semantic segmentation benchmarks. In contrast to CNN models, by applying hand-crafted features one can generally achieve rather limited accuracy.

C. Hazirbas and L. Ma—The authors contributed equally.

© Springer International Publishing AG 2017
S.-H. Lai et al. (Eds.): ACCV 2016, Part I, LNCS 10111, pp. 213–228, 2017.
DOI: 10.1007/978-3-319-54181-5_14

Fig. 1. An exemplar output of FuseNet. From left to right: input RGB and depth images, the predicted semantic labeling and the probability of the corresponding labels, where white and blue denote high and low probability, respectively. (Color figure online)

Utilizing depth additional to the appearance information (*i.e.* RGB) could potentially improve the performance of semantic segmentation, since the depth channel has complementary information to RGB channels, and encodes structural information of the scene. The depth channel can be easily captured with low cost RGB-D sensors. In general object classes can be recognized based on their color and texture attributes. However, the auxiliary depth may reduce the uncertainty of the segmentation of objects having similar appearance information. Couprie *et al.* [9] observed that the segmentation of classes having similar depth, appearance and location is improved by making use of the depth information too, but it is better to use only RGB information to recognize object classes containing high variability of their depth values. Therefore, the optimal way to fuse RGB and depth information has been left an open question.

In this paper we address the problem of indoor scene understanding assuming that both RGB and depth information simultaneously available (see Fig. 1). This problem is rather crucial in many perceptual applications including robotics. We remark that although indoor scenes have rich semantic information, they are generally more challenging than outdoor scenes due to more severe occlusions of objects and cluttered background. For example, indoor object classes, such as *chair*, *dining table* and *curtain* are much harder to recognize than outdoor classes, such as *car*, *road*, *building* and *sky*.

The contribution of the paper can be summarized as follows:

- We investigate a solution how to incorporate complementary depth information into a semantic segmentation framework. For this sake we propose an encoder-decoder type network, referred to as FuseNet, where the encoder part is composed of two branches of networks that simultaneously extract features from RGB and depth images and fuse depth features into the RGB feature maps as the network goes deeper (see Fig. 2).
- We propose and examine two different ways for fusion of the RGB and depth channels. We also analyze the proposed network architectures, referred to as dense and sparse fusion (see Fig. 3), in terms of the level of fusion.
- We experimentally show that our proposed method is successfully able to fuse RGB and depth information for semantic segmentation also on cluttered indoor scenes. Moreover, our method achieves competitive results with state-

of-the-art methods in terms of segmentation accuracy evaluated on the challenging SUN RGB-D dataset [10].

2   Related Work

A fully convolutional network (FCN) architecture has been introduced in [3] that combines semantic information from a deep, coarse layer with appearance information from a shallow, fine layer to produce accurate and detailed segmentations by applying end-to-end training. Noh *et al.* [6] have proposed a novel network architecture for semantic segmentation, referred to as DeconvNet, which alleviates the limitations of fully convolutional models (*e.g.*, very limited resolution of labeling). DeconvNet is composed of deconvolution and unpooling layers on top of the VGG 16-layer net [11]. To retrieve semantic labeling on the full image size, Zeiler *et al.* [12] have introduced a network composed of deconvolution and unpooling layers. Concurrently, a very similar network architecture has been presented [13] based on the VGG 16-layer net [11], referred to as *SegNet*. In contrast to DeconvNet, SegNet consists of smoothed unpooled feature maps with convolution instead of deconvolution. Kendall *et al.* [14] further improved the segmentation accuracy of SegNet by applying dropout [15] during test time [16].

Some recent semantic segmentation algorithms combine the strengths of CNN and conditional random field (CRF) models. It has been shown that the poor pixel classification accuracy, due to the invariance properties that make CNNs good for high level tasks, can be overcome by combining the responses of the CNN at the final layer with a fully connected CRF model [8]. CNN and CRF models have also been combined in [4]. More precisely, the method proposed in [4] applies mean field approximation as the inference for a CRF model with Gaussian pairwise potentials, where the mean field approximation is modeled as a recurrent neural network, and the defined network is trained end-to-end refining the weights of the CNN model. Recently, Lin *et al.* [7] have also combined CNN and CRF models for learning patch-patch context between image regions, and have achieved the current state-of-the-art performance in semantic segmentation. One of the main ideas in [7] is to define CNN-based pairwise potential functions to capture semantic correlations between neighboring patches. Moreover, efficient piecewise training is applied for the CRF model in order to avoid repeated expensive CRF inference during the course of back-propagation.

In [2] a feed-forward neural network has been proposed for scene labeling. The long range (pixel) label dependencies can be taken into account by capturing sufficiently large input context patch, around each pixel to be labeled. The method [2] relies on a recurrent convolutional neural networks (RCNN), *i.e.* a sequential series of networks sharing the same set of parameters. Each instance takes as input both an RGB image and the predictions of the previous instance of the network. RCNN-based approaches are known to be difficult to train, in particular, with large data, since long-term dependencies are vanished while the information is accumulated by the recurrence [5].

Byeon *et al.* [5] have presented long short term memory (LSTM) recurrent neural networks for natural scene images taking into account the complex spatial dependencies of labels. LSTM networks have been commonly used for sequence classification. These networks include recurrently connected layers to learn the dependencies between two frames, and then transfer the probabilistic inference to the next frame. This allows to easily memorize the context information for long periods of time in sequence data. It has been shown [5] that LSTM networks can be generalized well to any vision-based task and efficiently capture local and global contextual information with a low computational complexity.

State-of-the-art CNNs have the ability to perform segmentation on different kinds of input sources such as RGB or even RGB-D. Therefore a trivial way to incorporate depth information would be to stack it to the RGB channels and train the network on RGB-D data assuming a four-channel input. However, it would not fully exploit the structure of the scene encoded by the depth channel. This will be also shown experimentally in Sect. 4. By making use of deeper and wider network architecture one can expect the increase of the robustness and the accuracy. Hence, one may define a network architecture with more layers. Nevertheless, this approach would require huge dataset in order to learn all the parameter making the training infeasible even in the case when the parameters are initialized with a pre-trained network.

2.1   The State of the Arts on RGB-D Data

A new representation of the depth information has been presented by Gupta et al. [1]. This representation, referred to as HHA, consists of three channels: disparity, height of the pixels and the angle between of normals and the gravity vector based on the estimated ground floor, respectively. By making use of the HHA representation, a superficial improvement was achieved in terms of segmentation accuracy [1]. On the other hand, the information retrieved only from the RGB channels still dominates the HHA representation. As we shall see in Sect. 4, the HHA representation does not hold more information than the depth itself. Furthermore, computing HHA representation requires high computational cost. In this paper we investigate a better way of exploiting depth information with less computational burden.

Li et al. [17] have introduced a novel LSTM Fusion (LSTM-F) model that captures and fuses contextual information from photometric and depth channels by stacking several convolutional layers and an LSTM layer. The memory layer encodes both short - and long-range spatial dependencies in an image along vertical direction. Moreover, another LSTM-F layer integrates the contexts from different channels and performs bi-directional propagation of the fused vertical contexts. In general, these kinds of architectures are rather complicated and hence more difficult to train. In contrast to recurrent networks, we propose a simpler network architecture.

3   FuseNet: Unified CNN Framework for Fusing RGB and Depth Channels

We aim to solve the semantic segmentation problem on RGB-D images. We define the label set as $\mathcal{L} = \{1, 2, \ldots, K\}$. We assume that we are given a training set $\{(\mathbf{X}_i, \mathbf{Y}_i) \mid \mathbf{X}_i \in \mathbb{R}^{H \times W \times 4}, \mathbf{Y}_i \in \mathcal{L}^{H \times W}$ for all $i = 1, \ldots, M\}$ consisting of M four-channel RGB-D images $(\mathbf{X}_i)$, having the same size $H \times W$, along with the ground-truth labeling $(\mathbf{Y}_i)$. Moreover, we assume that the pixels are drawn as *i.i.d.* samples following a categorical distribution. Based on this assumption, we may define a CNN model to perform multinomial logistic regression.

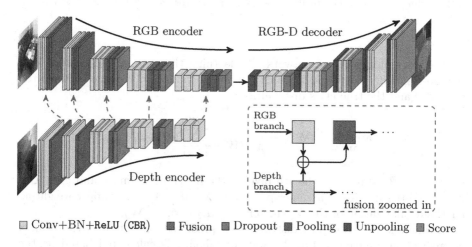

□ Conv+BN+ReLU (CBR)   ■ Fusion   ■ Dropout   ■ Pooling   ■ Unpooling   ■ Score

Fig. 2. The architecture of the proposed FuseNet. Colors indicate the layer type. The network contains two branches to extract features from RGB and depth images, and the feature maps from depth is constantly fused into the RGB branch, denoted with the red arrows. In our architecture, the fusion layer is implemented as an element-wise summation, demonstrated in the dashed box. (Color figure online)

The network extracts features from the input layer and through filtering provides classification score for each label as an output at each pixel. We model the network as a composition of functions corresponding to L layers with parameters denoted by $\mathbf{W} = [\mathbf{w}^{(1)}, \mathbf{w}^{(2)}, \ldots, \mathbf{w}^{(L)}]$, that is

$$f(\mathbf{x}; \mathbf{W}) = g^{(L)}(g^{(L-1)}(\cdots g^{(2)}(g^{(1)}(\mathbf{x}; \mathbf{w}^{(1)}); \mathbf{w}^{(2)}) \cdots ; \mathbf{w}^{(L-1)}); \mathbf{w}^{(L)}). \quad (1)$$

The classification score of a pixel $\mathbf{x}$ for a given class c is obtained from the function $f_c(\mathbf{x}; \mathbf{W})$, which is the cth component of $f(\mathbf{x}; \mathbf{W})$. Using the *softmax* function, we can map this score to a probability distribution

$$p(c \mid \mathbf{x}, \mathbf{W}) = \frac{\exp\left(f_c(\mathbf{x}; \mathbf{W})\right)}{\sum_{k=1}^{K} \exp\left(f_k(\mathbf{x}; \mathbf{W})\right)}. \quad (2)$$

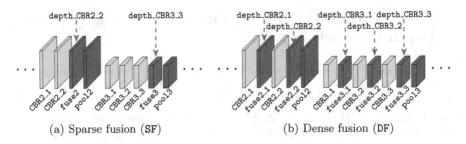

(a) Sparse fusion (SF)   (b) Dense fusion (DF)

Fig. 3. Illustration of different fusion strategies at the second (CBR2) and third (CBR3) convolution blocks of VGG 16-layer net. (a) The fusion layer is only inserted before each pooling layer. (b) The fusion layer is inserted after each CBR block. (Color figure online)

For the training of the network, *i.e.* learning the optimal parameters $\mathbf{W}^*$, the cross-entropy loss is used, which minimizes the KL-divergence between the predicted and the true class distribution:

$$\mathbf{W}^* = \arg\min_{\mathbf{W}} \frac{1}{2}\|\mathbf{W}\|^2 - \frac{\lambda}{MHW} \sum_{i=1}^{M} \sum_{j=1}^{HW} \log p(y_{ij} \mid \mathbf{x}_{ij}, \mathbf{W}),$$

where $\mathbf{x}_{ij} \in \mathbb{R}^4$ stands for the jth pixel of the ith training image and $y_{ij} \in \mathcal{L}$ is its ground-truth label. The hyper-parameter $\lambda > 0$ is chosen to apply weighting for the regularization of the parameters (*i.e.* L_2-norm of $\mathbf{W}$).

At inference, a probability distribution is predicted for each pixel via softmax normalization, defined in Eq. (2), and the labeling is calculated based on the highest class probability.

3.1   FuseNet Architecture

We propose an encoder-decoder type network architecture as shown in Fig. 2. The proposed network has two major parts: (1) the *encoder* part extracts features and (2) the *decoder* part upsamples the feature maps back to the original input resolution. This encoder-decoder style has been already introduced in several previous works such as DeconvNet [6] and SegNet [13] and has achieved good segmentation performance. Although our proposed network is based on this type of architecture, we further consider to have two encoder branches. These two branches extract features from RGB and depth images. We note that the depth image is normalized to have the same value range as color images, *i.e.* into the interval of [0,255]. In order to combine information from both input modules, we fuse the feature maps from the depth branch into the feature maps of the RGB branch. We refer to this architecture as *FuseNet* (see Fig. 2).

The encoder part of FuseNet resembles the 16-layer VGG net [11], except of the fully connected layers fc6, fc7 and fc8, since the fully connected layers reduce the resolution with a factor of 49, which increases the difficulty of the upsampling part. In our network, we always use batch normalization (BN) after convolution (Conv) and before rectified linear unit[1] (ReLU) to reduce the internal covariate shift [18]. We refer to the combination of convolution, batch normalization and ReLU as CBR block, respectively. The BN layer first normalizes the feature maps to have zero-mean and unit-variance, and then scales and shifts them afterwards. In particular, the scale and shift parameters are learned during training. As a result, color features are not overwritten by depth features, but the network learns how to combine them in an optimal way.

The decoder part is a counterpart of the encoder part, where memorized unpooling is applied to upsample the feature maps. In the decoder part, we again use the CBR blocks. We also did experiments with deconvolution instead of convolution, and observed very similar performance. As proposed in [14], we also apply dropout in both the encoder and the decoder parts to further boost the performance. However, we do not use dropout during test time.

The key ingredient of the FuseNet architecture is the fusion block, which combines the feature maps of the depth branch and the RGB branch. The fusion layer is implemented as element-wise summation. In FuseNet, we always insert the fusion layer after the CBR block. By making use of fusion the discontinuities of the features maps computed on the depth image are added into the RGB branch in order to enhance the RGB feature maps. As it can be observed in many cases, the features in the color domain and in the geometric domain complement each other. Based on this observation, we propose two fusion strategies: (a) dense fusion (DF), where the fusion layer is added after each CBR block of the RGB branch. (b) sparse fusion (SF), where the fusion layer is only inserted before each pooling. These two strategies are illustrated in Fig. 3.

3.2 Fusion of Feature Maps

In this section, we reason the fusion of the feature maps between the RGB and the depth branches. To utilize depth information a simple way would be just stacking the RGB and depth images into a four-channel input. However, we argue that by fusing RGB and depth information the feature maps are usually more discriminant than the ones obtained from the stacked input.

As we introduced before in Eq. (1), each layer is modeled as a function g that maps a set of input $\mathbf{x}$ to a set of output $\mathbf{a}$ with parameter $\mathbf{w}$. We denote the kth feature map in the lth layer by $g_k^{(l)}$. Suppose that the given layer operation consists of convolution and ReLU, therefore

$$\mathbf{x}_k^{(l+1)} = g_k^{(l)}(\mathbf{x}^{(l)}; \mathbf{w}_k^{(l)}) = \sigma(\langle \mathbf{w}_k^{(l)}, \mathbf{x}^{(l)} \rangle + b_k^{(l)}).$$

---

[1] The rectified linear unit is defined as $\sigma(x) = \max(0, x)$.

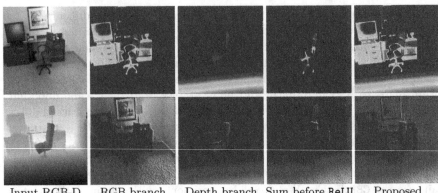

Input RGB-D   RGB branch   Depth branch Sum before ReLU   Proposed

Fig. 4. Comparison of two out of 64 feature maps produced at the CBR1_1 layer. The features from RGB and depth mostly compensate each other, where the textureless region usually have rich structure features and structureless regions usually present texture features. This visually illustrates that the proposed fusion strategy better preserves the informative features from color and depth than applying element-wise summation followed by ReLU. (Color figure online)

If the input is a four-channel RGB-D image, then the feature maps can be decomposed as $\mathbf{x} = [\boldsymbol{a}^\mathsf{T} \; \boldsymbol{b}^\mathsf{T}]^\mathsf{T}$, where $\boldsymbol{a} \in \mathbb{R}^{d_1}$, $\boldsymbol{b} \in \mathbb{R}^{d_2}$ with $d_1 + d_2 = d :=$ $\dim(\mathbf{x})$ are features learned from the color channels and from the depth channel, respectively. According to this observation, we may write that

$$
\begin{aligned}
\mathbf{x}_k^{(l+1)} &= \sigma(\langle \mathbf{w}_k^{(l)}, \mathbf{x}^{(l)} \rangle + b_k^{(l)}) = \sigma(\langle \mathbf{u}_k^{(l)}, \mathbf{a}^{(l)} \rangle + c_k^{(l)} + \langle \mathbf{v}_k^{(l)}, \mathbf{b}^{(l)} \rangle + d_k^{(l)}) \\
&= \max\left(\mathbf{0}, \langle \mathbf{u}_k^{(l)}, \mathbf{a}^{(l)} \rangle + c_k^{(l)} + \langle \mathbf{v}_k^{(l)}, \mathbf{b}^{(l)} \rangle + d_k^{(l)}\right) \\
&\leq \max(\mathbf{0}, \langle \mathbf{u}_k^{(l)}, \mathbf{a}^{(l)} \rangle + c_k^{(l)}) + \max(\mathbf{0}, \langle \mathbf{v}_k^{(l)}, \mathbf{b}^{(l)} \rangle + d_k^{(l)}) \qquad (3) \\
&= \sigma(\langle \mathbf{u}_k^{(l)}, \mathbf{a}^{(l)} \rangle + c_k^{(l)}) + \sigma(\langle \mathbf{v}_k^{(l)}, \mathbf{b}^{(l)} \rangle + d_k^{(l)}),
\end{aligned}
$$

where we applied the decomposition of $\mathbf{w}_k^{(l)} = [\mathbf{u}_k^{(l)^\mathsf{T}} \; \mathbf{v}_k^{(l)^\mathsf{T}}]^\mathsf{T}$ and $b_k^{(l)} = c_k^{(l)} + d_k^{(l)}$.

Based on the inequality in Eq. (3), we show that the fusion of activations of the color and the depth branches (*i.e.* their element-wise summation) produces a stronger signal than the activation on the fused features. Nevertheless, the stronger activation does not necessarily lead to a better accuracy. However, with fusion, we do not only increase the neuron-wise activation values, but also preserve activations at different neuron locations. The intuition behind this can be seen by considering low-level features (*e.g.*, edges). Namely, due to the fact that the edges extracted in RGB and depth images are usually complementary to each other. One may combine the edges from both inputs to obtain more information. Consequently, these low-level features help the network to extract better high-level features, and thus enhance the ultimate accuracy.

To demonstrate the advantage of the proposed fusion, we visualize the feature maps produced by CBR1_1 in Fig. 4, which corresponds to low-level feature extraction (*e.g.*, edges). As it can be seen the low-level features in RGB and depth are usually complementary to each other. For example, the textureless region can be distinguished by its structure, such as the lap against the wall, whereas the structureless region can be distinguished by the color, such as the painting on the wall. While combining the feature maps before the ReLU layer fail to preserve activations, however, the proposed fusion strategy, applied after the ReLU layer, preserves well all the useful information from both branches. Since low-level features help the network to extract better high-level ones, the proposed fusion thus enhances the ultimate accuracy.

4 Experimental Evaluation

In this section, we evaluate the proposed network through extensive experiments. For this purpose, we use the publicly available SUN RGB-D scene understanding benchmark [10]. This dataset contains 10335 synchronized RGB-D pairs, where pixel-wise annotation is available. The standard trainval-test split consists of 5050 images for testing and 5285 images for training/validation. This benchmark is a collection of images captured with different types of RGB-D cameras. The dataset also contains in-painted depth images, obtained by making use of multi-view fusion technique. In the experiments we used the standard training and test split with in-painted depth images. However, we excluded 587 training images that are originally obtained with RealSense RGB-D camera. This is due to the fact that raw depth images from the aforementioned camera consist of many invalid measurements, therefore in-painted depth images have many false values. We remark that the SUN RGB-D dataset is highly unbalanced in terms of class instances, where 16 out of 37 classes rarely present. To prevent the network from over-fitting towards unbalanced class distribution, we weighted the loss for each class with the median frequency class balancing according to [19]. In particular, the class *floormat* and *shower-curtain* have the least frequencies and they are the most challenging ones in the segmentation task. Moreover, approximately 0.25% pixels are not annotated and do not belong to any of the 37 target classes.

Training. We trained the all networks end-to-end. Therefore images were resized to the resolution of 224×224. To this end we applied bilinear interpolation on the RGB images and nearest-neighbor interpolation on the depth images and the ground-truth labeling. The networks were implemented with the Caffe framework [20] and were trained with stochastic gradient descent (SGD) solver [21] using a batch size of 4. The input data was randomly shuffled after each epoch. The learning rate was initialized to 0.001 and was multiplied by 0.9 in every 50,000 iterations. We used a momentum of 0.9 and set weight decay to 0.0005. We trained the networks until convergence, when no further decrease in the loss was observed. The parameters in the encoder part of the network were fine-tuned from the VGG 16-layer model [11] pre-trained on the ImageNet dataset [22]. The original VGGNet requires a three-channel color image. Therefore, for different input dimensions we processed the weights of first layer (*i.e.* conv1_1) as follows:

i) averaged the weights along the channel for a single-channel depth input;
ii) stacked the weights with their average for a four-channel RGB-D input;
iii) duplicated the weights for a six-channel RGB-HHA input.

Testing. We evaluated the results on the original 5050 test images. For quantitative evaluation, we used three criteria. Let TP, FP, FN denote the total number of true positive, false positive, false negative, respectively, and N denotes the total number of annotated pixels. We define the following three criteria:

i) Global accuracy, referred to as *global*, is the percentage of the correctly classified pixels, defined as

$$\text{Global} = \frac{1}{N} \sum_c \text{TP}_c, \quad c \in \{1...K\}.$$

ii) Mean accuracy, referred to as *mean*, is the average of classwise accuracy, defined as

$$\text{Mean} = \frac{1}{K} \sum_c \frac{\text{TP}_c}{\text{TP}_c + \text{FP}_c}.$$

iii) Intersection-over-union (IoU) is average value of the intersection of the prediction and ground truth regions over the union of them, defined as

$$\text{IoU} = \frac{1}{K} \sum_c \frac{\text{TP}_c}{\text{TP}_c + \text{FP}_c + \text{FN}_c}.$$

Among these three measures, the global accuracy is relatively less informative due to the unbalanced class distribution. In general, the frequent classes receive a high score and hence dominate the less frequent ones. Therefore we also measured the average class accuracy and IoU score to provide a better evaluation of our method.

Table 1. Segmentation results on the SUN RGB-D benchmark [10] in comparison to the state of the art. Our methods DF1 and SF5 outperforms most of the existing methods, except of the Context-CRF [7].

	Global	Mean	IoU
FCN-32s [3]	68.35	41.13	29.00
FCN-16s [3]	67.51	38.65	27.15
Bayesian SegNet [14] (RGB)	71.2	45.9	30.7
LSTM [17]	–	48.1	–
Context-CRF [7] (RGB)	78.4	53.4	42.3
FuseNet-**SF5**	76.27	48.30	37.29
FuseNet-**DF1**	73.37	50.07	34.02

4.1 Quantitative Results

In the first experiment, we compared our FuseNet to the state-of-the-art methods. The results are presented in Table 1. We denote the SparseFusion and Dense-Fusion by SF, DF, respectively, following by the number of fusion layers used in the network (*e.g.*, SF5). The results shows that our FuseNet outperforms most of the existing methods with a significant margin. FuseNet is not as competitive in comparison to the Context-CRF [7]. However, it is also worth noting that the Context-CRF trains the network with a different loss function that corresponds to piecewise CRF training. It also requires mean-field approximation at the inference stage, followed by a dense fully connected CRF refinement to produce the final prediction. Applying the similar loss function and post-processing, FuseNet is likely to produce on-par or better results.

In the second experiment, we compare the FuseNet to network trained with different representation of depth, in order to further evaluate the effectiveness of depth fusion and different fusion variations. The results are presented in Table 2. It can be seen that stacking depth and HHA into color gives slight improvements over network trained with only color, depth or HHA. In contrast, with the depth fusion of FuseNet, we improve over a significant margin, in particular with respect to the IoU scores. We remark that the depth fusion is in particular useful as a replacement for HHA. Instead of preprocessing a single channel depth

Table 2. Segmentation results of FuseNet in comparison to the networks trained with RGB, depth, HHA and their combinations. The second part of the table provides the results of variations of FuseNet. We show that FuseNet obtained significant improvements by extracting more informative features from depth.

Input	Global	Mean	IoU
Depth	69.06	42.80	28.49
HHA	69.21	43.23	28.88
RGB	72.14	47.14	32.47
RGB-D	71.39	49.00	31.95
RGB-HHA	73.90	45.57	33.64
FusetNet-SF1	75.48	46.15	35.99
FusetNet-SF2	75.82	46.44	36.11
FusetNet-SF3	76.18	47.10	36.63
FusetNet-SF4	76.56	48.46	37.76
FusetNet-SF5	76.27	48.30	37.29
FusetNet-DF1	73.37	50.07	34.02
FusetNet-DF2	73.31	49.39	33.97
FusetNet-DF3	73.37	49.46	33.52
FusetNet-DF4	72.83	49.53	33.46
FusetNet-DF5	72.56	49.86	33.04

Table 3. Classwise segmentation accuracy of 37 classes. We compare FuseNet-SF5, FuseNet-DF1 to the network trained with stacked RGB-D input.

	wall	floor	cabin	bed	chair	sofa	table	door	wdw	bslf	pic	cnter	blinds
RGB-D	77.19	93.90	**62.51**	74.62	71.22	59.09	**66.76**	42.27	62.73	29.51	64.66	**48.19**	48.80
SF5	**90.20**	**94.91**	61.81	**77.10**	**78.62**	66.49	65.44	46.51	62.44	**34.94**	67.39	40.37	43.48
DF1	82.39	93.88	56.97	73.76	78.02	62.85	60.60	45.43	**67.22**	28.79	**67.50**	39.89	44.73
	desk	shelf	ctn	drssr	pillow	mirror	mat	clthes	ceil	books	fridge	tv	paper
RGB-D	12.12	9.27	63.26	40.44	52.02	52.99	0.00	**38.38**	84.06	**57.05**	**34.90**	45.77	**41.54**
SF5	**25.63**	**20.28**	**65.94**	44.03	54.28	52.47	0.00	25.89	84.77	45.23	34.52	34.83	24.08
DF1	20.98	14.46	61.43	**48.63**	**58.59**	**55.96**	0.00	30.52	**86.23**	53.86	32.31	**53.13**	36.67
	towel	shwr	box	board	person	stand	toilet	sink	lamp	btub	bag	mean	
RGB-D	**27.92**	4.99	**31.24**	69.08	16.97	42.70	76.80	**69.41**	50.28	65.41	24.90	49.00	
SF5	21.05	**8.82**	21.94	57.45	19.06	37.15	76.77	68.11	49.31	73.23	12.62	48.30	
DF1	27.14	1.96	26.61	**66.36**	**30.91**	**43.89**	**81.38**	66.47	**52.64**	**74.73**	**25.80**	**50.07**	

Table 4. Classwise IoU scores of 37 classes. We compare FuseNet-SF5, FuseNet-DF1 to the network trained with stacked RGB-D input.

	wall	floor	cabin	bed	chair	sofa	table	door	wdw	bslf	pic	cnter	blinds
RGB-D	69.46	86.10	35.56	58.29	60.02	43.09	46.37	27.76	43.30	19.70	36.24	25.48	29.11
SF5	**74.94**	**87.41**	**41.70**	**66.53**	**64.45**	**50.36**	**49.01**	**33.35**	**44.77**	**28.12**	**46.84**	**27.73**	**31.47**
DF1	69.48	86.09	35.57	58.27	60.03	43.09	46.38	27.78	43.31	19.75	36.30	25.44	29.12
	desk	shelf	ctn	drssr	pillow	mirror	mat	clths	ceil	books	fridge	tv	paper
RGB-D	10.19	5.34	43.02	23.93	30.70	31.00	0.00	**17.67**	63.10	21.79	22.69	31.31	12.05
SF5	**18.31**	**9.20**	**52.68**	**34.61**	**37.77**	**38.87**	0.00	16.67	**67.34**	**27.29**	31.31	31.64	16.01
DF1	15.61	7.44	42.24	28.74	31.99	34.73	0.00	15.82	60.09	24.28	23.63	**37.67**	**16.45**
	towel	shwr	box	board	person	stand	toilet	sink	lamp	btub	bag	mean	
RGB-D	13.21	4.13	14.21	40.43	10.00	11.79	59.17	45.85	26.06	51.75	12.38	31.95	
SF5	**16.55**	**6.06**	**15.77**	49.23	14.59	**19.55**	**67.06**	**54.99**	**35.07**	**63.06**	9.52	**37.29**	
DF1	13.60	1.54	15.47	**45.21**	**15.49**	17.46	63.38	48.09	27.06	56.85	**12.92**	34.02	

images to obtain hand crafted three-channel HHA representation, FuseNet learns high dimensional features from depth end-to-end, which is more informative as shown by experiments.

In Table 2, we also analyzed the performance of different variations of FuseNet. Since the original VGG 16-layer network has 5 levels of pooling, we increase the number of fusion layers as the network gets deeper. The experiments show that segmentation accuracy gets improved from SF1 to SF5, however the increase appears saturated up to the fusion after the 4th pooling, *i.e.*, SF4. The possible reason behind the accuracy saturation is that depth already provides very distinguished features at low-level to compensate textureless regions in RGB, and we consistently fuse features extracted from depth into the RGB-branch. The same trend can be observed with DF.

In the third experiment, we further compare FuseNet-SF5, FuseNet-DF1 to the network trained with RGB-D input. In Table 3 and 4, we report the classwise accuracy and IoU scores of 37 classes, respectively. For class accuracy, all the three network architectures give very comparable results. However, for IoU scores, SF5 outperforms in 30 out of 37 classes in comparison to other two networks.

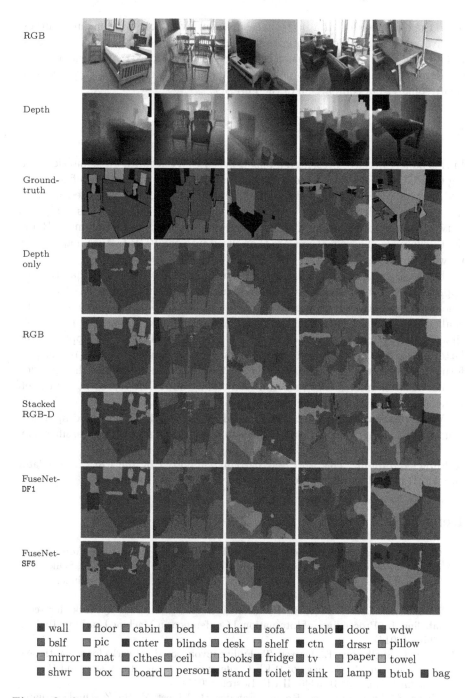

Fig. 5. Qualitative segmentation results for different architectures. The first three rows contain RGB and depth images along with the ground-truth, respectively, followed by the segmentation results. Last two rows contain the results obtained by our DF1 and SF5 approaches. (Color figure online)

Since the classwise IoU is a better measurement over global and mean accuracy, FuseNet obtains significant improvements over the network trained with stacked RGB-D, showing that depth fusion is a better approach to extract informative features from depth and to combine them with color features. In Fig. 5, we demonstrate some visual comparison of the FuseNet.

5   Conclusions

In this paper, we have presented a fusion-based CNN network for semantic labeling on RGB-D data. More precisely, we have proposed a solution to fuse depth information with RGB data by making use of a CNN. The proposed network has an encoder-decoder type architecture, where the encoder part is composed of two branches of networks that simultaneously extract features from RGB and depth channels. These features are then fused into the RGB feature maps as the network goes deeper.

By conducting a comprehensive evaluation, we may conclude that the our approach is a competitive solution for semantic segmentation on RGB-D data. The proposed FuseNet outperforms the current CNN-based networks on the challenging SUN RGB-D benchmark [10]. We have also investigated two possible fusion approaches, *i.e.* dense fusion and sparse fusion. By applying the latter one with a single fusion operation we have obtained a slightly better performance. Nevertheless we may conclude that both fusion approaches provide similar results. Interestingly, we can also claim that HHA representation itself provides a superficial improvement to the depth information.

We also remark that a straight-forward extension of the proposed approach can be applied for other classification tasks such as image or scene classification.

Acknowledgement. This work was partially supported by the ERC Consolidator Grant "3D Reloaded" and by the Alexander von Humboldt Foundation.

References

1. Gupta, S., Girshick, R., Arbeláez, P., Malik, J.: Learning rich features from RGB-D images for object detection and segmentation. In: Fleet, D., Pajdla, T., Schiele, B., Tuytelaars, T. (eds.) ECCV 2014. LNCS, vol. 8695, pp. 345–360. Springer, Cham (2014). doi:10.1007/978-3-319-10584-0_23
2. Pinheiro, P.O., Collobert, R.: Recurrent convolutional neural networks for scene labeling. In: Proceedings of International Conference on Machine Learning, Beijing, China (2014)
3. Long, J., Shelhamer, E., Darrell, T.: Fully convolutional networks for semantic segmentation. In: Proceedings of IEEE Conference on Computer Vision and Pattern Recognition, pp. 3431–3440. IEEE, Boston (2015)
4. Zheng, S., Jayasumana, S., Romera-Paredes, B., Vineet, V., Su, Z., Du, D., Huang, C., Torr, P.: Conditional random fields as recurrent neural networks. In: Proceedings of IEEE International Conference on Computer Vision, pp. 1529–1537. IEEE, Santiago (2015)

5. Byeon, W., Breuel, T.M., Raue, F., Liwicki, M.: Scene labeling with LSTM recurrent neural networks. In: Proceedings of IEEE Conference on Computer Vision and Pattern Recognition, pp. 3547–3555. IEEE, Boston (2015)
6. Noh, H., Hong, S., Han, B.: Learning deconvolution network for semantic segmentation. In: Proceedings of IEEE International Conference on Computer Vision (2015)
7. Lin, G., Shen, C., van den Hengel, A., Reid, I.: Exploring context with deep structured models for semantic segmentation. arXiv preprint arXiv:1603.03183 (2016)
8. Chen, L.C., Papandreou, G., Kokkinos, I., Murphy, K., Yuille, A.L.: Semantic image segmentation with deep convolutional nets and fully connected CRFs. In: Proceedings of International Conference on Learning Representations, San Diego (2015)
9. Couprie, C., Farabet, C., Najman, L., LeCun, Y.: Indoor semantic segmentation using depth information. In: Proceedings of International Conference on Learning Representations (2013)
10. Song, S., Lichtenberg, S.P., Xiao, J.: Sun RGB-D: a RGB-D scene understanding benchmark suite. In: Proceedings of IEEE Conference on Computer Vision and Pattern Recognition, pp. 567–576 (2015)
11. Simonyan, K., Zisserman, A.: Very deep convolutional networks for large-scale image recognition. In: Proceedings of International Conference on Learning Representations (2015)
12. Zeiler, M.D., Fergus, R.: Visualizing and understanding convolutional networks. In: Fleet, D., Pajdla, T., Schiele, B., Tuytelaars, T. (eds.) ECCV 2014. LNCS, vol. 8689, pp. 818–833. Springer, Cham (2014). doi:10.1007/978-3-319-10590-1_53
13. Badrinarayanan, V., Handa, A., Cipolla, R.: SegNet: a deep convolutional encoder-decoder architecture for robust semantic pixel-wise labelling. arXiv preprint arXiv:1505.07293 (2015)
14. Kendall, A., Badrinarayanan, V., Cipolla, R.: Bayesian SegNet: Model uncertainty in deep convolutional encoder-decoder architectures for scene understanding. arXiv preprint arXiv:1511.02680 (2015)
15. Srivastava, N., Hinton, G., Krizhevsky, A., Sutskever, I., Salakhutdinov, R.: Dropout: a simple way to prevent neural networks from overfitting. J. Mach. Learn. Res. **15**, 1929–1958 (2014)
16. Gal, Y., Ghahramani, Z.: Dropout as a Bayesian approximation: Representing model uncertainty in deep learning. Computing Research Repository (2015)
17. Li, L.Z., Yukang, G., Xiaodan, L., Yizhou, Y., Hui, C., Liang, L.: RGB-D Scene labeling with long short-term memorized fusion model. arXiv preprint arXiv:1604.05000v2 (2016)
18. Ioffe, S., Szegedy, C.: Batch normalization: accelerating deep network training by reducing internal covariate shift. In: Bach, F.R., Blei, D.M. (eds.): Proceedings of International Conference on Machine Learning, JMLR Proceedings, vol. 37, pp. 448–456. JMLR.org (2015)
19. Eigen, D., Fergus, R.: Predicting depth, surface normals and semantic labels with a common multi-scale convolutional architecture. In: Proceedings of IEEE International Conference on Computer Vision, pp. 2650–2658 (2015)
20. Jia, Y., Shelhamer, E., Donahue, J., Karayev, S., Long, J., Girshick, R., Guadarrama, S., Darrell, T.: Caffe: Convolutional architecture for fast feature embedding. arXiv preprint arXiv:1408.5093 (2014)

21. Bottou, Léon: Stochastic gradient descent tricks. In: Montavon, Grégoire, Orr, Geneviève, B., Müller, Klaus-Robert (eds.) Neural Networks: Tricks of the Trade. LNCS, vol. 7700, pp. 421–436. Springer, Heidelberg (2012). doi:10.1007/ 978-3-642-35289-8_25
22. Russakovsky, O., Deng, J., Su, H., Krause, J., Satheesh, S., Ma, S., Huang, Z., Karpathy, A., Khosla, A., Bernstein, M., Berg, A.C., Fei-Fei, L.: ImageNet large scale visual recognition challenge. Int. J. Comput. Vis. **115**, 211–252 (2015)

Point-Cut: Interactive Image Segmentation Using Point Supervision

Changjae Oh, Bumsub Ham, and Kwanghoon Sohn[✉]

Yonsei University, Seoul, Republic of Korea
{ocj1211,mimo,khsohn}@yonsei.ac.kr

Abstract. Interactive image segmentation is a fundamental task in many applications in graphics, image processing, and computational photography. Many leading methods formulate elaborated energy functionals, achieving high performance with reflecting human's intention. However, they show limitations in practical usage since user interaction is labor intensive to obtain segments efficiently. We present an interactive segmentation method to handle this problem. Our approach, called point cut, requires minimal point supervision only. To this end, we use off-the-shelf object proposal methods that generate object candidates with high recall. With the single point supervision, foreground appearance can be estimated with high accuracy, and then integrated into a graph cut optimization to generate binary segments. Intensive experiments show that our approach outperforms existing methods for interactive object segmentation both qualitatively and quantitatively.

1 Introduction

Image segmentation labels image pixels into discriminative regions such as an object and background. Unsupervised image segmentation (e.g., [1]) is a fundamental task for computer vision applications, e.g., object proposal [2–6], object tracking [7], and object detection [8]. This unsupervised approach, however, cannot reflect user's intention.

To overcome this limitation, many works have presented using manually provided initial labels, e.g., bounded box [9], snap [10], and scribble [11–14]. In other words, the user roughly labels parts of regions to segment, and the initial labels are automatically propagated to neighborhoods until the problem converges to the optimal solution. Such approaches broaden their applications to image editing [23–25]. Recently, interactive segmentation is incorporated to the framework of deep learning, overcoming the lack of densely annotated dataset in semantic segmentation [26,27].

Existing methods have concentrated on improving the accuracy of image segmentation [11–14]. They require lots of attention to label assignments, and give correct segments only with enough initial labels. This is because appearance models of foreground and background is critical to accurate segmentation. From the practical point of view, estimating foreground and background appearance model effectively and efficiently is important.

© Springer International Publishing AG 2017
S.-H. Lai et al. (Eds.): ACCV 2016, Part I, LNCS 10111, pp. 229–244, 2017.
DOI: 10.1007/978-3-319-54181-5_15

In order to address this problem, we present an interactive image segmentation method, called point-cut, that only uses single point (pixel) of user input as supervision. To this end, we use superpixels to encode spatial smoothness to an input image. The appearance model of foreground is then estimated by using the object proposals that are effectively generated from the superpixels. That is, the foreground appearance model is estimated by scoring the color distance between object proposals and a point seed. The point supervision enables us to easily estimate foreground appearance while suppressing background clutters. We incorporate the proposed appearance model into a graph-cut energy function, and perform segmentation by minimizing the energy function via the max flow/min cut algorithm.

The remainder of this paper is organized as follows. In Sect. 2, we review representative works on interactive image segmentation. The proposed model is then described in Sect. 3. In Sect. 4, we show experimental results, and compare our approach with the state of the art. Finally, we conclude our work with the future works in Sect. 5.

2   Related Works

There exist various interactive segmentation methods including contour-based methods, bounding box-based methods, and stroke-based methods. In this section we review bounding box-based methods and stroke-based approaches, which are most related to our work.

2.1   Bounding Box-Based Approaches

In general, bounding box-based methods estimate the appearance models of foreground and background iteratively, from initially provided bounding box that encloses a foreground object [9, 28–32]. The bounding box captures the spatial location of the foreground object and provides an initial appearance model for foreground (box inside) and background (box outside). The seminal work of [9] uses gaussian mixture models (GMMs) to estimate the foreground and background appearance models, and incorporates them into an energy function. The DenseCut (DC) accelerates this method using a fully connected conditional random field [31]. The PinPoint uses a topology prior from geometric properties of the bounding box, encouraging the segmented object to be tightly enclosed by the bounding box [28]. The OneCut (OC) presents an appearance model that measures L1 distance between the foreground and background histograms, which separates colors effectively [29]. The MILCut considers a segmentation problem as multiple instance learning, where the superpixels outside of the bounding box are regarded as negative samples [30]. The Loose cut relaxes the bounding box constraint, and uses a loosely bounding box to estimate the appearance model [32].

2.2 Stroke-Based Approaches

In stroke-based approaches, initial sparse seeds are provided in foreground and background, and the seed information is then propagated into other parts of regions [11–14].

The graph-cut based method is the seminal work [11]. Given initial strokes (seeds), it estimates the color variation of foreground and background and minimize total edge weights in the cut. However, this causes a small-cut problem, that an isolated region is split into many sub-regions. This problem can be handled by using a random walk (RW) model [12]. In RW, each pixel is regarded as a random walker and computes the probability that the random walker firstly reaches to each seed, avoiding the small-cut problem. However, this model does not segment out highly textured regions and weak boundaries, since the first arrival probability does not consider the relation between the starting location of each random walker and the locations inside the initial seeds. To handle this problem, the random walk with restart (RWR) model [13] considers the relationship between the random walker and all initial seeds. Recently, the Laplacian coordinate (LC) [14] has been presented using a higher-order graph model, which further regularizes segmentation results.

The contributions of our work are as follows: First, unlike existing methods, we perform image segmentation using point supervision while maintaining performance. Second, we use object proposals in interactive segmentation, which enables minimal supervision and alleviates the small-cut problem efficiently. Third, we present intensive experimental results to validate our approach to using object proposals in interactive segmentation.

Our work differs from several similar works, such as object segmentation with saliency [17], high-order cues [18,19], and topological constraints [20]. Saliency map highlights the visually attentive objects that correspond to human attention [16] by capturing various saliency cues. Saliency-based object segmentation, e.g. [17], thus can divide salient object from background, but its performance depends on the accuracy of saliency map and cannot reflect human prior. In [18], unsupervised segmentation is employed as high-order cues to simply facilitate propagation of local information to larger image areas. Appearance models in [19] incorporate edge, region texture and geometric information and divide an object with the level set framework. In [20], object segmentation is addressed by developing contour completion models with initial seeds. Contrary to the previous works, our work estimates foreground appearance by encoding object segments which imply object candidates for effective segmentation.

3 Point-Cut: Object Segmentation via Point Supervision

3.1 Motivation

In object segmentation, the general formulation of an energy function consists of an appearance model and a boundary regularization model [11]. The appearance model is commonly represented based on the intensity distributions,

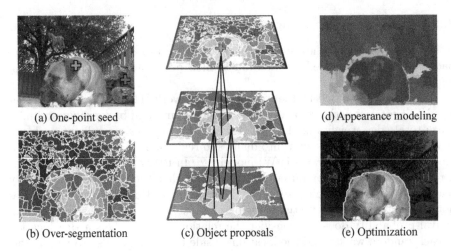

(a) One-point seed

(d) Appearance modeling

(b) Over-segmentation

(c) Object proposals

(e) Optimization

Fig. 1. Overview of the proposed method. (a) Point supervision on input image, (b) over-segmentation, (c) generating object proposals, (d) appearance modeling, and (e) optimization via max flow/min cut algorithm. Green and red crosses in (a,c) denote positive and negative points supervision, respectively, and the color is encoded blue to red as the value becomes higher in (d). (color figure online)

e.g., histogram [11] and Gaussian Mixture models (GMMs) [9], which are estimated from user inputs such as strokes or bounding box. Compared to the histogram, GMMs can better adapt to the colors of the image, while still being effective at capturing small appearance differences between foreground and background. However, GMMs require large amount of training samples to discriminate foreground/background appearance well. It thus requires enough initial labels for accurate segmentation.

To address this problem, we propose an appearance model by employing object proposals that greedily captures both appearance and spatial smoothness in an object. The proposed method only requires single point supervision to foreground and background for accurate segmentation. The overview of the proposed method is described in Fig. 1. First, under a single point of seed to foreground and background (Fig. 1(a)), we take an over-segmentation to form the basis of generating segment-based object proposals (Fig. 1(b)). Under the point supervision, we greedily extract the segments of object proposals which partially include an initial seed (Fig. 1(c)). We then model the foreground appearance which is estimated by accumulating segments of object proposals with color guided weight (Fig. 1(d)). Finally, the proposed appearance model is incorporated to the graph-cut energy function with the max flow/min cut algorithm (Fig. 1(e)).

3.2   Proposed Model

Let us suppose X be a set of binary labels for each pixel, where $x_i = 1$ and $x_i = 0$ if the pixel i is in foreground and background, respectively. We extract

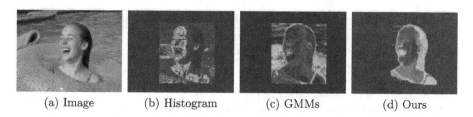

| (a) Image | (b) Histogram | (c) GMMs | (d) Ours |

Fig. 2. Comparison of the initial estimates of foreground appearances. (a) Input image, (b) histogram, (c) GMMs, and (d) the proposed model. Note that bounding-box is provided in (b) and (c), while the point supervision is given in (d). In (b)–(d), the color is encoded blue to red as the value becomes higher. (color figure online)

a set of object proposals $\mathcal{R}$ from an input image, where r_k and l_k in $\mathcal{R}$ are an average color value and a set of pixel indices of k-*th* object proposal (segment), respectively. Noting that any of segment-based object proposals [2,4–6] can be leveraged to our method, here we employed selective search method [3] for generating object proposals without loss of generality. Denoting a set of initial one-point seeds as $s_* = (s_+, s_-)$, where s_+ and s_- are positive and negative seeds, respectively, we estimate foreground appearance by a scoring function in the form:

$$f(x_i) = \begin{cases} \frac{1}{Z} \sum_{k \in \mathcal{N}(s_*)} d(x_i, r_k), & i \in l_k \\ 0, & i \notin l_k \end{cases}, \tag{1}$$

where Z is a normalization factor and $\mathcal{N}(s_*)$ indicates indices of object proposals which include a positive seed s_+, while excluding a negative seed s_-. Here, $d(x_i, r_k)$ measures a color distance between r_k and s_+ by a Gaussian kernel in three dimensional RGB space. For r_k, representative RGB values are computed by averaging values in r_k. By scoring segment-based object proposals as (1), the proposed model effectively captures object appearance which also imposes spatial regularization.

As described in Fig. 2, the proposed appearance model effectively captures foreground appearance under the point supervision. The proposed model can aggregate object-like parts since the task of generating object proposals greedily finds object candidates in high recall rate. Although object proposals generate a large number of outlier regions such as background regions, they can be excluded by simply considering s_*. In addition, since segment-based object proposals generate segments with various sizes, from a superpixel level to an image level, the proposed model naturally captures an object in various scales.

3.3   Graph-Cut Optimization

We then formulate the energy function by employing the proposed model as an appearance term and adding pairwise terms in the form:

$$E(X|\mathcal{R}) = \sum_i E_u(x_i|\mathcal{R}, s_*) + \sum_{i,j \in \mathcal{N}_8} E_p(x_i, x_j) + \sum_k \sum_{i,j \in t_k} E_h(x_i, x_j|\mathcal{T}), \tag{2}$$

Algorithm 1. Object segmentation with the point supervision.
___
1: **Input:** One-point seed s_+ and s_- on input image I.
2: **Output:** Binary labeling X to pixels in I.
3: Obtain initial superpixels t_k using [33].
4: Generate a set of object proposals $\mathcal{R}$.
5: Compute the scoring function f in (1).
6: Minimize E in (2) via the max flow/min cut algorithm.

---

where $\mathcal{N}_8$ is a set of 8 neighboring pixels, and $t_k \in \mathcal{T}$ is k-th super-pixel. Here we set the unary term E_u as an exponential loss of the proposed model as follows:

$$E_u\left(x_i | \mathcal{R}, s_*\right) = \begin{cases} 1 - \exp\left(-f(x_i)\right), & x_i = 1 \\ \exp\left(-f(x_i)\right), & x_i = 0. \end{cases} \tag{3}$$

A pairwise term penalizes the discontinuity between the object and background, which is defined by computing color distance between i and j as follows:

$$E_p\left(x_i, x_j\right) = \exp\left(-\frac{(I_i - I_j)^2}{2\sigma^2}\right), \tag{4}$$

where I_i denotes color values of pixel i, and σ controls the effect of the color difference between i and j.

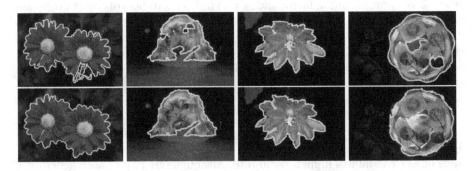

Fig. 3. Effect of the high-order term. (from top to bottom) Segmentation results without and with E_h.

Finally, motivated by [32], we further employ a high-order term E_h, which enforces labeling consistency among pixels in a superpixel t_k as follows:

$$E_h\left(x_i, x_j | \mathcal{T}\right) = \kappa \cdot \delta_k\left(x_i, x_j\right), \tag{5}$$

where κ is a constant that controls the effect of E_h, and $\delta_k\left(x_i, x_j\right) = 1$ for $i, j \in t_k$ and otherwise $\delta_k\left(x_i, x_j\right) = 0$. The high-order term E_h implies that pixels grouped with same segment are likely belong to the same label. As shown in Fig. 3, it alleviates the small-cut problem that usually occurs in graph-cut based segmentation.

Finally, the proposed energy function in (2) is solved via the max flow/min cut algorithm. The overall process is summarized in Algorithm 1.

4   Experimental Results

In order to validate the proposed model, we perform intensive experiments both qualitatively and quantitatively. We analyze the proposed method to describe the effectiveness of object proposals with the point supervision. We then compare our method against several state-of-the-art methods, including stroke-based methods [11,12,14] and bounding box-based methods [9,29,31].

4.1   Experimental Setup

Dataset. In experiments, we use two public benchmarks: the Grabcut dataset [9] and the MSRA-10K dataset [17,35]. Grabcut dataset contains 50 images and their ground truth, where 20 images are from the Berkeley Image Segmentation Benchmark Database [22]. The MSRA-10K dataset contains 10 K images and their ground truth, containing single and multiple objects in the images. From the MSRA-10K dataset, we selected 200 images which contains single object. In other words, totally 250 images are used for performance evaluation.

Parameter Settings. The proposed model has been implemented using Matlab 2014b on a desktop PC with an 3.4 GHz Intel processor. For generating object proposals using the selective search method [3], minimum size of segments is set to 50, 100, 200, and 300, and other parameters are fixed as in [3]. In graph-cut optimization, σ in (4) is set to 0.2, and κ in (5) is set to 0.3. Except for the experiment with varying the number of object proposals, the number of object proposals are set to 300, which shows the best performance. The processing time is about 13 s for a 300×400 image. Among all steps, generating object proposals takes up more than 70%, thus it can be alleviated by changing the number of object proposals.

4.2   Analysis of the Proposed Method

Effectiveness of the Point Supervision. In this section, we analyze the proposed approach, which demonstrates the effectiveness of the point supervision in estimating foreground appearance. As a baseline, as illustrated in Fig. 4(b), we measured scores of all the segments in object proposals that are generated without supervision (*no-supervision*). We also evaluate the results when only s_+ is provided to generate foreground appearance (s_+ *only*).

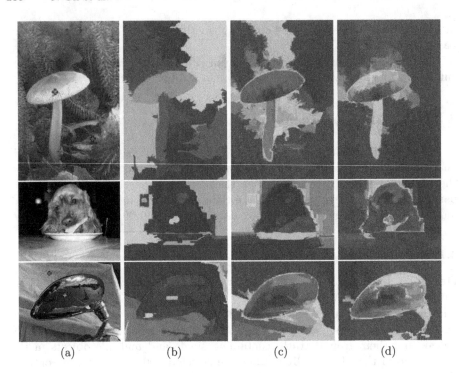

(a)   (b)   (c)   (d)

Fig. 4. Effect of the point supervision. (a) Input image, foreground appearance of (b) no-supervision, (c) s_+ only, and (d) the proposed method. In our method, foreground appearance can be effectively estimated while excluding background.

As shown in Fig. 4, the accuracy of foreground appearance is drastically enhanced when the point supervision is provided. By comparing Fig. 4(c) and (d), background clutter can be excluded by adding a negative seed s_- (*Ours*).

For quantitative comparisons, we show precision-recall (PR) curves in Fig. 5 of each case. Here, the PR curve is estimated by binarizing the final foreground appearance map using different thresholds (typically ranging from 0 to 255). This figure shows that the precision becomes drastically enhanced when the point supervision is added to the given object proposals compared to the baseline method. Since generating object proposals is designed to greedily extract object-like regions, it shows high recall. By adding a seed to the object region, we can easily aggregate parts of the foreground object while excluding redundant proposals, which enhances both precision and recall. Since s_- further excludes background clutters, the precision becomes more enhanced when considering both s_+ and s_- in modeling foreground appearance.

Effect of the Number of Proposals. Here we analyze the relationship between the segmentation accuracy and the number of object proposals. The experiments are performed by setting the number of proposals to 10, 50, 100,

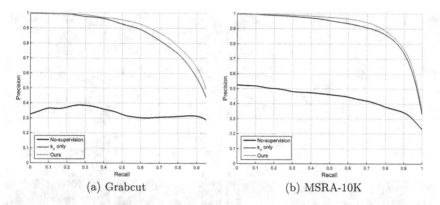

(a) Grabcut   (b) MSRA-10K

Fig. 5. Precision-recall evaluation to the foreground appearance of no-supervision, s_+ only, and ours: (a) Evaluation on the Grabcut dataset and (b) evaluation the MSRA-10K dataset.

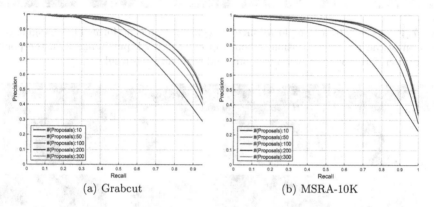

(a) Grabcut   (b) MSRA-10K

Fig. 6. Precision-recall evaluation of the proposed method by changing the number of object proposals: Results on (a) Grabcut and (b) MSRA-10K dataset.

200, and 300. We show PR curves in Fig. 6. Precision increases according to the number of object proposal, since various scales of object candidates are aggregated while excluding background clutters. Our results verify that the proposed approach effectively estimates foreground appearance, enabling accurate segmentation only with the single point supervision.

Robustness to Seeding Location. In order to verify the robustness to seeding location, we perform experiments with varying seed location. As shown in Fig. 7, the proposed method generates consistent results against varying seeding location. It is because the proposed model aggregates object proposals generated in various scales, which enables estimating foreground appearance even with the point supervision.

Fig. 7. Robustness to seeding location. The segmentation is performed by changing the location of the point supervision. (From top to bottom) Input image with the point supervision, foreground appearance map, and segmentation result.

(a)   (b)   (c)   (d)

Fig. 8. Examples of object selection. (a,c) Input images with the point supervision and (b,d) their segmentation results. An object can be selected based on the location of positive and negative seeds.

Seeding Adaptability. Our approach can easily divide a single object from multiple objects having similar appearance as shown in Fig. 8. Since each of object proposal is a subset of an object, we can segment an object part by providing the point supervision adaptively.

Table 1. Descriptions of existing methods

Method	GC	RW	LC	Grabcut	OC	DC	**Ours**
Model	Histogram	L2 distance	L2 distance	GMMs	Histogram	GMMs	**Object proposals**
Label	Stroke	Stroke	Stroke	Box	Box	Box	**Point**
Optimization	Discrete	Continuous	Continuous	Discrete	Discrete	Discrete	**Discrete**

4.3 Comparisons with State-of-the Art Methods

We compare the proposed method with the state-of-the-art: GC [11], RW [12], LC [14], Grabcut [9], OC [29], and DC [31]. The summary of the algorithms are represented in Table 1. Note that foreground/background seeds are provided in GC, RW, and LC, while the bounding box is given Grabcut, OC, and DC.

For a quantitative evaluation, we measure the Dice score [15, 21] between segmentation results S^x and ground truths GT^x for each label $x \in \{0, 1\}$ as follows:

$$Dice(S^x, GT^x) = \frac{2\,|S^x \cap GT^x|}{|S^x| + |GT^x|}, \tag{6}$$

where $|S^x \cap GT^x|$ indicates the overlapping area between the regions S^x and GT^x. We report an average Dice score over all labels.

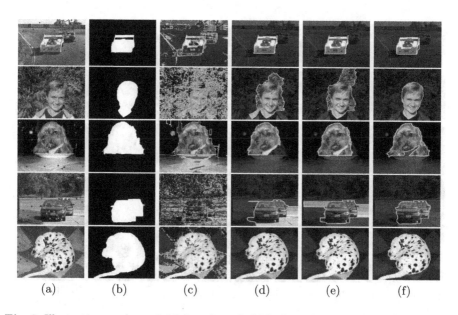

(a)　(b)　(c)　(d)　(e)　(f)

Fig. 9. Illustrative results on initial strokes of which the radius are 5 pixels. Upper two images are from the Grabcut dataset and others are the MSRA-10K dataset. (a) Input, (b) ground truth, (c) GC, (d) RW, (e) LC, and (f) ours.

Stroke-Based Approaches. We compare our method with stroke-based approaches: GC [11], RW [12], and LC [14]. We perform two experiments: stroke-based supervision and the point supervision. For stroke-based supervision, we randomly generated initial strokes similar to [34]. Namely, based on the ground truth, we randomly selected same amount of points on foreground and background regions. The selected pixels are then dilated by the radius of 5 pixels, which is employed as initial strokes. The number of points are set to 10 and 20, respectively. In our method, we set s_+ and s_- to include positive and negative strokes, respectively. For the point supervision, we randomly selected one pixel to each foreground and background region without any dilation.

Figure 9 shows the segmentation results with enough initial labels. It shows that the proposed method outperforms existing methods. GC severely suffers the small-cut problem while the proposed method effectively addresses the problem. Other methods show competitive performance since enough initial labels are provided. However, as presented in Fig. 10, the performance of existing methods are severely degraded in the point supervision since they cannot capture foreground appearance effectively. On the contrary, the proposed method coherently yields highly accurate results. GC still suffers the small-cut problem. RW and LC do not cause small-cut problem since they propagate initial seeds to spatially adjacent neighbors. However, they fail to estimate foreground appearance well when the seeds are not enough to describe foreground and background regions. Namely, GC, RW, and LC requires enough foreground and background

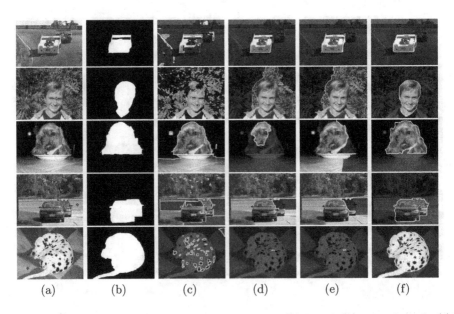

(a)   (b)   (c)   (d)   (e)   (f)

Fig. 10. Illustrative results on the point supervision. (a) Input, (b) ground truth, (c) GC, (d) RW, (e) LC, and (f) Ours. (Best viewed in electronic version.)

Table 2. Dice evaluation with stroke-based methods

	Strokes								Point supervision			
	GC		RW		LC		Ours		GC	RW	LC	**Ours**
	10	20	10	20	10	20	10	20				
Grabcut	0.500	0.518	0.852	0.853	0.904	0.897	0.897	0.901	0.608	0.496	0.577	**0.836**
MSRA-10K	0.632	0.644	0.902	0.916	0.928	0.941	0.940	0.962	0.686	0.548	0.640	**0.875**

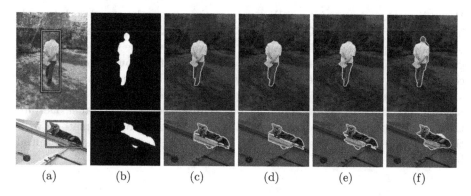

(a)   (b)   (c)   (d)   (e)   (f)

Fig. 11. Illustrative segmentation results of (top) the Grabcut and (bottom) MSRA-10K dataset with the bounding box as initial labels. (a) Input, (b) ground truth, (c) Grabcut, (d) OC, (e) DC, and (f) Ours.

seeds. The quantitative results are presented in Table 2, and demonstrate that the proposed method outperforms existing methods.

We also compare the segmentation result with respect to the dataset. As shown in Figs. 5 and 6 and Table 2, segmentation generally performs better on the MSRA-10K dataset than the Grabcut dataset. It is because the MSRA-10K dataset treats salient object which usually shows distinctive appearance from background region.

Further comparison with varying seeding location is shown in Table 3. One-point seeds were randomly generated in foreground and background regions based on the ground truth labels, and the experiment was repeated twenty times on MSRA-10K and Grabcut datasets. As shown in Table 3, the proposed method generates robust results while seeding location changes.

Table 3. Dice evaluation with varying seeding location

	GC		RW		LC		**Ours**	
	mean	var	mean	var	mean	var	mean	var
Grabcut	0.620	0.052	0.501	0.044	0.550	0.032	**0.818**	**0.020**
MSRA-10K	0.687	0.057	0.550	0.050	0.640	0.041	**0.872**	**0.021**

Table 4. Dice evaluation with bounding box-based methods

	Bounding box				One-point			
	Grabcut	OC	DC	Ours	Grabcut	OC	DC	**Ours**
Grabcut	0.940	0.916	0.903	0.901	-	-	-	**0.836**
MSRA-10K	0.960	0.956	0.939	0.941	-	-	-	**0.875**

Bounding Box Approaches. We compare the proposed method with bounding box-based approaches: Grabcut [9], OC [29], and DC [31]. Since these methods are not applicable under the point supervision, we use the bounding box generated from ground truth that tightly encloses the foreground object. For fair comparison, we use positive and negative initial labels of the bounding box as s_+ and s_-, respectively. As shown in Fig. 11 and Table 4, all methods yield high performance since the tight bounding box is provided as initial label. Despite the lack of supervision, as demonstrated in Table 4, the proposed method with the point supervision yields competitive performance to the results with initial bounding-box labels.

5   Discussions

We have introduced a point-cut that only requires the single point supervision on image segmentation. We design the appearance model of foreground by leveraging segment-based object proposals. Under the guidance of the single point label, located on the object and background, the hierarchically generated object proposals are aggregated to form foreground appearance. It is then incorporated to a graph-cut energy function with a high-order pairwise term. The intensive experimental results have shown that the proposed method is competitive to existing methods which use bounding box or strokes for initial labeling.

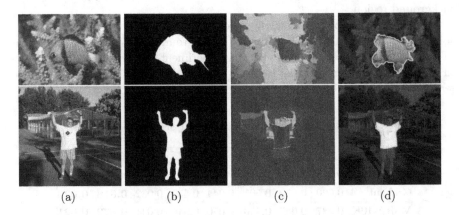

(a)   (b)   (c)   (d)

Fig. 12. Limitations on (top) highly textured and (bottom) human images. (a) Input with the point supervision, (b) ground truth, (c) foreground appearance maps, (d) segmentation results.

Our method uses segment based object proposals, and thus the segmentation performance largely depends on the quality of proposals. For example, our method does not handle highly textured foreground objects as shown in Fig. 12 (top). In addition, the point supervision might neglect parts of an object as shown in Fig. 12 (bottom). These limitations may be addressed by using additional supervision and object prior. For future work, combine our model with deep neural networks to leverage highly discriminative object prior.

Acknowledgement. This work was supported by Institute for Information and communications Technology Promotion (IITP) grant funded by the Korea government (MSIP) (No. R0115-15-1007, High quality 2d-to-multiview contents generation from large-scale RGB+D database).

References

1. Shi, J., Malik, J.: Normalized cuts and image segmentation. IEEE TPAMI **22**(8), 805–888 (2000)
2. Carreira, J., Sminchisescu, C.: CPMC: automatic object segmentation using constrained parametric min-cuts. IEEE TPAMI **34**(7), 1312–1328 (2012)
3. Uijlings, J.R., van de Sande, K.E., Gevers, T., Smeulders, A.W.: Selective search for object recognition. IJCV **104**(2), 154–171 (2013)
4. Manén, S., Guillaumin, M., Gool, L.: Prime object proposals with randomized prim's algorithm. In: ICCV (2013)
5. Arbeláez, P., Pont-Tuset, J., Barron, J., Marques, F., Malik, J.: Multiscale combinatorial grouping. In: CVPR (2014)
6. Krähenbühl, P., Koltun, V.: Geodesic object proposals. In: Fleet, D., Pajdla, T., Schiele, B., Tuytelaars, T. (eds.) ECCV 2014. LNCS, vol. 8693, pp. 725–739. Springer, Heidelberg (2014). doi:10.1007/978-3-319-10602-1_47
7. Ren, X., Malik, J.: Tracking as repeated figure/ground segmentation. In: CVPR (2007)
8. Cinbis, R., Verbeek, J., Schmid, C.: Segmentation driven object detection with Fisher vectors. In: ICCV (2013)
9. Rother, C., Kolmogorov, V., Blake, A.: Grabcut: Interactive foreground extraction using iterated graph cuts. In: ACM SIGGRAPH (2004)
10. Mortensen, E.N., Barret, W.A.: Intelligent scissors for image composition. In: ACM SIGGRAPH (1995)
11. Boykov, Y.Y., Jolly, M.P.: Interactive graph cuts for optimal boundary and region segmentation of objects in N-D images. In: ICCV (2001)
12. Grady, L.: Random walks for image segmentation. IEEE TPAMI **28**(11), 1768–1783 (2006)
13. Kim, T.H., Lee, K.M., Lee, S.U.: Generative image segmentation using random walks with restart. In: Forsyth, D., Torr, P., Zisserman, A. (eds.) ECCV 2008. LNCS, vol. 5304, pp. 264–275. Springer, Heidelberg (2008). doi:10.1007/978-3-540-88690-7_20
14. Casaca, W., Nonato, L.G., Taubin, G.: Laplacian coordinates for seeded image segmentation. In: CVPR (2014)
15. Santner, J., Pock, T., Bischof, H.: Interactive multi-label segmentation. In: Kimmel, R., Klette, R., Sugimoto, A. (eds.) ACCV 2010. LNCS, vol. 6492, pp. 397–410. Springer, Heidelberg (2011). doi:10.1007/978-3-642-19315-6_31

16. Borji, A., Cheng, M.M., Jiang, H., Li, J.: Salient object detection: a benchmark. IEEE TIP **24**(12), 5706–5722 (2015)

17. Cheng, M.M., Mitra, N.J., Huang, X., Torr, P.H., Hu, S.: Global contrast based salient region detection. IEEE TPAMI **37**(3), 569–582 (2015)

18. Kim, T., Lee, K., Lee, S.: Nonparametric higher-order learning for interactive segmentation. In: CVPR (2010)

19. Wang, T., Han, B., Collomosse, J.: TouchCut: fast image and video segmentation using single-touch interaction. CVIU **120**, 14–30 (2014)

20. Xu, J., Collins, M.D., Singh, V.: Incorporating user interaction and topological constraints within contour completion via discrete calculus. In: CVPR (2013)

21. Dice, L.R.: Measures of the amount of ecologic association between species. Ecology **26**(3), 297–302 (1945)

22. Martin, D., Fowlkes, C., Tal, D., Malik, J.: A database of human segmented natural images and its application to evaluating segmentation algorithms and measuring ecological statistics. In: ICCV (2001)

23. Levin, A., Lischinski, D., Weiss, Y.: Colorization using Optimization. In: ACM SIGGRAPH (2004)

24. Chuang, Y.Y., Curless, B., Salesin, D.H., Szeliski, R.: A bayesian approach to digital matting. In: CVPR (2001)

25. An, X., Pellacini, F.: AppProp: all-pairs appearance-space edit propagation. In: ACM SIGGRAPH (2008)

26. Dai, J., He, K., Sun, J.: BoxSup: exploiting bounding boxes to supervise convolutional networks for semantic segmentation. In: ICCV (2015)

27. Lin, D., Dai, J., Jia, J., He, K., Sun, J.: ScribbleSup: scribble-supervised convolutional networks for semantic segmentation. In: CVPR (2016)

28. Lempitsky, V., Kohli, P., Rother, C., Sharp, T.: Image segmentation with a bounding box prior. In: ICCV (2009)

29. Tang, M., Gorelick, L., Veksler, O., Boykov, Y.: Grabcut in one cut. In: ICCV (2013)

30. Wu, J., Zhao, Y., Zhu, J., Luo, S., Tu, Z.: MILCut: A sweeping line multiple instance learning paradigm for interactive image segmentation. In: CVPR (2014)

31. Cheng, M.M., Prisacariu, V.A., Zheng, S., Torr, P.H., Rother, C.: DenseCut: densely connected CRFs for realtime GrabCut. In: Pacific Graphics (2015)

32. Yu, H., Zhou, Y., Qian, H., Xian, M., Lin, Y., Guo, D., Zheng, K., Abdelfatah, K., Wang, S.: LooseCut: interactive image segmentation with loosely bounded boxes. arXiv preprint arXiv:1507.03060 (2015)

33. Felzenszwalb, P.F., Huttenlocher, D.P.: Efficient graph-based image segmentation. IJCV **59**(2), 167–181 (2004)

34. Bai, J., Wu, X.: Error-tolerant scribbles based interactive image segmentation. In: CVPR (2014)

35. Liu, T., Yuan, Z., Sun, J., Wang, J., Zheng, N., Tang, X., Shum, H.Y.: Learning to detect a salient object. IEEE TPAMI **33**(2), 353–367 (2011)

A Holistic Approach
for Data-Driven Object Cutout

Huayong Xu[1], Yangyan Li[3(✉)], Wenzheng Chen[1], Dani Lischinski[2],
Daniel Cohen-Or[3], and Baoquan Chen[1]

[1] Shandong University, Jinan, China
[2] Hebrew University of Jerusalem, Jerusalem, Israel
[3] Tel Aviv University, Tel Aviv, Israel
yangyan.lee@gmail.com

Abstract. Object cutout is a fundamental operation for image editing and manipulation, yet it is extremely challenging to automate it in real-world images, which typically contain considerable background clutter. In contrast to existing cutout methods, which are based mainly on low-level image analysis, we propose a more *holistic* approach, which considers the entire shape of the object of interest by leveraging higher-level image analysis and learnt global shape priors. Specifically, we leverage a deep neural network (DNN) trained for objects of a particular class (chairs) for realizing this mechanism. Given a rectangular image region, the DNN outputs a probability map (P-map) that indicates for each pixel inside the rectangle how likely it is to be contained inside an object from the class of interest. We show that the resulting P-maps may be used to evaluate how likely a rectangle proposal is to contain an instance of the class, and further process good proposals to produce an accurate object cutout mask. This amounts to an automatic end-to-end pipeline for catergory-specific object cutout. We evaluate our approach on segmentation benchmark datasets, and show that it significantly outperforms the state-of-the-art on them.

1 Introduction

Object cutout is a fundamental operation in image editing and manipulation [3,35,37], an operation which graphics artists perform routinely. Performing this operation in a completely automatic fashion involves solving two classical and challenging computer vision tasks: object detection and semantic segmentation. Furthermore, in some scenarios an automatic approach is infeasible, since the user's intent is difficult to predict. Thus, a variety of interactive cutout tools have been proposed over the years, e.g., [22,24,28]. A common approach is to let the user indicate the object of interest with a bounding box, and attempt to proceed automatically from this minimal input to obtain an accurate cutout mask [28].

Electronic supplementary material The online version of this chapter (doi:10. 1007/978-3-319-54181-5_16) contains supplementary material, which is available to authorized users.

© Springer International Publishing AG 2017
S.-H. Lai et al. (Eds.): ACCV 2016, Part I, LNCS 10111, pp. 245–260, 2017.
DOI: 10.1007/978-3-319-54181-5_16

(a)   (b)   (c)   (d)   (e)

Fig. 1. A cluttered scene with four chairs (a); an aggregated P-map visualizing object detection (b); local P-maps inside proposed rectangles (c); cutouts produced with the aid of our local P-maps (d); cutouts produced using GrabCut, for the same rectangles (e).

However, these tasks of detection and segmentation, which the human visual system accomplishes with ease, are notorious for being surprisingly hard for a computer program. They are especially challenging when the object of interest is located in front of a cluttered background, which may contain many distractions, such as other objects with similar low-level statistics to the foreground object. Such an example is demonstrated in Fig. 1(a), where the background contains chairs identical in appearance to those in the foreground.

Methods that are based mainly on low-level image analysis tend to fail when the foreground object and the background are not statistically separable, and when salient separating edges cannot be easily detected. Sparse user input, such as a bounding box [28] or a pair of scribbles [22], is not sufficient to overcome these difficulties, as demonstrated in Fig. 1(e). A more *holistic* approach, which considers the whole shape rather than its pieces by leveraging higher-level image analysis and global object shape priors, has a better chance of coping with these challenging scenarios.

Recent advances in deep neural networks (DNNs) have shown promising results in solving various image understanding tasks, such as classification, detection and segmentation [29]. However, object cutout presents DNNs with three additional challenges. Firstly, the network should learn a large variety of detailed shape priors, which differ significantly among different object classes. Secondly, the solution space is high-dimensional, since the images operated upon and the resulting cutout masks are required to be of high resolution. Thirdly, the cutout masks have sharp boundaries. For example, the state-of-the-art DNN-based instance-level object segmentation approach of [21] achieves 24.5% AP^r at 0.5 IoU^1, on the chair class, which is far from being useful for graphics applications.

---

[1] AP is short for *average precision*, which is the area under precision-recall (PR) curve. IoU is short for Intersection over Union, i.e., $A(P \bigcap G)/A(P \bigcup G)$, where P and G are segmentation prediction and ground truth, respectively, while $A(\bullet)$ indicates their areas. To measure the precision of segmentation, AP^r is used, which is *region* based AP. Here, a segmentation is considered to be positive when it reaches 0.5 IoU.

In this work, we leverage a Convolution-Deconvolution (DeconvNet) DNN [25]. However, we train it specifically using objects of a particular class (chairs). By focusing the training on a particular class, we reduce the learning difficulty and push it to learn more detailed shape priors. Moreover, to provide a more exhaustive coverage of the class in the training phase, we leverage synthetic imagery generated from ShapeNet [33]. Given a rectangular image region, the trained network generates a map (of the same resolution as the input region), where each pixel indicates the likelihood of belonging to the object. We refer to such maps as *P-maps* for short.

We show that the resulting P-maps are useful for a number of vision tasks and applications.

First, given a set of proposals (generated by any state-of-the-art method), we are able to evaluate and rank it better using the P-map. This capability enhances automatic location of chairs in an image. Second, getting back to the original motivation for our work, we are able to use the P-map to guide an iterative graphcut process [28] towards an accurate object cutout (see Fig. 1(d)). Thus, the approach described in this paper amounts to an end-to-end solution for automatic object cutout.

We use chairs as our running example, as they represent a family of shapes that have a rich variability of geometry and topology, and pose a challenge to state-of-the-art DNNs. Our technique is specifically designed to deal with cluttered images, learning to extract the foreground shape from a background that may contain objects with similar local statistics. We show that our holistic shape prior based approach considerably improves the accuracy of the resulting cutouts, compared to the current state-of-the-art, especially for cluttered images.

2   Related Work

Over the past few decades, tremendous amount of research have been devoted to studying how to faithfully perceive objects in images. Significant progress has been made on several sub-tasks towards this goal, including object recognition, object detection, and semantic segmentation, from which still only a coarse understanding of the scene can be established. In this section, we briefly review advances made in these directions and discuss their connections to the task of instance-level object cutout.

Image segmentation is the process of partitioning an image into multiple segments of similar appearance. The problem can be formulated as a clustering problem in color space [4]. To incorporate more spatial constrains into the process, the image may be modeled as a graph, converting image segmentation into a graph partition problem. The weights on the graph edges can either be inferred from pixel colors [10] or from sparse user input, as an addition [28]. Algorithms have been proposed for efficiently computing the partition, even when the pixels are densely connected (DenseCRF) [16]. Such methods are capable of inferring a sharp segmentation mask from sparse of fuzzy probabilities, and thus are widely used as a post-process for methods that produce segmentation probability maps.

Semantic segmentation. Instead of grouping pixels only by appearance, semantic segmentation forms segments by grouping pixels belonging to same semantic objects; thus, a single segment might contain heterogeneous appearances. Since such segmentation depends on semantic understanding of the image content, state-of-the-art methods operate by running classification neural networks on patches densely sampled from the image in order to predict the semantic label of their central pixels [23,26,36]. Instead, Noh et al. [25] proposed a DeconvNet to directly output a high resolution semantic segmentation. We leverage DeconvNet for solving the more challenging object cutout problem by adapting and training it extensively on objects from a specific class.

Object cutout. Object cutout further pushes semantic segmentation from category-level to instance-level. The additional challenge is that objects with similar appearance may hinder the cutout accuracy for individual instances. The state-of-the-art addresses the object cutout problem by solving it jointly with detection [13,21], object number prediction [20], or by explicitly modeling the occlusion interactions between different instances [2,30]. Though significant progress has been made recently, the performance on some object categories is still very low. In this work, we take advantage of being able to utilize training data synthesized from 3D models [31], and focus on leveraging rich holistic shape priors for addressing segmentation ambiguities.

3D object retrieval and view estimation. Recently, exciting advances in image based 3D object retrieval and object view estimation have made [1,19,31]. Such efforts are quite related to object cutout, as the retrieved 3D model can be rendered in the estimated view to approximate the object in the image, thus providing a strong prior for cutout. However, we found that the gap between projected proxies and accurate cutout masks cannot be easily bridged. One reason is that there are only few models in the existing shape databases that match well with real world objects. The inherent mismatch between 3D database and real world objects, plus the introduced retrieval and view estimation errors, render it infeasible to compute object cutout through such an approach, in general cases.

Object detection. Object detection is usually done in two steps: object bounding box proposal generation and proposal evaluation. Proposal generation yields a set of bounding boxes that potentially contain objects [17,34,38]. Proposal evaluation typically extracts features from the image patches contained in the proposed bounding rectangle, and estimates the confidence of the image patches to belong to objects of certain classes. R-CNN [11] is an representative work in object detection and several works extended it to further improve efficiency and accuracy [12,14,27]. We show that the P-maps generated by our category-specific DeconvNet can benefit proposal evaluation for improving object detection and subsequent object cutout.

3 Instance Probability Maps

In this section, we introduce our method for generating instance probability maps. The term "instance" indicates that the maps aim to locate specific instances of a particular object class, rather than only detect the presence of such an object in the image. These probability maps, which will be referred to as P-maps, specify for each pixel its likelihood of belonging to an object instance. As we show in later sections, they allow efficient detection and consequent cutouts of objects, as well as the retrieval of 3D shapes.

Our P-maps are based on the non-trivial observation that although an image of an object may be high-dimensional, the underlying object can often be represented by a compact feature vector. Dosovitskiy et al. [7] show that a DNN can be trained to generate object images from given object type, viewpoint, and color. This raises the expectation that neural networks can detect the presence of an object, encode it into a rather low-dimensional feature vector, from which it then should be possible to "reconstruct" the object, or its binary cutout mask. The premise of this approach is that the extraction of this low-dimensional representation in fact "peels off" the background clutter.

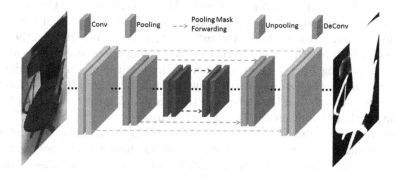

Fig. 2. A schematic illustration of the DeconvNet architecture. Records of the max pooling operations that occur during the first convolutional half, are forwarded to the subsequent deconvolution half of the network.

However, only rather fuzzy images can be reconstructed if the feature vector is extracted from real-world cluttered images, instead of a clean feature vector consisting of object type, viewpoint, and color [8]. To generate a sharper image or cutout mask, additional information must be passed into the reconstruction process, and we build our approach upon the DeconvNet architecture proposed by Noh et al. [25], which we found to be better suited for cluttered scenes. In this network, not only the feature vector, but also additional information about the feature extraction process is forwarded into the reconstruction process, which greatly improves the reconstruction sharpness.

More specifically, the feature extraction part of our network (see Fig. 2) is composed of convolutional layers and pooling layers, which gradually encode the

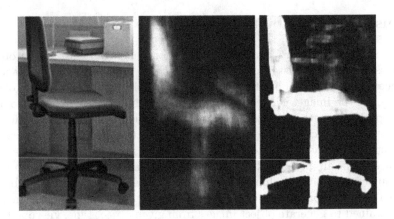

Fig. 3. Given an input image (left), DNN trained extensively with large amount of images from a particular class can learn to "reconstruct" a fuzzy image while ignoring background clutters (middle). Pooling mask forwarding in DeconvNet greatly improves the sharpness of output cutout probability maps (right).

input as a 4096-dimensional feature vector. This feature vector is then taken by the reconstruction part of the network composed of deconvolutional layers and unpooling layers, which gradually reconstruct the P-map. Importantly, the pooling masks, which record the full history of the pooling operations, are forwarded into the unpooling layers. The pooling mask forwarding relieves the difficulty in learning how to perform a sharp reconstruction, thus greatly outperforming approaches that only use the feature vector. See Fig. 3 for a visual comparison of results with and without the use of pooling masks.

The original DeconvNet was proposed for solving a semantic segmentation problem using 21 classes. We adapt it to solve our instance-level segmentation problem by changing its last layer to output only two channel images: one for foreground and one for background. Then a softmax function over these two channels gives the foreground/background probability for each pixel. DeconvNet was originally trained on PASCAL VOC 2012 [9] data, where the number of segmented images is not particularly high, since image segmentation is a hard task for crowd sourcing. When narrowing the data to a specific category, it is insufficient to inject enough shape priors into the trained model. Instead, we choose to train the network using a much larger number of synthetic images with ground truth cutout masks, which are generated completely automatically by rendering 3D models. In the reminder of the paper, we refer our adapted DeconvNet as DNCS (DeconvNet-Class-Specific). As we shall see, the amount and quality of our training data enables the trained network to learn a powerful shape prior, which makes it possible to perform well even in the presence of considerable background clutter.

Fig. 4. Comparison of instance probability maps and the resulting cutout masks generated by various baseline methods and by our approach. It may be seen that our DNCS is more successful at injecting the learnt shape priors into the probability map generation. Furthermore, our GrabCut+P cutout method makes more effective use of the probability maps to produce a cutout, compared to DenseCRF.

4 Proposal Evaluation

The ability to generate high-quality instance probability maps over rectangles of roughly the expected object size in the image is useful not only for generating accurate binary cutout masks (Sect. 5), but also helpful for locating object instances from a given scene image, referred to as *detection task* in computer vision. Given a proposal, we are able to evaluate and rank it better when using the corresponding P-map, thus improve detecting object out of an entire image.

Proposal evaluation on RGB-P images. A proposal is a rectangular region in a large image, which is deemed likely to contain an object of interest. There are many methods that generate proposals, whose objective is to avoid performing an exhaustive search over the entire image. We show that using an RGB-P image, where the fourth P channel is computed by the instance cutout DNN, benefits such proposal evaluation methods. More formally, let I_b be a rectangular proposal, its evaluation by a function $\mathcal{X} : I_b \to \mathcal{R}$, maps the input proposal to a real value that indicates the confidence of having an object of a specific class contained in it. In our case, the function $\mathcal{X}$ is no more than a binary classifier that tells how likely the proposal depicts a chair.

We train the classifier $\mathcal{X}$ with synthetic images, and we generate many rectangular proposals with any state-of-the-art methods. Since in our synthetic images we know the ground truth bounding boxes of the objects, we can easily generate positive and negative examples. We treat proposals with more than 80% overlap with the ground truth bounding boxes as positive samples, and the rest as negative samples. For each proposal, we also compute its P-channel.

We trained two classifiers: $\mathcal{X}_{SVM}$ and $\mathcal{X}_{CNN}$. For $\mathcal{X}_{SVM}$, we extract AlexNet [18] CNN features (4096 dimensions, the output of $fc7$ layer) from the RGB channels, and HoG [6] features (24304 dimensions) from the P-map, which are then reduced with PCA to 4096 dimensions. Then we concatenate the CNN features and the PCA reduced HoG features for training a linear Support Vector Machine (SVM). The $\mathcal{X}_{CNN}$ classifier is an end-to-end CNN approach, where we add a fourth channel to the filters of the first convolutional layer of

a   b   c

Fig. 5. An aggregated P-map for an entire image can be generated by accumulating instance probability map from bounding box proposals (b). By weighting the proposals with $\mathcal{X}_{CNN}$ an even better aggregated P-map can be generated (c).

Fig. 6. P-map enhanced chair detection results. Note that since our P-map "sees" the individual chairs, it can locate chairs well, even with heavy background clutter.

Table 1. IoU comparison of various instance cutout probability map generation methods with various post-processing methods. In *ShapeView*, the cutout probability map is generated by rendering images of similar shape retrieval in estimated viewpoint. *DeconvNet (original)* is the DeconvNet model trained on 21 classes of images from PASCAL VOC 2012. *GrabCut + P* is our method described in Sect. 5.

	GrabCut	ShapeView	DeconvNet (original)	DNCS	
				DenseCRF	GrabCut + P
PASCAL 2012	45.6	46.2	39.8	49.4	**52.1**
Our Benchmark	58.1	63.3	59.6	78.9	81.5

AlexNet to adapt the additonal P-channel, and fine tune the network to work as a binary classifier.

The effect of proposal evaluation is visualized in an aggregated P-map in Fig. 5, where we generate an aggregated P-map, by running instance cutout in all proposals, accumulating the resulting P-maps with weights from the confidences given by $\mathcal{X}_{CNN}$, and normalizing the result. Another example of such a map is shown in Fig. 1(b). It is clear that our P-map enhanced proposal evaluation can greatly narrow down attentions to chair regions. We compare the performance of our two classifiers with versions trained without using the P channel, and found that both classifiers perform better when P channels is used (see Table 2). This is a strong evidence that the P-channels are effectively improving the proposal evaluation. As can be seen from Fig. 6, chairs, even with heavy background clutters can be well located by our P-map powered detection.

5   Cutout Mask Extraction

Given a P-map generated by DNCS within a proposal rectangle, our goal is now to generate a binary cutout mask for the object of interest contained therein. We achieve this goal by adapting the iterative graphcut approach (GrabCut) of Rother et al. [28].

The original GrabCut algorithm uses the bounding rectangle to initialize two GMM color models, one for the background, based on colors outside the rectangle, and one for the foreground, based on colors inside the rectangle. The minimum graphcut is then computed [15], using the two color models to determine the unary (data) term for each pixel. The process is then repeated iteratively using the result from the previous iteration to update the background and foreground GMMs, instead of the initial rectangle.

The above process will generally fail to converge to an accurate cutout mask whenever there is a significant overlap between the background and foreground color models, which will happen if the background contains objects with similar colors to those of the foreground object, as demonstrated in Fig. 1(e). However, armed with our P-map we can initialize the background and foreground color models in a much more precise fashion.

Specifically, we first convert the continuous P-map into an initial binary foreground mask, by computing the minimum graphcut where the unary term at each pixel is determined by our P-map. Denoting by p_i the P-map value of pixel i, we set the foreground likelihood to $P_i^F = p_i^\alpha$ and the background likelihood to $P_i^B = (1 - p_i)^\alpha$, where $\alpha = 2.3$. The resulting binary mask is then used to initialize the two GMM color models, instead of the bounding rectangle. In subsequent iterations, we set the unary term to a weighted combination of the value predicted by the GMM color model and the P-map likelihood, with the latter's weight decreasing as the iterations progress:

$$CP_i^F = GMM_i^F \exp(-wP_i^B)$$
$$CP_i^B = GMM_i^B \exp(-wP_i^F), \tag{1}$$

where GMM^F and GMM^B are the color models for the foreground and background, respectively. The weight $w = b/k$, where k is the iteration number and $b = 25$ was empirically tuned to reduce the influence of P^F and P^B as the iterations progress.

Figure 1(d,e) compares two results produced using our P-map enhanced GrabCut (d) with those of the original GrabCut approach (e). It may be seen that the latter includes in the cutout mask parts of the background which have similar appearance to the foreground chair (in fact, these are parts of identical chairs in the background), while our approach produces a nearly perfect cutout mask.

6   Experiments

In this section, we quantitatively evaluate the performance of our instance cutout approach, and compare it against several other baseline methods. We also quantitatively evaluate the boost in object detection performance enabled by the use of our P-maps.

6.1   Evaluation of Instance Cutout

Dataset and evaluation metric. We evaluate our instance cutout performance on two chair image datasets. One is from PASCAL VOC 2012, which contains

175 chair images with ground truth cutout annotations. We found this dataset to be highly challenging for the cutout task, as it contains not only background clutter, but also heavy occlusion, thus many of the chair instances are only partially visible. Occlusion also makes it more challenging for object detection to providing reasonably good proposals, since a rather complete presence of the object of interest is expected. In addition, we have prepared another benchmark, with 418 chair images, which contains considerable background clutter, but fewer occlusions. We evaluate different approaches using the Intersection over Union (IoU) metric, which measures the ratio between the areas of intersection and union of ground truth and predicted cutout masks. Higher IoU score indicates better cutout accuracy.

Baseline methods. Recent advances in image based 3D object retrieval [19] and object view estimation [31] provide an potential solution for generating an instance probability map, by retrieving similar shapes and rendering them from the predicted viewpoint. The rendered images approximate the underlying object in the input image, and thus can be used as probability maps for instance cutout. More specifically, we pick top 5 retrievals, render them as binary images from the predicted viewpoints, weight the rendered images by the retrieval confidence, and then overlay them into a normalized instance cutout probability map. We refer to this approach as "ShapeView"; see Fig. 7 for examples of the resulting instance probability maps. Another baseline to our approach are the probability maps generated by the original DeconvNet, which was trained for

Fig. 7. Examples of instance probability maps generated by retrieving similar shapes and rendering them from the corresponding predicted viewpoints (ShapeView probability maps).

semantic segmentation with 21 classes. In our comparisons we use GrabCut [28] to generate a cutout mask directly from a given image with a proposal rectangle, while DenseCRF is used for generating a cutout mask from instance probability maps.

We compare our P-map enhanced GrabCut method (Sect. 5) applied on the P-maps generated by DNCS model against the original GrabCut, and DenseCRF applied on probability maps generated using the ShapeView approach and the original DeConvNet. The quantitative results are summarized in Table 1, while Fig. 4 shows a visual comparison using nine examples from our benchmark. Note that our method outperforms the baseline methods on both the PASCAL VOC 2012 dataset (by 5.9%) and on our benchmark (by 18.2%) (see Table 1). The full set of the test images and the results of these methods is included in our supplementary materials. The performance boost on our benchmark is much higher, since our network was trained with synthetic images that exhibit considerable background clutter, but no occlusions. This suggests an interesting future work direction on synthesizing images with realistic occlusion patterns for training occlusion-aware DNNs. Note that the ShapeView baseline method we proposed also consistently outperforms the original DeConvNet. This may be explained by the fact that it is trained on many classes, and thus cannot learn a sufficiently strong shape prior for each class.

6.2 Evaluation of Object Proposal Evaluation

We evaluate the performance of the $\mathcal{X}_{SVM}$ and $\mathcal{X}_{CNN}$ classifiers described in Sect. 4 on 35154 proposals generated by the Selective Search method [34]. These proposals were generated from 52 images from our benchmark, with each of the images containing a single chair. We measure the accuracy by the average recall on positive and negative samples.

We compare our P-map enhanced $\mathcal{X}_{SVM}$ and $\mathcal{X}_{CNN}$ classifiers against those trained without P-maps, and found that the use of P-maps greatly enhances proposal evaluation accuracy, as reported in Table 2. Our experiment suggests that the instance cutout task should be more tightly coupled with object detection tasks, as the improvement in one benefits the other.

Table 2. Object proposal evaluation accuracy of classifiers $\mathcal{X}_{SVM}$ and $\mathcal{X}_{CNN}$ on RGB images and RGB-P images. Augmenting the image with a P-channel boosts the performance of both classifiers.

	RGB images	RGB-P images
$\mathcal{X}_{SVM}$	69.6	87.9
$\mathcal{X}_{CNN}$	65.6	86.5

6.3   Comparison to Seeing 3D Chairs

We also compare chair detection performance based on Selective Search + $\mathcal{X}$ with that proposed in Seeing 3D Chairs [1]. Given an image, Seeing 3D Chairs outputs a ranked list of chair proposals. We generate chair proposals with Selective Search and then rank them with our classifiers. We compare the top-k detection accuracy of these approaches on the first 100 chair images from PASCAL VOC 2012. The results are reported in Table 3. Note that Seeing 3D Chairs is also an approach extensively trained on the chair class, yet we show that our P-map powered approach achieves better accuracy.

Table 3. Comparison of top-k detection accuracy between Seeing 3D Chairs, and our P-map powered detection pipeline.

	Top-1	Top-2	Top-3	Top-4	Top-5
Seeing 3D Chairs	13.86	24.67	28.11	28.78	30.61
Selective Search + $\mathcal{X}_{SVM}$	21.73	28.76	35.49	40.16	43.49
Selective Search + $\mathcal{X}_{CNN}$	20.29	31.58	38.41	44.86	49.37

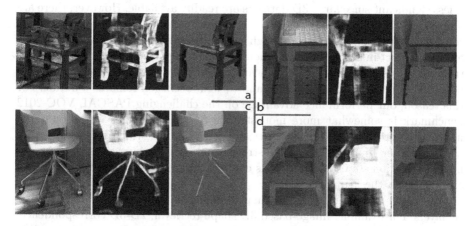

Fig. 8. Failure cases. We found several sources of errors in our cutout masks: (a) Chairs that are rarely seen in training data might be misunderstood by the DNN; (b) Occlusions pose additional challenges over background clutter; (c) The binary mask generation step sometimes eliminates thin structures even though they are preserved in the probability map; (d) Strong similarities between objects might result in highly confusing situation from specific view points.

7   Conclusions

Many computer graphics applications depend on accurate object cutouts. Facilitating automatic cutout extraction remains extremely challenging, since it cannot rely on low-level image analysis alone, and necessarily requires some degree of high-level semantic analysis. The P-maps that we introduced aim to provide some of the latent semantics to assist in the extraction of cutouts. The presented network aims to encode in the P-maps the essence of the shape prior with rich variability of geometry and topology.

The semantic information that P-maps carry was shown to be effective not only directly for cutouts, but also for locating the target object. We have shown that they significantly improve the evaluation of proposals, which are again means to enhance and accelerate a variety of applications that require image analysis.

The claim to fame of the P-maps is their competence to deal with cluttered images, where the target object has "rivals" in its background. Our network was designed explicitly to deal with these types of distractions, and together with our modified GrabCut approach makes a substantial step toward automatic and accurate instance cutout.

Nevertheless, our approach has its limitations. First, it is category specific, and requires training on the target class. It is intensively data-driven, which implies that a large amount of annotated data is required. For chairs, the problem is less significant since large 3D datasets are readily available. However, there are always peculiar shapes (see Fig. 8(a)). For many other object classes there is no comparable availability of rich enough 3D models, yet. Second, the relative size of target object in the input image should be in an expected range, defined by the training set. Arguably, a more significant limitation of our technique is occlusion (see Fig. 8(b)). While cluttering is handled well, occlusion remains a hurdle. For this reason, our performance advantage on the challenging PASCAL VOC 2012 benchmark is somewhat more modest. One of the challenges we encountered in training for occlusion is to realistically synthesize it, which is left for future work. Another limitation is demonstrated in Fig. 8(c), where the final binary mask generation step sometimes fails capture thin structures, even though they are present in the P-map.

We believe that more fundamental processing can benefit from similar semantic layers. For example, image-based 3D shape retrieval, 2D-3D correspondence, or fitting and registering 3D proxies into an image. The P-maps or possibly similar semantic layers have the potential to boost the performance of applications that link 2D to 3D. We would also like to explore the potential of P-maps for enhance other low-level image processing operations, such as edge detection, where the saliency of the edge is augmented or amplified by the P-channel.

Acknowledgement. We would first like to thank all the reviewers for their valuable comments and suggestions. This work is supported in part by grants from National 973 Program (2015CB352501), NSFC-ISF(61561146397), Shenzhen Knowledge innovation program for basic research (JCYJ20150402105524053).

References

1. Aubry, M., Maturana, D., Efros, A., Russell, B.C., Sivic, J.: Seeing 3D chairs: exemplar part-based 2D–3D alignment using a large dataset of CAD models. In: Proceedings of the CVPR, pp. 3762–3769. IEEE (2014)
2. Chen, Y.T., Liu, X., Yang, M.H.: Multi-instance object segmentation with occlusion handling. In: Proceedings of the CVPR, pp. 3470–3478 (2015)
3. Chen, T., Cheng, M.M., Tan, P., Shamir, A., Hu, S.M.: Sketch2photo: Internet image montage. ACM Trans. Graph. **28**, 124:1–124:10 (2009)
4. Comaniciu, D., Meer, P.: Mean shift: a robust approach toward feature space analysis. PAMI **24**, 603–619 (2002)
5. Dai, J., He, K., Sun, J.: Boxsup: Exploiting bounding boxes to supervise convolutional networks for semantic segmentation. In: Proceedings of the ICCV (2015)
6. Dalal, N., Triggs, B.: Histograms of oriented gradients for human detection. In: Proceedings of the CVPR, vol. 1, pp. 886–893. IEEE (2005)
7. Dosovitskiy, A., Springenberg, J.T., Brox, T.: Learning to generate chairs with convolutional neural networks. In: Proceedings of the CVPR, pp. 1538–1546. IEEE (2015)
8. Dosovitskiy, A., Brox, T.: Inverting visual representations with convolutional networks. arXiv preprint arXiv:1506.02753 (2015)
9. Everingham, M., Gool, L., Williams, C.K.I., Winn, J., Zisserman, A.: The pascal visual object classes (voc) challenge. IJCV **88**, 303–338 (2009)
10. Felzenszwalb, P.F., Huttenlocher, D.P.: Efficient graph-based image segmentation. IJCV **59**, 167–181 (2004)
11. Girshick, R., Donahue, J., Darrell, T., Malik, J.: Rich feature hierarchies for accurate object detection and semantic segmentation. In: Proceedings of the CVPR, pp. 580–587. IEEE (2014)
12. Girshick, R.: Fast R-CNN. In: Proceedings of the ICCV (2015)
13. Hariharan, B., Arbeláez, P., Girshick, R., Malik, J.: Simultaneous detection and segmentation. In: Fleet, D., Pajdla, T., Schiele, B., Tuytelaars, T. (eds.) ECCV 2014. LNCS, vol. 8695, pp. 297–312. Springer, Heidelberg (2014). doi:10.1007/978-3-319-10584-0_20
14. He, K., Zhang, X., Ren, S., Sun, J.: Spatial pyramid pooling in deep convolutional networks for visual recognition. In: Fleet, D., Pajdla, T., Schiele, B., Tuytelaars, T. (eds.) ECCV 2014. LNCS, vol. 8691, pp. 346–361. Springer, Heidelberg (2014). doi:10.1007/978-3-319-10578-9_23
15. Kolmogorov, V., Zabih, R.: What energy functions can be minimized via graph cuts? IEEE PAMI **26**, 147–159 (2004)
16. Krähenbühl, P., Koltun, V.: Efficient inference in fully connected CRFs with gaussian edge potentials. In: Shawe-Taylor, J., Zemel, R., Bartlett, P., Pereira, F., Weinberger, K. (eds.) NIPS, pp. 109–117. Curran Associates, Inc. (2011)
17. Krahenbuhl, P., Koltun, V.: Learning to propose objects. In: Proceedings of the CVPR, pp. 1574–1582 (2015)
18. Krizhevsky, A., Sutskever, I., Hinton, G.E.: Imagenet classification with deep convolutional neural networks. In: Proceedings of the NIPS, pp. 1097–1105 (2012)
19. Li, Y., Su, H., Qi, C.R., Fish, N., Cohen-Or, D., Guibas, L.J.: Joint embeddings of shapes and images via CNN image purification. ACM Trans. Graph. **34**(6), 234 (2015)
20. Liang, X., Wei, Y., Shen, X., Yang, J., Lin, L., Yan, S.: Proposal-free network for instance-level object segmentation. arXiv preprint arXiv:1509.02636 (2015)

21. Liang, X., Wei, Y., Shen, X., Jie, Z., Feng, J., Lin, L., Yan, S.: Reversible recursive instance-level object segmentation. arXiv preprint arXiv:1511.04517 (2015)
22. Li, Y., Sun, J., Tang, C.K., Shum, H.Y.: Lazy snapping. ACM Trans. Graph. **23**, 303–308 (2004)
23. Long, J., Shelhamer, E., Darrell, T.: Fully convolutional networks for semantic segmentation. In: Proceedings of the CVPR, pp. 3431–3440 (2015)
24. Mortensen, E.N., Barrett, W.A.: Intelligent scissors for image composition. In: Proceedings of the 22nd Annual Conference on Computer Graphics and Interactive Techniques, SIGGRAPH 1995, pp. 191–198. ACM, New York (1995)
25. Noh, H., Hong, S., Han, B.: Learning deconvolution network for semantic segmentation. In: Proceedings of the ICCV (2015)
26. Papandreou, G., Chen, L.C., Murphy, K., Yuille, A.L.: Weakly-and semi-supervised learning of a DCNN for semantic image segmentation. In: Proceedings of the ICCV (2015)
27. Ren, S., He, K., Girshick, R., Sun, J.: Faster R-CNN: Towards real-time object detection with region proposal networks. In: Proceedings of the NIPS (2015)
28. Rother, C., Kolmogorov, V., Blake, A.: "GrabCut": interactive foreground extraction using iterated graph cuts. ACM Trans. Graph. **23**, 309–314 (2004)
29. Russakovsky, O., Deng, J., Su, H., Krause, J., Satheesh, S., Ma, S., Huang, Z., Karpathy, A., Khosla, A., Bernstein, M., Berg, A., Fei-Fei, L.: Imagenet large scale visual recognition challenge. IJCV **115**, 211–252 (2015)
30. Silberman, N., Sontag, D., Fergus, R.: Instance segmentation of indoor scenes using a coverage loss. In: Fleet, D., Pajdla, T., Schiele, B., Tuytelaars, T. (eds.) ECCV 2014. LNCS, vol. 8689, pp. 616–631. Springer, Heidelberg (2014). doi:10.1007/978-3-319-10590-1_40
31. Su, H., Qi, C.R., Li, Y., Guibas, L.J.: Render for CNN: Viewpoint estimation in images using CNNs trained with rendered 3D model views. In: Proceedings of the ICCV (2015)
32. Su, H., Huang, Q., Mitra, N.J., Li, Y., Guibas, L.: Estimating image depth using shape collections. ACM Trans. Graph. **33**, 37:1–37:11 (2014)
33. Su, H., Yi, E., Savva, M., Chang, A., Song, S., Yu, F., Li, Z., Xiao, J., Huang, Q., Savarese, S., Funkhouser, T., Hanrahan, P., Guibas, L.: Shapenet: an ongoing effort to establish a richly-annotated, large-scale dataset of 3d shapes (2015). http://shapenet.org
34. Uijlings, J.R.R., Sande, K.E.A., Gevers, T., Smeulders, A.W.M.: Selective search for object recognition. IJCV **104**, 154–171 (2013)
35. Xu, K., Zheng, H., Zhang, H., Cohen-Or, D., Liu, L., Xiong, Y.: Photo-inspired model-driven 3D object modeling. ACM Trans. Graph. **30**, 80:1–80:10 (2011)
36. Zheng, S., Jayasumana, S., Romera-Paredes, B., Vineet, V., Su, Z., Du, D., Huang, C., Torr, P.H.S.: Conditional random fields as recurrent neural networks. In: Proceedings of the ICCV (2015)
37. Zheng, Y., Chen, X., Cheng, M.M., Zhou, K., Hu, S.M., Mitra, N.J.: Interactive images: cuboid proxies for smart image manipulation. ACM Trans. Graph. **31**, 1–11 (2012)
38. Zitnick, C.L., Dollár, P.: Edge boxes: locating object proposals from edges. In: Fleet, D., Pajdla, T., Schiele, B., Tuytelaars, T. (eds.) ECCV 2014. LNCS, vol. 8693, pp. 391–405. Springer, Heidelberg (2014). doi:10.1007/978-3-319-10602-1_26

Interactive Segmentation from 1-Bit Feedback

Ding-Jie Chen[(✉)], Hwann-Tzong Chen, and Long-Wen Chang

Department of Computer Science, National Tsing Hua University, Hsinchu, Taiwan
dj_chen_tw@yahoo.com.tw

Abstract. This paper presents an efficient algorithm for interactive image segmentation that responds to 1-bit user feedback. The goal of this type of segmentation is to propose a sequence of yes-or-no questions to the user. Then, according to the 1-bit answers from the user, the segmentation algorithm progressively revises the questions and the segments, so that the segmentation result can approach the ideal region of interest (ROI) in the mind of the user. We define a question as an event that whether a chosen superpixel hits the ROI or not. In general, an interactive image segmentation algorithm is better to achieve high segmentation accuracy, low response time, and simple manipulation. We fulfill these demands by designing an efficient interactive segmentation algorithm from 1-bit user feedback. Our algorithm employs techniques from over-segmentation, entropy calculation, and transductive inference. Over-segmentation reduces the solution set of questions and the computational costs of transductive inference. Entropy calculation provides a way to characterize the query order of superpixels. Transductive inference is used to estimate the similarity between superpixels and to partition the superpixels into ROI and region of uninterest (ROU). Following the clues from the similarity between superpixels, we design the query-superpixel selection mechanism for human-machine interaction. Our key idea is to narrow down the solution set of questions, and then to propose the most informative question based on the clues of the similarities among the superpixels. We assess our method on four publicly available datasets. The experiments demonstrate that our method provides a plausible solution to the problem of interactive image segmentation with merely 1-bit user feedback.

1 Introduction

Image segmentation is a building block of many applications in computer vision and image editing. It is typically used to partition images into several regions and thus to enable the subsequent high-level processing about image structure. However, the regions of interest may have semantic significance, or just have certain homogeneity. Since it is hard to define the region of interest selected by a human user, interactive segmentation provides a solution to this dilemma by invoking the

Electronic supplementary material The online version of this chapter (doi:10.1007/978-3-319-54181-5_17) contains supplementary material, which is available to authorized users.

© Springer International Publishing AG 2017
S.-H. Lai et al. (Eds.): ACCV 2016, Part I, LNCS 10111, pp. 261–274, 2017.
DOI: 10.1007/978-3-319-54181-5_17

aid of the user. In interactive segmentation, the user marks areas of the image as ROI or ROU, then the segmentation algorithm updates the segmentation according to the new marked areas. By iteratively providing more new marked areas, the user can guide the segmentation toward the ROI she or he prefers.

In interactive image segmentation, there are several types of the manipulation mechanisms for human-machine interactions, containing varying degrees of complexity. To a machine, a user can guide the segmentation by providing seed points [3], region selections [4,14,20], line segments [5,7,11,12], bounding boxes [6,17], contours [13,16], image set [18,21] and so on. There is no doubt that the high segmentation accuracy is the basic criterion of a segmentation algorithm. However, the user may be more sensitive to the manipulation mechanism and the response time of a segmentation algorithm. The manipulation mechanism is the aforementioned inputs from the user. The response time is the time a segmentation algorithm takes to react to a given new input. An interactive segmentation algorithm with simpler manipulation and lower response time makes a user more willing to interact with the algorithm.

Recently, Rupprecht *et al.* [19] introduce one kind of interactive image segmentation with very simple manipulation. In this kind of segmentation, the user only needs to decide whether the pixel queried by the machine hits the ROI or not. Their idea is inspired from the classical *Twenty Questions game*[1]. Twenty questions game is a kind of deductive questioning game with multiple players. One player, called *oracle*, selects an object in mind; the other players, called the *inquirers*, try to infer the selected object with a limited amount of questions. Each kind of player obeys one rule to play the twenty questions game. The rule for the oracle is that only 'yes' or 'no' can be used to response to inquirer; the rule for an inquirer is that only the questions that can be answered with 'yes' or 'no' are allowed to be asked. The twenty-questions-style interactive image segmentation proposed by Rupprecht *et al.* can be thus defined as follows. A user, playing the role as the oracle, chooses an ROI in a given image. Then the segmentation algorithm, i.e. the inquirer, proposes one pixel each round to query the user's response to fact that the pixel hits the ROI or not. The algorithm's goal is to infer the ROI in the user's mind with the information from those queried pixels. Since the user only provides 'yes' or 'no' for each question, we call the interaction as *1-bit feedback*. This is the simplest manipulation mechanism for an interactive image segmentation. In this paper, we focus on addressing the interactive image segmentation with 1-bit user feedback.

The interactive segmentation from 1-bit feedback has potential to provide a hands-free segmentation mechanism, because the users do not need to provide any scribbles on specific image locations. For example, in sterilized operating room, the physically touched computer control for medical image segmentation is inappropriate. Or, for instance, tiny screens on the wearable computers have limited interface capabilities. In such scenarios, 1-bit user feedback are more adequate since any input device that receives binary signals can be used to collect the responses.

---

[1] An online Twenty Questions game web site: http://www.20q.net/.

This paper describes an efficient algorithm to address the problem of interactive image segmentation with merely 1-bit user feedback. There are three advantages of the proposed interactive image segmentation method. First, the proposed method has a very simple human-machine interaction mechanism. In our implementation, user only needs to decide if the proposed superpixel hits the ROI or not. Second, the segmentation accuracy of our method is better than the competitor [19] in most datasets. Third, the average response time of our method is extreme short.

2   Related Work

The purpose of an interactive segmentation algorithm is to segment the region of interest in an image with the aid from the user. We briefly categorize some selected methods into two groups according to their interaction modes.

2.1   Passive Interaction Based Image Segmentation

In passive interaction based image segmentation [3,5–7,11–13,16–18,21], an interaction is triggered by the inputs from the user. The user directly defines various inputs to guide the machine to approach the segmentation of ROI.

In general, these algorithms allow the user to specify scribbles via seed points [3], line segments [5,7,11,12], bounding boxes [6,17], or contours [13,16]. Then, the segmentation algorithms minimize their energy functions use *level set, graph cuts, random walks,* or *geodesic distance* to segment the images. The segmentation results are updated according to the modified scribbles. Image co-segmentation [18,21] is a special case of interactive segmentation, which provides a way to implicitly define the region of interest via multiple images. The segmentation results may be different according to the different image sets.

To sum up, the manipulation methods of the passive interaction based image segmentation usually contain varying degrees of complexity. Different scribbles may get very different segmentation results. The user needs to take full responsibility on how to specify good scribbles to guide the segmentation algorithm.

2.2   Active Interaction Based Image Segmentation

Methods in the category of active interaction based image segmentation [4,14,20] usually propose several uncertain image regions to the user, and then the computer updates the segmentation results with the user selected regions.

To segment a lot of images simultaneously, Batra *et al.* [4] propose a co-segmentation algorithm which allows users to indicate where the ROI is. Based on the user indications, their system provides some uncertain image regions to ask for the real labels from the user. Kowdle *et al.* [14] propose an active-learning based segmentation algorithm for 3D reconstruction built on an energy minimization framework. They employs an active learning method to query whether the uncertain regions belong to ROI or not. Straehl *et al.* [20] provide non-local uncertainty measurements to suggest the uncertain regions for the user, and then apply the watershed cut to segment the large data sets guided by the user-selected locations. Rupprecht *et al.* [19] simulate the segmentation probability among the entire image by sampled segmentations. According to the probability distribution over the set of sampled segmentations, they pick the region that has the highest uncertainty to ask for the real label.

Comparing with the passive interaction based image segmentation, the active interaction based image segmentation has simpler human-machine interaction but unsatisfactory segmentation accuracy. This kind of segmentation algorithm needs to estimate the segmentation uncertainty of the entire image, and based on the estimation, it can query the user for reducing the segmentation uncertainty.

2.3 Yes-or-no Interaction

The most similar work to ours is the algorithm proposed by Rupprecht *et al.* [19]. Their segmentation model is based on the notion of the twenty questions game. In twenty questions game, the effective strategy for the inquirer is to come up with a question that can eliminate half the *all possible answers* at each iteration. In this way, a series of 20 questions can allow the inquirer to distinguish between 2^{20} potential answers. Based on this strategy, Rupprecht *et al.* use the MCMC sampling method to approximate the probability of *all possible segmentations*. Since the probability distributed among the entire image, they directly select the centroid pixel of the most uncertain region as the question.

In contrast to their method, we first reduce the solution set of all possible segmentations by over-segmentation. Then, we use transductive inference technique to directly explore the most informative superpixel for querying. Finally, we update the segmentation from the user feedback to approach the ROI in the user's mind. Figure 1 illustrates the outline of our approach.

We do the experiments on four datasets: Berkeley segmentation dataset (BSDS300) [9], Stanford Background Dataset (SBD) [10], Microsoft Research Asia Salient Object Dataset (MSRA1000) [1], and The PASCAL Visual Object Classes Challenge 2007 (VOC2007) [8]. The experimental results show that our method performs better than the recent approach [19] in response time and segmentation accuracy.

3 Our Approach

The proposed segmentation algorithm includes three phases: *Initialization*, *Adaptive Querying*, and *Information Updating*. The initialization phase aims to reduce the solution set of all possible segmentations. In addition, this phase also prepares the needed graph structure and feature descriptors for the subsequent phases. The adaptive querying phase and the information updating phase carry out interactive image segmentation with the aid of 1-bit feedback from the user. The adaptive querying phase adopts the transductive inference technique to explore the most informative superpixel for querying. Based on the new feedback, the information updating phase improves the segmentation uncertainty and thus pushes the segmentation toward the ROI that the user prefers.

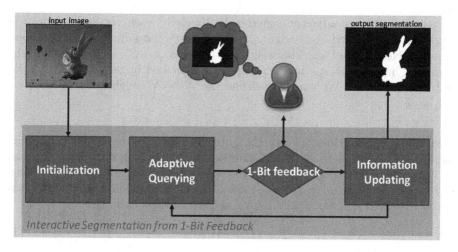

Fig. 1. The interaction pipeline of our method. The initialization phase aims to reduce the solution set of all possible segmentations. The adaptive querying phase and the information updating phase carry out interactive image segmentation with the aid of 1-bit feedback from the user. The adaptive querying phase adopts the transductive inference technique to explore the most informative superpixel for querying. Based on the user feedback, the information updating phase updates the needed information for next iteration.

3.1   Initialization

Given an image with height h and width w, the pixel-wise binary segmentation has $2^{h \times w}$ possible configurations. However, the discriminative power of twenty iterations in 1-bit feedback can only distinguish 2^{20} possible segmentations at most. Hence, our first goal is to reduce the solution set of all possible segmentations. One reasonable method is to over-segment the image into a superpixel set $\mathcal{S}$, thus the solution set could be greatly reduced to $2^{|\mathcal{S}|}$. That is the motivation we propose to consider superpixel-wise image segmentation in this work. In order to apply the transductive inference technique to estimate the similarity between any two superpixels, we need to model the given image as a graph for describing the neighborhood relationship among superpixels. Therefore, we first describe the steps of building the graph model of an image in this subsection, and then explain how to estimate the similarity between any two superpixels via the transductive inference technique.

Graph Model. For the given image, we use the efficient SLIC over-segmentation algorithm [2] to partition the image into a superpixel set $\mathcal{S} = \{s_1, s_2, \cdots, s_N\}$ with N elements. For each superpixel, we use the 3-D mean color in CIE-Lab space as its feature representation.

Given a superpixel set $\mathcal{S}$, we define a weighted connected graph $\mathcal{G} = (\mathcal{S}, \mathcal{E}, \omega)$, where the vertex set $\mathcal{S}$ contains all image superpixels and the edge set $\mathcal{E}$ consists all pairs of any two adjacent superpixels. Precisely, each vertex s_p denotes a single superpixel, and each edge $e_{pq} \in \mathcal{E}$ denotes the adjacent neighborhood of superpixels s_p and s_q. The weighting function $\omega : \mathcal{E} \to [0, 1]$ assigns the corresponding weight ω_{pq} to each edge e_{pq}, expressed in terms of mean color feature similarities. We can thus define the N-by-N weight matrix as $W = [\omega_{pq}]_{N \times N}$.

Similarity Estimation. The weight matrix W describes the similarity between any two *adjacent* superpixels. With the transductive inference method proposed by Zhou *et al.* [22], we can further estimate the transductive similarity between any two superpixels, no matter they are adjacent or not. The transductive similarity matrix T also has size N-by-N, and can be defined by

$$T = (D - \alpha W)^{-1} I , \tag{1}$$

where D is the diagonal matrix with each diagonal entry representing the row sum of W, α is a parameter in $(0, 1)$, and I is the N-by-N identity matrix.

3.2   Adaptive Querying

In the scenario of interactive image segmentation from 1-bit feedback, the goal of the segmentation algorithm is to guess the user's ROI. However, a region of interest may have semantic significance, or just have certain homogeneity. Figure 2 shows the diverse ground-truth segments of different datasets. In this paper, we set the goal of the adaptive querying phase as to explore the most informative superpixel for querying. We also design two strategies to deal with two cases existing in this phase. The two cases are categorized according to whether we obtain (i) only one of the ROI-superpixel or ROU-superpixel, or (ii) both of the ROI-superpixel and ROU-superpixel.

The first case corresponds to the situation of all queried superpixels belonging to the ROI or the ROU. In this case, the boundary of ROI is very difficult to define by transductive inference method. The first priority for this case is to find out the other label so that the second case can be applied. The second case corresponds to the situation of some queried superpixels belonging to the ROI and some queried superpixels belonging to the ROU. In this case, the boundary of ROI can be roughly described by transductive inference method. Now the goal is to find out the boundary superpixel with the most uncertainty to refine the ROI.

Case 1: Only One Label is Available. In this case, we aim to find out the other label. The most informative superpixel is defined as the superpixel that has the highest entropy so far. In this work, we diversify our queries using the entropy and the transductive similarity matrix T from Eq. 1. In transductive similarity matrix T, the n-th row represents the similarity between superpixel s_n and all other superpixels. Here we normalize the n-th row of T to make it sum to one.

We observe that if a superpixel s_n has more similar superpixels, then the normalized n-th row of T is more flattened. Hence we adopt the entropy function of n-th row of T to represent the proportion of similar superpixels that the superpixel s_n has. The higher entropy that a superpixel has also indicates that the superpixel is more informative, because no matter which label the superpixel has, there are large proportion of similar superpixels contain that label. Therefore, we choose the superpixel with the highest entropy as the query-superpixel in this case.

In practice, we define the query-superpixel selection function Q_1 in Case 1 by

$$Q_1(\mathcal{S}) = \arg\max_{s_n \in \mathcal{S}} \epsilon(s_n) = \arg\max_{s_n \in \mathcal{S}} \epsilon(T_{s_n, \cdot}) \,, \tag{2}$$

where $\epsilon(\cdot)$ is the entropy function, $T_{s_n, \cdot}$ is the normalized n-th row of T.

Case 2: Both Labels are Available. In this case, we aim to refine the ROI boundary. The most informative superpixel is defined as the superpixel which has the most uncertainty. We simulate the segmentation uncertainty with transductive inference and the known labels, and thus select the most uncertain superpixel to form the query question.

In practice, we use the following N-by-2 transductive similarity matrix to describe the similarity between each superpixel to ROI or ROU:

$$\hat{T} = (D - \beta W)^{-1} Y \,, \tag{3}$$

where β is a parameter in $(0, 1)$, $Y = [y_{ROI}, y_{ROU}]$ is a label matrix, y_{ROI} and y_{ROU} are both N-by-1 indicator vectors, in which the n-th element is 1 if the n-th superpixel has label ROI or ROU. The first column of $\hat{T}$ indicates the similarity of each superpixel to all ROI-superpixels, the second column of $\hat{T}$ indicates the similarity of each superpixel to all ROU-superpixels. The superpixel that has the highest uncertainty will have the smallest difference between these two columns. Hence, we define the query-superpixel selection function Q_2 in Case 2 by

$$Q_2(\mathcal{S}) = \arg\min_{s_n \in \mathcal{S}} \delta(s_n) = \arg\min_{s_n \in \mathcal{S}} |\hat{T}_{s_n, 1} - \hat{T}_{s_n, 2}| \,, \tag{4}$$

where $\delta(\cdot)$ is the absolute difference function, $\hat{T}_{p,q}$ is the entry of $\hat{T}$ that locates on p-th row and q-th column.

3.3   Information Updating

In the scenario of interactive image segmentation from 1-bit feedback, we can receive the 1-bit feedback between adaptive querying phase and the information updating phase. With one more certain label of the query-superpixel, we first generate the corresponding segmentation, and then update the needed information $\epsilon(s)$ and $\hat{T}$ for next iteration. Note that if the new feedback defines the counterpart label of Case 1, then we go into Case 2 and thus only need to update $\hat{T}$.

Case 1: Still only One Label is Available. To get the corresponding segmentation from Eq. (2), we define the current maximum entropy value among all superpixels $\{\epsilon(s_1), \epsilon(s_2), \cdots, \epsilon(s_N)\}$ as $\mathfrak{m}$. A superpixel with larger entropy value than $\mathfrak{m}/2$ is treated as an ROI-superpixel.

For preventing from selecting the high entropy superpixel that is similar to the one in the previous iteration, we have to alter the entropy value among all superpixels with the latest queried superpixels s_z. Here we define the new entropy value of superpixel s_n as

$$\epsilon(s_n) = \epsilon(s_n) - \frac{T_{s_z,s_n}}{T_{s_z,s_z}} \epsilon(s_z) . \tag{5}$$

This updated $\epsilon(s_n)$ will trigger the query-superpixel selection function Q_1 in Eq. (2).

Case 2: Both Labels are Available. To get the corresponding segmentation from Eq. (3), a superpixel with positive value of $\hat{T}_{s_n,1} - \hat{T}_{s_n,2}$ is treated as an ROI-superpixel.

Since we have one more new label, we have new indicator vector y'_{ROI} or y'_{ROU}. Hence we obtain new label matrix $Y' = [y'_{ROI}, y_{ROU}]$ or $Y' = [y_{ROI}, y'_{ROU}]$. This update, $Y = Y'$, will trigger the new transductive similarity matrix $\hat{T}$ in Eq. (3) and the new query-superpixel in Eq. (4).

4   Experimental Results

Since we deal with the interactive image segmentation problem, we aim to achieve faster response time, higher segmentation accuracy, and fewer queries. Hence we conduct the evaluations with respect to the response time and the qualitative and quantitative results. The experiments are performed on four datasets. The parameter settings are the same for the four datasets, we set the number of superpixels $N = 200$, the parameters $\alpha = 0.999$ and $\beta = 0.001$.

To evaluate our method on these datasets, every segment in each individual ground-truth annotation is selected as a region of interest (ROI). To measure the segmentation quality, we employ the median and the mean Dice scores as used in Rupprecht et al. [19], which measure the overlap between the segmentation and the ground truth. We separately compare the proposed interactive image segmentation with its several variants and the method of Rupprecht et al. [19] in following experiments.

Berkeley Segmentation DataSet (BSDS300) [9]: The BSDS300 dataset contains 300 natural images. Each image has several hand-labeled segmentations as the ground-truth human annotations. The example ground-truth of BSDS300 is shown in Fig. 2a. Note that we only show one human annotation for better visualization. This dataset contains various regions with several human annotators, thus provides difficult tasks of identifying regions of interest of a interactive

(a) BSDS300	(b) SBD	(c) MSRA1000	(d) VOC2007

Fig. 2. Testing examples from each dataset. Each color denotes an ground-truth segment. Black color in (b), (c), and (d) and creamy-white color in (d) are ignored segments in the experiments (Color figure online).

segmentation from 1-Bit Feedback algorithm. The average number of regions in each individual ground-truth is 20.37.

Stanford Background Dataset (SBD) [10]: The SBD dataset contains 715 natural images collected from other datasets. This dataset contains three types of ground-truth annotations. We select the *region* annotation as Rupprecht *et al.* [19] for comparison. Each image in the SBD dataset has one ground-truth human annotation. This dataset contains some semantic regions. Precisely, each image has eight possible semantic labels: building, foreground object, grass, mountain, road, sky, tree, and water. The example ground-truth of SBD is shown in Fig. 2b. The average number of semantic regions in each individual ground-truth is 4.22.

Microsoft Research Asia Salient Object Dataset (MSRA1000): The MSRA1000 dataset contains 1000 natural images collected from other datasets. The ground-truth human annotations are provided by Achanta *et al.* [1][2]. The natural images are provided by Liu *et al.* [15][3]. Each image has only one ground-truth human annotations. The example ground-truth of MSRA1000 is shown in Fig. 2c. This dataset contains only the ROI region, thus provides clearly-defined region of interest. The average number of regions in each individual ground truth is thus 1.0.

The PASCAL Visual Object Classes Challenge 2007 (VOC2007) [8]: From the VOC2007 dataset, we use the *trainval* data in segmentation subset for evaluation. The *trainval* image set in segmentation subset contains 422 natural images. This dataset also contains some semantic regions. Precisely, each image can has twenty possible semantic labels. Each image is partitioned into independent instances. The example ground-truth of VOC2007 is shown in Fig. 2d. The average number of regions in each individual ground-truth is 2.87.

---

[2] http://ivrlwww.epfl.ch/supplementary_material/RK_CVPR09/index.html.

[3] http://research.microsoft.com/en-us/um/people/jiansun/SalientObject/salient_object.htm.

4.1 Variants

This experiment compares some variant versions of our method. Figures 3 and 4 depict the mean Dice score and the median Dice score against the number of questions. In the legend blocks of these two figures, we use strategy1-strategy2 to represent the variant versions of our method. Specifically, strategy1 denotes the strategy used in Case 1 and strategy2 denotes the strategy in used Case 2. In addition, 'R' means selecting a random superpixel; 'T' means selecting the most uncertain superpixel according to the transductive inference similarity; 'W' means selecting a random superpixel weighted by its entropy value; 'F' means selecting the superpixel which has the most different feature to all previous selected superpixels; 'E' means selecting a superpixel with the highest entropy value. A dataset with a larger difference between Figs. 3 and 4 means the dataset is more difficult to do the interactive segmentation, because the median is helpful in ignoring several bad segmentations.

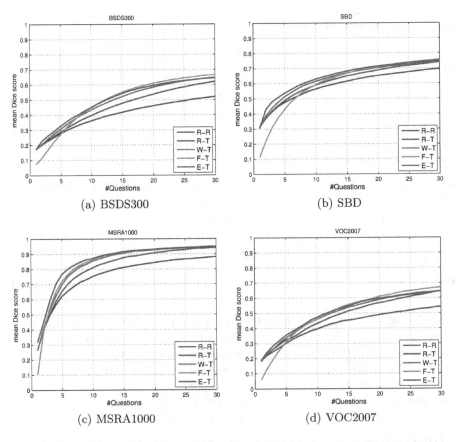

(a) BSDS300

(b) SBD

(c) MSRA1000

(d) VOC2007

Fig. 3. Performance comparison of the variant versions of our method on the mean Dice score against the number of questions.

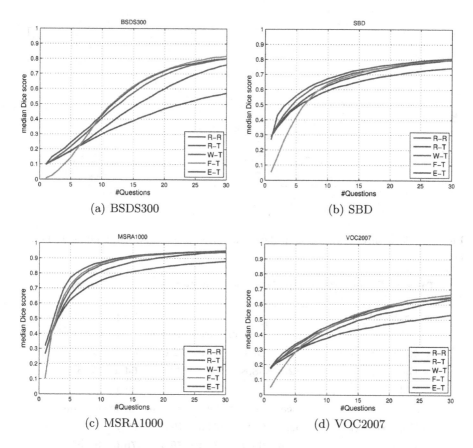

Fig. 4. Performance comparison the variant versions of our method on the median Dice score against the number of questions.

It is better to select the superpixel with the highest entropy value as the query-superpixel in the first case of the adaptive querying phase, because it contains the most information. On the other hand, it is better to select the most uncertain superpixel according to the transductive inference similarity in the second case of the adaptive querying phase, because it can find out the superpixel that locates on the object's boundary, and thus is better for segmentation refinement. Therefore, we select the 'E-T' version as our representative method in following experiments.

4.2 Comparison

Figure 5 illustrates the median Dice score against the number of questions. Since there is no released code of the method of Rupprecht *et al.*, we reproduce their results on Fig. 5 according to their experiments in [19]. It can bee seen that our method performs significantly better than the method of Rupprecht *et al.*

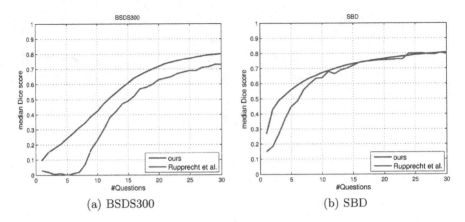

(a) BSDS300   (b) SBD

Fig. 5. Performance comparison on the median Dice score against the number of questions.

Table 1. Quantitative comparison: the mean and median Dice scores (%).

BSDS300	10Q		20Q		30Q	
	mean	median	mean	median	mean	median
Rupprecht *et al.* [19]	34.7	23.8	48.8	62.0	56.1	73.2
Ours	**42.8**	**38.8**	**58.6**	**70.0**	**64.1**	**79.8**
SBD	10Q		20Q		30Q	
	mean	median	mean	median	mean	median
Rupprecht *et al.* [19]	52.6	63.9	63.9	75.8	67.9	79.8
Ours	**61.4**	**65.4**	**70.9**	**76.1**	**75.0**	**80.5**

in BSDS300. In SBD, our method is much better than their method in the first fifteen rounds. Table 1 also shows that our method performs quite well. This experiment shows that our method is very competitive with respect to the criterion of segmentation accuracy.

4.3   Response Time

This experiment shows the efficiency of our method. In general, the average response time per iteration for our method is less than 1 ms on an Intel Core i7-4770 3.40 GHz CPU. For comparison, the average response time of the method of Rupprecht *et al.* is about 1 s on Intel Core i7-4820 3.70 GHz CPU. Their computation bottleneck is the step of MCMC sampling, which is used to approximate the segmentation probability among the entire image. In contrast, we directly use the transductive inference technique to explore the most informative superpixel for querying, thus prevent the complex sampling process. Another reason is because we use superpixels as the building blocks of our algorithm. Using superpixels

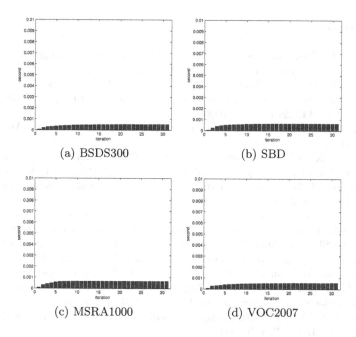

(a) BSDS300   (b) SBD

(c) MSRA1000   (d) VOC2007

Fig. 6. Average response time per iteration of our approach on the four datasets.

greatly reduces the number of graph nodes and speeds up the computation in transductive inference. Notice that it takes about 0.2 s in the initialization phase, which contains over-segmentation, feature extraction, and transductive inference. However, the initialization phase only needs to be done once. Figure 6 shows that our average time per iteration on the four datasets.

5   Conclusion

We have shown that the proposed method can efficiently solve the problem of interactive image segmentation from 1-bit feedback. Our interactive image segmentation algorithm achieves the preferable properties of high segmentation accuracy, low response time, and simple manipulation. We fulfill these requirements by designing an efficient algorithm that consists of over-segmentation, entropy estimation, and transductive inference. The experimental results show the good performance of our method, in particular, extremely short response time. Our key idea is to prune the solution space of possible segmentations, and then to propose the most informative question based on the clues of the similarity among the superpixels. The method helps to increase the probability of finding out the most uncertain image region to obtain its real label from the 1-bit user feedback.

Acknowledgement. This work was support in part by MOST Grants 103-2221-E-007-045-MY3 and 103-2218-E-007-017-MY3 in Taiwan.

References

1. Achanta, R., Hemami, S.S., Estrada, F.J., Süsstrunk, S.: Frequency-tuned salient region detection. In: CVPR, pp. 1597–1604 (2009)
2. Achanta, R., Shaji, A., Smith, K., Lucchi, A., Fua, P., Süsstrunk, S.: SLIC superpixels compared to state-of-the-art superpixel methods. IEEE Trans. Pattern Anal. Mach. Intell. **34**, 2274–2282 (2012)
3. Adams, R., Bischof, L.: Seeded region growing. IEEE Trans. Pattern Anal. Mach. Intell. **16**, 641–647 (1994)
4. Batra, D., Kowdle, A., Parikh, D., Luo, J., Chen, T.: iCoseg: Interactive co-segmentation with intelligent scribble guidance. In: CVPR, pp. 3169–3176 (2010)
5. Boykov, Y., Jolly, M.: Interactive graph cuts for optimal boundary and region segmentation of objects in N-D images. In: ICCV, pp. 105–112 (2001)
6. Cheng, M., Prisacariu, V.A., Zheng, S., Torr, P.H.S., Rother, C.: Densecut: Densely connected crfs for realtime grabcut. Comput. Graph. Forum **34**, 193–201 (2015)
7. Dong, X., Shen, J., Shao, L., Yang, M.: Interactive cosegmentation using global and local energy optimization. IEEE Trans. Image Process. **24**, 3966–3977 (2015)
8. Everingham, M., Van Gool, L., Williams, C.K.I., Winn, J., Zisserman, A.: The PASCAL Visual Object Classes Challenge (VOC 2007) Results (2007). http://www.pascal-network.org/challenges/VOC/voc2007/workshop/index.html
9. Fowlkes, C.C., Martin, D.R., Malik, J.: Local figure-ground cues are valid for natural images. J. Vis. **7**, 1–9 (2007)
10. Gould, S., Fulton, R., Koller, D.: Decomposing a scene into geometric and semantically consistent regions. In: ICCV, pp. 1–8 (2009)
11. Grady, L.: Random walks for image segmentation. IEEE Trans. Pattern Anal. Mach. Intell. **28**, 1768–1783 (2006)
12. Gulshan, V., Rother, C., Criminisi, A., Blake, A., Zisserman, A.: Geodesic star convexity for interactive image segmentation. In: CVPR, pp. 3129–3136 (2010)
13. Kass, M., Witkin, A.P., Terzopoulos, D.: Snakes: active contour models. Int. J. Comput. Vision **1**, 321–331 (1988)
14. Kowdle, A., Chang, Y., Gallagher, A.C., Chen, T.: Active learning for piecewise planar 3d reconstruction. In: CVPR, pp. 929–936 (2011)
15. Liu, T., Sun, J., Zheng, N., Tang, X., Shum, H.: Learning to detect a salient object. In: CVPR (2007)
16. Mortensen, E.N., Barrett, W.A.: Intelligent scissors for image composition. In: SIGGRAPH, pp. 191–198 (1995)
17. Rother, C., Kolmogorov, V., Blake, A.: "grabcut": interactive foreground extraction using iterated graph cuts. ACM Trans. Graph. **23**, 309–314 (2004)
18. Rother, C., Minka, T.P., Blake, A., Kolmogorov, V.: Cosegmentation of image pairs by histogram matching - incorporating a global constraint into MRFs. In: CVPR, pp. 993–1000 (2006)
19. Rupprecht, C., Peter, L., Navab, N.: Image segmentation in twenty questions. In: CVPR, pp. 3314–3322 (2015)
20. Straehle, C.N., Köthe, U., Knott, G., Briggman, K.L., Denk, W., Hamprecht, F.A.: Seeded watershed cut uncertainty estimators for guided interactive segmentation. In: CVPR, pp. 765–772 (2012)
21. Vicente, S., Rother, C., Kolmogorov, V.: Object cosegmentation. In: CVPR, pp. 2217–2224 (2011)
22. Zhou, D., Bousquet, O., Lal, T.N., Weston, J., Schölkopf, B.: Learning with local and global consistency. In: NIPS, pp. 321–328 (2003)

Geodesic Distance Histogram Feature for Video Segmentation

Hieu Le[1(✉)], Vu Nguyen[1], Chen-Ping Yu[2], and Dimitris Samaras[1]

[1] Stony Brook University, Stony Brook, USA
hle@cs.stonybrook.edu
[2] Harvard University, Cambridge, USA

Abstract. This paper proposes a geodesic-distance-based feature that encodes global information for improved video segmentation algorithms. The feature is a joint histogram of intensity and geodesic distances, where the geodesic distances are computed as the shortest paths between superpixels via their boundaries. We also incorporate adaptive voting weights and spatial pyramid configurations to include spatial information into the geodesic histogram feature and show that this further improves results. The feature is generic and can be used as part of various algorithms. In experiments, we test the geodesic histogram feature by incorporating it into two existing video segmentation frameworks. This leads to significantly better performance in 3D video segmentation benchmarks on two datasets.

1 Introduction

Video segmentation is an important pre-processing step for many high-level video applications such as action recognition [1], scene understanding [2], or 3D reconstruction [3]. A more compact representation not only reduces the subsequent processing space and time requirements, but also provides sets of visual segments that contain meaningful cues for higher-level computer vision tasks. However, generating supervoxels from videos is a significantly more difficult task than superpixel segmentation from images, due to the heavy computational cost and the extra temporal dimension. Specifically, well delineated spatio-temporal video segments can be used for tracking bounded regions, foreground moving objects, or semantic understanding. For example, locating the movement of hands is helpful for gesture or action recognition, and separating foreground/background can pin-point the region-of-interest for detecting moving objects. Therefore, these spatio-temporal segments should be temporally consistent in order to be beneficial for these computer vision tasks.

For video segmentations that are initialized from superpixels, the main goal is to consider the connections between neighboring superpixels and to decide which ones belong to the same spatio-temporal cluster. The connections are usually represented as a spatio-temporal graph, where the nodes are the superpixels and the edges connect superpixels that are adjacent to each other. The edges are weighted based on the similarity distances between pairs of superpixels.

© Springer International Publishing AG 2017
S.-H. Lai et al. (Eds.): ACCV 2016, Part I, LNCS 10111, pp. 275–290, 2017.
DOI: 10.1007/978-3-319-54181-5_18

Previous work [4,5] proposed a variety of features corresponding to a wide range of low and mid-level image cues from superpixels. For example, the within-frame similarities were computed from boundary magnitude, color, texture, and shape, and the temporal connections were defined by the direction of optical flow or motion trajectories. Importantly, the aforementioned features that were used for video segmentation encode only local information, extracted from within each superpixel. One would expect improved performance when combining local and global features, if the appropriate global features per superpixel were extracted.

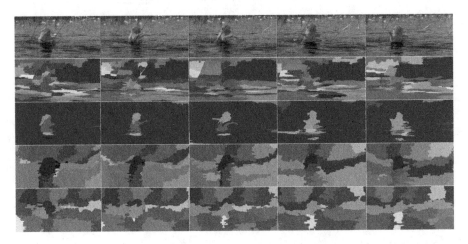

Fig. 1. The segmentation results on video "monkey" from Segtrack v2 dataset [6]. Top row: original frames with superimposed ground-truth (green). Second row: segmentation results of the PGP algorithm ([7]) using their four predefined features. Third row: result of PGP with our feature integrated. Fourth row: segmentation result of spectral clustering with the 6 features proposed in [8]. Bottom row: segmentation result of spectral clustering with our feature integrated. Our results show better temporal consistency and less over-segmentation. (Color figure online)

The geodesic distance has been shown to be effective for image segmentation problems [9,10] but its applications in the video domain have been limited [10–13]. In this work, we propose a complete methodology for the use of geodesic distance histogram features in the video segmentation problem. The histogram feature describes the superpixel-of-interest by the distribution of the geodesic distances from it, to all other superpixels in the same frame. The representation compactly encodes global similarity relations between segments. Thus, we want to use per-frame geodesic distance information to associate superpixels both within and across frames. However, the nature of this global representation, poses several challenges that need to be addressed, in order to successfully use geodesic distance histograms for video segmentation:

- The feature needs to be robust across frames in order to perform useful super-pixel association. That means if a superpixel has a unique representation in one frame, its representation in the next frame should be also unique, in order to facilitate matching.
- For relatively small segments, their similar relationship to global context can dwarf distinctive neighborhood information, which might make them hard to differentiate.
- The feature does not encode any spatial relationships between segments. Such relationships often offer constrains that allow otherwise similar segments to be distinguished from each other.

In this paper, we address these issues in order to derive a geodesic histogram feature that is appropriate for video segmentation tasks. In essence, we introduce the necessary local information in the global representation, in order to disambiguate associations across frames. For a given superpixel, we first extract the soft boundary map of the frame where it belongs, then we compute geodesic distances from the superpixel-of-interest to all other superpixels in the same frame using the boundary scores. If we were performing per frame segmentation, a 1D histogram of these scores would suffice [10]. However, due to motion, this 1D histogram is not robust across frames. As observed previously [13], a 2D joint histogram of intensity and geodesic distance is much more robust. To encode more spatial information into the feature, we compute multiple geodesic histograms in a spatial pyramid [14]. Finally, we weigh the bins with respect to their spatial distance from the superpixel-of-interest, in order to favor potentially discriminative neighborhood information. We show in experiments that when we add our complete geodesic histogram feature into existing frameworks, the resulting segmentations are greatly improved, especially in 3D segmentation accuracy and temporal consistency. The feature is also fast to compute, without increasing significantly processing time for the existing frameworks. The geodesic histogram features are added into two state-of-the-art video segmentation frameworks that are based on superpixel clustering, and tested on two popular datasets using standard 3D segmentation benchmarks.

The rest of paper is organized as follows: Sect. 2 discusses related work. Section 3 discusses the motivation, computation, and analysis of the proposed geodesic histogram features. Implementation details are described in Sect. 3.4. Section 4 presents the experimental results. Section 5 concludes the paper and discusses other possible applications.

2   Related Work

Many video segmentation works propose diverse features to capture various kinds of information in order to estimate the similarity between the components of the video. Appearance can be represented by features based on color [5,15], texture [16], and soft boundaries [17]. Motion related features have also been utilized often, including short-term motion features based on optical flow [18,19] and long-term motion features based on trajectories [20–23]. Superpixel shape is used to compute the similarities among superpixels across frames [15]. Some works discuss the choice of features to use [8] as well as the method to incorporate various kinds of features into affinity matrices [4].

Geodesic distances provide appearance-based similarity estimates. Geodesic distances have been applied widely on segmentation related problems on images [9,10,13]. A feature based on geodesic distance for matching images of deformed objects has been introduced in [13]. The authors showed that the geodesic distance could be invariant to object deformations, by encoding pixels as color histograms on the surrounding pixels that have the same geodesic distances. The geodesic distance is also used to propose object segments on images [9], which is based on the correlation between the object boundary and the change in the geodesic distance transform. Several video segmentation methods have employed geodesic distance for various purposes. The salient object segmentation framework uses a geodesic distance in each frame to estimate the objectness of superpixels [11] on a per frame basis. Further work further proposes a spatio-temporal geodesic distance [10] that extends image segmentation to video segmentation. However, the proposed spatio-temporal distance has to be constrained to be temporally non-decreasing to preserve the metric property, thus limiting the robustness of the method.

In this paper, we propose a feature based on geodesic distance to estimate the similarity between the superpixels in the video. We consider the frame-wise distribution of the geodesic distances, i.e., the histogram of geodesic distances from each superpixel to all other superpixels in the same frame. This representation compactly encodes the relative similarity distances between the segment containing the superpixel-of-interest to all the other segments on the frame. This global information therefore serves as a complement to the set of to the set of appearance, motion, and shape-based features which only encode information from the inner region of the superpixel-of-interest.

3   Geodesic Distance Histogram Feature

Given a frame of the video, let X be the set of superpixels: $X = \{x_1, \ldots, x_n\}$. The frame is then represented by a non-negative, undirected graph $G = (X, E)$, where each value in E is associated with a pair of neighboring superpixels in X, and the edge weight is computed as the boundary strength between the two superpixels. The geodesic distance between any two superpixels $x_i, x_j \in X$ is defined as the weight of the shortest path between the two superpixels in G.

Given a superpixel x_i on a frame, the geodesic distance between x_i and all other superpixels in the same frame is computed and pooled into a geodesic distance histogram. This histogram contains the global information of the frame with respect to x_i in terms of geodesic distance distribution, and can be used for computing pair-wise superpixel similarity both within and across frames.

3.1   1D Geodesic Distance Histogram

The simplest approach is to use an 1D histogram to describe the distribution of the geodesic distances, where a bin of the histogram represents the number of superpixels with a particular geodesic distance. This is similar to the concept of critical level sets [9], where each critical level defines a group of superpixels having their geodesic distances less than a certain threshold. Each bin of the histogram is then associated with a region in the image.

In order to keep our feature relatively constant across frames, the value of each bin should stay approximately the same. This means that the regions associated with each bin also remain relatively stable. Considering the superpixel (in red) shown in Fig. 2(a), two regions corresponding to the first two bins of the histogram are visualized in Fig. 2(b). The first bin collects the votes of all superpixels with the lowest geodesic distance interval, forming the region indicated by the leftmost arrow. However, the region corresponding to the second bin is the combination of superpixels from different semantic regions. The value of the second bin is therefore not robust since these regions could potentially move in different ways, and end up voting for different bins in subsequent frames.

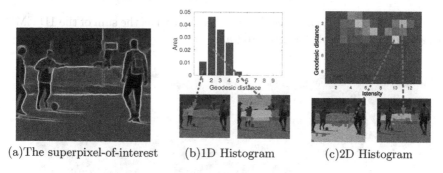

(a)The superpixel-of-interest   (b)1D Histogram   (c)2D Histogram

Fig. 2. The figure shows an example of 1D (geodesic distances) and 2D (intensity-geodesic distances) histogram features. (a): frame 1 of video "soccer" from Chen's Xiph.org dataset [24], with soft boundary scores highlighted, and a superpixel-of-interest marked in red. (b) and (c): the 1D and 2D histograms of the superpixel-of-interest, and the frame regions (green) that correspond to the selected bins and cells of the 1D and 2D histograms, respectively. (b) shows that the bins of the 1D histogram contain mixed information, while the cells in (c) contain regions that are more semantically homogeneous. (Color figure online)

3.2   2D Intensity-Geodesic Distance Histogram

We incorporate the intensity feature as an additional cue to complement the geodesic distance, on order to constrain bins to correspond to individual regions instead of disparate groups of regions. Thus the histogram becomes a 2D table where each cell is voted for by the superpixels that have a particular pair of geodesic distance and intensity. The joint distribution of intensity-geodesic distance was originally proposed in [13], where the joint distribution was expected to be stable and informative under a wide range of deformations.

Figure 2(c) visualizes the intensity-geodesic distance histogram of a superpixel-of-interest (shown in red in Fig. 2(a)). Notice that the second bin of the 1D histogram equals to the sum of all cells in the second row of the 2D histogram, and the region from the second bin in the 1D histogram is now separated into multiple smaller regions corresponding to these cells. This is a desired effect given that each of the cells in the 2D histogram contains superpixels from the same semantic region as the 1D case. We also visualized the cell with the highest value in Fig. 2(c), which corresponds to the superpixels within the entire grass field. Such a region is likely to be stable across frames and remain connected. This implies that as long as the intermediate boundaries remain the same, these regions would still contribute to the same cells in the histogram.

To compute the similarity distance between two histograms, we can use the χ^2 distance or the Earth Mover's Distance. Following [13], the χ^2 distance between two 2D histograms H_p and H_q with size $M \times N$ is defined by:

$$\chi^2(H_p, H_q) = \frac{1}{2} \sum_{k=1}^{K} \sum_{m=1}^{M} \frac{[H_p(k,m) - H_q(k,m)]^2}{H_p(k,m) + H_q(k,m)} \qquad (1)$$

The Earth Mover's Distance (EMD) is computed as the sum of the 1D EMDs at each intensity bin of the 2D histogram.

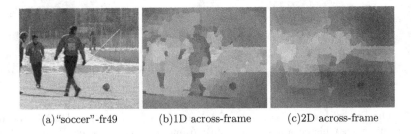

(a) "soccer"-fr49   (b) 1D across-frame   (c) 2D across-frame

Fig. 3. The figures visualize the similarities between the superpixel-of-interest in Fig. 2(a) on a later frame (frame 49) to all other superpixels. Warmer color represents higher similarity. (a): original frame. (b): the similarity map based on 1D geodesic distance histograms. (c): the similarity map based on 2D intensity-geodesic distance histograms. The figure shows that the 2D histogram is more robust than the 1D histogram for across-frame matching: there are multiple superpixels located in multiple regions that have similar 1D histograms with the superpixel-of-interest, while only the superpixels located within the upper-body region have the most similar 2D histograms.

Figure 3 visualizes the similarity values computed based on 1D and 2D feature histograms from the superpixel-of-interest in Fig. 2(a) on a later video frame. In the color scheme, higher similarity is represented by the warmer color. The figure shows that the 1D histogram is less robust than the 2D histogram: there are multiple regions having similar 1D histograms with the superpixel-of-interest, and the superpixel with the highest 1D histogram similarity is in the background. In contrast, the superpixel with the highest similarity using the 2D histogram falls within the same upper-body region, a desirable result.

3.3 Spatial Information

Pooling methods such as histograms discard spatial information, such as image distance relationships or local neighborhood patterns. We encode spatial cues in two ways: (1) by embedding spatial distances into the voting weight of each superpixel, and (2) by adopting a commonly used spatial pyramid scheme [14].

Spatial Distance Voting Weight. For a given superpixel x, its histogram feature is constructed by its intensity and geodesic distances to all other pixels in the same frame. To take the spatial location of these other superpixels into account, the geodesic distances are weighted by the spatial distance of those superpixels to x. In particular, the weighting of superpixel y to the histogram bins of superpixel x in frame f is defined by:

$$weight_y = \frac{|y|}{|f|} \times \exp(-\mu \times L_2(x, y)) \tag{2}$$

where $|\cdot|$ is the area and $L_2(\cdot)$ is the Euclidean distance between two superpixels' center locations.

The area component normalizes the influence of superpixels of different sizes. The exponential ensures that nearby superpixels contribute more to the geodesic histogram of x. This is especially helpful for superpixels that belong to smaller segments, for which most other superpixels have large geodesic distances, that would dominate the histogram. Hence two small regions that are locally different would have very similar histograms. The parameter μ of the exponential controls the trade-off between global and local information.

Spatial Pyramid Histogram. Inspired by the popularity of spatial pyramids [14], we incorporated the pyramid scheme into the construction of our feature histogram to encode more spatial information into the features. We implemented two scales of the spatial pyramid: 1×1 and 2×2 grids over a given frame. A histogram is extracted from each cell of the grid. Histograms from the same scale are concatenated.

3.4 Implementation Details

Our features are constructed from the intensity and boundary probability maps. For more robust boundary extraction, we also experiment with two different boundary map methods: spatial edge maps using structured forests [25], and motion boundary maps using the method proposed in [26].

Given the combined edge map and the superpixel graph, the geodesic distance feature for each superpixel is computed using Dijkstra's algorithm in

$O(|X||E|log|X|)$, with the cost of a path being the accumulated boundary scores between one superpixel to another.

We empirically set the intensity dimension of the feature histogram at 13 bins, and the geodesic dimension at 9 bins.

4   Experiments

In this section, we describe our experiments using the geodesic histogram features for video segmentation. We incorporated our features into two existing frameworks that are based on different clustering algorithms: spectral clustering [8] and parametric graph partitioning [7]. Spectral clustering performs dimensionality reduction on an affinity matrix based on eigenvalues, while parametric graph partitioning directly performs the clustering on the superpixel graph by modeling L_p affinity matrices probabilistically. Also, the method in [8] generates coarse-to-fine hierarchical segmentation results, while [7] only outputs a single level of segmentation.

The experiments were conducted on the Segtrack V2 [6] and Chen's Xiph.org [24] datasets, covering a wide range of scenarios for evaluating video segmentation algorithms. We evaluate our segmentation results using the metrics proposed in [27], including 3D Accuracy (AC), 3D Under-segmentation Error (UE), 3D Boundary Recall (BR), and 3D Boundary Precision (BP). All experiments were conducted with the exact same set of initial superpixels and other parameter settings.

4.1   Video Segmentation Using Spectral Clustering

We first evaluate the performance of the framework by adding our feature to spectral clustering [8]. We use the same 6 features as [8]: short term temporal, long term temporal, spatio temporal appearance, spatio temporal motion, across boundary appearance, and across boundary motion. The affinity matrix was computed by combining the 6 affinity matrices computed from each feature. We combined the original computed affinity matrix with the geodesic histogram features in order to preserve the algorithm settings and superpixel configurations. The similarity distances based on our features were computed using the χ^2 distance.

Figure 4 shows the evaluation results of spectral clustering with and without our feature on Segtrack v2 and Chen Xiph.org datasets. We tested four settings of our feature: (i) 2D histogram using only spatial edge maps to compute geodesic distances and without spatial distance voting weight (2D−0), (ii) 2D histogram using spatial edge maps and spatial distance voting weight with $\mu = 0.02$ (2D−0.02), (iii) 2D histogram using both spatial edge and motion boundary maps with $\mu = 0.02$ (2D + 0.02) and, (iv) 2D histograms with spatial pyramid (2D + 0.02 sp). Compared to the baseline, our feature significantly improved segmentation performance. The improvement was most significant in 3D accuracy: increased by 5% for Segtrack v2 and 10% for Chen Xiph.org. For Segtrack v2 dataset, our feature was able to improve the segmentation results on all four metrics. For Chen Xiph.org dataset, the feature gave a strong boost to 3D accuracy and 3D boundary precision. For all settings tested, we noticed that motion boundary maps did not affect performance much. Given that motion boundary map generation requires optic flow computation, which can be time consuming, its omission might result in faster implementations. The spatial distance voting weights had a strong impact on the results and clearly improved segmentation.

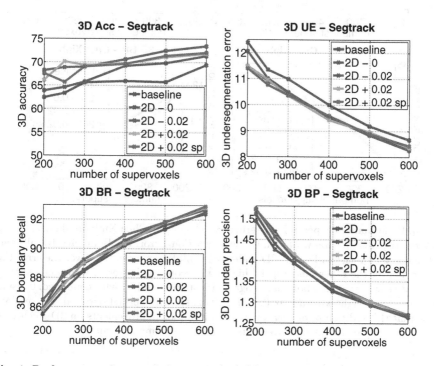

Fig. 4. Performance of spectral clustering (SC) [8] on the Segtrack v2 dataset, using four metrics: 3D Accuracy, 3D Under Segmentation Error, 3D Boundary Recall, and 3D Boundary Precision. For the 3D under-segmentation metric, the lower the error the better. For all the other metrics, the higher the score the better. -: using only spatial boundary edge. +: spatial boundary edge and motion boundary edge combined. 0: using spatial voting weight with $\mu = 0$. 0.02: $\mu = 0.02$. sp: with spatial pyramid. These plots show that the addition of our features result in major improvements on 3D Accuracy, and minor but consistent improvements on the three remaining metrics.

In addition to these improvements, Fig. 6 shows that the average temporal length of supervoxels consistently increased for all parameter settings of our feature by 10% for Segtrack v2 dataset and 5% for Chen Xiph.org dataset, showing that the segmentation results acquired better temporal consistency. Having both longer supervoxels and improved segmentation metrics indicate that our feature provides additional information for more reliable temporal consistency. This is

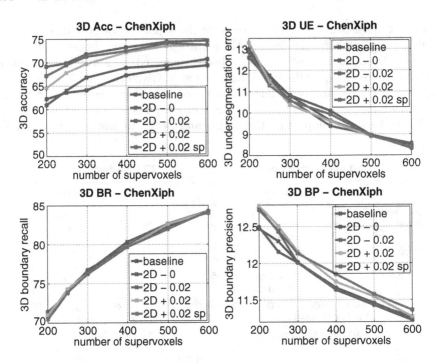

Fig. 5. Performance of spectral clustering (SC) [8] on the Chen Xiph.org dataset, using four metrics: 3D Accuracy, 3D Under Segmentation Error, 3D Boundary Recall, and 3D Boundary Precision. For the 3D under-segmentation metric, the lower the error the better. For all the other metrics, the higher the score the better. -: using only spatial boundary edge. +: spatial boundary edge and motion boundary edge combined. 0: using spatial voting weight with $\mu = 0$. 0.02: $\mu = 0.02$. sp: with spatial pyramid. These plots show that the addition of our features result in major improvements on 3D Accuracy, and minor but consistent improvements on the three remaining metrics.

significant, since connecting more corresponding superpixels temporally is a crucial and challenging part of the video segmentation task.

An interesting qualitative example is shown in Fig. 7, showing the segmentation results for video "soldier" with only two clusters. The second row visualizes the two clusters generated by [8] using the 6 predefined features with only local information, only capturing the lower leg of the moving soldier. In contrast, the segmentation results improved with the addition of our geodesic feature. The global information that is encoded by our feature seems to have provided better information to the spectral clustering algorithm to segment the main object out of the background. Another qualitative example is shown in the 4th and 5th row of Fig. 1. The segment of the baseline shown in the 4th row shows some under-segmentation over the main moving object. This issue however, is less pronounced with our feature.

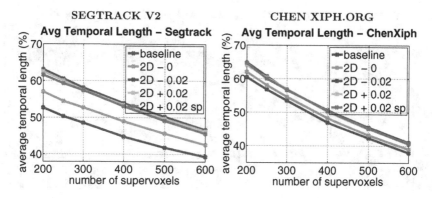

Fig. 6. Average temporal length of supervoxels generated by spectral clustering (SC) [8] on the Segtrack v2 and Chen Xiph.org datasets. The results show significant improvements on the temporal consistency with the addition of our feature on Segtrack v2 dataset, and minor but consistent improvement on the Chen Xiph.org dataset.

Fig. 7. The figure shows the segmentation results for the video "soldier" from the Segtrack v2 dataset using spectral clustering [8] with and without our feature. We set the number of output clusters at 2 for this example. The top row shows the original frames with the ground truth highlighted in green. The second row shows the results of spectral clustering with 6 features, as originally proposed in [8]. The third row shows the results of the algorithm when using the 6 original features plus our feature (2D histogram with spatial information). All other settings were set to be exactly the same. (Color figure online)

4.2   Video Segmentation Using Parametric Graph Partitioning

Parametric Graph Partitioning (PGP) [7] is a recent graph-based unsupervised method that generates a single level of video segmentation. The method models edge weights by a mixture of Weibull distributions, and requires that an L_p-norm based similarity distance to be utilized. Therefore, we conduct experiments in this section using Earth Mover's Distance as in [7]. The baseline is the setting originally proposed in [7] which uses four feature types: intensity, the hue of the HSV color space, the AB component of LAB color space, and gradient orientation. We did not use the motion feature since it did not contribute significantly toward PGP performance as suggested in the original paper.

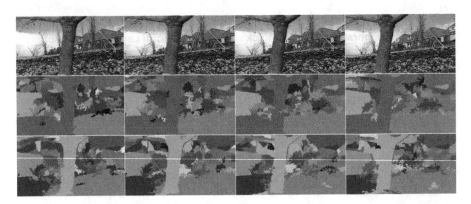

Fig. 8. The segmentation results of PGP on video "garden" from the Chen Xiph.org dataset with and without our feature. The top row shows the original frames. The second row shows the segmentation results of PGP using the 4 features proposed in [7]. The bottom row is the segmentation results of PGP using the 4 features plus our feature (2D histogram with spatial information).

Tables 1 and 2 report the quantitative evaluation of PGP with and without our feature on the two datasets. We evaluated the 1D histogram feature on the Chen Xiph.org dataset, shown in Table 2. While PGP with the 1D feature outperforms the baseline in general, the benchmarks of 3 out of 8 videos decreased. On the other hand, the 2D feature significantly improved the segmentation performance of PGP. For the Segtrack v2 dataset, quantitative results in Table 1 show clear improvements of our feature for PGP, as well as the additional benefits from the spatial pyramid configuration.

Table 1. Quantitative evaluation on the Chen Xiph.org dataset. Best values are shown in bold. The table shows the evaluation results of the segmentation generated from the method proposed in [7] with and without our feature in two configurations: 1D geodesic distance histogram and 2D intensity-geodesic distance histogram. All videos are initialized with 300 superpixels.

Metrics	3D ACC			UE 3D			BR 3D			BP 3D		
Methods	[7]	1D	2D	[7]	1D	2D	[7]	1D	2D	[7]	1D	2D
Bus_fa	70.72	70.58	**70.98**	6.22	10.31	**5.75**	80.22	81.64	**82.46**	37.64	38.60	**38.98**
Container_fa	88.68	86.69	**89.05**	3.66	7.54	**3.45**	**71.24**	70.38	70.74	8.68	**16.28**	8.55
Garden_fa	81.69	83.72	**85.46**	1.80	1.68	**1.47**	72.46	77.48	**79.91**	**12.83**	12.73	12.41
Ice_fa	86.71	**87.54**	77.83	**26.70**	42.58	58.59	**83.29**	80.82	67.47	30.99	29.54	**44.48**
Paris_fa	40.46	51.37	**61.44**	13.50	**12.99**	13.15	47.17	53.73	**56.68**	4.22	4.70	**4.73**
Soccer_fa	85.79	83.95	**87.04**	4.84	5.46	**2.74**	31.37	30.47	**43.35**	**5.51**	5.20	5.49
Salesman_fa	83.39	72.54	**84.69**	40.48	54.33	**12.41**	73.01	72.76	**79.88**	**22.41**	19.93	13.47
Stefan_fa	83.56	81.57	**90.14**	6.76	19.80	**4.87**	80.66	74.62	**83.30**	10.98	**15.16**	11.04
Mean	77.62	77.25	**80.83**	12.99	19.34	**12.80**	67.43	67.74	**70.47**	16.66	**17.77**	17.40

Table 2. Quantitative evaluation on the SegTrack v2 dataset. Best values are shown in bold. The table shows the evaluation results of the segmentation generated from the algorithm proposed in [7], and two of our feature configurations: basic 2D histogram (2D) and 2D histogram with spatial information (2Dsp). The algorithms are all initialized with 300 superpixels per frame.

Metrics	3D ACC			UE3D			BR3D			BP3D		
Methods	[7]	2D	2Dsp	[7]	2D	2Dsp	[7]	2D	2Dsp	[7]	2D	2Dsp
B.o.paradise	96.77	**96.81**	96.79	**2.74**	3.62	3.90	93.12	94.47	**94.80**	6.83	**6.98**	6.71
Birdfall	58.61	**67.54**	62.22	24.42	11.15	**10.52**	77.99	90.92	**92.36**	**0.61**	0.45	0.47
Bmx-1	**94.60**	94.50	94.56	5.49	6.44	**5.40**	98.31	98.32	**98.58**	4.05	**4.66**	4.23
Bmx-2	78.00	78.39	**81.40**	**11.43**	13.33	12.48	94.00	91.49	**95.01**	3.72	**4.17**	3.92
Cheetah-1	73.26	75.76	**76.35**	30.62	6.59	**5.46**	92.19	97.54	**98.62**	**1.65**	1.09	1.10
cheetah-2	63.84	**73.68**	69.38	34.64	**6.95**	8.73	97.85	98.54	**98.66**	**2.19**	1.38	1.38
Drift-1	**93.85**	93.20	93.34	3.77	**3.29**	3.42	92.70	**94.54**	94.53	1.22	1.20	**1.22**
Drift-2	**92.43**	92.41	92.06	3.31	2.98	**2.96**	90.52	**92.53**	92.13	0.94	0.93	**0.94**
Frog	56.92	64.72	**86.67**	16.32	14.01	**11.60**	59.28	76.14	**83.26**	**10.42**	3.84	2.25
Girl	87.71	**89.18**	**89.18**	10.76	**10.18**	10.27	90.18	94.59	**94.68**	**5.46**	5.39	5.32
Hum.bird-1	65.07	**73.32**	73.27	9.41	**9.16**	9.20	**88.50**	88.48	87.10	3.14	3.26	**3.76**
Hum.bird-2	77.71	84.95	**85.52**	**6.35**	7.04	9.06	**94.64**	94.26	94.58	5.00	5.18	**6.09**
Monkey	86.86	89.06	**89.62**	13.66	3.84	**3.73**	93.07	98.32	**98.37**	**2.79**	1.59	1.62
M.dog-1	88.09	88.70	**88.97**	9.50	**9.30**	9.38	95.74	97.44	**98.50**	1.40	**1.42**	**1.42**
M.dog-2	62.57	**65.60**	64.79	5.82	5.36	**5.15**	86.80	**91.13**	90.56	0.91	**0.95**	0.94
Parachute	92.54	92.31	**92.31**	19.54	18.29	**5.65**	95.24	95.71	**97.34**	**1.27**	1.13	0.76
Penguin-1	**95.72**	23.36	93.45	3.38	**3.56**	**3.56**	**49.25**	44.57	44.53	**0.89**	0.83	0.66
Penguin-2	95.51	95.77	**95.79**	3.39	**3.28**	**3.28**	73.19	71.41	**74.78**	1.38	**1.39**	1.17
Penguin-3	96.49	**96.79**	96.48	3.87	3.87	**3.83**	67.89	68.13	74.44	1.28	**1.32**	1.16
Penguin-4	**95.72**	94.50	94.74	**3.87**	3.95	3.92	73.54	73.82	**73.44**	1.16	**1.21**	0.96
Penguin-5	**93.27**	92.25	91.63	8.38	**8.21**	8.22	**74.01**	72.87	71.14	1.03	**1.05**	0.82
Penguin-6	92.37	92.64	**93.09**	**3.73**	4.02	4.03	62.33	**63.52**	59.50	1.02	**1.08**	0.81
Soldier	89.81	**90.19**	**90.19**	4.71	**4.11**	4.40	92.38	93.29	**93.48**	1.87	1.86	**1.89**
Worm	92.21	92.71	**92.75**	**10.31**	15.18	14.95	89.28	92.72	**93.48**	1.01	**1.19**	1.17
Average	84.16	83.26	**86.86**	10.39	7.40	**6.80**	84.25	86.45	**87.24**	**2.55**	2.23	2.12

Two example cases of PGP are shown in Fig. 8, and the 2nd and 3rd row of Fig. 1. For the over-segmented scenario in Fig. 1, the water was unfavorably divided into many spurious segments by the PGP baseline. Adding our feature did not only help merging the background into one segment, but also enhanced temporal consistency and boundary awareness. Given the under-segmented baseline result on the lower part of the tree shown in Fig. 8, our feature helped to segment the entire tree and also reduced over-segmentation in other parts of the video.

4.3   Feature Extraction Running Time

All experiments were conducted on an Intel Core i7 CPU with 3.5 Ghz, and 16 Gb of memory. When adding our feature into the framework of [8], the average additional running time was increased by 67 s on a 85-frame video using the default parameter settings, which is a just small fraction of the total running time of several hours. The additional running time increase for the PGP framework was on average 48 s, with 300 initial superpixels per frame. These results show that the computational cost of our feature is low, and adds very little overhead to existing frameworks.

5   Conclusion

In this paper, we introduced a novel feature for video segmentation based on geodesic distance histograms. The histogram is computed as a spatially-organized distribution of accumulated boundary costs between superpixels, which is a representation that includes more global information than conventional features. We validated the efficacy of our feature by adding it into two recent frameworks for video segmentation using spectral clustering and parametric graph partitioning, and showed that the proposed feature improved the performance of both frameworks in 3D video segmentation benchmarks, as well as the temporal consistency of the resulting supervoxels. We believe that the encoded global information can be further applied to other video related tasks such as moving object tracking, object proposals, and foreground background segmentation.

Acknowledgement. Partially supported by the Vietnam Education Foundation, NSF IIS-1161876, FRA DTFR5315C00011, the Stony Brook SensonCAT, the SubSample project from the DIGITEO Institute, France, and a gift from Adobe Corporation

References

1. Taralova, E.H., Torre, F., Hebert, M.: Motion words for videos. In: Fleet, D., Pajdla, T., Schiele, B., Tuytelaars, T. (eds.) ECCV 2014. LNCS, vol. 8689, pp. 725–740. Springer, Heidelberg (2014). doi:10.1007/978-3-319-10590-1_47
2. Jain, A., Chatterjee, S., Vidal, R.: Coarse-to-fine semantic video segmentation using supervoxel trees. In: ICCV, pp. 1865–1872. IEEE Computer Society (2013)
3. Kundu, A., Li, Y., Dellaert, F., Li, F., Rehg, J.M.: Joint semantic segmentation and 3D reconstruction from monocular video. In: Fleet, D., Pajdla, T., Schiele, B., Tuytelaars, T. (eds.) ECCV 2014. LNCS, vol. 8694, pp. 703–718. Springer, Heidelberg (2014). doi:10.1007/978-3-319-10599-4_45
4. Khoreva, A., Galasso, F., Hein, M., Schiele, B.: Classifier based graph construction for video segmentation. In: 2015 IEEE Conference on Computer Vision and Pattern Recognition (CVPR), pp. 951–960 (2015)
5. Grundmann, M., Kwatra, V., Han, M., Essa, I.: Efficient hierarchical graph-based video segmentation. In: 2010 IEEE Conference on Computer Vision and Pattern Recognition (CVPR), pp. 2141–2148 (2010)
6. Li, F., Kim, T., Humayun, A., Tsai, D., Rehg, J.M.: Video segmentation by tracking many figure-ground segments. In: 2013 IEEE International Conference on Computer Vision, pp. 2192–2199 (2013)

7. Yu, C.P., Le, H., Zelinsky, G., Samaras, D.: Efficient video segmentation using parametric graph partitioning. In: The IEEE International Conference on Computer Vision (ICCV) (2015)

8. Galasso, F., Cipolla, R., Schiele, B.: Video segmentation with superpixels. In: Lee, K.M., Matsushita, Y., Rehg, J.M., Hu, Z. (eds.) ACCV 2012. LNCS, vol. 7724, pp. 760–774. Springer, Heidelberg (2013). doi:10.1007/978-3-642-37331-2_57

9. Krähenbühl, P., Koltun, V.: Geodesic object proposals. In: Fleet, D., Pajdla, T., Schiele, B., Tuytelaars, T. (eds.) ECCV 2014. LNCS, vol. 8693, pp. 725–739. Springer, Heidelberg (2014). doi:10.1007/978-3-319-10602-1_47

10. Bai, X., Sapiro, G.: A geodesic framework for fast interactive image and video segmentation and matting. In: 2007 IEEE 11th International Conference on Computer Vision, pp. 1–8 (2007)

11. Wang, W., Shen, J., Porikli, F.: Saliency-aware geodesic video object segmentation. In: 2015 IEEE Conference on Computer Vision and Pattern Recognition (CVPR), pp. 3395–3402 (2015)

12. Price, B.L., Morse, B., Cohen, S.: Geodesic graph cut for interactive image segmentation. In: 2010 IEEE Conference on Computer Vision and Pattern Recognition (CVPR), pp. 3161–3168 (2010)

13. Ling, H., Jacobs, D.W.: Deformation invariant image matching. In: Tenth IEEE International Conference on Computer Vision (ICCV 2005), vols. 1, 2, pp. 1466–1473 (2005)

14. Lazebnik, S., Schmid, C., Ponce, J.: Beyond bags of features: spatial pyramid matching for recognizing natural scene categories. In: 2006 IEEE Computer Society Conference on Computer Vision and Pattern Recognition (CVPR 2006), vol. 2, pp. 2169–2178 (2006)

15. Cheng, H.T., Ahuja, N.: Exploiting nonlocal spatiotemporal structure for video segmentation. In: 2012 IEEE Conference on Computer Vision and Pattern Recognition (CVPR), pp. 741–748 (2012)

16. Leung, T., Malik, J.: Representing and recognizing the visual appearance of materials using three-dimensional textons. Int. J. Comput. Vision 43, 29–44 (2001)

17. Galasso, F., Keuper, M., Brox, T., Schiele, B.: Spectral graph reduction for efficient image and streaming video segmentation. In: The IEEE Conference on Computer Vision and Pattern Recognition (CVPR) (2014)

18. Galasso, F., Iwasaki, M., Nobori, K., Cipolla, R.: Spatio-temporal clustering of probabilistic region trajectories. In: Metaxas, D.N., Quan, L., Sanfeliu, A., Gool, L.J.V. (eds.) ICCV, pp. 1738–1745. IEEE Computer Society (2011)

19. Tsai, Y.H., Yang, M.H., Black, M.J.: Video segmentation via object flow. In: The IEEE Conference on Computer Vision and Pattern Recognition (CVPR) (2016)

20. Brox, T., Malik, J.: Object segmentation by long term analysis of point trajectories. In: Daniilidis, K., Maragos, P., Paragios, N. (eds.) ECCV 2010. LNCS, vol. 6315, pp. 282–295. Springer, Heidelberg (2010). doi:10.1007/978-3-642-15555-0_21

21. Lezama, J., Alahari, K., Sivic, J., Laptev, I.: Track to the future: Spatio-temporal video segmentation with long-range motion cues. In: 2011 IEEE Conference on Computer Vision and Pattern Recognition (CVPR) (2011)

22. Palou, G., Salembier, P.: Hierarchical video representation with trajectory binary partition tree. In: Computer Vision and Pattern Recognition (CVPR), Portland, Oregon (2013)

23. Brox, T., Malik, J.: Object segmentation by long term analysis of point trajectories. In: Daniilidis, K., Maragos, P., Paragios, N. (eds.) ECCV 2010. LNCS, vol. 6315, pp. 282–295. Springer, Heidelberg (2010). doi:10.1007/978-3-642-15555-0_21

24. Chen, A.Y.C., Corso, J.J.: Propagating multi-class pixel labels throughout video frames. In: 2010 Western New York Image Processing Workshop (WNYIPW), pp. 14–17 (2010)
25. Dollár, P., Zitnick, C.L.: Structured forests for fast edge detection. In: ICCV, International Conference on Computer Vision (2013)
26. Weinzaepfel, P., Revaud, J., Harchaoui, Z., Schmid, C.: Learning to detect motion boundaries. In: The IEEE Conference on Computer Vision and Pattern Recognition (CVPR) (2015)
27. Xu, C., Corso, J.J.: Evaluation of super-voxel methods for early video processing. In: 2012 IEEE Conference on Computer Vision and Pattern Recognition (CVPR), pp. 1202–1209 (2012)

HF-FCN: Hierarchically Fused Fully Convolutional Network for Robust Building Extraction

Tongchun Zuo, Juntao Feng, and Xuejin Chen[✉]

CAS Key Laboratory of Technology in Geo-spatial Information Processing
and Application System, University of Science and Technology of China, Hefei, China
{ztcustc,fjt}@mail.ustc.edu.cn, xjchen99@ustc.edu.cn

Abstract. Automatic building extraction from remote sensing images
plays an important role in a diverse range of applications. However, it is
significantly challenging to extract arbitrary-size buildings with largely
variant appearances or occlusions. In this paper, we propose a robust
system employing a novel hierarchically fused fully convolutional network (HF-FCN), which effectively integrates the information generated
from a group of neurons with multi-scale receptive fields. Our architecture takes an aerial image as the input without warping or cropping it
and directly generates the building map. The experiment results tested
on a public aerial imagery dataset demonstrate that our method surpasses state-of-the-art methods in the building detection accuracy and
significantly reduces the time cost.

1 Introduction

With the rapid development of remote sensing technologies and popularization
of geospatial related commercial software, high resolution satellite images are
easily accessible. These valuable data provide a huge fuel for interpreting real
terrestrial scenes. The building rooftop is one of the most important types of
terrestrial objects because it is essential for a wide range of technologies, such as,
urban planning, automated map making, 3D city modelling, disaster assessment,
military reconnaissance, etc. However, it is very costly and time-consuming to
manually delineate the footprint of buildings even for human experts.

In recent decades, many researchers have made massive attempts to extract
buildings automatically. Much of the past work defines criteria according to
the particular characteristics of rooftop, such as, polygonal boundary [1–4],
homogeneous color or texture [5], surrounding shadow [6–9], and their combinations [10,11]. However, such approaches are weakly capable of handling real-
world data because hand-coded rules or probabilistic models learned from a small
set of samples are heavily dependent on data. For example, they usually assume
that the building rooftop is a polygon. However, stadiums typically have circle
or oval shapes. Mnih [12] proposed a patch-based convolutional neural network
to extract location of objects automatically and provided a huge public dataset

© Springer International Publishing AG 2017
S.-H. Lai et al. (Eds.): ACCV 2016, Part I, LNCS 10111, pp. 291–302, 2017.
DOI: 10.1007/978-3-319-54181-5_19

including large-scale aerial images and their corresponding human-labeled maps. Based on Mnih's work, Saito *et al.* improved the extraction accuracy further by developing new cost function and model averaging techniques [13]. Though these methods achieve high performance, they still have limited ability to deal with two frequently appearing cases: (1) buildings are occluded by shadows or trees and (2) buildings possess moderately variant appearances.

Extracting buildings from aerial image is essentially a problem of semantic segmentation. Recent work suggests a number of methods in processing natural images. Long *et al.* [14] firstly proposed an effective architecture for semantic image segmentation, namely, fully convolutional network (FCN). Chen *et al.* [15] presented a system which combines the responses at the final convolutional layer with a fully connected conditional random field (CRF). The system is able to accurately segment semantic objects. Zheng *et al.* [15] introduced an end-to-end network which integrates CRF with CNNs to avoid off-line post-processing for object delineation. Noh *et al.* [16] applied a deconvolution network to each proposal in an input image, and constructed the final semantic segmentation map by combining the results from all proposals.

Although these methods show good performance in natural image segmentation, they have components not suited for building extraction in aerial image in three aspects. Firstly, each image in the PASCAL VOC 2012 dataset [17] has a handful of targets, while our source image is a complex scene, which has a number of targets with significant occlusions, variant appearances, and low contrast, as shown in Fig. 1(a)(b)(c), respectively. We directly integrate the coarse but strong semantic response into the output, instead of using CRF post-processing [12,15]. Secondly, the image in our remote sensing dataset contains many tiny buildings, as Fig. 1(d) shows. Noh *et al.* [16] indicated that FCNs [14] have less abilities in processing small objects. Thirdly, building extraction has much higher demand in precision of structure. The output of FCNs [14] has lower resolution which sacrifices precise structures severely.

In this paper, we present a robust building extraction system by developing a hierarchically fused fully convolutional network (HF-FCN). We trained our network on the large aerial image dataset [12]. In our architecture (HF-FCN),

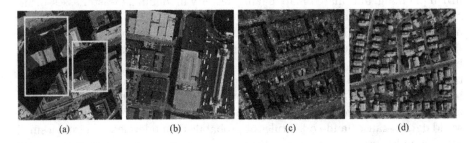

(a)   (b)   (c)   (d)

Fig. 1. Examples of aerial images with different type of challenges. (a) Occlusions in red boxes. (b) Variant appearances. (c) Low contrast. (d) A large number of tiny buildings. (Color figure online)

we design a new scheme to integrate multi-level semantic information generated from the convolutional layers with a group of increasing receptive fields, which capture context information of neighborhoods in different size. Therefore, it is more effective to handling buildings with arbitrary sizes, variant appearances or occlusions. Compared with the previous methods using convolution neural network [12,13], our HF-FCN does not require overlapped cropping and model averaging. Taking the whole image as input, it directly outputs the segmentation map by one pass of forward propagation. Therefore, the computational complexity is reduced significantly. In conclusion, our contributions include:

1. A new architecture is developed for building extraction, which has a strong ability in processing appearance variations, varying building sizes and occlusions. The overall accuracy exceeds the state-of-the-art algorithms.
2. Our approach leads to a notable reduction of computation cost compared with previous solutions.

The rest of this article is organized as follows. In Sect. 2, we summarize the related work for building extraction. Section 3 provides details of our neural network architecture. Section 4 introduces the dataset and training strategies of our proposed network, and experimental results while comparing our results to two state-of-the-art methods.

2 Related Work

In previous studies, extracting buildings by employing their shape information is a dominant method. It is observed that rooftops have more regular shapes, which usually are rectangular or combinations of several rectangles. Several studies [1–4] exploited a graph-based search to establish a set of rooftop hypotheses through examining the relationship of lines and line intersections, and then removing the fake hypotheses using a series of manually designed criteria. Cote and Saeedi [5] generated the rooftop outline from selected corners in multiple color and color-invariance spaces, further refine the outline by the level-set curve evolution algorithm. Though these methods based on geometric primitives achieved good performance in high contrast remote sensing imagery, they suffer from three shortcomings. Firstly, they lack the ability of detecting arbitrarily shaped building rooftop. Secondly, they fail to extract credible geometric features in buildings with inhomogeneous color distribution or low contrast with surroundings. Thirdly, it is time-consuming to process large-scale scenes because of their high computational complexity.

Apart from using shape information, spectral information is a distinctive feature for terrestrial object extraction. For instance, shadows are commonly dark grey or black, vegetations are usually green or yellow with particular textures, and main roads are dim gray in most cases. According to these prior knowledge, Ghaffarian et al. [18] split aerial scenes into three components (respectively, shadows and the vegetation, roads and the bare soil, buildings) using a group of manually established rules. Afterwards, a purposive fast independent component analysis technique is employed to separate building area in remote sensing image. However, their results are significantly sensitive to parameter choice. A feasible alternative strategy is to learn the appearance representation using supervised learning algorithm [8–10,19]. Firstly, an aerial image is divided into superpixels. Secondly, hand-crafted features, such as color histograms or local binary patterns, are extracted from each over-segmented region. Finally, each region is classified using machine learning tools and a gallery of training descriptors. Since it is inevitable for machine learning methods to mislabel regions with similar appearance, additional information is utilized to refine previous results. Ngo et al. [9] removed false rooftops using the assumption that buildings are surrounded by shadows because of illumination. Baluyan et al. [10] devised a "histogram method" to detect missed rooftops. Li et al. [11] selected probable rooftops after pruning out blobs using shadows, light direction, a series of shape criteria, and then these rooftops are refined by high order conditional random field. The drawbacks of these algorithms are threefold. (1) It is problematic to recognize an over-segmented region as building because terrestrial objects have hugely variant appearances in real scene. (2) Hand-craft features are less expressive to tremendous shape or appearance difference of buildings. Therefore, it is not robust to process large-scale remote sensing images. (3) Additional information is unreliable in many cases. For instance, some low buildings have no shadow in its neighborhood, and many buildings have unique structures that do not satisfy the hand-coded criteria.

As mentioned above, traditional methods are weakly capable of adapting to real scenes with huge variant appearances, occlusions or low contrast. Mnih, a pioneer, presented a patch-based framework for learning to label aerial images [12]. A neural network architecture is carefully designed for predicting buildings in aerial imagery, and the output of this network is processed by conditional random fields (CRFs). Satito et al. [13] improved Mnih's networks for extracting multiple kinds of objects simultaneously, two techniques consisting of model averaging with spatial displacement (MA) and channel-wise inhibited softmax (CIS) are introduced to enhance the performance. However, these methods need to crop test image to a fixed size, which not only increases the time cost, but also breaks the integrity of buildings. Our system takes whole images as inputs without overlapped cropping or wrapping and directly outputs labelling images. It is much beneficial to preserve the whole structure of buildings and shorten computation time.

3   Algorithm

In this section, we introduce our hierarchically fused fully convolutional network (HF-FCN) for extracting rooftops, and the implementation in the training stage.

3.1   Network Architecture

Given an input aerial image **S**, our goal is to predict a label image $\hat{M}$ where 1 for the pixel belonging to a building and 0 otherwise. We use similar strategy with semantic segmentation. We modify the VGG16 Net [20] by hierarchically fusing the response of all layers together, as shown in Fig. 2. The VGG16 Net [20] has 16 convolutional layers and five 2-stride down-sampling layers, from which we can acquire enough multi-level information. Its network parameters pre-trained on ImageNet are helpful for initializing our network because our aerial data are essentially optical imagery. We made the following modifications to detect buildings more effectively. Firstly, the sixth and seventh fully connected layers and the fifth pooling layer in VGG16 Net are cut, because they are at 1/32 of the resolution of the input image. As a result, the interpolated prediction map will be too fuzzy to utilize. Meanwhile, the number of neurons in the sixth and seventh convolutional layers is too large to cost intensive computation. The trimmed VGG16 Net is denoted as Level 1 in our HF-FCN. Secondly, the feature map from each convolutional layer in Level 1 are fed into a convolutional layer with a filter of 1×1 kernel. The outputs of these convolutional layers are upsampled and cropped to the size of input image. Upsampling is implemented via deconvolution which is initialized by bilinear interpolation. These upsampled feature maps compose the Level 2 in our HF-FCN. Finally, all the feature maps in Level 2 are stacked and put into a convolutional layer with a filter of kernel size of 1×1 to yield final predicted map, denoted as Level 3 in our HF-FCN. The size of the feature map in last stage of Level 1 is 1/16 of input image, which is too small to use. Thus, the input images are padded with all-zero band to enlarge the size of feature maps, similar as [14].

Table 1. The receptive field (RF) and the stride size of Level 2 in our architecture.

Layer	F1_1	F1_2	F2_1	F2_2	F3_1	F3_2	F3_3	F4_1	F4_2	F4_3	F5_1	F5_2	F5_3
RF	3	5	10	14	24	32	40	60	76	92	124	164	196
Stride	1	1	2	2	4	4	4	8	8	8	16	16	16

In Level 2, the feature maps with increasing receptive field (see Table 1) capture local information in larger neighbourhood sizes at higher semantic levels. The shallow layers generate feature maps with fine spatial resolution but low level semantic information. In contrast, the deep layers generate coarse feature maps with high-level semantic information. The feature maps at middle layers correspond to certain intermediate-level features. Integrating all these feature maps, buildings with variant appearances or occlusions are effectively extracted. An example is shown in Fig. 3. Given an aerial image, the U1_1 in Fig. 3(b) with small receptive field extracts low-level features like edges and corners. In Fig. 3(c), the U1_2 functions like an over-segmentation which groups pixels with similar color or texture into a subregion. In the U2_1 as Fig. 3(d) shows, shape

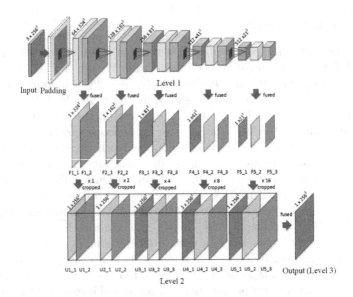

Fig. 2. Our network architecture. F1_1 means the fusion of feature maps generated from its corresponding convolutional layer conv1_1, U1_1 means the upsampling of F1_1, and so forth.

information is augmented. From the U3_3 as Fig. 3(e) shows, we can see that regions with significantly varying appearances are merged into an integrated building by considering high-level features. In U4_2 and U5_2 (see Fig. 3(f)(g)), our network learns strong semantic knowledge to distinguish dark rooftops with dim shadows and dark-green water area. In Level 3, we show that HF-FCN obtains a reliable prediction by combining multi-level semantic information and spatial information, as Fig. 3(h) shows.

3.2   Network Training

In the training stage, we train our network to directly generate a prediction map $\hat{\mathbf{M}}$ from raw pixels in the input aerial image $\mathbf{S}$ to approach a true label image $\tilde{\mathbf{M}}$. Figure 4 shows an example of $\mathbf{S}$, $\tilde{\mathbf{M}}$, $\hat{\mathbf{M}}$. We denote our input training data set as $\mathbf{I} = \{(\mathbf{S}_i, \tilde{\mathbf{M}}_i), i = 1, \dots, N\}$, N is the number of aerial image and labeled map pairs.

Taking account of each input image holistically and independently, the subscript i is ignored for notational simplicity in the following definition. In our image-to-image training stage, the loss function is computed over all pixels in a training image $\mathbf{S} = \{s_j, j = 1, \dots, |\mathbf{S}|\}$ and building map $\tilde{\mathbf{M}} = \{\tilde{m}_j, j = 1, \dots, |\mathbf{S}|\}$, $\tilde{m}_j \in \{0, 1\}$, where $|S|$ is the number of pixels in $\mathbf{S}$. For simplicity, we denote the collection of all standard network layer parameters as $\mathbf{W}$. For each pixel j in a training image, the probability that assigns it to building is

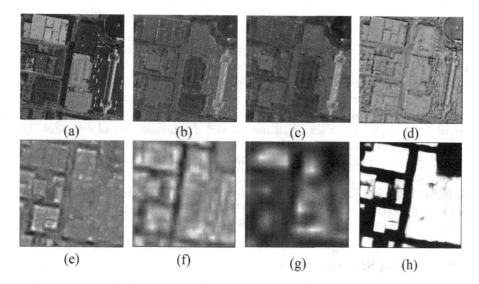

Fig. 3. (a) Input aerial image. (b–g) Feature maps generated from U1_1, U1_2, U2_1, U3_3, U4_2, U5_2, respectively. (h) Predicted labelling map.

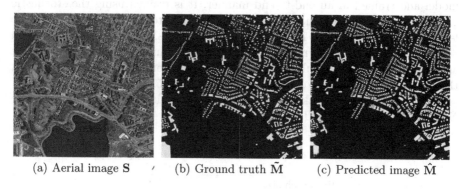

(a) Aerial image **S**   (b) Ground truth **M̃**   (c) Predicted image **M̂**

Fig. 4. An example of the predicted image.

denoted as its probability as a building $\hat{m}_j$. We use the sigmoid cross-entropy loss function defined as

$$\mathcal{L} = -\frac{1}{|\mathbf{S}|} \sum_{s_j \in \mathbf{S}} \left[\tilde{m}_j \log \hat{m}_j + (1 - \tilde{m}_j) \log (1 - \hat{m}_j) \right]. \tag{1}$$

4   Experiments

In this section, we introduce our detailed implementation and report the performance of our proposed algorithm.

4.1 Dataset

In our experiments, we use Massachusetts Buildings Dataset (*Mass. Buildings*) proposed by Mnih [12]. The dataset consists of 151 aerial images of the Boston area, with each image being 1500×1500 pixels for an area of 2.25 square kilometers. The entire dataset covers roughly 340 square kilometers. The intensity of each aerial image is scaled into the range of $[0, 1]$. The data is split into a training set of 137 images, a test set of 10 images and a validation set of 4 images. To train the network, we create a set of image tiles for training and validation by cropping each aerial image using a sliding window with size of 256×256 pixels and stride of 64 pixels. After scanning, the training and validation datasets include 75938 tiles and 2500 tiles respectively, with their corresponding building masks. For testing, we use ten 1500×1500 images excluded from the training images.

4.2 Training Settings

The implementation of our network is based on the *Caffe* Library [21]. Our HF-FCN is fine-tuned from an initialization with the pre-trained VGG16 Net model and trained in an end-to-end manner. It is trained using the stochastic gradient descent algorithm, with the hyper-parameters listed in Table 2. The learning rate is divided by 10 for each 8000 iterations. We find that the learned deconvolutions provide no noticeable improvements in our experiments, similar as [14,22]. Therefore, lr_mult is set to zero for all deconvolutional layers. Except that the pad of first convolutional layer is set to 35, the others are set to 1, same as VGG16 Net. It takes about six hours to train our network on a single NVIDIA Titan 12 GB GPU.

Table 2. Parameters for network training.

Mini-batch size	18
Initial learning rate	10^{-5}
Momentum	0.9
Weight decay	0.02
Clip_gradients	10000
The number of training iterations	16000

4.3 Results

To show the effectiveness of HF-FCN, we compare our method with two state-of-the-art approaches [12,13]. Three common metrics are used to evaluate the performance of our algorithm: (1) the relaxed precision and recall scores with $\rho = 3$; (2) the standard precision and recall scores ($\rho = 0$); (3) the time cost. The relaxed precision is defined as the fraction of detected pixels that are within

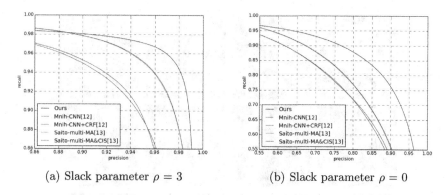

(a) Slack parameter $\rho = 3$ (b) Slack parameter $\rho = 0$

Fig. 5. The relaxed precision-recall curves from different methods with two slack parameters.

ρ pixels of a true pixel, while the relaxed recall is defined as the fraction of the true pixels that are within ρ pixels of a detected pixel. The slack parameter ρ is set to 3, which is the same value as used in [12,13]. The relaxed precision-recall curves generated from different methods are shown in Fig. 5(a). As can be seen, all curves of ours are located above others in building prediction obviously. More strictly, we set slack parameter ρ as 0, that is to say, it becomes a standard precision and recall scores. The precision-recall curves generated from different methods are shown in Fig. 5(b). We can see that our approach is more appropriate for detecting rooftops in complex scene, which significantly outperforms [12,13]. To compare the system efficiency, we calculate the average time of processing ten test images in the same computer using different methods. Table 3 shows that our method is able to not only significantly improve the performance, but also dramatically reduces the time cost.

To prove that our network has strong ability in extracting buildings with variant appearances, arbitrary sizes, occlusions, a new experiment is designed. We crop seven 256×256 image patches that have buildings with variant appearances or occlusions from the test images. Corresponding predictions are directly

Table 3. Performance comparison with [12,13]. Recall here means recall at breakeven points. Time is computed in the same computer with a single NVIDIA Titan 12 GB GPU.

	Recall ($\rho = 3$)	Recall ($\rho = 0$)	Time (s)
Mnih-CNN [12]	0.9271	0.7661	8.70
Mnih-CNN+CRF [12]	0.9282	0.7638	26.60
Saito-multi-MA [13]	0.9503	0.7873	67.72
Saito-multi-MA&CIS [13]	0.9509	0.7872	67.84
Ours (HF-FCN)	**0.9643**	**0.8424**	**1.07**

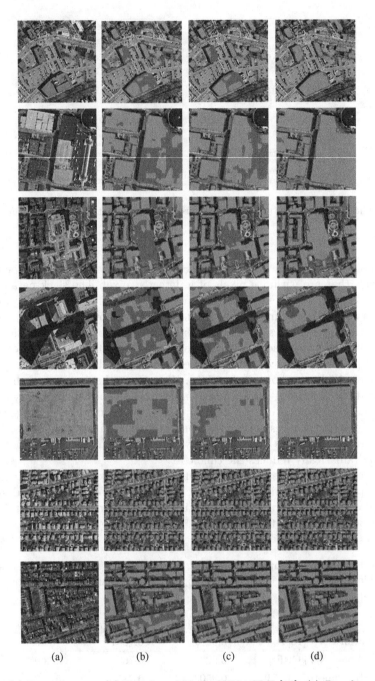

Fig. 6. (a) Input images. (b) Results of Mnih-CNN+CRF [12]. (c) Results of Saito-multi-MA&CIS [13]. (d) Our results. Correct results (TP) are shown in green, false positives (FP) are shown in blue, and false negatives (FN) are shown in red. (Color figure online)

Table 4. Recall at the selected regions of the test images.

Image ID	01	02	03	04	05	06	07	Mean
Mnih-CNN+CRF [12]	0.784	0.869	0.769	0.653	0.893	0.764	0.800	0.784
Saito-multi-MA&CIS [13]	0.773	0.915	0.857	0.789	0.945	0.773	0.830	0.851
Ours (HF-FCN)	**0.874**	**0.964**	**0.899**	**0.901**	**0.986**	**0.840**	**0.851**	**0.911**

cropped from the predicted images generated by three approaches, including Mnih-CNN+CRF [12], Saito-multi-MA&CIS [13] and ours. We binarize the probability map using a threshold of 0.5. A series of examples is shown in Fig. 6. In addition, Table 4 shows the resulting recalls at the breakeven points of the standard precision recall curve for each patch. The accuracy of our approach is 12.7%, 6.0% higher than [12,13], receptively.

5   Conclusions

In this article, we propose a novel fully convolutional network which is strongly capable of extracting buildings of arbitrary sizes, variant appearances or occlusions without any post-processing. Meanwhile, it further improves the overall accuracy. The proposed network can take arbitrary-size image as the input as long as the GPU memory allows. Compared with patch-based methods, there is no need to label a whole image by cropping the image into small patches. As consequence, inconsistent border caused by cropping would not occur in our system. Moreover, the time cost is tremendously reduced using our HF-FCN. The proposed method is demonstrated robust to various types of aerial scenes selected from real-world data. Furthermore, our architecture can be easily extended to extract multi-objects in remote sensing imagery. Consequently, we believe that our technique potentially provides a generic solution to understand complex aerial scenes.

Acknowledgement. We would like to thank the anonymous reviewers. This work was supported by the National Natural Science Foundation of China (NSFC) under Nos. 61472377 and 61331017, and the Fundamental Research Funds for the Central Universities under No. WK2100060011.

References

1. Noronha, S., Nevatia, R.: Detection and modeling of buildings from multiple aerial images. IEEE Trans. Pattern Anal. Mach. Intell. **23**, 501–518 (2001)
2. Nosrati, M.S., Saeedi, P.: A novel approach for polygonal rooftop detection in satellite/aerial imageries. In: 2009 16th IEEE International Conference on Image Processing (ICIP), pp. 1709–1712 (2009)
3. Izadi, M., Saeedi, P.: Three-dimensional polygonal building model estimation from single satellite images. IEEE Trans. Geosci. Remote Sens. **50**, 2254–2272 (2012)

4. Wang, J., Yang, X., Qin, X., Ye, X., Qin, Q.: An efficient approach for automatic rectangular building extraction from very high resolution optical satellite imagery. IEEE Geosci. Remote Sens. Lett. **12**, 487–491 (2015)

5. Cote, M., Saeedi, P.: Automatic rooftop extraction in nadir aerial imagery of suburban regions using corners and variational level set evolution. IEEE Trans. Geosci. Remote Sens. **51**, 313–328 (2013)

6. Sirmacek, B., Unsalan, C.: Building detection from aerial images using invariant color features and shadow information. In: 23rd International Symposium on Computer and Information Sciences, ISCIS 2008, pp. 1–5 (2008)

7. Manno-Kovcs, A., Ok, A.O.: Building detection from monocular VHR images by integrated urban area knowledge. IEEE Geosci. Remote Sens. Lett. **12**, 2140–2144 (2015)

8. Chen, D., Shang, S., Wu, C.: Shadow-based building detection and segmentation in high-resolution remote sensing image. J. Multimedia **9**, 181–188 (2014)

9. Ngo, T.T., Collet, C., Mazet, V.: Automatic rectangular building detection from VHR aerial imagery using shadow and image segmentation. In: 2015 IEEE International Conference on Image Processing (ICIP), pp. 1483–1487 (2015)

10. Baluyan, H., Joshi, B., Al Hinai, A., Woon, W.L.: Novel approach for rooftop detection using support vector machine. ISRN Mach. Vis. **2013** (2013)

11. Li, E., Femiani, J., Xu, S., Zhang, X., Wonka, P.: Robust rooftop extraction from visible band images using higher order CRF. IEEE Trans. Geosci. Remote Sens. **53**, 4483–4495 (2015)

12. Mnih, V.: Machine learning for aerial image labeling. Doctoral (2013)

13. Saito, S., Yamashita, Y., Aoki, Y.: Multiple object extraction from aerial imagery with convolutional neural networks. J. Imaging Sci. Technol. **60** (2016)

14. Long, J., Shelhamer, E., Darrell, T.: Fully convolutional networks for semantic segmentation. In: 2015 IEEE Conference on Computer Vision and Pattern Recognition (CVPR), pp. 3431–3440 (2015)

15. Zheng, S., Jayasumana, S., Romera-Paredes, B., Vineet, V., Su, Z., Du, D., Huang, C., Torr, P.H.S.: Conditional random fields as recurrent neural networks. In: 2015 IEEE International Conference on Computer Vision (ICCV), pp. 1529–1537 (2015)

16. Noh, H., Hong, S., Han, B.: Learning deconvolution network for semantic segmentation. In: 2015 IEEE International Conference on Computer Vision (ICCV), pp. 1520–1528 (2015)

17. Everingham, M., Van Gool, L., Williams, C.K.I., Winn, J., Zisserman, A.: The PASCAL Visual Object Classes Challenge 2012 (VOC2012) Results. http://www.pascal-network.org/challenges/VOC/voc2012/workshop/index.html

18. Ghaffarian, S., Ghaffarian, S.: Automatic building detection based on purposive FastICA (PFICA) algorithm using monocular high resolution google earth images. ISPRS J. Photogramm. Remote Sens. **97**, 152–159 (2014)

19. Dornaika, F., Moujahid, A., Bosaghzadeh, A., El Merabet, Y., Ruichek, Y.: Object classification using hybrid holistic descriptors: application to building detection in aerial orthophotos. Polibits **51**, 11–17 (2015)

20. Simonyan, K., Zisserman, A.: Very deep convolutional networks for large-scale image recognition. Computer Science (2015)

21. Jia, Y., Shelhamer, E., Donahue, J., Karayev, S., Long, J., Girshick, R., Guadarrama, S., Darrell, T.: Caffe: Convolutional architecture for fast feature embedding. Eprint Arxiv, pp. 675–678 (2014)

22. Xie, S., Tu, Z.: Holistically-nested edge detection. In: The IEEE International Conference on Computer Vision (ICCV) (2015)

Dictionary Learning, Retrieval, and Clustering

Dictionary Reduction: Automatic Compact Dictionary Learning for Classification

Yang Song$^{(\boxtimes)}$, Zhifei Zhang, Liu Liu, Alireza Rahimpour, and Hairong Qi

Department of Electrical Engineering and Computer Science,
University of Tennessee, Knoxville, TN 37996, USA
ysong18@vols.utk.edu

Abstract. A complete and discriminative dictionary can achieve superior performance. However, it also consumes extra processing time and memory, especially for large datasets. Most existing compact dictionary learning methods need to set the dictionary size manually, therefore an appropriate dictionary size is usually obtained in an exhaustive search manner. How to automatically learn a compact dictionary with high fidelity is still an open challenge. We propose an automatic compact dictionary learning (ACDL) method which can guarantee a more compact and discriminative dictionary while at the same time maintaining the state-of-the-art classification performance. We incorporate two innovative components in the formulation of the dictionary learning algorithm. First, an indicator function is introduced that automatically removes highly correlated dictionary atoms with weak discrimination capacity. Second, two additional constraints, namely, the sum-to-one and the nonnegative constraints are imposed on the sparse coefficients. On one hand, this achieves the same functionality as the L_2-normalization on the raw data to maintain a stable sparsity threshold. On the other hand, this effectively preserves the geometric structure of the raw data which would be otherwise destroyed by the L_2-normalization. Extensive evaluations have shown that the preservation of geometric structure of the raw data plays an important role in achieving high classification performance with smallest dictionary size. Experimental results conducted on four recognition problems demonstrate the proposed ACDL can achieve competitive classification performance using a drastically reduced dictionary (https://github.com/susanqq/ACDL.git).

1 Introduction

Given a set of measurements $\{\mathbf{y}_i\}_{i=1}^n$, dictionary learning (DL) is designed to learn an overcomplete dictionary $\{\mathbf{d}_i\}_{i=1}^d$, which is utilized to represent the measurement in a sparse manner. Most DL algorithms fall into one of the two categories: unsupervised learning and supervised learning. Many unsupervised DL approaches [1–5] aim to minimize the reconstruction error between observations and their sparse representation, making them suitable to solve problems like image denosing, image decoding and inpainting.

© Springer International Publishing AG 2017
S.-H. Lai et al. (Eds.): ACCV 2016, Part I, LNCS 10111, pp. 305–320, 2017.
DOI: 10.1007/978-3-319-54181-5_20

Driven by the classification task, supervised DL approaches [6–13] aim to learn a dictionary as discriminative as possible by exploiting the label information. In the big data era, such a large and overcomplete dictionary brings challenges to both processing time and storage. Obviously, a large dictionary requires large memory, and it is more time-consuming in solving sparse coefficient. To reduce the dictionary size, compact dictionary learning techniques have been studied to learn a dictionary with less redundancy and high distinguishability [8,9,14–19]. DLSI [6] learns each class-wise dictionary with less coherence to ensure compactness. FDDL [9] simultaneously learns the discriminative class-wise dictionary and sparse coefficient which satisfies the Fisher criterion. DL-COPAR [16] fixed this problem by separating the dictionary into the common part and class-specific part via a predefined threshold. However, it is difficult to find an appropriate threshold that balances these two parts. ITDL [19] enforces the incoherence of selected dictionary atoms by maximizing the mutual information measurement on the dictionary. All of above methods need the user to specify the size of dictionary. Therefore, given a dataset, an appropriate dictionary size could only be obtained by exhaustively experimenting with different dictionary sizes, which is a tedious procedure for different applications and is often not desirable. LCKSVD [8] represents state-of-the-art DL and has shown superior performance in various computer vision related applications, but it does not have a mechanism to automatically determine the dictionary size. SADL [20] adaptively learns a compact codebook by involving the row sparsity, but it is an unsupervised method which is not for classification purpose. LDL [18] is also an automatic compact DL algorithm, where a latent matrix is used to control the redundancy, but it is a class-wise learning algorithm which cannot guarantee the most discriminative and compact dictionary. Another drawback of existing works is that they require L_2-normalization on the raw data to facilitate the determination of the sparsity threshold. However, this normalization also destroys the geometric structure of the raw data that should be utilized as an important and discriminate feature.

In this paper, we propose an automatic compact dictionary learning (ACDL) method globally. Inspired by DKSVD [7], a classification error term is also introduced to learn a linear classifier jointly with the dictionary. The contribution of ACDL is two-fold. First, an indicator function is designed to automatically remove highly correlated dictionary atoms with weak discrimination capacity. By alternatively updating the dictionary and classifier, the indicator function automatically identifies those common and redundant atoms, and an appropriate dictionary is obtained until the function output is stable. Second, instead of using the common L_2-normalization to maintain a stable sparsity threshold which, on the other hand, unavoidably destroys the geometric structure of the raw data set, ACDL introduces two constraints, namely, the sum-to-one (S2O) and the non-negative (NN) constraints, on the sparse coefficients. These two constraints achieve the same effect as L_2-normalization in terms of maintaining a stable sparsity threshold. However, they also effectively preserve the structure of the raw data in the original space. We show through extensive experiments as well as the toy example (Sect. 2) that the geometric structure of raw data is essential in achieving high classification performance.

The rest of the paper is organized as follows: Sect. 2 overviews representative works and compares them with the propose method using a toy example. Sect. 3 elaborates on the proposed automatic compact dictionary learning (ACDL). Experimental results are shown in Sect. 4. Section 5 concludes the paper.

2 Motivation

Most state-of-the-art DL works designed for classification, e.g., SDL [17], DL-COPAR [16], FDDL [9] and LCKSVD [8], will first normalize (i.e., L_2 normalization) the dataset and then learn a dictionary of certain size that is manually set in ahead. However, normalization will destroy the geometric structure of the data and may mix the data together that leads misclassification. For example, two 2-D points of different classes locate at $(0,0)$ and $(2,0)$ will overlap at $(1,0)$ after L_2-normalization. In addition, manual setting of the dictionary size is difficult to achieve the optimal because the latent optimal dictionary size will vary with applications. A common way to get an appropriate size is exhaustive searching that is time-consuming.

To preserve the geometric structure of the raw dataset, we propose the automatic compact dictionary learning (ACDL) method that does not adopt any normalization methods. Therefore, the data from different classes will not overlap by mistake. The non-negative and sum-to-one constraints are incorporated to force the learned dictionary items to represent the skeleton of geometric structure. A toy example is shown in Fig. 1 to illustrate the effect of non-negative and sum-to-one constraints.

Two classes of data samples are constructed based on the Gaussian distribution with the first class using a uni-modal distribution and the second class using a bi-modal distribution. The purpose of the experiment is to see the effect of normalization and the two constraints. Figure 1 shows the data set and learned dictionary, where red squares and blue circles indicate samples from two classes. The hexagrams in magenta and cyan, denote the learned dictionary atoms belonging to the red and blue class, respectively. It mainly demonstrates two advantages of the proposed ACDL: preservation of the geometric structure and automatic determination of dictionary size.

Figure 1(a) has illustrated the importance of geometric structure because the normalization used in most algorithms (e.g., LCKSVD) will overlap the two clusters on the diagonal direction of the coordinate system. Therefore, the dictionary learned by LCKSVD cannot well distinguish the samples located along the diagonal direction. Figure 1(b) shows the learned dictionary without normalization. The dictionary atoms tends to approach the clusters. However, the atoms cannot well represent the geometric structure, and the location of atoms are unstable in each learning process. Compared to Fig. 1(c) and (d), ACDL (Fig. 1(e)) well preserves the geometric structure through the non-negative and sum-to-one constraints and automatically learns a dictionary with three atoms.

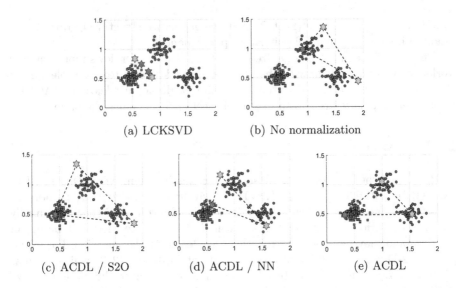

Fig. 1. A toy example to illustrate the learning results of different conditions. Red squares and blue circles are samples from two classes. Hexagrams denote the learned dictionary atoms, and magenta and cyan have the same label with red and blue, respectively. The L_2-normalization is required and the dictionary size needs to be set manually in all methods except the ACDL. (a) The result of LCKSVD, and the dictionary size is set to be 3. (b) Illustration of LCKSVD without L_2-normalization. (c) and (d) The results of ACDL without sum-to-one (S2O) and non-negative (NN) constraints, respectively. (e) The result of the proposed ACDL method, which incorporates the S2O and NN, relaxing the requirement of normalization and preserving the geometric structure of the data and dictionary items. (Color figure online)

To summarize, the key advantage of ACDL over the other methods is that it can automatically determine the dictionary size according the dataset without human intervention. In addition, the geometric structure of the raw data set is preserved to potentially improve classification performance.

3   Approach

In this section, we elaborate on the proposed ACDL approach for the automatic determination of the size of a compact dictionary while at the same time maintaining competitive performance as compared to the state-of-the-art. Assume a set of observations $\mathbf{Y} = [\mathbf{y}_1, \mathbf{y}_2, \cdots, \mathbf{y}_n] \in \mathbb{R}^{m \times n}$ that can be represented by linear composition of a few atoms in a dictionary $\mathbf{D} = [\mathbf{d}_1, \mathbf{d}_2, \cdots, \mathbf{d}_d] \in \mathbb{R}^{m \times d}$, with the corresponding weight of each atom (or sparse coefficients) being $\mathbf{X} = [\mathbf{x}_1, \mathbf{x}_2, \cdots, \mathbf{x}_n] \in \mathbb{R}^{d \times n}$. Note that $n > d$, and each column of $\mathbf{X}$ is sparse. Equation 1 shows the sparse representation.

$$\mathbf{Y} \approx \mathbf{DX} \tag{1}$$

For classification purpose, each atom in the dictionary $\mathbf{D}$ is learned to distinguishably represent certain class. Therefore, the label of an observation can be decided based on the corresponding sparse coefficients. For simplicity, the dictionary atom with the largest sparse coefficient indicates that the observation belongs to the same class as the atom. Suppose the label of observation is $\mathbf{G} = [\mathbf{g}_1, \mathbf{g}_2, \cdots, \mathbf{g}_n] \in \mathbb{R}^{c \times n}$ (assume there are c classes), and each $\mathbf{g}_i$ is a column vector whose elements are zero except for the one having the same row index as the class index. For example, if $c = 2$, then $\mathbf{g}_1 = [1, 0]^T$ means the label of the first observation is 1. By the same token, a linear classifier can be written as $\mathbf{W} = [\mathbf{w}_1, \mathbf{w}_2, \cdots, \mathbf{w}_d] \in \mathbb{R}^{c \times d}$, where for a $\mathbf{w}_i = [w_1, w_2, \cdots, w_c]^T$, certain element w_j indicates the probability that $\mathbf{d}_i$ belongs to the jth class. Therefore, class information of the observations could be represented sparsely as in Eq. 2.

$$\mathbf{G} \approx \mathbf{WX} \qquad (2)$$

Equations 1 and 2 consider reconstruction fidelity and classification error, receptively. By combining these two terms, a global dictionary learning algorithm can be formulated.

3.1 Automatic Compact Dictionary Learning

A preliminary objective function to globally learn a dictionary and the corresponding classifier can be written in Eq. 3,

$$\arg \min_{\mathbf{D}, \mathbf{W}} \|\mathbf{Y} - \mathbf{DX}\|_F^2 + \gamma \|\mathbf{G} - \mathbf{WX}\|_F^2 + \lambda \|\mathbf{X}\|_1$$
$$s.t. \ \mathbf{X} \succeq 0, \ \sum \mathbf{x}_i = 1 \qquad (3)$$

where the first term denotes the reconstruction error, the second term represents the classification error, and the last term indicates the sparsity constraint. Different from existing compact dictionary learning methods, in ACDL, we constrain the sparse coefficients $\mathbf{X}$ under two additional conditions, i.e., the non-negative condition and the column-wise sum-to-one condition in order to preserve the geometric structure of the dataset. Specifically, these two constraints together force the learned dictionary to form a polyhedron upon the dataset, whose vertices (dictionary atoms) have the tendency to be placed in the middle of the data samples with dense distributions. Thus, the polyhedron more explicitly reflects the distribution of the dataset in its raw spatial scale.

Given a set of observations and their labels, as well as predefined dictionary size, minimizing Eq. 3 could obtain a dictionary with appropriate size along with the classifier. However, we aim to learn a dictionary with its size automatically determined. Ideally, the dictionary $\mathbf{D}$ is initialized to be the observation matrix $\mathbf{Y}$, and then $\mathbf{D}$ is updated and reduced automatically according to the dataset given. From this perspective, both the size of the dictionary and the atom need to be updated during the learning procedure. To achieve this goal, we rewrite

Eqs. 3 and 4 by adding an operator $\mathcal{F}(\mathbf{D}, \mathbf{W})$ which is a function of $\mathbf{D}$ and $\mathbf{W}$ to indicate the atoms that need to be removed from the dictionary.

$$\arg\min_{\mathbf{D},\mathbf{W}} \|\mathbf{Y} - \mathbf{D}\mathcal{F}(\mathbf{D}, \mathbf{W})\mathbf{X}\|_F^2 + \gamma\|\mathbf{G} - \mathbf{W}\mathcal{F}(\mathbf{D}, \mathbf{W})\mathbf{X}\|_F^2 + \lambda\|\mathbf{X}\|_1$$
$$s.t.\ \mathbf{X} \succeq 0,\ \sum \mathbf{x}_i = 1 \tag{4}$$

Specifically, $\mathcal{F}(\mathbf{D}, \mathbf{W})$ yields a diagonal matrix of dimension $d \times d$. The elements on the diagonal are either 0 or 1 with 0 indicating the corresponding atom in $\mathbf{D}$ should be removed. Assume $\mathcal{D}_i = \mathbf{D}\mathcal{F}(\mathbf{D}, \mathbf{W})diag(\mathcal{B}(\mathbf{W})_{i*})$, where i is the class index, $\mathbf{W}_{i*}$ denotes the ith row of $\mathbf{W}$, $\mathcal{B}$ is a column-wised binary non maximum suppression operator, for example, $\mathbf{W} = [0.8, 0.2, 0.1; 0.7, 0.2, 0.1]^T$, $\mathcal{B}(\mathbf{W}) = [1, 0, 0; 1, 0, 0]^T$. $diag(\mathbf{W}_{i*})$ forms a diagonal matrix indicating the probability that each atom of $\mathbf{D}$ belongs to the ith class. That is, $\mathcal{D}_i$ is the same matrix as $\mathbf{D}$ but the atoms that are indicated as removed or do not belong to the ith class are set to zero. By the same token, $\mathcal{W}_i = \mathbf{W}\mathcal{F}(\mathbf{D}, \mathbf{W})diag(\mathbf{W}_{i*})$. Based on $\mathcal{D}_i$'s and $\mathcal{W}_i$'s, $\mathcal{F}(\mathbf{D}, \mathbf{W})$ is updated by Eq. 5.

$$\mathcal{F}(\mathbf{D}, \mathbf{W})_{jj} = \begin{cases} \mathcal{F}(\mathbf{D}, \mathbf{W})_{jj}, & \mathbf{d}_j = 0, \mathbf{d}_j \in \mathcal{D}_i \\ 0, & \dfrac{\min_k \|\mathbf{d}_j - \mathbf{d}_k\|_2}{\max_{p,q} \|\mathbf{d}_p - \mathbf{d}_q\|_2} < \epsilon \\ & \text{and } \mathbf{w}_j \ln \mathbf{w}_j < \mathbf{w}_k \ln \mathbf{w}_k \\ & \mathbf{w}_j, \mathbf{w}_k \in \mathcal{W}_i \\ & \mathbf{d}_* \in \mathcal{D}_i, \mathbf{d}_* \neq 0 \\ & or\ \mathcal{F}(\mathbf{D}, \mathbf{W})_{jj} = 0 \\ 1, & \text{otherwise} \end{cases} \tag{5}$$

where $\mathcal{F}(\mathbf{D}, \mathbf{W})_{jj}$ denotes the jth element on the diagonal of $\mathcal{F}(\mathbf{D}, \mathbf{W})$. ϵ is a threshold from 0 to 1. Generally speaking, if a dictionary atom is close to another from the same class as measured in Euclidian distance or other distance metrics (we use Euclidian distance in this paper) and less discriminative (the corresponding column in $\mathbf{W}$ has lower entropy), it will be removed (or set to zero) in the subsequent iterations. $\mathcal{F}(\mathbf{D}, \mathbf{W})$, $\mathbf{D}$ and $\mathbf{W}$ are updated alternatively. Usually, $\mathbf{D}$ and $\mathbf{W}$ are initialized to be $\mathbf{Y}$ and $\mathbf{G}$, respectively. The initial $\mathbf{X}$ and $\mathcal{F}(\mathbf{D}, \mathbf{W})$ are set to be an identity matrix. Then, iterating Eqs. 4 and 5 will result in a compact dictionary with its size (i.e., number of non-zero columns) automatically and optimally determined. Finally, the dictionary and classifier are generated using the non-zero columns of $\mathbf{D}\mathcal{F}(\mathbf{D}, \mathbf{W})$ and $\mathbf{W}\mathcal{F}(\mathbf{D}, \mathbf{W})$, respectively.

We will discuss more on solving Eq. 4. Since $\mathcal{F}(\mathbf{D}, \mathbf{W})$ is updated after $\mathbf{D}$ and $\mathbf{W}$, it can be considered as a constant when $\mathbf{D}$ and $\mathbf{W}$ are being updated. Because $\mathcal{F}(\mathbf{D}, \mathbf{W})$ forces certain atoms in $\mathbf{D}$ and $\mathbf{W}$ to be zero, the exact dictionary and classifier that need to be updated are $\widehat{\mathbf{D}} = \{\mathbf{D}\mathcal{F}(\mathbf{D}, \mathbf{W})_{*j} | \mathcal{F}(\mathbf{D}, \mathbf{W})_{*j} \neq 0\}$ and $\widehat{\mathbf{W}} = \{\mathbf{W}\mathcal{F}(\mathbf{D}, \mathbf{W})_{*j} | \mathcal{F}(\mathbf{D}, \mathbf{W})_{*j} \neq 0\}$, respectively, where $\mathcal{F}(\mathbf{D}, \mathbf{W})_{*j}$ denotes the jth column of $\mathcal{F}(\mathbf{D}, \mathbf{W})$. Correspondingly, the equivalent sparse coefficients $\widehat{\mathbf{X}} = \{\mathcal{F}(\mathbf{D}, \mathbf{W})_{j*}\mathbf{X} | \mathcal{F}(\mathbf{D}, \mathbf{W})_{j*} \neq \mathbf{0}\}$. Thus, Eq. 4 can be rewritten as Eq. 6.

$$\arg\min_{\widehat{\mathbf{D}}, \widehat{\mathbf{W}}} \|\mathbf{Y} - \widehat{\mathbf{D}}\widehat{\mathbf{X}}\|_F^2 + \gamma\|\mathbf{G} - \widehat{\mathbf{W}}\widehat{\mathbf{X}}\|_F^2 + \lambda\|\widehat{\mathbf{X}}\|_1$$

$$s.t. \ \widehat{\mathbf{X}} \succeq 0, \ \sum \widehat{\mathbf{x}}_i = 1 \tag{6}$$

Combining the first two terms and augmenting the sum-to-one constraint, Eq. 6 is simplified as Eq. 7.

$$\arg\min_{\widehat{\mathbf{D}}, \widehat{\mathbf{W}}} \left\| \begin{pmatrix} \mathbf{Y} \\ \sqrt{\gamma}\mathbf{G} \\ \delta\mathbf{1}_{1\times n} \end{pmatrix} - \begin{pmatrix} \widehat{\mathbf{D}} \\ \sqrt{\gamma}\widehat{\mathbf{W}} \\ \delta\mathbf{1}_{1\times d} \end{pmatrix} \widehat{\mathbf{X}} \right\|_F^2 + \lambda\|\widehat{\mathbf{X}}\|_1$$

$$s.t. \ \widehat{\mathbf{X}} \succeq 0 \tag{7}$$

Assume $\widetilde{\mathbf{Y}} = \begin{pmatrix} \mathbf{Y} \\ \sqrt{\gamma}\mathbf{G} \\ \delta\mathbf{1}_{1\times n} \end{pmatrix}$ and $\widetilde{\mathbf{D}} = \begin{pmatrix} \widehat{\mathbf{D}} \\ \sqrt{\gamma}\widehat{\mathbf{W}} \\ \delta\mathbf{1}_{1\times d} \end{pmatrix}$, where $\mathbf{1}$ denotes the matrix of all 1's and δ balances the effect of the sum-to-one constraint. Equation 7 is further simplified to Eq. 8.

$$\arg\min_{\widetilde{\mathbf{D}}, \widehat{\mathbf{X}}} \|\widetilde{\mathbf{Y}} - \widetilde{\mathbf{D}}\widehat{\mathbf{X}}\|_F^2 + \lambda\|\widehat{\mathbf{X}}\|_1$$

$$s.t. \ \widehat{\mathbf{X}} \succeq 0 \tag{8}$$

Given $\widetilde{\mathbf{Y}}$ and applying the proximal gradient descent [21], $\widetilde{\mathbf{D}}$ and $\widehat{\mathbf{X}}$ are updated alternatively. Then, new $\widehat{\mathbf{D}}$ and $\widehat{\mathbf{W}}$ can be obtained. Let $\mathbf{D} = \widehat{\mathbf{D}}$, $\mathbf{W} = \widehat{\mathbf{W}}$ and $\mathbf{X} = \widehat{\mathbf{X}}$, then $\mathcal{F}(\mathbf{D}, \mathbf{W})$ is updated through Eq. 5. Note that the size of $\mathbf{D}$ and $\mathbf{W}$ may be smaller than their original size now. Iterating the above procedure, a compacted dictionary, as well as the corresponding classifier, is learned in an automatic and joint manner. Algorithm 1 provides the pseudo-code for the propose ACDL method.

3.2 Proximal Gradient Descent

The proximal mapping or proximal operator [21] of a convex function $g(x)$ is defined in Eq. 9.

$$\mathbf{prox}_g(x) = \arg\min_u \left(g(u) + \frac{1}{2}\|u - x\|_2^2 \right) \tag{9}$$

If $g(x) = \lambda\|x\|_1$, then $\mathbf{prox}_g(x)$ is the shrinkage or soft threshold operation as shown in Eq. 10.

$$\mathbf{prox}_{\lambda, g}(x)_i = \begin{cases} x_i - \lambda, & x_i \geq \lambda \\ x_i + \lambda, & x_i \leq -\lambda \\ 0, & \text{otherwise} \end{cases} \tag{10}$$

Algorithm 1. Automatic Compact Dictionary Learning (ACDL)

1: **Input:** $\mathbf{Y}_{m \times n}$, $\mathbf{G}_{c \times n}$, γ, λ, δ, ϵ
2: **Output:** $\mathbf{D}$ and $\mathbf{W}$
3: **Initialization:** $\mathbf{D} = \mathbf{Y}$, $\mathbf{W} = \mathbf{G}$, $\mathbf{X} = \mathbf{I}_{n \times n}$, $\mathcal{F}(\mathbf{D}, \mathbf{W}) = \mathbf{I}_{n \times n}$, $stop$ = false
4: **while** not $stop$ **do**
5:     $\widehat{\mathbf{D}} = \{\mathbf{D}\mathcal{F}(\mathbf{D}, \mathbf{W})_{*j} | \mathcal{F}(\mathbf{D}, \mathbf{W})_{*j} \neq \mathbf{0}\}$
6:     $\widehat{\mathbf{W}} = \{\mathbf{W}\mathcal{F}(\mathbf{D}, \mathbf{W})_{*j} | \mathcal{F}(\mathbf{D}, \mathbf{W})_{*j} \neq \mathbf{0}\}$
7:     $\widehat{\mathbf{X}} = \{\mathcal{F}(\mathbf{D}, \mathbf{W})_{j*} \mathbf{X} | \mathcal{F}(\mathbf{D}, \mathbf{W})_{j*} \neq \mathbf{0}\}$
8:     $\widetilde{\mathbf{Y}} = \begin{pmatrix} \mathbf{Y} \\ \sqrt{\gamma}\mathbf{G} \end{pmatrix}$, $\widetilde{\mathbf{D}} = \begin{pmatrix} \widehat{\mathbf{D}} \\ \sqrt{\gamma}\widehat{\mathbf{W}} \end{pmatrix}$
9:     Solve Eq. 8 using proximal gradient descent (Eqs. 12 and 13)
10:    Get updated $\widehat{\mathbf{D}}$, $\widehat{\mathbf{W}}$ and $\widehat{\mathbf{X}}$
11:    **if** $size(\mathbf{D}) = size(\widehat{\mathbf{D}})$ **then**
12:        $stop$ = true
13:    **end if**
14:    $\mathbf{D} = \widehat{\mathbf{D}}$, $\mathbf{W} = \widehat{\mathbf{W}}$, $\mathbf{X} = \widehat{\mathbf{X}}$
15:    Compute $\mathcal{F}(\mathbf{D}, \mathbf{W})$ through Eq. 5
16: **end while**

$\widetilde{\mathbf{D}}$ and $\widehat{\mathbf{X}}$ are updated alternatively by minimizing one while keeping the other fixed. The update of $\widetilde{\mathbf{D}}$ can adopt the basic gradient descent method. Since the function with respect to $\widehat{\mathbf{X}}$ is not continuously differentiable because of the L_1-norm, the proximal mapping is employed to update $\widehat{\mathbf{X}}$. As illustrated in Eq. 11, the two terms in Eq. 8 correspond to the two functions $f(\widehat{\mathbf{X}}, \widetilde{\mathbf{D}})$ and $g(\widehat{\mathbf{X}})$.

$$\underset{\widetilde{\mathbf{D}}, \widehat{\mathbf{X}}}{\arg\min} \ \underbrace{\|\widetilde{\mathbf{Y}} - \widetilde{\mathbf{D}}\widehat{\mathbf{X}}\|_F^2}_{f(\widehat{\mathbf{X}}, \widetilde{\mathbf{D}})} + \underbrace{\lambda \|\widehat{\mathbf{X}}\|_1}_{g(\widehat{\mathbf{X}})} \tag{11}$$

Then, updating $\widetilde{\mathbf{D}}$ and $\widehat{\mathbf{X}}$ can be expressed in Eq. 12.

$$\widehat{\mathbf{X}}^{k+1} := \max \left\{ \mathbf{prox}_{\lambda, \eta_1^k, g} \left(\widehat{\mathbf{X}}^k - \eta_1^k \nabla f(\widehat{\mathbf{X}}^k, \widetilde{\mathbf{D}}) \right), \ 0 \right\} \tag{12}$$

$$\widetilde{\mathbf{D}}^{k+1} := \widetilde{\mathbf{D}}^k - \eta_2^k \nabla f(\widehat{\mathbf{X}}, \widetilde{\mathbf{D}}^k) \tag{13}$$

where

$$\nabla_{\widehat{\mathbf{X}}} f(\widehat{\mathbf{X}}, \widetilde{\mathbf{D}}) = \widetilde{\mathbf{D}}^T (\widetilde{\mathbf{D}}\widehat{\mathbf{X}} - \widetilde{\mathbf{Y}}) \tag{14}$$

$$\nabla_{\widetilde{\mathbf{D}}} f(\widehat{\mathbf{X}}, \widetilde{\mathbf{D}}) = (\widetilde{\mathbf{D}}\widehat{\mathbf{X}} - \widetilde{\mathbf{Y}})\widehat{\mathbf{X}}^T \tag{15}$$

Iterating Eqs. 12 and 13 until certain criterion is satisfied, the optimal $\widetilde{\mathbf{D}}$ and $\widehat{\mathbf{X}}$ can be obtained. Specifically, the optimal step size are calculated through ($\beta = 1$ [22]) as shown in Eqs. 16 and 17, where $\|\cdot\|_2$ denotes the spectral norm.

$$\eta_1^k = \frac{1}{\beta \|\widetilde{\mathbf{D}}^{kT} \widetilde{\mathbf{D}}^k\|_2} \tag{16}$$

$$\eta_2^k = \frac{1}{\beta \|\widehat{\mathbf{X}}^k \widehat{\mathbf{X}}^{kT}\|_2} \tag{17}$$

3.3 Classification

Given the learned dictionary $\mathbf{D}$ and classifier $\mathbf{W}$, new observations $\mathbf{Y}$ can be sparsely represented by $\mathbf{X}$ via Eq. 18, which can be solved by combining the sum-to-one constraint into the objective function like Eq. 7 and applying the proximal gradient descent.

$$\arg\min_{\mathbf{X}} \|\mathbf{Y} - \mathbf{DX}\|_F^2 + \lambda\|\mathbf{X}\|_1$$
$$s.t. \ \mathbf{X} \succeq 0, \ \sum \mathbf{x}_i = 1 \tag{18}$$

Then the predicted labels $\mathbf{L}$ of the observations $\mathbf{Y}$ can be obtained through Eq. 19,

$$\mathbf{L} = [\mathbf{l}_1, \mathbf{l}_2, \cdots, \mathbf{l}_n] = \mathbf{WX} \tag{19}$$

where $\mathbf{L}$ has the same number of columns as $\mathbf{Y}$ (assume n), and each column of $\mathbf{L}$ is the label vector of the corresponding observation in $\mathbf{Y}$. Note that $\mathbf{W}$ is L_2-normalized as the following, assuming the dictionary size is d.

$$\mathbf{W} = \left[\frac{\mathbf{w}_1}{\|\mathbf{w}_1\|_2}, \frac{\mathbf{w}_2}{\|\mathbf{w}_2\|_2}, \cdots, \frac{\mathbf{w}_d}{\|\mathbf{w}_d\|_2}\right] \tag{20}$$

Finally, the label of certain observation is the index corresponding to the largest element in its label vector as expressed in Eq. 21.

$$label(\mathbf{y}_i) = \arg\max_j \{l_j\}$$
$$s.t. \ l_j \in \mathbf{l}_i \tag{21}$$

where $\mathbf{y}_i$ denotes the ith column of $\mathbf{Y}$, and $\mathbf{l}_i = [l_1, l_2, \cdots]^T$ is the ith column of $\mathbf{L}$. $label(\cdot)$ yields an integer that indicates the class label.

4 Experimental Evaluation

We evaluate the proposed ACDL algorithm on four classification tasks, including multi-view classification using the Berkeley multiview wireless (BMW) dataset [23] (Sect. 4.1), scene recognition using the 15 scene categories dataset [24] (Sect. 4.2), object recognition using Caltech101 [25] (Sect. 4.3), and handwritten digit recognition using MNIST (Sect. 4.4). We compare ACDL with SDL [17], FDDL [9], DL-COPAR [16], KSVD [3], DKSVD [7], LCKSVD [8], and SRC [12] in terms of classification accuracy and dictionary size. In addition, some state-of-the-art approaches, such as spatial pyramid matching [24], sparse PCA [26], CNN [27,28], and LeNet [29], are cited to compare to our approach in terms of classification accuracy. Finally, Sect. 4.5 discusses the parameter setting.

4.1   The Berkeley Multiview Wireless Dataset

The Berkeley multiview wireless (BMW) [23] is a dataset of 20 landmark build-ings on the Berkeley campus under multiple views. For each building, wide-baseline images were captured from 16 different vantage points. At each vantage point, 5 short-baseline images were taken by five cameras simultaneously. And thereby there are 80 images per category. All images are 640 × 480 RGB color images. We follow the common setting, the training dataset are images cap-tured at the even vantage by camera #2 (320 images), and the rest are the testing dataset. The SURF features are postprocessed by Sparse PCA [26] in this experiment to filter noisy and background points.

In this experiment, the parameters are set as $\lambda = 0.1$, $\gamma = 20$, $\delta = 2$, $\epsilon = 0.6$ (see details in Sect. 4.5). The learned dictionary size is 109, and the average accuracy is 90.2%. The confusion matrix is shown in Fig. 2(b). Similarly, we compare our results with FDDL [9], SDL [17], DL-COPAR [16], LCKSVD [8] and a baseline method Naikal [26] as listed in Table 1. To illustrate the effectiveness of dictionary reduction, the dictionary-based algorithms except for ACDL are performed with two dictionary sizes, respectively. One is around 109 that is learned automatically from ACDL, and the other is 200. Our approach achieves a competitive performance while using a much smaller dictionary. This, again, demonstrates that ACDL can automatically learn a compact dictionary with appropriate size without any prior knowledge. In addition, the learned dictionary is discriminative enough to yield state-of-the-art classification performance. At the end of the table, the results of ACDL without sum-to-one (S2O) and non-negative (NN) constraints are reported to illustrate the significance of S2O and NN constraints.

Table 1. Classification results of different algorithms on the BMW dataset (the results marked by * are directly cited from corresponding papers)

Method	Accuracy(%)	Dictionary size
Naikal [26]	84.5*	–
FDDL [9]	74.0	320
SDL [17]	**90.8**	200
SDL [17]	90.1	109
DL-COPAR [16]	69.5	200
DL-COPAR [16]	73.8	100
LCKSVD1 [8]	89.9	200
LCKSVD1 [8]	89.8	109
LCKSVD2 [8]	90.4	200
LCKSVD2 [8]	89.7	109
ACDL	**90.2**	**109**
ACDL/S2O	81	127
ACDL/NN	76	80

4.2   The Fifteen Scene Categories Dataset

This dataset [24] contains 15 outdoor and indoor scene categories. The average image size is about 250×300 pixels, and each category has 200 to 400 images. The spatial pyramid features algorithm [24] was the first applied on this dataset and achieved an impressive classification accuracy of 81.4%. Following the same experimental setup in the spatial pyramid features [24] and LCKSVD [8], we randomly select 100 images per category as training data and the rest are testing data.

In ACDL, we initialize the dictionary with the whole training set (1500 atoms), and the parameters are set as $\lambda = 0.1$, $\gamma = 5$, $\delta = 2$, $\epsilon = 0.6$ (see details in Sect. 4.5). The learned dictionary size is drastically reduced to 122, with an average classification accuracy as 93.6%. The confusion matrix is shown in Fig. 2(a). We also compare our results with FDDL [9], DL-COPAR [16], LCKSVD [8], DKSVD [7], SDL [17] and some common baseline works as listed in Table 2. The results marked with * are the best results directly cited from corresponding papers. Among the baseline works, Zhou [30] obtained the highest accuracy of 91.6%.

For comparison purpose, the dictionary size of SDL and DKSVD are set to be 450 that is the same as used in LCKSVD. Under this setting, LCKSVD performs better than SDL, but SDL improves its performance to a great extent when we set the dictionary size according to the automatically learned size from ACDL. However, if there is no prior knowledge, it will be difficult to manually set an appropriate dictionary size. This experiment has demonstrated that ACDL can automatically learn a much more compact dictionary with appropriate size, while at the same time achieving higher recognition accuracy than state-of-the-art works.

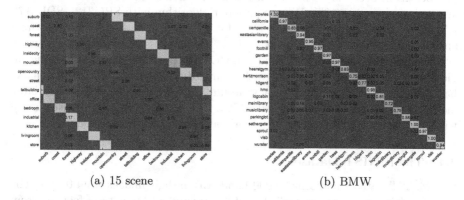

(a) 15 scene   (b) BMW

Fig. 2. (a) Confusion matrix using the dictionary (122 atoms) learned from ACDL on the 15 scene categories. (b) Confusion matrix using the dictionary (109 atoms) learned from ACDL on the BMW dataset.

Table 2. Classification results of different algorithms on 15 scene categories (the results marked by * are directly cited from corresponding papers)

Method	Accuracy(%)	Dictionary size
Lazebnik [24]	81.4*	–
Gao [31]	89.7*	–
Zhou [30]	**91.6***	–
FDDL [9]	73.5	317
FDDL [9]	54.1	170
DL-COPAR [16]	85.4*	1024*
LCKSVD1 [8]	90.4*	450*
LCKSVD2 [8]	92.9*	450*
DKSVD [7]	89.1	450
SDL [17]	84.7	450
SDL [17]	**93.0**	122
ACDL	**93.6**	**122**

4.3   The Caltech101 Dataset

The Caltech101 dataset [25] consists of 9144 images from 101 objects such as animals, cars, planes, etc. Each category has 31 to 800 images with significant shape variability. Following the common setting, the spatial pyramid features are used as the input and the feature dimension is 3000. We randomly choose 30% (2743 samples) to build the training dataset and the rest forms the testing dataset.

In this experiment, the parameters are set as $\lambda = 0.1$, $\gamma = 10$, $\delta = 2$, $\epsilon = 0.6$ (see details in Sect. 4.5). The dictionary size learned from ACDL is 161 and the average accuracy is 76.3%. We compare our algorithm with SRC [12], SDL [17], KSVD [3], DKSVD [7], LCKSVD [8], DL-COPAR [16] and some deep learning approaches as shown in Table 3. DeCAF [27] using convolutional neural network achieves the highest accuracy, and DL-COPAR performs the best among all dictionary-based approaches, but with a large dictionary size. Most dictionary-based methods achieve an accuracy around 70% or higher with a very large dictionary size, while the proposed ACDL can reach competitive performance with an extremely small dictionary size (more than 100 times smaller).

4.4   The MNIST Dataset

The MNIST dataset contains 70,000 handwritten digit images from 0 to 9. The size of each image is 28×28 pixels. LeNet [29] is employed to extract features from each image. The LeNet we used consists of two convolution-pooling layers and two fully connected networks, and we use the output of the last fully connected layer as the feature, which is a 10 dimensional vector. We randomly choose 4% (2400 samples) from the original training dataset which has 60000 samples, and the testing dataset has 10000 samples.

Table 3. Classification results of different algorithms on the Caltech dataset (the results marked by * are directly cited from corresponding papers)

Method	Accuracy(%)	Dictionary size
Lazebnik [24]	64.6*	–
Zeiler [28]	86.5*	–
DeCAF [27]	**86.9***	–
SRC [12]	70.7*	3060*
SDL [17]	75.3*	3060*
KSVD [3]	73.0*	3060*
DKSVD [7]	73.0*	3060*
LCKSVD1 [8]	73.4*	3060*
LCKSVD2 [8]	73.6*	3060*
DL-COPAR [16]	**83.3***	2048*
ACDL	**76.3**	**161**

In this experiment, the parameters are set as $\lambda = 0.1$, $\gamma = 20$, $\delta = 2$, $\epsilon = 0.4$ (see details in Sect. 4.5). The learned dictionary size from ACDL is 21, and the average classification accuracy is 98.5%. We compare our approach with FDDL [9], SDL [17], LCKSVD [8], DL-COPAR [16] and some deep learning methods as listed in Table 4. Similar to the experiment on the BMW dataset, two dictionary sizes are set for each algorithm, one is similar to the size automatically learned from ACDL and the other is a larger size. Among the dictionary-based methods, ACDL performs best and achieves a compact and discriminative dictionary size with consistent performance as observed from previous experiments.

Table 4. Classification results of different algorithms on the MNIST dataset (the results marked by * are directly cited from corresponding papers)

Method	Accuracy(%)	Dictionary size
LeNet-4 [29]	98.9*	–
Jarrett [32]	99.5*	–
Ciresan [33]	**99.7***	–
FDDL [9]	96.2	77
SDL [17]	98.5	100
SDL [17]	94.7	20
LCKSVD [8]	98.2	100
LCKSVD [8]	98.4	20
DL-COPAR [16]	**98.5**	100
DL-COPAR [16]	98.3	20
ACDL	**98.5**	**21**

4.5   Parameter Setting

From the above experiments conducted on four different datasets, the parameters need to be set manually are listed in Table 5. The parameter δ that balances the sum-to-one constraint and λ that dominates sparse penalty remain as constant for different datasets. The parameters for reduction threshold ϵ do not vary a lot. Only the parameter γ that controls the classification error has significant change when using different datasets. In practice, we adjust γ with step size of 5 within a narrow range, i.e. from 5 to 30. Usually, an appropriate dictionary would be constructed within several trials.

Table 5. Parameter settings on different datasets

Dataset	λ	γ	δ	ϵ
15 scenes	0.1	5	2	0.6
BMW	0.1	20	2	0.6
Caltech101	0.1	10	2	0.6
MNIST	0.1	20	2	0.4

Specifically, γ and ϵ affect the discrimination capability and the size, respectively, of the final dictionary. Therefore, γ reflects the distinguishability of the dataset, and ϵ, in a sense, reflects the geometric structure. For example, if the samples from the same class fall into two clusters that locate far from each other, then ϵ should be smaller. If the samples from different classes have large overlap, then γ needs to pick a larger value.

5   Conclusion

This paper proposed an automatic compact dictionary learning (ACDL) method, which learns a more compact and discriminative dictionary as compared with existing works, while maintaining a competitive classification performance. While most literature requires manual intervention to construct a dictionary with appropriate size, ACDL automatically removes common and redundant atoms by introducing the class error and indicator function into the objective function. The former ensures global learning that facilitates the removal of common atoms between classes, and the latter guarantees automatic dictionary reduction. In addition, the geometric structure of the dataset is preserved and utilized during the learning procedure by the non-negative and sum-to-one constraints enforced on the sparse coefficient, which may increase the classification performance. Experimental results demonstrated the effectiveness of the propose ACDL method.

References

1. Olshausen, B.A., Field, D.J.: Sparse coding with an overcomplete basis set: a strategy employed by V1? Vis. Res. **37**, 3311–3325 (1997)
2. Elad, M., Aharon, M.: Image denoising via sparse and redundant representations over learned dictionaries. IEEE Trans. Image Process. **15**, 3736–3745 (2006)
3. Aharon, M., Elad, M., Bruckstein, A.: K-SVD: an algorithm for designing overcomplete dictionaries for sparse representation. IEEE Trans. Sig. Process. **54**, 4311 (2006)
4. Mairal, J., Bach, F., Ponce, J., Sapiro, G.: Online dictionary learning for sparse coding. In: Proceedings of the 26th Annual International Conference on Machine Learning, pp. 689–696. ACM (2009)
5. Yang, J., Wright, J., Huang, T., Ma, Y.: Image super-resolution as sparse representation of raw image patches. In: IEEE Conference on Computer Vision and Pattern Recognition (CVPR), pp. 1–8. IEEE (2008)
6. Ramirez, I., Sprechmann, P., Sapiro, G.: Classification and clustering via dictionary learning with structured incoherence and shared features. In: IEEE Conference on Computer Vision and Pattern Recognition (CVPR), pp. 3501–3508. IEEE (2010)
7. Zhang, Q., Li, B.: Discriminative K-SVD for dictionary learning in face recognition. In: IEEE Conference on Computer Vision and Pattern Recognition (CVPR), pp. 2691–2698. IEEE (2010)
8. Jiang, Z., Lin, Z., Davis, L.S.: Label consistent K-SVD: learning a discriminative dictionary for recognition. IEEE Trans. Pattern Anal. Mach. Intell. **35**, 2651–2664 (2013)
9. Yang, M., Zhang, L., Feng, X., Zhang, D.: Fisher discrimination dictionary learning for sparse representation. In: IEEE International Conference on Computer Vision (ICCV), pp. 543–550. IEEE (2011)
10. Mairal, J., Ponce, J., Sapiro, G., Zisserman, A., Bach, F.R.: Supervised dictionary learning. In: Advances in Neural Information Processing Systems, pp. 1033–1040 (2009)
11. Mairal, J., Bach, F., Ponce, J.: Task-driven dictionary learning. IEEE Trans. Pattern Anal. Mach. Intell. **34**, 791–804 (2012)
12. Wright, J., Yang, A.Y., Ganesh, A., Sastry, S.S., Ma, Y.: Robust face recognition via sparse representation. IEEE Trans. Pattern Anal. Mach. Intell. **31**, 210–227 (2009)
13. Rahimpour, A., Taalimi, A., Luo, J., Qi, H.: Distributed object recognition in smart camera networks. In: IEEE International Conference on Image Processing, Phoenix, Arizona, USA. IEEE (2016)
14. Lazebnik, S., Raginsky, M.: Supervised learning of quantizer codebooks by information loss minimization. IEEE Trans. Pattern Anal. Mach. Intell. **31**, 1294–1309 (2009)
15. Liu, J., Shah, M.: Learning human actions via information maximization. In: IEEE Conference on Computer Vision and Pattern Recognition (CVPR), pp. 1–8. IEEE (2008)
16. Kong, S., Wang, D.: A dictionary learning approach for classification: separating the particularity and the commonality. In: Fitzgibbon, A., Lazebnik, S., Perona, P., Sato, Y., Schmid, C. (eds.) ECCV 2012. LNCS, vol. 7572, pp. 186–199. Springer, Heidelberg (2012). doi:10.1007/978-3-642-33718-5_14
17. Jiang, Z., Zhang, G., Davis, L.S.: Submodular dictionary learning for sparse coding. In: IEEE Conference on Computer Vision and Pattern Recognition (CVPR), pp. 3418–3425. IEEE (2012)

18. Yang, M., Dai, D., Shen, L., Van Gool, L.: Latent dictionary learning for sparse representation based classification. In: IEEE Conference on Computer Vision and Pattern Recognition (CVPR), pp. 4138–4145. IEEE (2014)
19. Qiu, Q., Patel, V.M., Chellappa, R.: Information-theoretic dictionary learning for image classification. IEEE Trans. Pattern Anal. Mach. Intell. **36**, 2173–2184 (2014)
20. Lu, C., Shi, J., Jia, J.: Scale adaptive dictionary learning. IEEE Trans. Image Process. **23**, 837–847 (2014)
21. Parikh, N., Boyd, S.: Proximal algorithms. Found. Trends Optim. **1**, 123–231 (2013)
22. Rakotomamonjy, A.: Direct optimization of the dictionary learning problem. IEEE Trans. Sig. Process. **61**, 5495–5506 (2013)
23. Naikal, N., Yang, A.Y., Sastry, S.S.: Towards an efficient distributed object recognition system in wireless smart camera networks. In: 13th Conference on Information Fusion (FUSION), pp. 1–8. IEEE (2010)
24. Lazebnik, S., Schmid, C., Ponce, J.: Beyond bags of features: spatial pyramid matching for recognizing natural scene categories. In: IEEE Computer Society Conference on Computer Vision and Pattern Recognition (CVPR), vol. 2, pp. 2169–2178. IEEE (2006)
25. Fei-Fei, L., Fergus, R., Perona, P.: Learning generative visual models from few training examples: an incremental Bayesian approach tested on 101 object categories. In: CVPR Workshop on Generative-Mode Based Vision. IEEE (2004)
26. Naikal, N., Yang, A.Y., Sastry, S.S.: Informative feature selection for object recognition via sparse PCA. In: IEEE International Conference on Computer Vision (ICCV), pp. 818–825. IEEE (2011)
27. Donahue, J., Jia, Y., Vinyals, O., Hoffman, J., Zhang, N., Tzeng, E., Darrell, T.: DECAF: a deep convolutional activation feature for generic visual recognition. In: Proceedings of the 31st International Conference on Machine Learning (ICML-2014), pp. 647–655 (2014)
28. Zeiler, M.D., Fergus, R.: Visualizing and understanding convolutional networks. In: Fleet, D., Pajdla, T., Schiele, B., Tuytelaars, T. (eds.) ECCV 2014. LNCS, vol. 8689, pp. 818–833. Springer, Heidelberg (2014). doi:10.1007/978-3-319-10590-1_53
29. LeCun, Y., Bottou, L., Bengio, Y., Haffner, P.: Gradient-based learning applied to document recognition. Proc. IEEE **86**, 2278–2324 (1998)
30. Zhou, B., Lapedriza, A., Xiao, J., Torralba, A., Oliva, A.: Learning deep features for scene recognition using places database. In: Advances in Neural Information Processing Systems, pp. 487–495 (2014)
31. Gao, S., Tsang, I.W.H., Chia, L.T.: Laplacian sparse coding, hypergraph laplacian sparse coding, and applications. IEEE Trans. Pattern Anal. Mach. Intell. **35**, 92–104 (2013)
32. Jarrett, K., Kavukcuoglu, K., Ranzato, M., LeCun, Y.: What is the best multistage architecture for object recognition? In: IEEE International Conference on Computer Vision (ICCV), pp. 2146–2153. IEEE (2009)
33. Ciregan, D., Meier, U., Schmidhuber, J.: Multi-column deep neural networks for image classification. In: IEEE Conference on Computer Vision and Pattern Recognition (2012)

A Vote-and-Verify Strategy for Fast Spatial Verification in Image Retrieval

Johannes L. Schönberger[1]([✉]), True Price[2], Torsten Sattler[1],
Jan-Michael Frahm[2], and Marc Pollefeys[1,3]

[1] ETH Zürich, Zürich, Switzerland
jsch@inf.ethz.ch
[2] UNC Chapel Hill, Chapel Hill, USA
[3] Microsoft, Redmond, USA

Abstract. Spatial verification is a crucial part of every image retrieval system, as it accounts for the fact that geometric feature configurations are typically ignored by the Bag-of-Words representation. Since spatial verification quickly becomes the bottleneck of the retrieval process, runtime efficiency is extremely important. At the same time, spatial verification should be able to reliably distinguish between related and unrelated images. While methods based on RANSAC's hypothesize-and-verify framework achieve high accuracy, they are not particularly efficient. Conversely, verification approaches based on Hough voting are extremely efficient but not as accurate. In this paper, we develop a novel spatial verification approach that uses an efficient voting scheme to identify promising transformation hypotheses that are subsequently verified and refined. Through comprehensive experiments, we show that our method is able to achieve a verification accuracy similar to state-of-the-art hypothesize-and-verify approaches while providing faster runtimes than state-of-the-art voting-based methods.

1 Introduction

Image retrieval, *i.e.*, finding relevant database images for a given query picture, is a fundamental problem in computer vision with applications in object retrieval [1,2], location recognition [3–5], image-based localization [6,7], automatic photo annotation [8], view clustering [9,10], loop-closure [11], and Structure-from-Motion [12–14]. Although methods that use compact image representations [15–17] have gained popularity, especially in combination with deep learning [18–20], state-of-the-art systems [2,21–25] still follow the *Bag-of-Words* (BoW) paradigm [1] proposed over a decade ago. The BoW model represents each image as a set

---

J.L. Schönberger, T. Price and T. Sattler—These authors contributed equally to the paper.

Electronic supplementary material The online version of this chapter (doi:10. 1007/978-3-319-54181-5_21) contains supplementary material, which is available to authorized users.

S.-H. Lai et al. (Eds.): ACCV 2016, Part I, LNCS 10111, pp. 321–337, 2017.
DOI: 10.1007/978-3-319-54181-5_21

of *visual words* obtained by quantizing the local feature space. Visually similar database images can then be found by searching for photos with similar visual words, usually implemented efficiently using inverted files [1]. For the sake of computational efficiency, BoW models generally only consider the presence and absence of visual words in an image and largely ignore their spatial configuration. Thus, a subsequent *spatial verification* phase [26] is typically used to filter retrieved photos whose visual words are not spatially consistent with the words in the query image. If not implemented efficiently, this spatial verification step quickly becomes the bottleneck of an image retrieval pipeline.

Spatial verification computes a geometric transformation, *e.g.*, a similarity or affine transformation [26], from feature correspondences between the query and database images. The correspondences are obtained from common visual word assignments of the query and database features. This often leads to many wrong correspondences, especially in the presence of repetitive structures or when using small vocabularies to reduce quantization artifacts. This makes traditional RANSAC-based approaches [27] that estimate transformations from multiple matches infeasible, as their runtime grows exponentially with the percentage of outliers. A key insight for fast spatial verification is to leverage the local feature geometry to hypothesize a geometric transformation from a single correspondence [26,28]. This significantly reduces the number of hypotheses that need to be verified. Spatial verification can be accelerated by replacing the hypothesize-and-verify framework with a Hough voting scheme based on quantizing the space of transformations [28,29]. These methods approximate the similarity between two images by the number of matches falling into the same bin in the voting space. Quantization artifacts are typically handled by using hierarchical voting schemes [29] or by allowing each match to vote for multiple bins [30].

Recent work demonstrates that a better verification accuracy can be obtained by explicitly incorporating verification into voting schemes, *e.g.*, as a more detailed verification step [31] or when casting multiple votes [30]. However, in this paper, we show that even such advanced voting schemes still achieve lower accuracy than classic hypothesize-and-verify approaches [26]. To close this gap, we propose a novel spatial verification approach that incorporates voting into a hypothesize-and-verify framework. In detail, we propose a hierarchical voting approach to efficiently identify promising transformation hypotheses that are subsequently verified and refined on all matches. Instead of finding the correct hypothesis through random sampling, we use a progressive sampling strategy on the most probable hypotheses. Furthermore, our approach offers multiple advantages over voting-based methods: First, rather than explicitly handling quantization artifacts in the voting space, our approach only requires that a reasonable estimate for the true transformation can be obtained from some matches falling into the same bin. Quantization artifacts are then automatically handled by the subsequent verification and refinement stages. As a result, our approach is rather insensitive to the quantization of the voting space and the visual vocabulary size. Second, in contrast to voting-based methods, which usually return only a similarity score, our approach explicitly returns a transformation and a set of inliers. Hence, it can be readily combined with query expansion (QE) schemes [2,21,32,33] as well as further reasoning based on the detected inliers [4,34]. Experimental evaluation on existing and new datasets show that our approach achieves accuracy equivalent to state-of-the-art hypothesize-and-verify methods while providing runtimes faster than state-of-the-art voting-based methods. The new query and distractor image datasets and the source code for our method are released to the public at https://github.com/vote-and-verify.

2   Related Work

In the following, we discuss prior work on spatial verification in the context of image retrieval. We classify these works based on whether they employ *weak geometric models*, use RANSAC's [27] *hypothesize-and-verify framework*, or follow a *Hough voting*-based approach. In this context, our method can be seen as a hybrid between the latter two types of approaches, as it replaces RANSAC's hypothesis generation stage with a voting scheme.

Weak geometric models. Instead of explicitly estimating a geometric transformation between two images, methods based on weak geometric models either use partial transformations [25,31] or local consistency checks [35–37]. For a feature match between two images, both Sivic & Zisserman [35] and Sattler *et al.* [36] define a consistency score based on the number of matches shared in the spatial neighborhoods of the two features. A threshold on this score is then used to prune matches. The geometry of the local features, *i.e.*, their position, orientation, and scale, can be used to hypothesize a similarity transformation from a single correspondence [26,28]. Jegou *et al.* [25] focus on the change in scale and orientation predicted by a correspondence. They quantize the space of changes into a fixed set of bins and use Hough voting to determine a subset of matches that are all similar in terms of either scale or orientation change. Pairwise Geometric Matching (PGM) of Li *et al.* [31] uses a two-stage procedure to handle noise in the voting process caused by inaccurate feature frames. First, voting provides a rough estimate for the orientation and scale change between the images, as well as putative matches. PGM then checks whether the transformations obtained from correspondence pairs are consistent with the initial estimate. To improve the runtime, Li *et al.* perform a pruning step that enforces 1-to-1 correspondences. PGM's computational complexity is $\mathcal{O}(n + m^2)$, where n is the total number of matches and m is the number of matches falling into the best bin. $O(n)$ is required for voting for the best bin and $O(m^2)$ for pairwise verification. If all matches are correct, $m = n$ holds, and the complexity is $O(n^2)$. This worst case actually happens in practice, *e.g.*, when the query is a crop of a database image. In comparison, while we also apply pruning, our vote-and-verify strategy has $\mathcal{O}(n)$ complexity and achieves both faster runtimes and better verification accuracy. A drawback to methods based on weak geometric models is they only determine a similarity score and do not identify individual feature correspondences. Thus, query expansion (QE) [21,32], which transforms features from the database into the query image, cannot be directly applied.

Hypothesize-and-verify methods. Probably the most popular approach to compute a geometric transformation in the presence of outliers is RANSAC [27] or one of its many variants [36,38–41]. However, matching features through their visual word assignments usually generates many wrong correspondences. This makes it impractical to use any RANSAC variant that samples multiple matches in each iteration, as the runtime grows exponentially with the outlier ratio. Philbin *et al.* [26] therefore propose a more efficient spatial verification approach, termed Fast Spatial Matching (FSM), exploiting the fact that a single correspondence already defines a transformation hypothesis. Consequently, they generate and evaluate all possible n transformations and apply local optimization [38,39] whenever a new best model is found. FSM is the *de facto* standard for spatial verification and is used in most state-of-the-art retrieval pipelines [2,21–25]. Its main drawback is the $\mathcal{O}(n^2)$ computational complexity due to evaluating all n hypotheses on all n correspondences. In practice, FSM can be accelerated by integrating it into a RANSAC framework and using early termination once the probability of finding a better model falls below a given threshold. Still, the worst-case complexity remains $\mathcal{O}(n^2)$. Our approach uses voting to efficiently identify promising transformation hypotheses in time $\mathcal{O}(n)$, and we show that it is sufficient to progressively sample a constant number of these hypotheses, resulting in an overall complexity of $\mathcal{O}(n)$. As such, our approach can be seen as borrowing the hypothesis prioritization of PROSAC [40] and using voting to achieve linear complexity. PROSAC orders matches based on matching quality and initially only generates transformation hypotheses from matches more likely to be correct. However, PROSAC is not directly applicable in our scenario, since matching via visual words does not provide an estimate of the matching quality. The output of FSM is a set of inliers and a geometric transformation, which can be used as input to QE. Our approach provides the same output and can thus directly be combined with existing QE methods.

Hough voting-based approaches. Lowe [28] uses Hough voting to identify a subset of all putative matches, consisting of all matches whose corresponding similarity transformation belongs to the largest bin in the voting space. Next, they apply RANSAC on this consistent subset of matches to estimate an affine transformation. To reduce the complexity to $\mathcal{O}(n)$, Avrithis & Tolias [29] propose to restrict spatial verification to the voting stage. To mitigate quantization artifacts, their Hough Pyramid Matching (HPM) uses a hierarchical voting space, where every match votes for a single transformation on each level. For each match, they then compute a strength score based on the number of other correspondences falling in the the same bins and aggregate these strengths into an overall similarity score for the two images. Wu & Kashino propose to handle quantization artifacts by having each match vote for multiple bins [30]. For each feature match m, their Adaptive Dither Voting (ADV) scheme finds neighboring correspondences, similar to [35,36]. If m is consistent with the transformation hypothesis of its neighbor, a vote is cast in the neighbor's bin. This additional verification step helps ADV to avoid casting unrelated votes, resulting in better accuracy compared to HPM. Since ADV finds a fixed number of nearest neighbors in the image space, its computational complexity is $\mathcal{O}(n \log n)$. Similar to methods based on weak geometric models, both HPM and ADV only provide an overall similarity score and thus cannot be directly combined with standard QE. In addition, our approach achieves superior verification accuracy at faster runtimes than ADV.

Verification during retrieval. All previously discussed methods operate as a post-processing step after image retrieval. Ideally, spatial verification should be performed during the retrieval process to detect and reject incorrect votes. While there exist a variety of approaches that directly integrate geometric information [42–46], this usually comes at the price of additional memory requirements or longer retrieval times. Thus, most state-of-the-art approaches still apply spatial verification only as a post-processing step [2, 21–23].

3   Vote-and-Verify for Fast Spatial Verification

One inherent problem of hypothesize-and-verify methods is that finding a good hypothesis through either random sampling or exhaustive search often takes quite some time. In this paper, we propose to solve this problem by finding promising transformation hypotheses through voting. Starting with the most promising ones, we verify a fixed number of these hypotheses on all matches, *i.e.*, perform inlier counting. As in FSM [26], local optimization [38, 39] is applied every time a new best transformation is found.

At first glance, the voting stage of our approach may seem identical to existing voting-based approaches [29–31], but there are two important differences between previous methods and our approach: (i) Existing works [29–31] use voting with the aim of identifying *all* geometrically consistent matches. As such, handling quantization artifacts in the voting space is very important and the methods are rather sensitive to the number of bins in the voting space. In contrast, we only require to find transformations that need to be geometrically consistent with *some* of the matches. Matches missed due to quantization are then automatically detected during inlier counting and are subsequently used during local optimization to refine the model. (ii) Instead of only using the number of matches associated with the same bin, we are also interested in the transformation induced by these matches. To this end, we find that simply taking the transformation defined by the center of a bin does not provide an accurate enough hypothesis, even for high levels of quantization. Consequently, we propose a refinement process to obtain more accurate transformation hypotheses. As a result, our approach is rather insensitive to the quantization level.

In the following, we first recapitulate the process of computing a similarity transformation from a single feature match (Sect. 3.1). We next detail the voting procedure (Sect. 3.2) and explain how to derive transformation hypotheses from the voting space (Sect. 3.3). Section 3.4 then describes the verification and local optimization stage, while Sect. 3.5 analyses the computational complexity of our method. Finally, Algorithm 1 gives an overview of our proposed method.

Algorithm 1. Proposed Vote-and-Verify algorithm: $\text{VERIFY}(\text{VOTE}(m_{i=1...n}))$.

1: **procedure** $\text{VOTE}(m_{i=1...n})$
2:     **for** $i = 1...n$ **do**
3:         **for** $l = 1...L$ **do**
4:             Vote for $\text{C}(m_i, l)$ with $w(l)$ in Hough space
5:     Find bins $b_{t=1...T}$ with highest votes $w(b_t)$
6:     **return** Hypotheses $\text{T}(b_{t=1...T})$ ordered in decreasing number of votes
7: **procedure** $\text{VERIFY}(\text{T}(b_{t=1...T}))$
8:     Set initial score $\hat{s} = 0$ and transformation $\hat{\text{T}}$ as invalid
9:     **for** $t = 1...T$ **do**
10:         Verify $\text{T}(b_t)$ and count the number of inliers s_t
11:         **if** $s_t > \hat{s}$ **then**
12:             Refine $\text{T}(b_t)$ with local optimization and update s_t
13:             Update best score $\hat{s} = s_t$ and transformation $\hat{\text{T}} = \text{T}(b_t)$
14:             Update probability p of finding better $\hat{\text{T}}$
15:             **if** $p < \hat{p}$ **then**
16:                 **break**
17:     **return** Effective inlier count $\hat{s}_{\text{eff}}$ for $\hat{\text{T}}$

3.1   From Local Features to Similarity Transformation

Consider a local image feature f, e.g., a SIFT feature [28], defined by its local feature frame $f = (f_x, f_y, f_\sigma, f_\theta)$. Here, $(f_x, f_y)^T$ and f_σ are the location and scale of the detected feature in the image, while f_θ denotes the feature orientation. Following [29], each feature f is associated with a canonical coordinate frame in which the feature is located at the origin with unit scale and zero orientation. The transformation $\text{M}_f(\mathbf{x})$ maps a location $\mathbf{x} = (x, y)^T$ in the image to a position in the canonical frame as

$$\text{M}_f(\mathbf{x}) = \frac{1}{f_\sigma} \text{R}(f_\theta) \left(\begin{bmatrix} x \\ y \end{bmatrix} - \begin{bmatrix} f_x \\ f_y \end{bmatrix} \right) = \frac{1}{f_\sigma} \begin{bmatrix} \cos f_\theta & \sin f_\theta \\ -\sin f_\theta & \cos f_\theta \end{bmatrix} \left(\begin{bmatrix} x \\ y \end{bmatrix} - \begin{bmatrix} f_x \\ f_y \end{bmatrix} \right), \quad (1)$$

where the rotation matrix $\text{R}(f_\theta)$ performs a clockwise rotation by f_θ degrees. Consequently, a feature match $m = (f^\mathcal{Q}, f^\mathcal{D})$ between a query image $\mathcal{Q}$ and a database image $\mathcal{D}$ defines a similarity transformation between the two images:

$$\text{M}_{(f^\mathcal{Q}, f^\mathcal{D})}(\mathbf{x}) = M_{f^\mathcal{Q}}^{-1} \left(M_{f^\mathcal{D}}(\mathbf{x}) \right) \quad (2)$$

$$= \frac{f_\sigma^\mathcal{Q}}{f_\sigma^\mathcal{D}} \text{R}(f_\theta^\mathcal{Q})^T \text{R}(f_\theta^\mathcal{D}) \left(\begin{bmatrix} x \\ y \end{bmatrix} - \begin{bmatrix} f_x^\mathcal{D} \\ f_y^\mathcal{D} \end{bmatrix} \right) + \begin{bmatrix} f_x^\mathcal{Q} \\ f_y^\mathcal{Q} \end{bmatrix} = \sigma \text{R}(\theta)\mathbf{x} + \mathbf{t}. \quad (3)$$

In Eq. 3, $\sigma = f_\sigma^\mathcal{Q}/f_\sigma^\mathcal{D}$, $\theta = f_\theta^\mathcal{Q} - f_\theta^\mathcal{D}$, and $\mathbf{t} = (t_x, t_y)^T$ define the relative scale, rotation angle, and translation of the similarity transformation. Thus, each match $m = (f^\mathcal{Q}, f^\mathcal{D})$ can be associated with a 4-dimensional coordinate

$$\text{C}(m) = [\sigma, \theta, t_x, t_y] . \quad (4)$$

3.2   From Similarity Transformation to Hough Voting

Each of the n feature matches between a query and database image defines a transformation hypothesis. We are interested in determining a fixed-sized set of transformations that is consistent with as many of these n hypotheses as possible. This can be done very efficiently using Hough voting, as similar transformations are likely to fall into the same voting bin. This allows us to obtain a set of promising transformation hypotheses from the best scoring bins. However, standard Hough voting suffers from quantization artifacts, caused by inaccuracies in the detected feature frames. This is especially problematic when only few matches are correct, in which case it becomes harder to distinguish between bins corresponding to the underlying geometric transformation and bins receiving votes from wrong matches. Thus, we use a hierarchical voting scheme similar to [29], which we describe in the following.

Following Avrithis & Tolias [29], we quantize each of the four similarity transformation parameters independently. The Hough voting space of transformations is then defined as the product space of the four individual quantizations. We discretize the space at different resolution levels $l = 0...L$ and use n_x, n_y, n_σ, and n_θ bins for translation, scale, and orientation at the finest resolution $l = 0$. Each successive resolution level divides the number of bins in half until only two bins are left for each dimension. Out of the four dimensions, only the rotation space is naturally bounded by $[0, 2\pi)$, while translation and scale in theory can take any value from $\mathbb{R}^2$ and $\mathbb{R}^+$, respectively. In practice, the space of possible scale changes σ is bounded, since feature detectors only consider a few octaves of scale space [29]. A feature match inducing a large scale change between two images can usually be safely discarded as an incorrect correspondence. Consequently, we only consider scale changes in the range $[1/\sigma_{\max}, \sigma_{\max}]$ [29]. In addition, we bound the translation parameters by $\max(|t_x|, |t_y|) \leq \max(W, H)$, where W and H are the width and height of the query image. Matches violating at least one constraint are ignored [29].

Given a feature match $m = (f^Q, f^D)$ with transformation parameters σ, θ, and $\mathbf{t}$ as defined above, we obtain the corresponding Hough space coordinate as

$$C_x(t_x, l) = 2^{-l} \lfloor n_x (t_x + \max(W, H)) / (2 \max(W, H)) \rfloor, \tag{5}$$

$$C_y(t_y, l) = 2^{-l} \lfloor n_y (t_y + \max(W, H)) / (2 \max(W, H)) \rfloor, \tag{6}$$

$$C_\sigma(\sigma, l) = 2^{-l} \lfloor n_\sigma (\log_2(\sigma) + \log_2(\sigma_{\max})) / (2 \log_2(\sigma_{\max})) \rfloor, \tag{7}$$

$$C_\theta(\theta, l) = 2^{-l} \lfloor n_\theta (\theta + \pi) / (2\pi) \rfloor. \tag{8}$$

For uniform sampling in the scale space, we linearize the scale change using the logarithmic function. The factor 2^{-l} normalizes the Hough coordinates to the respective resolution level of the voting space. Here, each match $m = (f^Q, f^D)$ determines a coordinate $\mathbf{C}(m, l)$ at level l in the voting space, with

$$\mathbf{C}(m, l) = [C_x(t_x, l), C_y(t_y, l), C_\sigma(\sigma, l), C_\theta(\theta, l)]. \tag{9}$$

The match m then contributes a level-dependent weight $w(l) = 2^{-l}$ to the score of its corresponding voting bin at level l. Next, we describe how to use these scores to detect the T most promising transformation hypotheses.

3.3   Hypothesis Generation

The goal of our voting scheme is to provide a set of transformation hypotheses for subsequent verification. As described in the last section, the center of any bin in the hierarchical representation defines a transformation, and we could simply pick the transformations corresponding to the bins with the highest scores. For coarser levels in the hierarchy, however, it is unlikely that the center of the bin is close to the actual transformation between the images. As such, we only hypothesize transformations corresponding to bins at level $l = 0$. To mitigate quantization artifacts, we propagate the scores from coarser levels to the bins at level 0. Each bin b at level 0 uniquely defines a path through the hierarchy to the coarsest level L. The total score for this bin is computed by summing the scores of all bins along this path as $w(b) = \sum_{l=0...L} w(b, l)$. Finally, we simply select the T bins that received the highest scores $w(b)$.

As we will show in Sect. 4, the naive approach of associating each bin at level 0 with the transformation defined by the bin's center coordinate does not perform well, even when using a reasonably deep hierarchy. In other words, the center coordinate of a bin can be rather far away from the true image transformation. In order to obtain a better estimate, we use the mean transformation of all matches falling into the bin instead. Let $\mathcal{M}(b)$ be the matches falling into bin b. Following Eq. 4, the mean transformation $\mathbf{T}(b)$ is defined as $\mathbf{T}(b) = \frac{1}{|\mathcal{M}(b)|} \sum_{m \in \mathcal{M}(b)} \mathbf{C}(m)$ and can be computed efficiently during voting without a significant memory overhead by maintaining a running average. Intuitively, one can think of this as local optimization (*cf.* [38]) on the level of hypothesis generation.

It is well-known that outliers, *i.e.*, wrong matches falling into a bin, significantly impact the computation of the mean. As a more robust alternative, one could use the median transformation instead. However, the mean can be computed much more efficiently, and experiments in Sect. 4.3 show that its performance is very robust to the choice of the quantization resolution.

3.4   Accurate and Efficient Hypothesis Verification

The scores associated with each of the T similarity transformation hypotheses only provides an estimate on how well the transformation explains the matches. As a next step, we thus perform detailed hypothesis verification using inlier counting. For this stage, we follow FSM [26] and consider a match an inlier to a transformation if its two-way reprojection error is below a threshold and if the scale change between two corresponding features induced by the transformation is consistent with the scales of the two features. The $t = 1...T$ transformation hypotheses are verified in decreasing order of their scores, and we apply local optimization [38,39] every time a new best model is found. The latter step refines the transformation by drawing a constant number of non-minimal sets of inliers to obtain a least squares estimate for the transformation. If there are at least $s = 3$ inliers, we estimate an affine transformation in local optimization. Affine transformations have been shown to perform better than similarity transformations [26], as they can handle more general geometric configurations.

We progressively verify the hypotheses ordered in decreasing number of votes until the probability $p = (1 - e)^t$ of finding a better model at the current inlier ratio e falls below a threshold. Our method is a variant of PROSAC [40] using Hough voting for ordering and 1-point sampling with a fixed-size hypothesis set.

One advantage over voting-based methods is that we explicitly determine the set of inliers. This allows us to use image similarity functions that are more discriminative than simply counting the number of inliers [4]. One such function is the *effective inlier count* [34], which has been shown to outperform raw inlier counting for image-based localization [7] and image retrieval [4] tasks. The effective inlier count is defined as

$$s_{\text{eff}} = \frac{|\cup_{i=1}^{\hat{n}} A_i|}{\sum_{i=1}^{\hat{n}} |A_i|} \hat{n}. \tag{10}$$

Here, $\hat{n}$ is the number of inlier matches, A_i is a region centered around the i-th inlier feature in the query image, $|\cup_{i=1}^{\hat{n}} A_i|$ is the area of the union of all regions, and $\sum_{i=1}^{\hat{n}} |A_i|$ is the maximum area that could be covered if none of the regions would overlap. In our experiments, we use square regions of size 24×24 px.

3.5 Computational Complexity

For a fixed number of levels in the hierarchy, each match votes for a fixed number of bins, where each transformation bin can be computed in time $\mathcal{O}(1)$. Performing a single vote is also a constant time operation, because it only requires incrementing a counter and updating the running mean for the transformation. Consequently, voting can be done in time $\mathcal{O}(n)$ for n matches. Since T is a constant, finding and evaluating the T best hypotheses also requires time $\mathcal{O}(n)$. Each hypothesis is evaluated on all matches, which can again be done in time $\mathcal{O}(n)$. Local optimization is performed at most once for each hypothesis using a fixed-sized non-minimal set, *i.e.*, each least squares transformation can be computed in constant time as well. Thus, the overall computational complexity is linear in the number n of matches.

4 Experimental Evaluation

In this section, we first introduce the datasets and describe our experimental setup. Next, we perform an ablation study and analyze the impact of the different parameters on the performance of our approach. Finally, we provide an extensive comparison with state-of-the-art spatial verification methods.

4.1 Query and Distractor Datasets

Following standard procedure [30,31], we primarily evaluate on the Oxford5k [26] and Paris6k [47] datasets. These image sets, collected from Flickr, contain $\sim$5 k and $\sim$6 k images, respectively, each consisting of 11 distinct landmarks, with 5 query images per landmark. In addition, we created the new *World5k* dataset consisting of 5320 images from 61 landmarks around the world, where the images were obtained from the Yahoo 100 M images dataset [48]. The landmark images were selected based on geo-tags and using overlap information from Heinly *et al.* [49] for ground-truth. Each landmark has between 30 and 100 database images and 1 to 3 associated query photos, resulting in a total of 163 query images. Different from Oxford5k and Paris6k, which represent an object retrieval task where query photos are obtained by cropping regions from database images, our query images are full-resolution and are not contained in the database.

To simulate larger datasets and thus create harder retrieval scenarios, it is common to combine the individual datasets with an additional "distractor" dataset of ~100k unrelated images collected from Flickr [26]. It has been shown that adding this Flickr100 k distractor set significantly impacts the retrieval performance. However, the Flickr100k (F100k) dataset mostly contains very unrelated photos, obtained by searching for generic terms, such as "graffiti", "uk", or "vacation". We thus collected four additional distractor sets from the Yahoo 100M images dataset. The first distractor set consists of 140 k images taken between 2 km and 50 km from the center of the University of Oxford. The other three distractor sets consist of images taken from 30 cities around the UK (171 k images), 30 cities throughout Europe (179 k), and 30 cities across the US (233 k). The collections were formed using the set of geo-tagged images within a 1 km radius (2 km for the US images) of the geographic center of the city. We expect to find more geometrically consistent matches for distractor images, *e.g.*, due to buildings with similar architectural styles, on the new distractor datasets compared to the Flickr100k set. As such, we expect that our new distractor sets represent more challenging scenarios for spatial verification. In the following, we refer to the distractor sets as *Ox* (Oxford), *UK*, *EU* (Europe), and *US*. Further information about our new datasets is available in the supplementary material.

4.2 Experimental Setup

Retrieval system. We employ a state-of-the-art retrieval system using Hamming embedding [25] and visual burstiness weighting [24]. We use a vocabulary containing 200k words[1] to quantize RootSIFT descriptors [2,28] extracted from keypoints provided by an upright Hessian affine feature detector [50]. Following standard procedure [23], we ensure that the vocabulary used for each dataset has been trained on another image collection. Correspondingly, we use a vocabulary trained on Oxford5k for the experiments on Paris6k and our new dataset. A vocabulary trained on Paris6k is then used for all experiments performed on the Oxford5k dataset. After retrieval, the top-1000 ranked database images are considered for spatial verification and re-ranked based on the similarity scores computed during verification. We enforce 1-to-1 matches prior to verification [31], since initial experiments showed that this significantly improved verification efficiency and quality for all spatial verification methods.

---

[1] Results obtained with 20k and 1M words can be found in the supplementary material.

Table 1. The impact of the number of voting bins (n_σ, n_θ, n_x, n_y) and the number T of verified transformation hypotheses on the performance and efficiency of our method. Results, obtained on Oxford5k, with the highest mAP (80.1%) are highlighted in green.

n_σ	n_θ	T=10, 16 mAP	time	32 mAP	time	64 mAP	time	T=20, 16 mAP	time	32 mAP	time	64 mAP	time	T=30, 16 mAP	time	32 mAP	time	64 mAP	time
8	8	**80.1**	0.7	79.7	0.7	79.8	0.8	79.9	0.8	79.9	0.9	79.9	0.9	80.0	0.9	79.9	0.9	80.0	1.0
	16	79.9	0.7	79.7	0.8	79.8	0.8	80.0	0.8	79.8	0.9	79.9	0.9	80.0	0.8	79.9	0.9	80.0	1.0
	32	79.7	0.7	79.8	0.8	79.9	0.8	80.0	0.9	79.9	0.9	79.9	0.9	79.9	0.9	79.9	0.9	80.0	1.0
16	8	79.9	0.7	79.7	0.7	79.8	0.8	80.0	0.8	79.9	0.9	79.9	0.9	**80.1**	0.8	80.0	0.9	**80.1**	0.9
	16	79.9	0.8	79.8	0.7	79.8	0.8	80.0	0.9	79.9	0.9	79.9	0.9	80.0	1.0	80.0	0.9	80.0	0.9
	32	79.7	0.7	79.9	0.8	79.8	0.8	80.0	0.8	79.9	0.9	80.0	0.9	80.0	0.9	80.0	0.9	80.0	0.9
32	8	79.9	0.7	80.0	0.8	79.8	0.8	80.0	0.8	80.0	0.9	79.9	0.9	80.0	0.9	80.0	0.9	**80.1**	0.9
	16	79.7	1.0	79.9	0.7	79.7	0.8	80.0	0.9	80.0	0.9	79.9	0.9	80.0	0.8	80.0	0.9	**80.1**	0.9
	32	79.8	0.7	79.7	0.9	80.0	0.8	79.9	0.8	80.0	1.0	80.0	0.9	80.0	0.9	80.0	1.1	**80.1**	0.9

Evaluation protocol. We follow the standard evaluation procedure and assess the verification performance using mean average precision (mAP), which essentially averages the area under the precision-recall curves. For each verification approach, we report the total time in seconds required to verify the 1000 top-ranked retrievals for all query images. We ignore retrieval and setup time, *e.g.*, the time required to enforce a 1-to-1 matching, since this is separate from verification. To facilitate comparability, we run single-threaded implementations on an Intel E5-2697 2.7GHz CPU with 256 GB RAM.

4.3   Ablation Study

We evaluate the impact of the different parameters of our approach on its verification performance and efficiency. All experiments presented in this section have been performed on the Oxford5k dataset without any distractor images.

Impact of the level of quantization. As a first experiment, we evaluate the impact of the number of bins for rotation (n_θ), scale (n_σ) and translation (n_x, n_y) as well as the number T of transformation hypotheses that are verified. Table 1 shows the results obtained for different parameter configurations. For this experiment, we used the mean transformation per bin to generate the hypotheses (*cf.* Sect. 3.3). As can be seen from the table, the verification performance of our approach is rather insensitive against the number of bins and verified transformations, although increasing T has a slightly positive impact on the measured mAP. Naturally, increasing the number of bins and T also increases the overall runtime, but the increase is rather small. We also experimented with fewer (2 and 4) and more (128) bin sizes but found that the former resulted in a significant drop in mAP while the latter did not noticeably improve mAP. For all following experiments, we use $T = 30$ transformation proposals, $n_x = 64$ and $n_y = 64$ translation bins, $n_\sigma = 32$ scale bins, and $n_\theta = 8$ rotation bins.

Impact of refining the transformation hypotheses. In Sect. 3.3, we proposed to use the mean transformation for the bins in the voting space, arguing that the transformation defined by the center coordinate is not accurate enough.

We measure an mAP of 76.2 when using the center coordinate and an mAP of 80.1 when using the mean transformation. At the same time, we do not observe an increase in runtime when computing the running mean. This clearly confirms our approach for refining the transformation hypotheses. In addition, we also experimented with using the median transformation per bin. As expected, the measured mAP increases to 80.3 since the median is less affected by outliers than the mean. However, this increase comes at significantly slower runtimes of 2.5 s, compared to 0.9 s when using the mean. This increase is caused by the fact that computing the median requires the individual transformations to be stored and then partially sorted.

4.4   Comparison with State-of-the-Art Spatial Verification Methods

In the next experiments, we verify the claim that our approach achieves a verification accuracy similar to hypothesize-and-verify methods while obtaining faster runtimes than voting-based methods. Towards this goal, we compare our approach against state-of-the-art methods for spatial verification. Hypothesize-and-verify approaches are represented by different variants of FSM [26]: The original *FSM* method exhaustively evaluates each transformation hypothesis obtained from a single feature match. The 1-point-RANSAC version of FSM (*FSM-R*) randomly samples from the hypotheses and terminates spatial verification once the probability of finding a better hypothesis falls below a threshold of $\hat{p} = 0.99$. For both FSM and FSM-R, we use two variants that either estimate an affine (*Aff.*) or a similarity (*Sim.*) transformation from each correspondence. All variants use local optimization to estimate an affine transformation from the inlier matches, independent of the type of transformation estimated from the individual correspondences. Besides ranking transformation hypotheses based on their numbers of inliers, we also evaluate FSM and FSM-R in combination with the effective inlier count (*cf.* Eq. 10). We again evaluate two variants. The first variant uses the effective inlier count instead of the standard inlier count during verification (*Eff. Inl. Eval*). The second variant simply applies the effective inlier count as a post-processing step (*Eff. Inl. Post*) on the best transformation found by FSM and FSM-R. The effective inlier count of this hypothesis is then used for re-ranking after spatial verification.

We also compare our approach against the current state-of-the-art approaches for voting-based verification: HPM [29], ADV [30], and PGM [31]. Since the three methods do not return inlier matches, they cannot be combined with the effective inlier count. Notice that the results reported in [30,31] are not directly comparable due to using different types of features and vocabularies of different sizes trained on different datasets. Thus, our results were obtained with our own implementations of HPM (without idf-weighting), ADV, and PGM.

Tables 2, 3, and 4 present the accuracy and runtimes on the Oxford5k, Paris5k, and new datasets, respectively. There are multiple interesting insights to be gained from our results: Both FSM and FSM-R outperform HPM, ADV, and PGM in terms of mAP. The result is especially pronounced on the Paris6k dataset (*cf.* Table 3). Using early stopping (FSM-R) rather than evaluating all

Table 2. Verification accuracy and efficiency measured on the Oxford5k dataset using a vocabulary of 20 k words. The best , second-best , and third-best results are highlighted for each column.

	-	F100k	Ox	UK	EU	US	Ox+UK	Ox+EU	Ox+US	UK+EU	UK+US	EU+US	Ox+UK+EU	Ox+UK+US	Ox+EU+US	UK+EU+US	Ox+UK+EU+US	All
mAP [%]																		
Pure Retrieval	76.2	66.4	64.1	60.3	60.3	59.6	57.6	57.4	57.2	56.3	55.9	55.8	54.4	54.3	54.1	53.7	52.3	51.7
FSM Aff	79.9	75.1	72.9	70.3	70.9	71.4	68.6	69.2	69.7	68.3	68.6	69.1	67.2	67.3	67.9	67.5	66.4	66.3
+ Eff. Inl. Eval	79.8	75.9	73.7	71.0	71.8	72.5	69.7	70.3	71.0	69.6	69.9	70.5	68.6	68.8	69.5	69.1	68.1	67.9
+ Eff. Inl. Post	79.6	75.2	73.3	70.1	71.1	71.7	68.8	69.6	70.2	68.5	68.8	69.5	67.5	67.7	68.4	67.8	66.8	66.6
FSM-R Aff	79.9	75.3	72.9	70.3	70.9	71.4	68.6	69.2	69.6	68.3	68.6	69.1	67.2	67.3	67.9	67.4	66.4	66.3
+ Eff. Inl. Eval	79.8	75.9	73.7	71.0	71.8	72.5	69.7	70.3	70.9	69.6	69.9	70.5	68.6	68.8	69.4	69.0	68.1	67.7
+ Eff. Inl. Post	79.6	75.2	73.1	70.0	71.0	71.7	68.6	69.5	70.0	68.3	68.7	69.4	67.3	67.5	68.3	67.6	66.6	66.3
FSM Sim	79.8	74.8	72.3	69.4	70.2	70.9	67.6	68.3	68.9	67.3	67.7	68.3	66.1	66.4	67.0	66.5	65.5	65.3
+ Eff. Inl. Eval	79.9	75.4	73.5	70.3	71.2	72.1	69.0	69.8	70.5	68.6	69.0	69.7	67.7	68.0	68.7	68.1	67.2	66.7
+ Eff. Inl. Post	79.1	74.4	72.5	69.1	70.0	71.0	67.7	68.5	69.2	67.2	67.8	68.5	66.2	66.6	67.4	66.6	65.7	65.3
FSM-R Sim	79.8	74.9	72.3	69.3	70.2	70.9	67.6	68.3	68.9	67.3	67.7	68.3	66.1	66.4	67.0	66.5	65.5	65.3
+ Eff. Inl. Eval	79.8	75.3	73.4	70.1	71.0	71.9	68.8	69.5	70.2	68.3	68.8	69.5	67.4	67.8	68.5	67.8	67.0	66.4
+ Eff. Inl. Post	79.0	74.3	72.4	69.0	69.9	70.8	67.6	68.3	69.0	67.1	67.6	68.3	66.1	66.5	67.2	66.5	65.5	65.2
HPM	73.4	65.4	63.3	59.7	59.6	60.0	58.1	58.1	58.2	57.1	57.1	57.4	56.0	56.0	56.3	55.8	54.9	54.4
ADV	78.5	73.7	71.9	68.2	68.8	70.0	66.8	67.4	68.3	66.1	66.5	67.0	65.1	65.4	65.9	65.2	64.3	64.0
PGM	76.0	64.6	62.5	58.7	59.1	57.9	55.5	55.3	54.9	54.6	53.9	53.8	52.6	52.3	52.0	51.4	50.2	49.4
Ours	80.1	74.5	71.9	68.7	69.5	69.7	67.0	67.6	67.8	66.6	66.7	67.2	65.3	65.4	65.9	65.4	64.4	64.1
+ Eff. Inl. Post	79.8	75.7	73.5	70.5	71.1	72.3	69.2	69.7	70.6	68.6	69.2	69.7	67.6	68.1	68.6	67.9	67.0	66.8
Runtime [s]																		
FSM Aff	31.1	43.4	44.1	46.3	45.9	49.7	68.0	68.1	70.1	69.9	70.5	68.7	73.5	71.2	69.9	69.8	73.3	73.9
+ Eff. Inl. Eval	309.6	356.4	381.0	456.5	446.8	408.6	476.8	467.1	438.0	500.7	488.6	475.3	523.4	495.0	484.7	512.1	522.7	506.2
+ Eff. Inl. Post	31.0	43.2	44.5	46.8	46.1	49.3	68.9	68.4	71.0	70.4	71.3	71.1	73.9	71.2	71.5	71.0	75.5	77.2
FSM-R Aff	4.3	5.7	6.6	8.4	8.0	7.5	10.4	9.9	9.7	11.2	11.1	10.6	12.1	11.2	10.7	11.3	12.0	12.0
+ Eff. Inl. Eval	193.1	237.1	258.2	333.9	324.8	282.5	347.9	339.2	303.9	370.1	357.5	344.1	387.0	363.2	354.4	381.0	390.0	378.7
+ Eff. Inl. Post	4.4	5.9	6.9	8.9	8.4	7.9	11.3	10.7	10.4	12.2	11.8	11.4	13.2	12.2	11.7	12.4	14.2	15.0
FSM Sim	33.0	45.0	47.5	48.8	47.8	51.0	70.9	70.9	73.5	72.2	73.3	72.1	75.9	73.4	74.2	73.4	76.3	76.1
+ Eff. Inl. Eval	317.3	367.3	389.6	476.3	454.6	416.4	485.7	476.2	446.9	508.1	497.5	482.6	531.2	502.1	491.8	520.8	530.7	516.1
+ Eff. Inl. Post	33.1	45.6	47.4	53.7	48.2	50.2	72.2	71.6	75.0	73.4	74.9	72.9	77.4	74.6	73.6	75.3	78.4	79.2
FSM-R Sim	4.2	5.5	6.3	8.1	7.6	7.2	10.1	9.5	9.5	10.8	10.6	10.1	11.6	10.8	10.2	10.9	11.6	11.5
+ Eff. Inl. Eval	192.5	236.6	257.0	332.4	323.2	281.2	346.0	336.4	301.7	368.7	356.2	341.3	385.9	361.4	352.9	378.1	386.7	376.2
+ Eff. Inl. Post	4.3	5.8	6.7	9.0	8.1	7.7	10.8	10.3	10.4	11.8	11.5	11.0	12.6	11.7	11.2	11.9	13.6	14.3
HPM	1.1	1.4	1.7	1.9	1.8	1.6	2.2	2.2	1.9	2.3	2.2	2.2	2.3	2.2	2.1	2.4	2.5	2.6
ADV	2.3	3.0	3.4	4.6	4.3	3.9	5.1	4.9	4.6	5.5	5.5	5.3	6.0	5.4	5.0	5.1	5.4	5.3
PGM	15.0	15.2	15.4	15.6	15.6	15.9	15.8	15.7	16.2	16.0	16.2	15.9	16.7	16.1	16.1	16.0	16.1	15.5
Ours	0.9	1.6	1.7	2.2	2.0	1.8	2.4	2.3	2.2	2.5	2.4	2.4	2.8	2.7	2.9	2.9	3.1	2.8
+ Eff. Inl. Post	1.2	1.8	1.9	2.5	2.3	2.0	2.7	2.7	2.5	3.0	2.8	2.9	3.4	3.4	3.6	3.5	5.6	5.9

possible transformations (FSM) significantly accelerates the verification without a significant impact on mAP. In fact, FSM-R is not more than a factor-of-4 slower than ADV, which is surprising given that one of the main arguments [29] for voting-based methods is that they are about an order of magnitude faster than FSM. Moreover, there is little difference between using an affine or similarity transformation for both FSM and FSM-R, likely due to the local optimization step. Compared to both ADV and PGM, our method achieves faster runtimes, which is most pronounced on our new dataset, where more features are found in each image. At the same time, our method also achieves a better accuracy. Especially on the Oxford5k and Paris6k datasets, our approach performs nearly as well as FSM and FSM-R, which are significantly slower than our method.

The influence of the distractor sets. Tables 2, 3, and 4 report the impact of combining each dataset with various combinations of the distractor sets. While using the effective inlier count provides little benefits without distractors, we observe a noticeable gain when adding distractors and thus making the problem harder. Naturally, the best results are obtained by directly incorporating the count into the verification stage of FSM and FSM-R. However, this comes at significant runtime costs since it needs to be evaluated often. Yet, using the count

Table 3. Verification accuracy and efficiency measured on the Paris6k dataset.

Table 4. Verification accuracy and efficiency measured on our new dataset.

	−	F100k	EU	US	EU+US	UK+EU+US	Ox+UK+EU+US	All
mAP [%]								
Pure Retrieval	71.2	60.2	56.2	54.6	51.2	49.5	48.7	47.7
FSM Aff	74.2	66.0	62.1	62.0	59.1	57.8	57.1	56.2
+ Eff. Inl. Eval	74.5	66.4	62.5	62.5	59.5	58.3	57.6	56.6
+ Eff. Inl. Post	73.9	65.6	61.5	61.4	58.3	57.1	56.4	55.6
FSM-R Aff	74.2	66.0	62.1	62.1	59.1	57.8	57.2	56.2
+ Eff. Inl. Eval	74.5	66.2	62.3	62.3	59.3	58.1	57.4	56.4
+ Eff. Inl. Post	73.7	65.3	61.3	61.2	58.2	56.9	56.3	55.4
FSM Sim	74.1	65.7	61.8	61.8	58.8	57.5	56.9	55.9
+ Eff. Inl. Eval	74.4	66.0	62.2	62.1	59.1	57.8	57.2	56.2
+ Eff. Inl. Post	73.8	65.2	61.3	61.2	58.2	56.8	56.2	55.3
FSM-R Sim	73.9	65.4	61.5	61.5	58.6	57.3	56.7	55.7
+ Eff. Inl. Eval	74.1	65.6	61.8	61.7	58.8	57.5	56.9	55.9
+ Eff. Inl. Post	73.6	64.9	60.9	60.9	57.9	56.6	55.9	55.0
HPM	70.8	60.8	56.1	55.5	52.2	50.8	50.2	49.3
ADV	71.5	62.9	60.1	60.1	57.3	56.0	55.3	54.2
PGM	70.7	59.3	54.8	53.6	50.0	48.1	47.1	46.2
Ours	73.4	64.9	60.9	60.7	57.8	56.6	56.0	54.9
+ Eff. Inl. Post	73.9	65.6	61.9	61.8	58.9	57.5	57.0	55.9
Runtime [s]								
FSM Aff	55.9	64.7	82.2	87.2	125.1	131.6	137.6	118.6
+ Eff. Inl. Eval	545.1	668.6	782.8	800.9	901.4	964.0	987.3	982.0
+ Eff. Inl. Post	55.6	67.4	83.1	89.5	127.5	133.5	144.9	124.2
FSM-R Aff	10.8	17.5	19.5	21.0	26.7	30.5	31.6	33.8
+ Eff. Inl. Eval	314.0	463.0	529.1	541.2	614.8	671.1	688.1	723.2
+ Eff. Inl. Post	11.0	17.8	20.1	21.7	27.8	32.0	35.2	39.2
FSM Sim	61.7	68.5	91.4	97.2	137.4	142.8	151.6	120.9
+ Eff. Inl. Eval	570.0	674.0	816.6	834.8	925.8	1005.1	1013.7	983.4
+ Eff. Inl. Post	62.0	68.4	93.0	98.3	140.0	155.9	152.8	126.7
FSM-R Sim	10.3	16.6	18.7	20.1	25.4	29.4	30.3	32.3
+ Eff. Inl. Eval	310.2	459.5	523.6	536.2	608.7	663.2	679.3	715.2
+ Eff. Inl. Post	10.6	17.2	19.3	20.9	26.7	31.0	33.9	37.9
HPM	2.2	2.3	2.8	2.8	3.8	4.2	4.4	4.5
ADV	3.7	5.4	6.0	6.0	7.7	8.5	8.6	9.1
PGM	19.1	19.8	19.7	19.9	20.3	20.7	22.1	23.5
Ours	2.5	2.8	3.3	3.3	4.4	4.6	4.9	5.4
+ Eff. Inl. Post	2.9	3.2	3.7	3.8	5.0	5.6	8.4	10.0

	−	EU	US	EU+US	UK+EU+US	Ox+UK+EU+US
mAP [%]						
Pure Retrieval	97.1	92.3	92.5	90.1	89.4	88.9
Aff	98.2	95.8	96.6	95.2	94.9	94.7
+ Eff. Inl. Eval	98.2	95.9	96.8	95.3	95.0	94.9
+ Eff. Inl. Post	97.9	95.5	96.4	94.9	94.6	94.4
Aff RANSAC	98.2	95.8	96.6	95.2	94.8	94.7
+ Eff. Inl. Eval	98.1	95.8	96.6	95.2	94.9	94.7
+ Eff. Inl. Post	97.8	95.4	96.3	94.7	94.4	94.3
Sim	98.1	95.8	96.6	95.1	94.8	94.6
+ Eff. Inl. Eval	98.1	95.9	96.7	95.3	95.0	94.8
+ Eff. Inl. Post	97.8	95.4	96.2	94.8	94.5	94.3
Sim RANSAC	98.0	95.7	96.4	95.0	94.7	94.5
+ Eff. Inl. Eval	98.0	95.8	96.6	95.2	94.9	94.7
+ Eff. Inl. Post	97.6	95.2	96.0	94.6	94.3	94.1
HPM	96.0	91.6	92.0	90.3	89.9	89.6
ADV	97.5	94.2	94.9	93.2	92.9	92.7
PGM	96.4	90.8	90.4	88.0	87.3	86.7
Ours	97.8	95.2	95.8	94.4	94.1	93.9
+ Eff. Inl. Post	97.8	95.4	96.1	94.7	94.4	94.3
Runtime [s]						
Aff	255.6	341.4	323.1	437.6	450.1	450.7
+ Eff. Inl. Eval	4234.8	6009.0	5599.8	6372.5	6629.2	6744.8
+ Eff. Inl. Post	256.0	345.2	325.9	441.8	455.7	457.0
Aff RANSAC	55.7	99.5	92.1	123.6	132.1	136.5
+ Eff. Inl. Eval	2485.8	4103.6	3757.2	4361.7	4574.0	4671.2
+ Eff. Inl. Post	57.8	104.0	96.3	128.7	138.6	143.8
Sim	270.5	360.7	339.7	466.5	478.3	475.2
+ Eff. Inl. Eval	4378.2	6146.8	5732.3	6501.5	6758.4	6864.2
+ Eff. Inl. Post	273.5	366.1	344.3	471.9	485.8	483.3
Sim RANSAC	54.8	96.2	88.9	119.3	128.1	132.2
+ Eff. Inl. Eval	2479.8	4087.6	3741.5	4338.6	4544.2	4639.7
+ Eff. Inl. Post	57.2	101.3	93.5	124.9	134.9	139.5
HPM	12.4	17.1	16.4	21.6	21.9	22.8
ADV	24.8	35.7	32.4	39.9	42.1	46.4
PGM	54.5	57.5	54.9	57.9	58.5	59.9
Ours	13.7	19.8	19.7	26.7	27.5	23.4
+ Eff. Inl. Post	16.5	24.3	23.8	29.7	30.8	28.0

as a post-processing step incurs only negligible cost. Combining the effective inlier count with our method further increases the accuracy of our method.

The distractor set showing the largest decrease in mAP in combination with the Oxford5k dataset was the image set from 30 cities around the UK and not *Ox*. This is somewhat counter-intuitive, since it might be expected that the distractor set consisting of images taken close to Oxford would be likely to contain images similar to the query. However, since there are fewer cities in this area, it turns out that those (typically non-urban) images are more easily discarded during spatial verification, compared to image sets targeted toward city centers. At the same time, the UK distractor set proved harder than the Europe and US sets, despite its smaller size. The Paris6k dataset had a similar drop in accuracy using our targeted distractor sets, compared to the Flickr100k distractor set.

5   Conclusion

In this work, we presented a novel method for fast spatial verification in image retrieval. Our method is a hybrid of voting-based approaches for efficient hypotheses generation and inlier counting-based methods for accurate hypothesis verification. Comprehensive experiments demonstrate high robustness to the choice of parameters. Our method achieves superior performance in balancing precision and efficiency versus the state of the art. Furthermore, we studied the impact of distractor image distribution and introduced a new query image set, which is released to the public alongside the source code of our method.

Acknowledgement. True Price and Jan-Michael Frahm were supported in part by the NSF No. IIS-1349074, No. CNS-1405847.

References

1. Sivic, J., Zisserman, A.: Video google: a text retrieval approach to object matching in videos. In: ICCV (2003)
2. Arandjelović, R., Zisserman, A.: Three things everyone should know to improve object retrieval. In: CVPR (2012)
3. Arandjelović, R., Zisserman, A.: DisLocation: scalable descriptor distinctiveness for location recognition. In: Cremers, D., Reid, I., Saito, H., Yang, M.-H. (eds.) ACCV 2014. LNCS, vol. 9006, pp. 188–204. Springer, Heidelberg (2015). doi:10. 1007/978-3-319-16817-3_13
4. Sattler, T., Havlena, M., Schindler, K., Pollefeys, M.: Large-scale location recognition and the geometric burstiness problem. In: CVPR (2016)
5. Torii, A., Arandjelovic, R., Sivic, J., Okutomi, M., Pajdla, T.: 24/7 place recognition by view synthesis. In: CVPR (2015)
6. Sattler, T., Weyand, T., Leibe, B., Kobbelt, L.: Image retrieval for image-based localization revisited. In: BMVC (2012)
7. Sattler, T., Havlena, M., Radenovic, F., Schindler, K., Pollefeys, M.: Hyperpoints and fine vocabularies for large-scale location recognition. In: ICCV (2015)
8. Gammeter, S., Quack, T., Van Gool, L.: I know what you did last summer: object-level auto-annotation of holiday snaps. In: ICCV (2009)
9. Weyand, T., Leibe, B.: Discovering favorite views of popular places with iconoid shift. In: ICCV (2011)
10. Weyand, T., Leibe, B.: Discovering details and scene structure with hierarchical iconoid shift. In: ICCV (2013)
11. Lee, G.H., Fraundorfer, F., Pollefeys, M.: Structureless pose-graph loop-closure with a multi-camera system on a self-driving car. In: IROS (2013)
12. Schönberger, J.L., Radenović, F., Chum, O., Frahm, J.M.: From single image query to detailed 3d reconstruction. In: IEEE Conference on Computer Vision and Pattern Recognition (CVPR) (2015)
13. Radenović, F., Schönberger, J.L., Ji, D., Frahm, J.M., Chum, O., Matas, J.: From dusk till dawn: modeling in the dark. In: CVPR (2016)
14. Schönberger, J.L., Frahm, J.M.: Structure-from-motion revisited. In: IEEE Conference on Computer Vision and Pattern Recognition (CVPR) (2016)
15. Jégou, H., Douze, M., Schmid, C., Pérez, P.: Aggregating local descriptors into a compact image representation. In: CVPR (2010)
16. Perronnin, F., Dance, C.: Fisher kernels on visual vocabularies for image categorization. In: CVPR, pp. 1–8 (2007)
17. Jégou, H., Zisserman, A.: Triangulation embedding and democratic aggregation for image search. In: CVPR (2014)
18. Arandjelović, R., Gronat, P., Torii, A., Pajdla, T., Sivic, J.: NetVLAD: CNN architecture for weakly supervised place recognition. In: IEEE Conference on Computer Vision and Pattern Recognition (2016)

19. Radenović, F., Tolias, G., Chum, O.: CNN image retrieval learns from BoW: unsupervised fine-tuning with hard examples. In: Leibe, B., Matas, J., Sebe, N., Welling, M. (eds.) ECCV 2016. LNCS, vol. 9905, pp. 3–20. Springer, Heidelberg (2016). doi:10.1007/978-3-319-46448-0_1

20. Gordo, A., Almazan, J., Revaud, J., Larlus, D.: Deep image retrieval: learning global representations for image search. arXiv:1604.01325 (2016)

21. Chum, O., Mikulik, A., Perdoch, M., Matas, J.: Total recall II: query expansion revisited. In: CVPR (2011)

22. Mikulík, A., Perdoch, M., Chum, O., Matas, J.: Learning vocabularies over a fine quantization. IJCV (2013)

23. Tolias, G., Avrithis, Y., Jégou, H.: To aggregate or not to aggregate: selective match kernels for image search. In: ICCV (2013)

24. Jégou, H., Douze, M., Schmid, C.: On the burstiness of visual elements. In: CVPR (2009)

25. Jegou, H., Douze, M., Schmid, C.: Hamming embedding and weak geometric consistency for large scale image search. In: Forsyth, D., Torr, P., Zisserman, A. (eds.) ECCV 2008. LNCS, vol. 5302, pp. 304–317. Springer, Heidelberg (2008). doi:10.1007/978-3-540-88682-2_24

26. Philbin, J., Chum, O., Isard, M., Sivic, J., Zisserman, A.: Object retrieval with large vocabularies and fast spatial matching. In: CVPR (2007)

27. Fischler, M., Bolles, R.: Random sample consensus: a paradigm for model fitting with applications to image analysis and automated cartography. Comm. ACM (1981)

28. Lowe, D.: Distinctive image features from scale-invariant keypoints. IJCV (2004)

29. Avrithis, Y., Tolias, G.: Hough pyramid matching: speeded-up geometry re-ranking for large scale image retrieval. IJCV (2014)

30. Wu, X., Kashino, K.: Adaptive dither voting for robust spatial verification. In: ICCV (2015)

31. Li, X., Larson, M., Hanjalic, A.: Pairwise geometric matching for large-scale object retrieval. In: CVPR (2015)

32. Chum, O., Philbin, J., Sivic, J., Isard, M., Zisserman, A.: Total recall: automatic query expansion with a generative feature model for object retrieval. In: ICCV (2007)

33. Mikulík, A., Radenović, F., Chum, O., Matas, J.: Efficient image detail mining. In: Cremers, D., Reid, I., Saito, H., Yang, M.-H. (eds.) ACCV 2014. LNCS, vol. 9004, pp. 118–132. Springer, Heidelberg (2015). doi:10.1007/978-3-319-16808-1_9

34. Irschara, A., Zach, C., Frahm, J.M., Bischof, H.: From structure-from-motion point clouds to fast location recognition. In: CVPR (2009)

35. Sivic, J., Zisserman, A.: Efficient visual search cast as text retrieval. PAMI (2009)

36. Sattler, T., Leibe, B., Kobbelt, L.: SCRAMSAC: improving RANSAC's efficiency with a spatial consistency filter. In: ICCV (2009)

37. Wu, X., Kashino, K.: Robust spatial matching as ensemble of weak geometric relations. In: BMVC (2015)

38. Chum, O., Matas, J., Kittler, J.: Locally optimized RANSAC. In: Michaelis, B., Krell, G. (eds.) DAGM 2003. LNCS, vol. 2781, pp. 236–243. Springer, Heidelberg (2003). doi:10.1007/978-3-540-45243-0_31

39. Lebeda, K., Matas, J., Chum, O.: Fixing the locally optimized ransac. In: BMVC (2012)

40. Chum, O., Matas, J.: Matching with prosac-progressive sample consensus. In: CVPR (2005)

41. Raguram, R., Chum, O., Pollefeys, M., Matas, J., Frahm, J.: Usac: a universal framework for random sample consensus. PAMI (2013)
42. Chum, O., Perdoch, M., Matas, J.: Geometric min-hashing: finding a (thick) needle in a Haystack. In: CVPR (2009)
43. Zhang, Y., Jia, Z., Chen, T.: Image retrieval with geometry-preserving visual phrases. In: CVPR (2011)
44. Johns, E.D., Yang, G.-Z.: Pairwise probabilistic voting: fast place recognition without RANSAC. In: Fleet, D., Pajdla, T., Schiele, B., Tuytelaars, T. (eds.) ECCV 2014. LNCS, vol. 8690, pp. 504–519. Springer, Heidelberg (2014). doi:10.1007/978-3-319-10605-2_33
45. Tolias, G., Kalantidis, Y., Avrithis, Y., Kollias, S.: Towards large-scale geometry indexing by feature selection. CVIU (2014)
46. Shen, X., Lin, Z., Brandt, J., Wu, Y.: Spatially-constrained similarity measure for large-scale object retrieval. PAMI (2014)
47. Philbin, J., Chum, O., Isard, M., Sivic, J., Zisserman, A.: Lost in quantization: improving particular object retrieval in large scale image databases. In: CVPR (2008)
48. Thomee, B., Shamma, D.A., Friedland, G., Elizalde, B., Ni, K., Poland, D., Borth, D., Li, L.J.: Yfcc100m: the new data in multimedia research. Comm. ACM (2016)
49. Heinly, J., Schönberger, J.L., Dunn, E., Frahm, J.M.: Reconstructing the world* in six days *(as captured by the yahoo 100 million image dataset). In: CVPR (2015)
50. Perdoch, M., Chum, O., Matas, J.: Efficient representation of local geometry for large scale object retrieval. In: CVPR (2009)

SSP: Supervised Sparse Projections for Large-Scale Retrieval in High Dimensions

Frederick Tung[(✉)] and James J. Little

Department of Computer Science, University of British Columbia, Vancouver, Canada
{ftung,little}@cs.ubc.ca

Abstract. As "big data" transforms the way we solve computer vision problems, the question of how we can efficiently leverage large labelled databases becomes increasingly important. High-dimensional features, such as the convolutional neural network activations that drive many leading recognition frameworks, pose particular challenges for efficient retrieval. We present a novel method for learning compact binary codes in which the conventional dense projection matrix is replaced with a discriminatively-trained sparse projection matrix. The proposed method achieves two to three times faster encoding than modern dense binary encoding methods, while obtaining comparable retrieval accuracy, on SUN RGB-D, AwA, and ImageNet datasets. The method is also more accurate than unsupervised high-dimensional binary encoding methods at similar encoding speeds.

1 Introduction

In the past few years, "big data" has transformed how we approach computer vision problems. The emergence and ongoing development of today's leading architecture for many recognition tasks, deep convolutional neural networks, has been enabled by the creation of massive labelled databases such as ImageNet [1]. With unprecedented amounts of labelled training examples available for tasks ranging from image classification [1], to semantic segmentation and captioning [2], to 3D reconstruction [3], increasingly complex deep models have been trained to achieve exciting new results [4].

As our models grow in complexity and labelled data sources grow in scale, an important question demands our attention: how can we leverage these massive data sources efficiently? In particular, how can we search labelled "big data" efficiently while maintaining economy of storage?

One traditionally highly successful approach to efficient large-scale retrieval is compact binary encoding [5–12]. The idea is to encode each database feature into a short binary signature (or binary code) using a generated projection matrix. The projection matrix may be randomly generated, or it may be optimized to minimize a quantization loss or classification loss on training examples. Given a novel feature as a query, the query is similarly projected into a binary code and compared to the binary codes in the database. Binary encoding is time-efficient in that binary code distances can be computed directly in modern hardware:

© Springer International Publishing AG 2017
S.-H. Lai et al. (Eds.): ACCV 2016, Part I, LNCS 10111, pp. 338–352, 2017.
DOI: 10.1007/978-3-319-54181-5_22

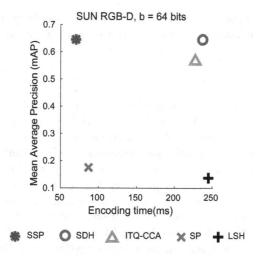

Fig. 1. Retrieval accuracy versus encoding time on the SUN RGB-D benchmark [3] using 4096-dimensional convolutional neural network features. The proposed supervised sparse projections (SSP) method achieves two to three times faster encoding speeds than conventional dense binary encoding methods, while obtaining comparable retrieval accuracy. SSP also achieves higher accuracy than unsupervised high-dimensional binary encoding methods at similar encoding speeds. Binary encoding methods: supervised discrete hashing (SDH) [12]; iterative quantization with canonical correlation analysis (ITQ-CCA) [8]; sparse projections (SP) [10]; locality-sensitive hashing (LSH) [17,18]. See Sect. 4 for complete results.

the Hamming distance between two binary codes can be computed by taking an XOR and counting the ones. Binary encoding is also storage-efficient as the generated binary codes are typically at least an order of magnitude more compact to store than the original features.

With the growth of big data, computer vision algorithms are relying on increasingly complex models and producing increasingly high dimensional features. From second-order pooling of traditional hand-crafted features [13], to Fisher Vectors [14], to the activations of deep neural networks [15], features with thousands or tens of thousands of dimensions have become the norm. However, scaling traditional binary encoding (also known as hashing) algorithms to handle high dimensional features is challenging. For example, traditional algorithms lack an effective regularizer for the learned projections, which sometimes leads to overfitting in high dimensions [10]. Moreover, encoding in high dimensions is computationally expensive due to the large projection matrix [10,16].

We present a novel binary encoding method called Supervised Sparse Projections (SSP), in which the conventional dense projection matrix is replaced with a discriminatively-trained sparse projection matrix. This design decision enables SSP to achieve two to three times faster encoding in comparison to conventional dense methods, while obtaining comparable retrieval accuracy. SSP is also more accurate than unsupervised high-dimensional binary encoding methods while providing similar encoding speeds. Figure 1 shows sample experimental results using 4096-dimensional convolutional neural network features.

2   Related Work

A broad range of binary encoding algorithms has been developed for large-scale retrieval. Locality-sensitive hashing (LSH) [17,18] is a traditional, data-independent binary encoding method that uses randomly generated projections. LSH offers theoretical asymptotic guarantees and can be kernelized [19,20], however a large number of bits (i.e. a large random projection matrix) is typically required for good performance in practice. This performance limitation motivated the development of data-driven binary encoding methods, which rely on training or optimization to learn the projection matrix [5–12]. For example, iterative quantization [8] learns a rotation of PCA (unsupervised) or CCA (supervised) projected data that minimizes the quantization error. Binary reconstructive embedding [5] uses coordinate descent to minimize the error between the original pairwise distances and the Hamming pairwise distances. Minimal loss hashing [6] applies structured prediction techniques to minimize a hinge-like error. Kernel-based supervised hashing [7] sequentially learns a projection in kernel space. Graph cuts coding [9] integrates the binary codes into the optimization as auxiliary variables. Supervised discrete hashing [12] learns a projection that optimizes the linear classification performance of the binary codes. All of these binary encoding methods learn *dense* projection matrices.

Gong et al. [16] showed that modern binary encoding methods are infeasible for searching high-dimensional feature vectors such as VLAD [21] and Fisher Vectors [14]. Their bilinear projections (BP) method replaces the usual expensive projection matrix with two smaller bilinear projections. Circulant binary embedding (CBE) [22] learns a projection by a circulant matrix, which can be accelerated using the Fast Fourier Transform. Xia et al. [10] recently proposed an unsupervised sparse projection (SP) method for searching high-dimensional feature vectors. The sparse projection is obtained by an iterative, ITQ-like optimization with an additional L0 regularization constraint. SP achieves an order of magnitude faster encoding than unsupervised dense methods while offering comparable accuracy. In addition, SP produces much more accurate results than BP or CBE, while having similar encoding speed. Concurrent with SP, Rastegari et al. [23] proposed sparse binary embedding, which finds a sparse projection matrix via a different SVM-based optimization. Kronecker binary embedding [24] computes fast projections using a structured matrix that is the Kronecker product of small orthogonal matrices.

3   Method

We begin by formulating the binary encoding problem. Let $\mathbf{X} \in \mathbf{R}^{n \times d}$ denote a database containing n feature vectors, with each feature vector being d-dimensional. The goal of a binary encoding algorithm is to learn a projection matrix $\mathbf{P} \in \mathbf{R}^{d \times b}$ that can be applied to $\mathbf{X}$ to produce similarity-preserving binary codes $\mathbf{B} \in \{-1, 1\}^{n \times b}$, where b is the number of bits in the binary codes. In the case of supervised binary encoding, we also have semantic class labels $\mathbf{Y} \in \mathbf{R}^{n \times C}$, where C is the number of classes, and $y_{ki} = 1$ if $\mathbf{x}_i$ belongs to class k and 0 otherwise.

Given a novel query $\mathbf{x}_q \in \mathbf{R}^d$, the learned projection $\mathbf{P}$ is applied to obtain the corresponding binary code $\mathbf{b} \in \{-1, 1\}^b$:

$$\mathbf{b}_q = \text{sgn}(\mathbf{x}_q \mathbf{P}) \tag{1}$$

where $\text{sgn}(\cdot)$ denotes the sign function applied to a vector, which returns $+1$ for positive values and -1 otherwise. It is also possible to learn and apply the projection to a non-linear mapping of $\mathbf{x}_q$: that is, $\mathbf{b}_q = \text{sgn}(\phi(\mathbf{x}_q)\mathbf{P})$ where ϕ is a non-linear mapping such as an RBF kernel mapping. For ease of formulation and experimentation we develop the simple linear case in Eq. 1.

Once the query has been encoded, it is often feasible to compare $\mathbf{b}_q$ to the database codes $\mathbf{B}$ by a linear search. The Hamming distance between two binary codes can be computed by taking an XOR and counting the ones, which is an operation supported in modern hardware. Algorithmic accelerations for efficient Hamming search can also be applied [25]. The fast comparison, combined with the compact storage requirements of the binary codes (storage requirements are typically less than the original feature vectors by at least an order of magnitude), make binary encoding a time and storage efficient option for large-scale search.

What makes for an effective projection matrix $\mathbf{P}$? A good projection matrix should preserve semantic relationships, as revealed in the supervisory signal $\mathbf{Y}$. It should be generalizable to new, previously unseen queries. Finally, we argue that it should be fast to apply at test time – an important criterion as today's increasingly complex models produce increasingly high-dimensional features.

To obtain an effective projection matrix, the Supervised Sparse Projections (SSP) method solves the following optimization problem:

$$\min_{\mathbf{B}, \mathbf{W}, \mathbf{P}} ||\mathbf{Y} - \mathbf{B}\mathbf{W}||_F^2 + \lambda ||\mathbf{W}||_F^2 \tag{2}$$

$$\text{s.t. } \mathbf{B} = \text{sgn}(\mathbf{X}\mathbf{P}), \ ||\mathbf{P}||_1 \leq \tau$$

where $\mathbf{W} \in \mathbf{R}^{b \times C}$, $||\cdot||_F$ denotes the Frobenius norm, λ is a regularization parameter, and τ is a sparsity parameter. The objective (2) can be seen as a standard L2 classification loss fitting the binary codes $\mathbf{B}$ to the supervisory signal $\mathbf{Y}$, with an additional constraint that $\mathbf{B}$ is generated by applying a sparse projection $\mathbf{P}$ to the feature vectors $\mathbf{X}$. We solve this optimization problem following a sequential approach. In the first subproblem, we find $\hat{\mathbf{B}}$ and $\hat{\mathbf{W}}$ satisfying

$$\min_{\mathbf{B}, \mathbf{W}} ||\mathbf{Y} - \mathbf{B}\mathbf{W}||_F^2 + \lambda ||\mathbf{W}||_F^2 \tag{3}$$

$$\text{s.t. } \mathbf{B} \in \{-1, 1\}^{n \times b}$$

In the second subproblem, given $\hat{\mathbf{B}}$ from the first subproblem, we solve for the projection matrix $\mathbf{P}$ satisfying

$$\min_{\mathbf{P}} ||\hat{\mathbf{B}} - \mathbf{XP}||_F^2 \tag{4}$$

$$\text{s.t. } ||\mathbf{P}||_1 \leq \tau$$

An alternative approach would be to incorporate the constraint $\mathbf{B} = \text{sgn}(\mathbf{XP})$ as a penalty term in a single objective,

$$\min_{\mathbf{B,W,P}} ||\mathbf{Y} - \mathbf{BW}||_F^2 + \lambda||\mathbf{W}||_F^2 + \nu||\mathbf{B} - \mathbf{XP}||_F^2$$

$$\text{s.t. } \mathbf{B} \in \{-1,1\}^{n \times b}, \ ||\mathbf{P}||_1 \leq \tau,$$

and then iterate the following until convergence, similar to [12]: (i) fix $\mathbf{B}$, $\mathbf{P}$, and update $\mathbf{W}$, (ii) fix $\mathbf{W}$, $\mathbf{P}$, and update $\mathbf{B}$, (iii) fix $\mathbf{W}$, $\mathbf{B}$, and update $\mathbf{P}$. However, the sequential approach is much faster to optimize and produces similar results in practice. Concretely, on the SUN RGB-D benchmark with $b = 32$ bits, the sequential approach is over 800% faster than iterating (i),(ii),(iii), with less than 0.1% difference in mAP (averaged over ten trials). Training time is dominated by the subproblem of solving for $\mathbf{P}$ given the sparsity constraint and fixed $\mathbf{W}$ and $\mathbf{B}$. We therefore optimize the original objective (2) by solving subproblems (3) and (4) sequentially.

Our formulation is similar to that of [12], with two main differences: our method computes a sparse projection, which allows for two to three times faster encoding with comparable accuracy; and our optimization procedure is different.

3.1   Solving for $\hat{\mathbf{B}}$ and $\hat{\mathbf{W}}$

We solve for $\hat{\mathbf{B}}$ and $\hat{\mathbf{W}}$ in subproblem (3) via an iterative alternating optimization. First, we keep $\mathbf{B}$ fixed and update $\mathbf{W}$; then, we keep $\mathbf{W}$ fixed and update $\mathbf{B}$. These steps are repeated until convergence or a maximum number of iterations is reached.

(i) Fix B, update W. With $\mathbf{B}$ fixed in subproblem (3), the problem of solving for $\mathbf{W}$ becomes standard L2-regularized least squares regression (ridge regression), which admits a closed form solution. In particular, we update $\mathbf{W}$ as

$$\mathbf{W} = (\mathbf{B}^\top \mathbf{B} + \lambda \mathbf{I})^{-1} \mathbf{B}^\top \mathbf{Y} \tag{5}$$

(ii) Fix W, update B. With $\mathbf{W}$ fixed in subproblem (3), the resulting problem for $\mathbf{B}$ admits an iterative solution via the discrete cyclic coordinate descent (DCC) method [12]. DCC allows us to solve for $\mathbf{B}$ bit by bit – in other words, one column of $\mathbf{B}$ at a time. Each column is solved in closed form holding the other columns fixed. In particular, with the other columns fixed, the kth bit (column) of $\mathbf{B}$ has the optimal solution

$$\mathbf{b}_{:k} = \text{sgn}(\mathbf{q}_{:k} - \mathbf{B}'\mathbf{W}'\mathbf{w}_k^\top) \tag{6}$$

Algorithm 1. Learning Supervised Sparse Projections (SSP)

Input: Database feature vectors $\mathbf{X} \in \mathbf{R}^{n \times d}$; Supervisory labels $\mathbf{Y} \in \mathbf{R}^{n \times C}$; λ, τ
Output: Projection matrix $\mathbf{P} \in \mathbf{R}^{d \times b}$; Binary codes $\mathbf{B} \in \{-1, 1\}^{n \times b}$
    Initialize $\mathbf{P}$ as a random projection and $\mathbf{B}$ as sgn($\mathbf{XP}$).
    repeat
        Update $\mathbf{W}$ by Eq. 5.
        Update $\mathbf{B}$ by discrete cyclic coordinate descent (Eq. 6).
    until converged or maximum number of iterations reached.
    Solve L1-regularized least squares (lasso) problem (Eq. 4) to obtain $\mathbf{P}$.

---

where $\mathbf{q}_{:k}$ is kth column of $\mathbf{Q} = \mathbf{YW}^{\top}$, $\mathbf{B}'$ is $\mathbf{B}$ excluding column k, $\mathbf{w}_k$ is the kth row of $\mathbf{W}$, and $\mathbf{W}'$ is $\mathbf{W}$ excluding row k. We invite the interested reader to refer to [12] for the complete derivation.

3.2 Solving for P

Given $\hat{\mathbf{B}}$ from the previous section, we next solve for $\mathbf{P}$ in subproblem (4). With $\hat{\mathbf{B}}$ fixed, we can recognize subproblem (4) as an L1-regularized least squares (lasso) problem. The L1 regularizer $||P||_1 \leq \tau$ induces sparsity in the solution $\mathbf{P}$, where the degree of sparsity is controlled by the user parameter τ.

L1-regularized least squares is a well studied problem and we apply one of the standard algorithms in the optimization community, SPGL1 [26], to solve for $\mathbf{P}$ iteratively. We use the publicly available implementation of SPGL1 with the default parameter settings.

The entire learning algorithm is summarized in Algorithm 1.

4 Experiments

We compared SSP with several representative binary encoding methods:

1. Locality-sensitive hashing (LSH) [17,18] is an unsupervised binary encoding method that has been applied in a variety of computer vision tasks, including image retrieval, feature matching, and object classification [27].
2. Iterative quantization [8] with canonical correlation analysis (ITQ-CCA) and the recently introduced supervised discrete hashing (SDH) method [12] are supervised binary encoding methods.
3. Sparse projections (SP) [10] is a state-of-the-art method for fast unsupervised high-dimensional binary encoding.

Our method can be extended to a non-linear embedding by RBF kernel in the same way as SDH [12]. For a fair comparison with ITQ-CCA, SP, and LSH, we evaluated all five methods without RBF embedding.

We performed experiments on three large vision datasets from different task domains:

1. The SUN RGB-D dataset [3] is a state-of-the-art benchmark for 3D scene understanding. It consists of 10,335 RGB-D images of indoor scenes and supports a wide range of scene understanding tasks, including scene categorization, semantic segmentation, object detection, and room layout estimation. We followed the recommended protocol for scene categorization, which uses the scene categories having more than 80 images (a total of 9,862 images over 21 categories). The SUN RGB-D benchmark provides convolutional neural network features as a baseline. These convolutional neural network features have $d = 4096$ dimensions and are generated using Places-CNN [28], a leading architecture for scene categorization on the SUN database [29]. We conducted ten trials, randomly setting aside 1,000 images as queries and using the remainder as the database in each trial.

2. The Animals with Attributes (AwA) dataset [30] was originally introduced for evaluating zero-shot learning. The task in zero-shot learning is to perform image classification for categories that are described in terms of their semantic attributes [30,31], but which have no training examples. Here we use AwA simply for multi-class retrieval. The dataset consists of 30,475 images covering 50 animal categories, with a minimum of 92 images in each category. AwA includes as a baseline 4096-dimensional convolutional neural network features generated using the VGG-19 "very deep" network [32]. We conducted ten trials, randomly setting aside 1,000 images as queries and using the remainder as the database in each trial.

3. The ILSVRC (ImageNet Large Scale Visual Recognition Challenge) 2012 dataset [1] is a standard benchmark for image classification algorithms. It consists of 1.2 million images spanning 1,000 semantic categories. We extracted 4096-dimensional convolutional neural network features from each image using the VGG CNN-M network [15]. To hold the database in memory for training, we conducted ten trials in which 100 classes were randomly selected each time. Within the subset of 100 classes, we randomly set aside 1,000 images as queries and used the remainder as the database. We sampled 10,000 images from the database to train the binary encoding methods.

We measure retrieval accuracy and encoding speed for all methods. Retrieval accuracy is measured by the mean average precision (mAP). With labelled data, we are interested in preserving semantic similarity. The retrieval ground truth for computing the mean average precision consists of database examples sharing the same semantic category label as the query. We compute the encoding time as the processor time required for the matrix multiplication $\mathbf{b}_q = \mathrm{sgn}(\mathbf{x}_q \mathbf{P})$ (Eq. 1), where $\mathbf{x}_q$ is the query vector and the projection matrix $\mathbf{P}$ may be sparse or dense, depending on the binary encoding method. In particular, SP and the proposed SSP produce sparse projections $\mathbf{P}$; LSH, ITQ-CCA, and SDH produce dense projections $\mathbf{P}$. Both sparse and dense matrix-vector multiplications are executed using the Intel Math Kernel Library (MKL). All timings are performed on a desktop computer with a 3.60 GHz CPU.

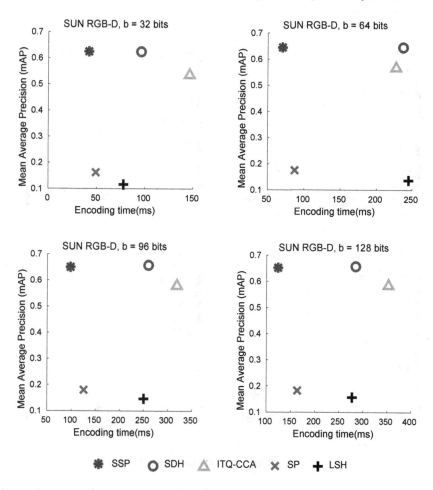

Fig. 2. Experimental results on the SUN RGB-D dataset [3] using 4096-dimensional convolutional neural network features (Places-CNN [28]) and binary code lengths $b = 32, 64, 96,$ and 128. Results are averaged over ten trials. Binary encoding methods: supervised sparse projections (SSP, this paper); supervised discrete hashing (SDH) [12]; iterative quantization with canonical correlation analysis (ITQ-CCA) [8]; sparse projections (SP) [10]; locality-sensitive hashing (LSH) [17,18].

Figure 2 summarizes the experimental results on SUN RGB-D for $b = 32,$ 64, 96, and 128 bits per feature vector. The vertical axis shows the mean average precision; the horizontal axis shows the encoding time in milliseconds for 1,000 queries. Compared to the unsupervised high-dimensional binary encoding method, SP, the proposed SSP obtains substantially higher retrieval accuracy at similar encoding speed. For example, at $b = 64$ bits SSP achieves 64.6% mAP compared to 17.4% for SP. These results demonstrate the advantage of

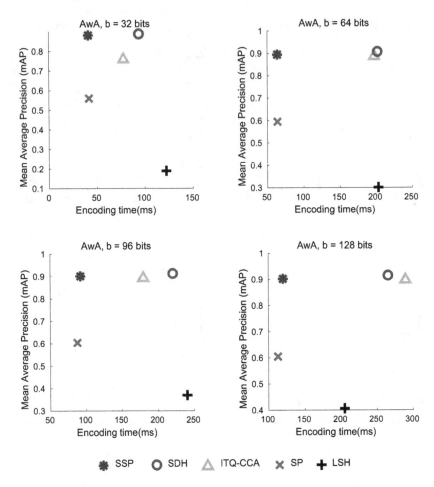

Fig. 3. Experimental results on the AwA dataset [30] using 4096-dimensional convolutional neural network features (VGG-19 "very deep" [32]) and binary code lengths $b = 32, 64, 96,$ and 128. Results are averaged over ten trials. Binary encoding methods: supervised sparse projections (SSP, this paper); supervised discrete hashing (SDH) [12]; iterative quantization with canonical correlation analysis (ITQ-CCA) [8]; sparse projections (SP) [10]; locality-sensitive hashing (LSH) [17,18].

taking into consideration the supervisory information available in many large-scale databases.

Because of the sparsity of the projection matrix, SSP provides significantly faster encoding speed than modern dense methods such as LSH, ITQ-CCA, and SDH. At the same time, retrieval accuracy is competitive with the dense methods. SSP achieves higher accuracy than ITQ-CCA and is comparable with SDH across all bit lengths. In particular, at $b = 32, 64, 96,$ and 128 bits, SSP is 2.3 to 3.4 times faster than SDH to encode, while matching SDH's accuracy within

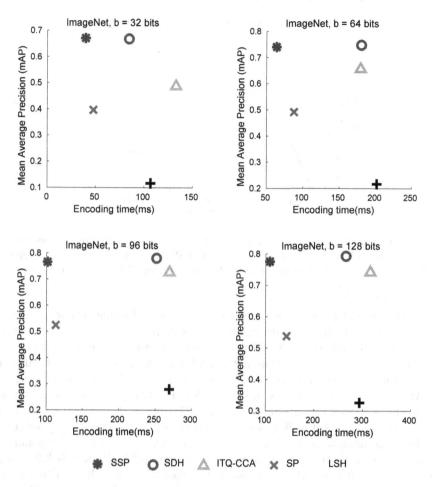

Fig. 4. Experimental results on the ImageNet dataset [1] using 4096-dimensional convolutional neural network features (VGG CNN-M [15]) and binary code lengths $b =$ 32, 64, 96, and 128. Results are averaged over ten trials. Binary encoding methods: supervised sparse projections (SSP, this paper); supervised discrete hashing (SDH) [12]; iterative quantization with canonical correlation analysis (ITQ-CCA) [8]; sparse projections (SP) [10]; locality-sensitive hashing (LSH) [17,18].

1% mAP. SSP is similarly faster than ITQ-CCA to encode, while obtaining 6 to 9% higher mAP.

In fact, these experimental results likely underestimate the practical improvement in encoding speed. The recent work on SP [10] reports encoding speeds that are linearly related to the proportion of nonzeros. For example, encoding with a projection matrix with 20% nonzeros is reportedly 5 times faster than encoding with a dense projection matrix. On the other hand, in Fig. 2, the sparse projection matrix for SSP has 20% nonzeros but we observed the encoding to be only 2.3 to 3.4 times faster than SDH, for instance. Since the authors of [10] also

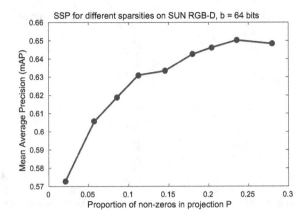

Fig. 5. Retrieval accuracy versus percentage of non-zeros in the SSP projection matrix **P**, for SUN RGB-D at $b = 64$ bits. The sparsity of **P** is determined by the parameter τ in Eq. 2. We set τ to obtain approximately 20% non-zeros in **P** in all other experiments.

use MKL for timing evaluation[1], this difference may be due to optimizations not fully supported in our implementation. In summary, comparisons with LSH, ITQ-CCA, and SDH on SUN RGB-D demonstrate that L1 regularization can be used to sparsify the projection matrix and improve the encoding efficiency of high-dimensional feature vectors, with only a minor cost in retrieval accuracy.

We observed similar results on the AwA dataset. Figure 3 shows the mean average precision versus encoding time for 1,000 queries, for $b = 32$, 64, 96, and 128 bits per feature vector. SSP achieves significantly faster encoding speed than the dense methods while providing comparable retrieval accuracy. For example, in comparison with the best performing dense method SDH, SSP is 2.2 to 3.2 times faster to encode and within 2% mAP across all bit lengths. SSP is significantly more accurate than the fast high-dimensional encoding method SP, with a relative improvement of 49 to 57% mAP across all bit lengths.

Figure 4 summarizes the experimental results on ImageNet (ILSVRC 2012) for $b = 32$, 64, 96, and 128 bits per feature vector. Experimental results on ImageNet follow the general trends observed on the SUN RGB-D and AwA datasets. Compared to fast high-dimensional binary encoding with SP, SSP obtains more accurate results at similar encoding speeds, with relative improvements of 43 to 69% mAP across all bit lengths. Compared to the best performing dense method SDH, SSP is 2.1 to 2.8 times faster to encode and is within 2% mAP across all bit lengths.

The sparsity of the learned projection matrix in SSP can be controlled by setting the τ parameter. In the preceding experiments we set τ to obtain approximately 20% non-zeros in the projection matrix **P**. Figure 5 shows how retrieval accuracy changes as we vary τ to obtain sparser or denser projection matrices. Experiments were conducted on the SUN RGB-D dataset with 64-bit codes,

---

[1] Personal communication.

Fig. 6. Qualitative results on the SUN RGB-D dataset at $b = 64$ bits. The left column shows query images. The right column shows ten of the top retrieved neighbors (sampled from multi-way ties) using SSP. The results indicate that SSP reliably preserves semantic neighbors in the learned binary codes.

averaging over ten random trials. A tradeoff between projection sparsity and retrieval accuracy can be seen in the figure. In general, making the projection matrix more sparse enables faster encoding but at a cost in retrieval accuracy.

On the SUN RGB-D dataset, training requires 5.9 min for $b = 32$ bits; 10.3 min for $b = 64$ bits; 15.9 min for $b = 96$ bits; and 21.0 min for $b = 128$ bits, averaging over ten random trials.

Figure 6 shows typical qualitative results on the SUN RGB-D dataset using SSP with $b = 64$ bits. SSP reliably preserves the supervisory information in the learned binary codes. We observed that even in failure cases, SSP returns semantically related neighbors. For example, the ground truth class label of the third query image from the top is 'lab', but the top retrieved nearest neighbors are 'office' images. In summary, SSP allows for scalable, fast, and semantics-preserving retrieval in large labelled databases. Replacing the original $d = 4096$ dimensional convolutional neural network features with 64-bit SSP binary codes results a reduction in storage costs by three orders of magnitude (131,072 or 262,144 bits reduced to 64 bits).

5   Conclusion

We have presented Supervised Sparse Projections, a novel supervised binary encoding method in which the conventional dense projection matrix is replaced with a discriminatively-trained sparse projection matrix. Experiments using 4096-dimensional convolutional neural network features on three large computer vision datasets confirm the validity of the proposed approach: SSP achieves two to three times faster encoding than modern dense binary encoding methods, while obtaining comparable retrieval accuracy, and is more accurate than unsupervised high-dimensional binary encoding methods while providing similar encoding speed. As big data continues to enable increasingly complex recognition models in computer vision, we believe that solutions for efficiently searching high-dimensional labelled data will be increasingly important.

Acknowledgements. We thank Yan Xia for helpful discussion. This work was funded in part by the Natural Sciences and Engineering Research Council of Canada.

References

1. Russakovsky, O., Deng, J., Su, H., Krause, J., Satheesh, S., Ma, S., Huang, Z., Karpathy, A., Khosla, A., Bernstein, M., Berg, A.C., Fei-Fei, L.: ImageNet large scale visual recognition challenge (2014). arXiv:1409.0575
2. Lin, T.Y., Maire, M., Belongie, S., Hays, J., Perona, P., Ramanan, D., Dollár, P., Zitnick, C.L.: Microsoft COCO: common objects in context. In: Proceedings of European Conference on Computer Vision (2014)
3. Song, S., Lichtenberg, S., Xiao, J.: SUN RGB-D: a RGB-D scene understanding benchmark suite. In: Proceedings of IEEE Conference on Computer Vision and Pattern Recognition (2015)

4. He, K., Zhang, X., Ren, S., Sun, J.: Deep residual learning for image recognition. In: Proceedings of IEEE Conference on Computer Vision and Pattern Recognition (2016)
5. Kulis, B., Darrell, T.: Learning to hash with binary reconstructive embeddings. In: Advances in Neural Information Processing Systems (2009)
6. Norouzi, M., Fleet, D.J.: Minimal loss hashing for compact binary codes. In: Proceedings of International Conference in Machine Learning (2011)
7. Liu, W., Wang, J., Ji, R., Jiang, Y.G., Chang, S.F.: Supervised hashing with kernels. In: Proceedings of IEEE Conference on Computer Vision and Pattern Recognition (2012)
8. Gong, Y., Lazebnik, S., Gordo, A., Perronnin, F.: Iterative quantization: a procrustean approach to learning binary codes for large-scale image retrieval. IEEE Trans. Pattern Anal. Mach. Intell. **35**, 2916–2929 (2013)
9. Ge, T., He, K., Sun, J.: Graph cuts for supervised binary coding. In: Proceedigs of European Conference on Computer Vision (2014)
10. Xia, Y., He, K., Kohli, P., Sun, J.: Sparse projections for high-dimensional binary codes. In: Proceedings of IEEE Conference on Computer Vision and Pattern Recognition (2015)
11. Cakir, F., Sclaroff, S.: Adaptive hashing for fast similarity search. In: Proceedings of IEEE International Conference on Computer Vision (2015)
12. Shen, F., Shen, C., Liu, W., Shen, H.T.: Supervised discrete hashing. In: Proceedings of IEEE Conference on Computer Vision and Pattern Recognition (2015)
13. Carreira, J., Caseiro, R., Batista, J., Sminchisescu, C.: Semantic segmentation with second-order pooling. In: Proceedings of European Conference on Computer Vision (2012)
14. Perronnin, F., Sánchez, J., Mensink, T.: Improving the Fisher kernel for large-scale image classification. In: Proceedings of European Conference on Computer Vision (2010)
15. Chatfield, K., Simonyan, K., Vedaldi, A., Zisserman, A.: Return of the devil in the details: delving deep into convolutional nets. In: Proceedings of British Machine Vision Conference (2014)
16. Gong, Y., Kumar, S., Rowley, H.A., Lazebnik, S.: Learning binary codes for high-dimensional data using bilinear projections. In: Proceedings of IEEE Conference on Computer Vision and Pattern Recognition (2013)
17. Gionis, A., Indyk, P., Motwani, R.: Similarity search in high dimensions via hashing. In: Proceedings of International Conference on Very Large Data Bases (1999)
18. Charikar, M.S.: Similarity estimation techniques from rounding algorithms. In: Proceedings of ACM Symposium on Theory of Computing (2002)
19. Kulis, B., Grauman, K.: Kernelized locality-sensitive hashing. IEEE Trans. Pattern Anal. Mach. Intell. **34**, 1092–1104 (2012)
20. Jiang, K., Que, Q., Kulis, B.: Revisiting kernelized locality-sensitive hashing for improved large-scale image retrieval. In: Proceedings of IEEE Conference on Computer Vision and Pattern Recognition (2015)
21. Jégou, H., Douze, M., Schmid, C., Pérez, P.: Aggregating local descriptors into a compact image representation. In: Proceedings of IEEE Conference on Computer Vision and Pattern Recognition (2010)
22. Yu, F.X., Kumar, S., Gong, Y., Chang, S.F.: Circulant binary embedding. In: Proceedings of International Conference in Machine Learning (2014)
23. Rastegari, M., Keskin, C., Kohli, P., Izadi, S.: Computationally bounded retrieval. In: Proceedings of IEEE Conference on Computer Vision and Pattern Recognition (2015)

24. Zhang, X., Yu, F.X., Guo, R., Kumar, S., Wang, S., Chang, S.F.: Fast orthogonal projection based on Kronecker product. In: Proceedings of IEEE International Conference on Computer Vision (2015)
25. Norouzi, M., Punjani, A., Fleet, D.J.: Fast search in Hamming space with multi-index hashing. In: Proceedings of IEEE Conference on Computer Vision and Pattern Recognition (2012)
26. van den Berg, E., Friedlander, M.P.: Probing the pareto frontier for basis pursuit solutions. SIAM J. Sci. Comput. **31**, 890–912 (2008)
27. Dean, T., Ruzon, M.A., Segal, M., Shlens, J., Vijayanarasimhan, S., Yagnik, J.: Fast, accurate detection of 100,000 object classes on a single machine. In: Proceedings of IEEE Conference on Computer Vision and Pattern Recognition (2013)
28. Zhou, B., Lapedriza, A., Xiao, J., Torralba, A., Oliva, A.: Learning deep features for scene recognition using Places database. In: Advances in Neural Information Processing Systems (2014)
29. Xiao, J., Hays, J., Ehinger, K., Oliva, A., Torralba, A.: SUN database: large-scale scene recognition from abbey to zoo. In: Proceedings of IEEE Conference on Computer Vision and Pattern Recognition, pp. 3485–3492 (2010)
30. Lampert, C.H., Nickisch, H., Harmeling, S.: Learning to detect unseen object classes by between-class attribute transfer. In: Proceedings of IEEE Conference on Computer Vision and Pattern Recognition, pp. 951–958 (2009)
31. Farhadi, A., Endres, I., Hoiem, D., Forsyth, D.: Describing objects by their attributes. In: Proceedings of IEEE Conference on Computer Vision and Pattern Recognition, pp. 1778–1785 (2009)
32. Simonyan, K., Zisserman, A.: Very deep convolutional networks for large-scale image recognition (2014). arXiv:1409.1556

An Online Algorithm for Efficient
and Temporally Consistent Subspace Clustering

Vasileios Zografos[1(✉)], Kai Krajsek[2], and Bjoern Menze[1]

[1] Department of Informatics, Institute for Advanced Study,
Technical University of Munich, Munich, Germany
vasileios@zografos.org, bjoern.menze@tum.de
[2] IEK-8, Forschungszentrum Jülich, 52425 Jülich, Germany
k.krajsek@fz-juelich.de

Abstract. We present an online algorithm for the efficient clustering of data drawn from a union of arbitrary dimensional, non-static subspaces. Our algorithm is based on an online min-Mahalanobis distance classifier, which simultaneously clusters and is updated from subspace data. In contrast to most existing methods, our algorithm can cope with large amounts of batch or sequential data and is temporally consistent when dealing with time varying data (i.e. time-series). Starting from an initial condition, the classifier provides a first estimate of the subspace clusters in the current time-window. From this estimate, we update the classifier using stochastic gradient descent. The updated classifier is applied back onto the data to refine the subspace clusters, while at the same time we recover the explicit rotations that align the subspaces between time-windows. The whole procedure is repeated until convergence, resulting in a fast, efficient and accurate algorithm. We have tested our algorithm on synthetic and three real datasets and compared with competing methods from literature. Our results show that our algorithm outperforms the competition with superior clustering accuracy and computation speed.

1 Introduction

Given a set of N points $\mathbf{x} = \{x_j \in \mathbb{R}^D\}_{j=1}^N$ that is drawn from a union of K subspaces $\cup_{i=1}^K L_i$, we may define *subspace clustering* [1] as the problem of recovering the number of subspaces K, their parameters as well as the membership (clustering) of the data points $\mathbf{x}$ to each subspace. In practice it is sufficient to recover only the membership of points to subspaces and afterwards estimate the parameters of each subspace individually using PCA [2]. Additionally, K is a problem-specific parameter that is usually provided in each case, or can be estimated with existing techniques such as MDL-based pruning [3]. Subspace clustering has been applied successfully to many different problems in computer vision and machine learning, with a number of scientific publications now available [4–8], proposing a wide range of elegant solutions.

Electronic supplementary material The online version of this chapter (doi:10. 1007/978-3-319-54181-5_23) contains supplementary material, which is available to authorized users.

© Springer International Publishing AG 2017
S.-H. Lai et al. (Eds.): ACCV 2016, Part I, LNCS 10111, pp. 353–368, 2017.
DOI: 10.1007/978-3-319-54181-5_23

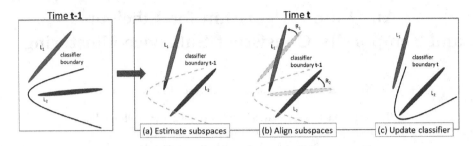

Fig. 1. OCRS overview: First we use the classifier from time-window $t-1$ to get an estimate of the new subspaces L_1, L_2 at time-window t (a). We then find the rotations R_1, R_2 that align the respective subspaces between time-windows in order to improve our estimates (b). Finally from the improved estimates we update the current classifier using stochastic gradient descent (c).

However, the vast majority of existing approaches are *batch* methods, in that they produce a clustering solution once all the data points have been observed. Because of their offline nature, batch methods have some fundamental limitations: **First** of all they cannot deal efficiently with sequential data. That is, data that arrives one or a few points at a time. Using a batch method to cluster sequential data means that the algorithm needs to be applied afresh to the whole data, for each new point arriving. This is obviously very inefficient and costly. The **second** limitation is that batch methods cannot cope with large datasets. This is because most batch methods perform an eigen-decomposition of the data matrix, due to the often used spectral clustering approach [9] and so there is a physical memory restriction on the amount of data that can be processed simultaneously by a computer. The **third** limitation is that batch methods assume that the parameters of the subspaces do not change as a function of time. What this means is that when dealing with dynamic processes and time-series/temporal data, batch methods are typically applied using a moving time-window, with the assumption that the subspaces are static within that time-window. Each time-window is treated independently without the use of past information. This may lead to temporal inconsistencies in the clustering, such as *concept drift* and *label switching*, i.e. although the clustering may be correct withing each window, there can be an arbitrary cluster label permutation between time-windows. In addition, ignoring past information is inefficient since a new subspace clustering problem is solved separately at each time-window.

Because of these fundamental limitations with batch methods, recent research has been focusing instead on *online subspace clustering*. This usually means either dealing with sequential data; or time-series data; or sometimes both. The distinction between sequential and time-series data is that in the former the data points are fed into the algorithm one/a few at a time. The number of points the algorithm sees is increased (i.e. *more observations*) and the algorithm must update the subspace clusters for each new point arriving. Sequential data usually, but not always, implies static subspaces. Conversely, time-series data do not increase over time, but instead the dimensionality of the data points increase (i.e. *longer* observations). Typically, clustering algorithms use a moving time-window inside which they solve the clustering, ensuring that the solution is consistent between time-windows. Varying time-series data is the result of a dynamic process and always implies non-static subspaces.

Notable examples of sequential methods are the works by [10–13], which are essentially computationally and memory efficient subspace clustering solutions that can cope with large data. Some of them, for example [12,13], can be adapted for use with non-static subspaces by considering new time-windows as out-of-sample data. Time-series methods, include the seminal work by [14] on non-static subspaces and the motion segmentation-specific approach by [15], which uses incremental PCA and results from previous time-windows to initialise a sparse solution in the current time-window. Both of these methods however cannot deal with sequential data by construction. It is also possible under certain conditions, such as normally distributed clusters, to address online subspace clustering via incremental Gaussian mixture models (GMMs). Of particular interest is the work by [16] on incremental learning of temporally consistent GMMs, or the IGMM method by [17] or that by [18], which although are incremental, do not enforce temporal consistency. Just like batch GMMs though, incremental GMMs can have problems when dealing with degenerate covariance matrices, which are common in subspace data.

In this paper we propose a new approach called Online Clustering of Rotating Subspaces (OCRS). OCRS uses a simple iterative *clustering-update* scheme. At each time-window t we apply a classifier that has been trained from the previous time-window $t-1$, in order to get a rough estimate of the current subspace clusters. We then use this estimate to iteratively: (i) recover the subspace rotations between current and previous time-windows and (ii) use the rotations to re-align the subspaces and online update the classifier using stochastic gradient descent. An overview of the OCRS method is illustrated in Fig. 1.

As a result, OCRS is capable of clustering any number of points lying on low-dimensional subspaces of arbitrary configurations. In addition, OCRS is efficient and fast, and can handle both sequential and large batch datasets, because of its stochastic gradient descent step. Time-series data are also dealt with in a temporally consistent manner by explicit recovery of the subspace rotations, which avoids the problem of label switching between time-windows. Finally, since OCRS is a fully online algorithm, it does not need to store already seen data. We have tested our method on a number of real and synthetic datasets and against state-of-the-art methods from literature. Our experiments demonstrate that OCRS outperforms competing methods in terms of convergence speed and recovers superior clustering solutions. Our method has very few parameters and is very stable across a large range of parameter values.

2   Our Approach

2.1   Batch Subspace Clustering

We begin with a batch algorithm for subspace clustering, which we will use as a foundation to construct our online approach later on. Our formulation is similar

to the gradient-descend K-means construction by [19] but has been extended
to work with anisotropic covariance matrices, bearing some resemblance to K-
subspaces type of methods [20–22]. Given therefore a distribution $dP(x)$ of points
$x \in \mathbb{R}^D$, the algorithm will compute K subspace clusters s_i, $i \in \{1, 2, \ldots, K\}$,
which minimise the expectation value of the Mahalanobis distance between each
point and its closest subspace. This can be considered as the equivalent task of
assigning cluster labels to the points using a min-Mahalanobis distance classifier.
More formally, we define the clusters by the minimum of a cost function $E(\mu, \mathbf{C})$:
$\mathbb{R}^D \times Sym^+ \to \mathbb{R}^+$, where $\mu := \{\mu_i\}$, $\mathbf{C} := \{\mathbf{C}_i\}$ denote the sets of means and
covariance matrices of each subspace cluster distribution, and Sym^+ denotes the
space of positive definite symmetric matrices. The cost function is defined as

$$E(\mu, \mathbf{C}) = \int Q(x, \mu, \mathbf{C})dP(x), \tag{1}$$

where

$$Q(x, \mu, \mathbf{C}) = \min_i \left((x - \mu_i)^T \mathbf{C}_i^{-1}(x - \mu_i) + \log(|\mathbf{C}_i|) \right), \tag{2}$$

is the loss function. If we write i^* as the index of the closest subspace to point
x and denote $\hat{x}^i = x - \mu_i$ then (1) becomes

$$E(\mu, \mathbf{C}) = \int (\hat{x}^{i^*})^T \mathbf{C}_{i^*}^{-1} \hat{x}^{i^*} + \log(|\mathbf{C}_{i^*}|)dP(x). \tag{3}$$

A local minimum of the cost function can then be obtained by classical gradient
descent:

$$\begin{aligned} \mu_i^t &= \mu_i^{t-1} - \gamma^t \nabla_{\mu_i} E = \mu_i^{t-1} + \gamma^t \Delta_{\mu_i} \\ \mathbf{C}_i^t &= \mathbf{C}_i^{t-1} - \gamma^t \nabla_{\mathbf{C}_i} E = \mathbf{C}_i^{t-1} + \gamma^t \Delta_{\mathbf{C}_i} \end{aligned}, \tag{4}$$

where γ^t is the learning rate. Given thus N samples from $dP(x)$ we can estimate
the integral in (3) and its gradient by the sample average. Note that (3) is not
differentiable for points lying on the Voronoi set w.r.t. the Mahalanobis distance
(i.e. points lying on the classifier boundary). In such cases, we randomly choose
one of the clusters and take the corresponding one-sided derivative. We then
obtain the batch update equations for the classifier's parameters μ, $\mathbf{C}$ given the
data points $\hat{x}^{i^*}$:

$$\Delta_{\mu_i} = \sum_j \begin{cases} \mathbf{C}_{i^*}^{-1}(\hat{x}^{i^*}), & \text{if} \quad i = i^* \\ 0, & \text{otherwise} \end{cases}$$

$$\Delta_{\mathbf{C}_i} = \sum_j \begin{cases} -\mathbf{C}_{i^*}^{-1} \hat{x}^{i^*} (\hat{x}^{i^*})^T \mathbf{C}_{i^*}^{-1} - \mathbf{C}_{i^*}^{-1}, & \text{if} \quad i = i^* \\ 0, & \text{otherwise.} \end{cases} \tag{5}$$

for $j = 1, \ldots, N$.

2.2   Online Subspace Clustering

Stochastic gradient descent for sequential data
However as we can see, the batch algorithm needs to iterate through all the points for estimating the gradients in (5), before the classifier can be updated with (4). In order to turn the batch method into an online approach and be able to deal with sequential data, we can use the *general stochastic gradient descent approach* introduced by [23]. Instead of computing the exact gradient of (3), at each iteration we calculate:

$$\mu_i^t = \mu_i^{t-1} - \gamma^t H_{\mu_i}$$
$$\mathbf{C}_i^t = \mathbf{C}_i^{t-1} - \gamma^t H_{\mathbf{C}_i} \tag{6}$$

where $H_\mu, H_{\mathbf{C}}$ are the update functions, with the explicit constraints:

$$\int H_\mu dP(x) = \nabla_\mu E$$
$$\int H_{\mathbf{C}} dP(x) = \nabla_{\mathbf{C}} E \tag{7}$$

In the stochastic gradient descent approach, we deal with points on the Voronoi set (i.e. classifier boundary) by setting the corresponding gradients to zero. Since the Voronoi set has zero probability mass, (7) are still fulfilled. We then obtain the corresponding *online* update steps:

$$H_{\mu_i} = \begin{cases} \mathbf{C}_{i^*}^{-1} \hat{x}_j^{i^*}, & \text{if } i = i^* \\ 0, & \text{otherwise} \end{cases}$$

$$H_{\mathbf{C}_i} = \begin{cases} \mathbf{C}_{i^*}^{-1} \hat{x}_j^{i^*} (\hat{x}_j^{i^*})^T \mathbf{C}_{i^*}^{-1} - \mathbf{C}_{i^*}^{-1}, & \text{if } i = i^* \\ 0, & \text{otherwise.} \end{cases} \tag{8}$$

It can be shown that the update functions in (7) correspond to a valid gradient approach on the cost function, since:

$$\nabla E = \nabla \int Q(x, \mu, \mathbf{C}) dP(x) = \int \nabla Q(x, \mu, \mathbf{C}) dP(x) = \int H dP(x). \tag{9}$$

The step from the second to third term in (9) is possible according to Lebesque monotone convergence theorem. The algorithm is now able to process data on the fly (sequential data) without the need to remember/store previously seen data.

Improving effciency: Inv. covariance matrix and pre-conditioning
Still however, the algorithm is far from efficient since it needs to invert the covariance matrix at each iteration of the stochastic gradient descent. This can be computationally demanding especially for high dimensional data. Furthermore, the matrix $\mathbf{C}^t$ computed from (6) is not guaranteed to be positive definite. We can fix both of these problems by working instead with the inverse covariance

matrix $\mathbf{B} = \mathbf{C}^{-1}$. The iterative update of the covariance matrix at time t may be written alternatively as:

$$\mathbf{C}^t = \frac{1}{t}\mathbf{C}^0 + \frac{1}{t}\sum_{\tau=1}^{t-1}\hat{x}_\tau^\tau(\hat{x}_\tau^\tau)^T + \frac{1}{t}\hat{x}_t^t(\hat{x}_t^t)^T \tag{10}$$

$$= \mathbf{C}^{t-1} + \frac{1}{t}(\hat{x}_t^t(\hat{x}_t^t)^T - \mathbf{C}^{t-1}) \tag{11}$$

$$= \mathbf{C}^{t-1} + \frac{1}{t}\mathbf{C}^{t-1}H_{\mathbf{C}^t}\mathbf{C}^{t-1}. \tag{12}$$

Equation (12) is simply a pre-conditioning of the estimated gradient at time t that guarantees that $\mathbf{C}^t$ is positive definite, provided that $\mathbf{C}^{t-1}$ is also positive definite. This is achieved by initializing the algorithm with a scaled identity matrix $\mathbf{C}^0 := \eta\mathbf{I}$, for some small $\eta \in \mathbb{R}^+$. We may now use the Sherman-Morrison formula [24]

$$(\mathbf{P} + uu^T)^{-1} = \mathbf{P}^{-1} + \frac{\mathbf{P}^{-1}uu^T\mathbf{P}^{-1}}{1 + u^T\mathbf{P}^{-1}u} \tag{13}$$

which computes the rank-1 update of the inverse of a matrix and set $\mathbf{P} := (t-1)\mathbf{C}^{t-1}$ and $u := \hat{x}_t^t$ in (13) to obtain the recursive update:

$$\mathbf{B}^t = \frac{t\mathbf{B}^{t-1}}{t-1} + \frac{t\mathbf{B}^{t-1}\hat{x}_t^t(\hat{x}_t^t)^T\mathbf{B}^{t-1}}{\left(1 + (\hat{x}_t^t)^T\frac{\mathbf{B}^{t-1}}{t-1}\hat{x}_t^t\right)(t-1)^2}. \tag{14}$$

Similarly for the mean, we use the recursive formula:

$$\mu^t = \mu^{t-1} + \frac{1}{t}(x_t - \mu^{t-1}). \tag{15}$$

The convergence of our online clustering algorithm is guaranteed, and is formally stated as:

Proposition 1. *The general stochastic online clustering algorithm converges to a stationary point. More precisely the cost function and its gradient converge almost surely*

$$E(\mu^t, \mathbf{C}^t) \xrightarrow[t\to\infty]{a.s.} E(\mu^\infty, \mathbf{C}^\infty),$$
$$\nabla_{\mu,\mathbf{C}}E(\mu^t, \mathbf{C}^t) \xrightarrow[t\to\infty]{a.s.} 0. \tag{16}$$

Proof. The proof follows the work of [23] on general stochastic gradients. It is sufficient to show that our algorithm fulfills the three following requirements: *a) The cost function is three times differentiable with continuous derivatives, except for a zero probability set; b) The infinite sum of the learning rates diverges and the infinite sum of the square of the learning rates is finite; and c) The parameter values are bounded.* Due to lack of space, a detailed proof has been omitted but is available upon request.

Temporal consistency: incorporating rotations

The online algorithm we have presented so far is able to update the classifier in light of new data. However, when dealing with dynamic processes and by extension, varying time-series data, the subspaces will no longer be static but will exhibit motions between time-windows. These motions will depend on the temporal variation of the data, on how the time-windows are defined and also on the subspace projection basis used. For instance, in a linear dynamic process we may consider the data matrix $\mathbf{W}(t) \in \mathbb{R}^{N \times \zeta}$ at time t, where N is the number of data points and ζ the time-window size (e.g. the number of image frames in a video sequence). To obtain a data point $x_j(t) \in \mathbb{R}^D$ we may project the jth row $w_j^T(t)$ of $\mathbf{W}(t)$ onto a subspace of dimension D, i.e. $x_j(t) = \Pi(t)w_j(t)$ where $\Pi(t) \in \mathbb{R}^{D \times \zeta}$ is a projection matrix. Note that although the projection of the data points to a lower-dimensional basis is not essential for our algorithm to function, it is desirable since having fewer dimensions to process means a speed-up in the algorithm's performance.

Treating each time-window independently with a new projection matrix at every t, would project the data onto different basis thereby inducing large motions on the subspaces and possibly breaking temporal consistency. Instead we can consider overlapping time-windows and in addition obtain a fixed set of projection basis, by using the D principal components of the first time-window. So if $\mathbf{W}(1) = \mathbf{U\Sigma V}^T$ is a rank-D approximation of $\mathbf{W}(1)$ computed by SVD, we can define the projection matrix $\Pi = \mathbf{\Sigma}^{-1}\mathbf{V}^T$, which remains fixed for all t so that $x_j(t) = \mathbf{\Sigma}^{-1}\mathbf{V}^T w_j(t)$.

Overlapping the time-windows and a fixed projection matrix should induce smoother motions that in the majority of cases can be effectively addressed by Eqs. (14) and (15) alone. However, if we wish for additional robustness against larger temporal variations in the data, one option is to recover also the explicit subspace motions between time-windows. In this work we will consider cases where the intrinsic dimensions of the subspaces do not change between time-windows and so the only significant motion components are the rotations and translations in $\mathbb{R}^D$. Any other transformation will keep the data on the same subspace and therefore leave the parameters of the subspaces unchanged. What we mean by this is that any two d-dim subspaces with orthonormal basis and fixed d, will appear identical from a geometric point of view, up to a rotation and a translation in $\mathbb{R}^D$. Obviously from a statistical viewpoint the data distribution on the subspaces might change but the data will not leave the subspace. We further assume that the translation components are small enough that can be effectively captured when the classifier is updated inside each time-window. Therefore, we only need to deal with the rotation components since the rotations are more likely to induce large changes in the subspaces from one time-window to another.

If a subspace is rotating between two time-windows t_1 and t_2 then the following holds:

$$\mathbf{B}^{t_2} = \mathbf{R}^T \mathbf{B}^{t_1} \mathbf{R}, \tag{17}$$

where $\mathbf{R} \in \mathbb{R}^D$ is the rotation matrix and $\mathbf{B}^{t_1}, \mathbf{B}^{t_2}$ the inverse covariance matrices at each respective time-window. To recover the rotation we may solve:

$$\underset{\mathbf{A}}{\arg\min} \|\mathbf{B}^{t_2} - \mathbf{R}^T \mathbf{B}^{t_1} \mathbf{R}\|_F, \tag{18}$$

where we parameterize the rotation matrices by means of the Cayley transformation of a skew symmetric matrix $\mathbf{A}$:

$$\mathbf{R} = (\mathbf{I} + \mathbf{A})^{-1}(\mathbf{I} - \mathbf{A}). \tag{19}$$

The norm in (18), is the Euclidean metric in the ambient space of symmetric matrices and we use it here because of its tractable derivatives. Other choices for proper geodesic metrics are possible (e.g. affine-invariant metric) that work directly on the symmetric positive-definite manifold, but their derivatives are not so straightforward to compute for arbitrary high dimensions. Let us denote a_{mn} the components of matrix $\mathbf{A}$. The derivative of the rotation matrix $\mathbf{R}$ can be easily computed using basic matrix identities

$$-\frac{\partial \mathbf{R}}{\partial a_{ij}} = (\mathbf{I} + \mathbf{A})^{-1}\frac{\partial \mathbf{A}}{\partial a_{ij}} + (\mathbf{I} + \mathbf{A})^{-1}\frac{\partial \mathbf{A}}{\partial a_{ij}}(\mathbf{I} + \mathbf{A})^{-1}(\mathbf{I} - \mathbf{A}), \tag{20}$$

leading to the derivative of the objective function (18):

$$\frac{\partial \mathcal{L}(\mathbf{R}, \mathbf{B}^{t_1}, \mathbf{B}^{t_2})}{\partial a_{ij}} = \mathrm{Tr}\left(\left(\frac{\partial \mathbf{R}}{\partial a_{ij}}\right)^T (\mathbf{B}^{t_1})^T \mathbf{B}^{t_1} \mathbf{R}\right) + \mathrm{Tr}\left(\mathbf{R}^T (\mathbf{B}^{t_1})^T \mathbf{B}^{t_2}\left(\frac{\partial \mathbf{R}}{\partial a_{ij}}\right)\right)$$
$$-2\mathrm{Tr}\left(\left(\frac{\partial \mathbf{R}}{\partial a_{ij}}\right)^T \mathbf{B}^{t_1} \mathbf{R}(\mathbf{B}^{t_2})^T\right) - 2\mathrm{Tr}\left((\mathbf{B}^{t_2})^T \mathbf{R}^T \mathbf{B}^{t_1}\left(\frac{\partial \mathbf{R}}{\partial a_{ij}}\right)\right).$$
$$\tag{21}$$

(18) can be minimised using (21) and standard optimisation tools.

Full algorithm and extensions

The full algorithm for the temporally consistent, online clustering of non-static subspaces is presented in pseudocode in Algorithm 1. There are three nested loops: the outer loop over all time-windows (ln. 1), the intermediate loop to recover the rotations that align the subspaces (ln. 3) and the innermost loop (ln. 4) that uses stochastic gradient descent to cluster the data points and update the classifier.

In more detail, the algorithm works as follows: For each time-window $t = 1, ..., \mathcal{T}$ (ln. 1) we first initialise the rotations to $\mathbf{I}$ and take the classifier trained in the previous time-window $t - 1$ (ln. 2). Then we start the rotation recovery loop (ln. 3), which terminates when the rotated subspaces of the previous time-window $t - 1$ are closely aligned to the subspaces of the current time window t. Inside the rotation recovery loop, we initiate the stochastic gradient descent loop (ln. 4), which iterates over a few random data points $P << N$. As it is common with stochastic methods, the convergence depends on the data points that were drawn. For each data point, we apply the min-Mahalanobis classifier to assign that point to its closest subspace and simultaneously update the subspace clusters s_i (ln. 6). Given the clusters, we then update the parameters of the

Algorithm 1. Online clustering of non-stationary subspaces

Input: Initial $\{\mu_i^0, \mathbf{B}_i^0\}$, **Output:** Clusters s_{i*}
1: **for** $t = 1 : T$ **do**   //time-window loop
2:    $\mathbf{A}, \mathbf{R} = \mathbf{I}$, $\mu^t = \mu^{t-1}$, $\mathbf{B}^t = \mathbf{B}^{t-1}$
3:    **while** $\|\mathbf{B}^t - \mathbf{R}^T \mathbf{B}^{t-1} \mathbf{R}\|_F > \epsilon$ **do**   //rotation recovery loop
4:       **for** $j = 1 : P$ **do**   //stochastic gradient descent loop
5:          Pick random x_j
6:          $s_{i*} = \min \sum_i^K (\hat{x}_j^i)^T \mathbf{B}_i \hat{x}_j^i - \log(|\mathbf{B}_i|)$
7:          $n_{i*}^j = n_{i*}^{j-1} + 1$
8:          $\mu_{i*}^j = \mu_{i*}^{j-1} + \frac{1}{n_{i*}^j}(x_j - \mu_{i*}^{j-1})$
9:          $\mathbf{B}_{i*}^j = \frac{n_{i*}^j \mathbf{B}_{i*}^{j-1}}{n_{i*}^j - 1} + \frac{n_{i*}^j \mathbf{B}_{i*}^{j-1} \hat{x}_j^{i*} (\hat{x}_j^{i*})^T \mathbf{B}_{i*}^{j-1}}{\left(1 + (\hat{x}_j^{i*})^T \frac{\mathbf{B}_{i*}^{j-1}}{n_{i*}^j - 1} \hat{x}_j^{i*}\right)(n_{i*}^j - 1)^2}$
10:       **end for**
11:       $\mu^t = \mu^j$, $\mathbf{B}^t = \mathbf{B}^j$
12:       $\text{argmin}_{\mathbf{A}} \|\mathbf{B}^t - \mathbf{R}^T \mathbf{B}^{t-1} \mathbf{R}\|_F$
13:       $\mathbf{R} = (\mathbf{I} + \mathbf{A})^{-1}(\mathbf{I} - \mathbf{A})$
14:       $\mathbf{B}^t = \mathbf{R}^T \mathbf{B}^{t-1} \mathbf{R}$
15:    **end while**
16: **end for**

classifier (ln. 7–9). Note that we use n_i in (ln. 7) since not every subspace cluster contains the same number of data points and so the normalisations in (ln. 8, 9) will be different for each cluster. Once the stochastic gradient descent loop is finished, we take the best estimate of the current subspace clusters, as given by the classifier (ln. 11) and use it to recover the rotations that align the subspaces between time-windows $t-1$ and t (ln. 12, 13). The subspaces are rotated (ln. 14) and the process (cluster-update-rotate) is repeated until convergence. In practice we only require a few iterations of the intermediate loop (ln. 3). In the end, the algorithm outputs the subspace clusters s_i for each time-window.

Optionally, we may further improve the convergence properties of our method by increasing smoothly the $D - d$ largest eigenvalues of $\mathbf{B}$ at each time-window. d is the intrinsic dimensionality of the subspaces and could be provided as problem-specific information (e.g. for 3D rigid motion segmentation $d \leq 4$). Such a modification allows for better behaviour in certain problems by suppressing the recovered subspace clusters to the desired dimensions and avoiding "over-inflation" (i.e. where one cluster grows uncontrollably to include all data points). We may also incorporate a forgetting factor λ in Eq. (14), depending on our specific problem requirements (e.g. faster/slower subspace changes between time-windows). Thus (14) becomes:

$$\mathbf{B}^t = \frac{(\lambda(t-1) + 1)\mathbf{B}^{t-1}}{\lambda(t-1)} + \frac{(\lambda(t-1) + 1)\mathbf{B}^{t-1}\hat{x}_t^t(\hat{x}_t^t)^T \mathbf{B}^{t-1}}{\left(1 + (\hat{x}_t^t)^T \frac{\mathbf{B}^{t-1}}{\lambda(t-1)}\hat{x}_t^t\right)(\lambda(t-1))^2}. \tag{22}$$

For $\lambda > 1$ we increase the weight of older data and for $\lambda < 1$ that of new data. For $\lambda = 1$ we get back (14). A similar forgetting factor for the mean in (15) can be included as:

$$n^t = \rho n^{t-1} + 1, \quad \mu^t = \frac{1}{n^t}(x_t + \rho n^{t-1}\mu^{t-1}), \tag{23}$$

for $0 \leq \rho \leq 1$. For $\rho = 1$ then we get back (15).

3   Experiments

In this section, we present our experiments on real and synthetic subspace data. We compared against the state-of-the-art methods dGPCA [14], SSSC [13] and IGMM [17]. Other methods were not considered due to unavailability of computer source code. From the tested methods, only dGPCA cannot deal with sequential data, while SSSC was modified to become fully online by considering data in new time-windows as out-of-sample data. In order to keep our experiments fair and comprehensive we split our tests into two types: *sequential-data* and *non-static subspaces*. OCRS does not make any distinction between the two types and has been applied unmodified in every case. For our experiments we quote the clustering error:

$$\text{Error} = \text{Number of misclassified points}/N \cdot 100\%. \tag{24}$$

Synthetic experiments
(Sequential data): We generated $K = [2,...,8]$ subspaces each with $N = 500$ uniform random data points and intrinsic dimensions d ranging from 1 to 10. We then used a random 25% of the data as ground truth training to initialise the different methods and used the rest as sequential testing data to be clustered. So in the case of OCRS, we initialised $\{\mu, \mathbf{B}\}$ with the sample mean and inverse

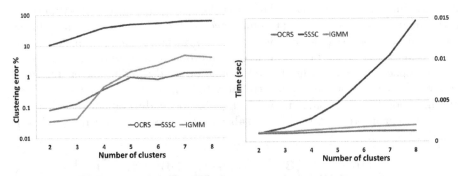

Fig. 2. Sequential subspace clustering. Comparison between OCRS, IGMM and SSSC with clustering error results shown on the left and computation time on the right, for increasing number of clusters. We use a logarithmic scale for the clustering error for ease of comparison. Best viewed in colour. (Color figure online)

sample covariance of the ground truth data respectively. The clustering error and computation time w.r.t. the number of clusters are shown in Fig. 2 (left) and (right) respectively. We see that OCRS obtains more accurate and stable solutions over an increasing number of subspaces, followed by IGMM. SSSC obtains the worst results by comparison. In terms of speed, both OCRS and IGMM are fast while SSSC is the most inefficient method since it is affected by the additional out-of-sample points as the number of clusters increase. (Non-static subspaces): We generated uniform random data on $K = 3$ subspaces in $\mathbb{R}^7$ with intrinsic dimensions $d = 2$, on which we applied a fixed translation vectors and rotations of increasing magnitudes. Given a random 25% of the data as ground truth initialisation, we applied OCRS, dGPCA, IGMM and SSSC on the remainder 75% of the data. Each method was allowed to run for 15 iterations and the final converged results are shown in Fig. 3. We see that OCRS and IGMM obtain good results aided by the low levels of noise in the data. OCRS is able to produce more stable solutions as the subspace rotation magnitude increases. dGPCA and SSSC on the other hand fail to converge to good solutions, indicating that they are much more sensitive to subspace rotations.

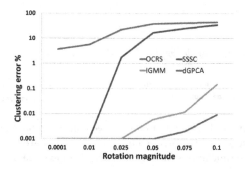

Fig. 3. Clustering of rotating subspaces. Comparison between OCRS, dGPCA, IGMM and SSSC on the accuracy of segmenting rotating subspaces, for increasing rotation magnitudes. We use a logarithmic scale for the clustering error for ease of comparison. Best viewed in colour. (Color figure online)

In Fig. 4 (left), we examine the behaviour of OCRS for different values of the forgetting factor λ from (22), in a synthetic setup with a fixed subspace rotation of 0.001 in every time-window. We see how the different convergence profiles become steeper for $\lambda < 1$ as the method downweighs past observations in favour of new ones. Also on the same figure (right), we see a more challenging synthetic example with significant temporal variations in the data. dGPCA and OCRS without the explicit rotation search, both fail to converge. However, when we optimise for the rotations we see just how the behaviour of OCRS improves and manages to converge to the correct solution within a few time-windows.

In our final synthetic example we compare OCRS against the batch method SCC [25], on the task of sequential subspace clustering. Although both methods

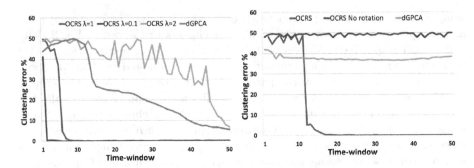

Fig. 4. The effects of the forgetting factor λ on OCRS convergence (left) and the effects of explicitly optimising for the rotations in cases with significant temporal variations in the data (right). We include the dGPCA method as a reference. Best viewed in colour. (Color figure online)

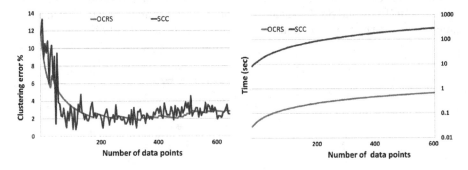

Fig. 5. Accuracy (left) and associated cumulative computational cost (right) of using a batch method for sequential subspace clustering, as compared to those of an online method. Note that the time axis is displayed in logarithmic scale. Best viewed in colour. (Color figure online)

have similar convergence profiles (Fig. 5 left), the batch method is considerably more expensive to use (Fig. 5 right). This is because SCC, an otherwise fast method, has to be applied afresh to the whole, accumulated data, for every new point arriving. In the end, the total computational cost of SCC explodes to more than 200 times that of the online method. This experiment highlights just why batch methods cannot cope with sequential data or very large datasets, as efficiently as online methods can.

Real dataset 1: Online motion segmentation - (Non-static subspaces)
We examined the problem of online motion segmentation that can be solved as the equivalent problem of clustering non-static 4-dimensional subspaces. We used the Hopkins155 dataset [26], which contains 159 sequences of 3D rigidly moving objects with approximately 200 data points and 30–40 frames in each sequence. The data was first projected to $\mathbb{R}^5$ and the methods were initialised from the first ζ frames, either on their own (SSSC, IGMM) or by using the batch

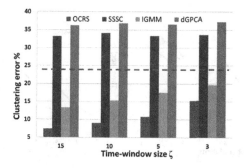

Fig. 6. Segmentation errors comparison of OCRS, dGPCA, IGMM and SSSC on the Hopkins155 dataset for variable size time-windows. The horizontal dashed line indicates the performance of the batch GPCA approach and is included here as a reference. Best viewed in colour. (Color figure online)

GPCA algorithm [27] on the first time-window (dGPCA, OCRS). We then used an overlapping time-window of size ζ and segmented the different subspaces in an online fashion. The segmentation results on the final time-window (i.e. after convergence) are shown in Fig. 6. We see that all methods have better accuracy with larger time-windows since there is more temporal information (i.e. longer trajectories) available to solve the problem. The accuracy drops as the size ζ of the time-windows decreases. OCRS is significantly more accurate than the competing methods. IGMM shows some promising results but not nearly as good as OCRS. SSSC and dGPCA come last, since they are unable to fully capture the dynamics of the moving subspaces.

Note that in general, the online methods have worse performance than state-of-the-art offline methods that see all the data. This is not unusual especially when there are very few time-windows to ensure good convergence. This is certainly not a characteristic of our method alone but mirrored across all online methods. Direct comparison with state-of-the-art offline methods in very short motion sequences might be biased so it has been omitted.

Real dataset 2: Online handwritten digit recognition - (Sequential data) Next we compared OCRS, SSSC and IGMM at online recognition of handwritten digits, which can be solved as the equivalent problem of sequential subspace clustering, where each digit belongs to one of 10 different subspaces. We use the MNIST dataset [28] that contains 10000 examples of 10 handwritten digits, with considerable noise present due to the large variation between handwriting styles. We projected the data to $\mathbb{R}^{10}$ and used 100 samples from each digit as ground truth training. The rest of the samples were used for testing. The resulting recognition accuracy and computation time plots are illustrated in Fig. 7. We see that similarly to the synthetic experiments, OCRS and IGMM exhibit the best performance in terms of accuracy and speed, with OCRS being more accurate as the number of clusters (i.e. digits) increases. We also observe that the accuracy

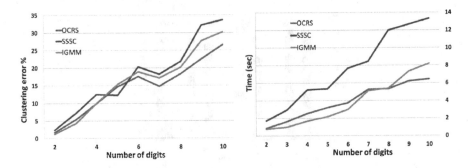

Fig. 7. Online handwritten digit segmentation.. Comparison between OCRS, IGMM and SSSC on the MNIST dataset, with accuracy results on the left and computation time on the right, for increasing number of digits. Best viewed in colour. (Color figure online)

gap between OCRS, IGMM and SSSC has now decreased, indicating that all methods are affected by the significant levels of noise present in the data.

Real dataset 3: Dynamic texture segmentation - (Non-static subspaces) Finally, we compared OCRS and dGPCA on the problem of segmenting video sequences of dynamic textures [29]. This can be equivalently solved as non-static subspace clustering with each dynamic texture lying in a different subspace. The set-up of our experiment was identical to that described in [14], where for each video sequence we used an overlapping time-window of $\zeta = 5$ frames, and the data in each time-window was projected to $\mathbb{R}^5$ with a fixed projection matrix. Both methods were initialised by GPCA [27] using the first time-window. We illustrate example segmentation results from 4 different video sequences in Fig. 8. We see that even though both methods start with the same initial estimate, OCRS obtains segmentations with less noise and better delineation of the different textures.

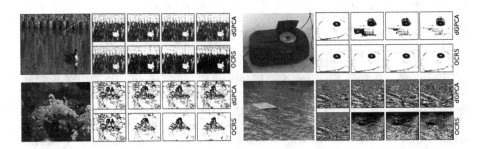

Fig. 8. Dynamic texture segmentation. Example frames of dynamic textures segmentations for both OCRS and dGPCA. The red frames indicate the initial segmentations provided by batch GPCA. Best viewed in colour. (Color figure online)

4 Conclusion

We have presented a new, online algorithm for the efficient clustering of non-static subspaces, which addresses key gaps in existing literature. Our method, OCRS, uses an online min-Mahalanobis distance classifier to sequentially cluster the data, while it deals with temporal variations by recovering the subspace rotations between time-windows. Our contributions in this paper may be summarised as: (i) a novel combination of stochastic gradient descent with rotation search that allows us to deal with sequential data and non-static subspaces simultaneously; (ii) working directly with the inverse covariance matrix B, which is considerably more efficient and avoids degeneracies; and (iii) closed-form derivatives for subspace rotation minimisation for arbitrary $\mathbb{R}^D$. We have tested our method on real and synthetic data and compared with competing methods from literature. Our algorithm consistently outperforms existing approaches both in segmentation accuracy and computation speed.

Acknowledgements. This research was supported by the TU Munchen - IAS (funded by the German Excellence Initiative and the EU 7th Framework Programme under grant agreement no. 291763, the Marie Curie COFUND program of the EU).

References

1. Vidal, R.: Subspace clustering. IEEE Signal Process. Mag. **28**, 52–68 (2011)
2. Jolliffe, I.: Principal Component Analysis. Springer, New York (1986)
3. Agrawal, R., Gehrke, J., Gunopulos, D., Raghavan, P.: Automatic subspace clustering of high dimensional data. Data Min. Knowl. Disc. **11**, 5–33 (2005)
4. Elhamifar, E., Vidal, R.: Sparse subspace clustering. In: CVPR (2009)
5. Liu, G., Lin, Z., Yu, Y.: Robust subspace segmentation by low-rank representation. In: ICML (2010)
6. Zografos, V., Ellis, L., Mester, R.: Discriminative subspace clustering. In: CVPR (2013)
7. Vidal, R., Favaro, P.: Low rank subspace clustering (LRSC). Pattern Recogn. Lett. **43**, 47–61 (2014)
8. Patel, V.M., Nguyen, H.V., Vidal, R.: Latent space sparse and low-rank subspace clustering. IEEE J. Sel. Top. Sign. Proces. **9**, 691–701 (2015)
9. Ng, A.Y., Jordan, M.I., Weiss, Y.: On spectral clustering: analysis and an algorithm. In: NIPS (2001)
10. Wang, S., Fan, Y., Zhang, C., Xu, H., Hao, X., Hu, Y.: Subspace clustering of high dimensional data streams. In: 2008 Seventh IEEE/ACIS International Conference on Computer and Information Science, ICIS 2008, pp. 165–170 (2008)
11. Park, N.H., Lee, W.S.: Memory efficient subspace clustering for online data streams. In: Proceedings of the 2008 International Symposium on Database Engineering and Applications, IDEAS 2008, pp. 199–208 (2008)
12. Peng, X., Zhang, L., Yi, Z.: Scalable sparse subspace clustering. In: 2013 IEEE Conference on Computer Vision and Pattern Recognition (CVPR), pp. 430–437 (2013)
13. Peng, X., Zhang, L., Yi, Z.: Inductive sparse subspace clustering. Electron. Lett. **49**, 1222–1224 (2013)

14. Vidal, R.: Online clustering of moving hyperplanes. In: Advances in Neural Information Processing Systems 19, pp. 1433–1440. MIT Press (2006)

15. Wang, J., Fu, Z.: Online motion segmentation based on sparse subspace clustering. J. Inf. Comput. Sci. **12**, 1293–1300 (2015)

16. Arandjelovic, O., Cipolla, R.: Incremental learning of temporally-coherent gaussian mixture models. In: BMVC (2005)

17. Engel, P.M., Heinen, M.R.: Incremental learning of multivariate gaussian mixture models. In: Rocha Costa, A.C., Vicari, R.M., Tonidandel, F. (eds.) SBIA 2010. LNCS (LNAI), vol. 6404, pp. 82–91. Springer, Heidelberg (2010). doi:10.1007/978-3-642-16138-4_9

18. Declercq, A., Piater, J.: Online learning of gaussian mixture models - a two-level approach. In: VISAPP (2008)

19. Bottou, L., Bengio, Y.: Convergence properties of the K-Means algorithm. In: Advances in Neural Information Processing Systems (1995)

20. Bradley, S., Mangasarian, O.: K-plane clustering. J. Global Optim. **16**(1), 23–32 (2000)

21. Tseng, P.: Nearest q-flat to M points. J. Optim. Theory Appl. **105**, 249–252 (2000)

22. Wang, D., Ding, C., Li, T.: K-subspace clustering. Mach. Learn. Knowl. Disc. Databases **5782**, 506–521 (2009)

23. Bottou, L.: Online algorithms and stochastic approximations. In: Online Learning and Neural Networks. Cambridge University Press (1998)

24. Sherman, J., Morrison, W.J.: Adjustment of an inverse matrix corresponding to a change in one element of a given matrix. Ann. Math. Stat. **21**, 124–127 (1950)

25. Chen, G., Lerman, G.: Spectral curvature clustering (SCC). IJCV **81**, 317–330 (2009)

26. Tron, P., Vidal, R.: A benchmark for the comparison of 3-D motion segmentation algorithms. In: CVPR (2007)

27. Vidal, R., Ma, Y., Sastry, S.: Generalized principal component analysis (GPCA). IEEE PAMI **27**, 1945–1959 (2005)

28. Lecun, Y., Bottou, L., Bengio, Y., Haffner, P.: Gradient-based learning applied to document recognition. In: Proceedings of the IEEE, vol. 86 (1998)

29. Péteri, R., Fazekas, S., Huiskes, M.J.: Dyntex: a comprehensive database of dynamic textures. Pattern Recogn. Lett. **31**, 1627–1632 (2010)

Sparse Gradient Pursuit for Robust Visual Analysis

Jiangxin Dong[1], Risheng Liu[1(✉)], Kewei Tang[2], Yiyang Wang[1], Xindong Zhang[1], and Zhixun Su[1]

[1] Dalian University of Technology, Dalian, China
rsliu@dlut.edu.cn
[2] Liaoning Normal University, Dalian, China

Abstract. Many high-dimensional data analysis problems, such as clustering and classification, usually involve the minimization of a Laplacian regularization, which is equivalent to minimize square errors of the gradient on a graph, i.e., the disparity among the adjacent nodes in a data graph. However, the Laplacian criterion usually preserves the locally homogeneous data structure but suppresses the discrimination among samples across clusters, which accordingly leads to undesirable confusion among similar observations belonging to different clusters. In this paper, we propose a novel criterion, named Sparse Gradient Pursuit (SGP), to simultaneously preserve the within-class homogeneity and the between-class discrimination for unsupervised data clustering. In addition, we show that the proposed SGP criterion is generic and can be extended to handle semi-supervised learning problems by incorporating the label information into the data graph. Though this unified semi-supervised learning model leads to a nonconvex optimization problem, we develop a new numerical scheme for the SGP related nonconvex optimization problem and analyze the convergence property of the proposed algorithm under mild conditions. Extensive experiments demonstrate that the proposed algorithm performs favorably against the state-of-the-art unsupervised and semi-supervised methods.

1 Introduction

The main idea of graph-based methods is to construct an informative data graph that explores the pairwise similarity between data points to capture the data structure. As is well known, the Laplacian matrix can be used to describe many properties of graphs. For instance, the multiplicity of the eigenvalue 0 of the graph Laplacian equals the number of connected components in the graph [1]. Therefore, Laplacian-based methods have been widely exploited in graph-based applications in computer vision, information retrieval, and exploratory data analysis. In the Laplacian related literatures [1–11], this central idea is formulated by first defining a distance-based adjacency graph on the data set

Electronic supplementary material The online version of this chapter (doi:10. 1007/978-3-319-54181-5_24) contains supplementary material, which is available to authorized users.

© Springer International Publishing AG 2017
S.-H. Lai et al. (Eds.): ACCV 2016, Part I, LNCS 10111, pp. 369–384, 2017.
DOI: 10.1007/978-3-319-54181-5_24

and then learning an informative similarity graph to explore the data structure by Laplacian-based methods. Inspired by this idea, many Laplacian-based approaches have been proposed in various contexts, such as dimensionality reduction [6,7], unsupervised clustering [1–5], and semi-supervised classification [8,9].

1.1 Summary of Notations

Hereafter, bold upper-case letters (e.g., $\mathbf{Z}$) and bold lower-case letters (e.g., $\mathbf{z}$) denote matrices and column vectors, respectively. $\mathbf{z}_i$ represents the i-th column of $\mathbf{Z}$, and z_{ij} is the (i,j)-th entry of $\mathbf{Z}$. $\|\mathbf{Z}\|_1 = \sum_{i,j} |z_{ij}|$ is the matrix ℓ_1 norm. $\|\cdot\|$ indicates the ℓ_2 norm of a vector and the Frobenius norm of a matrix. $\|\mathbf{Z}\|_{2,1} = \sum_i \|\mathbf{z}_i\|$ denotes the matrix $\ell_{2,1}$ norm. $\mathbf{Z} \geq 0$ indicates that all $z_{ij} \geq 0$. $\mathscr{S}^n$ is the space of $n \times n$ symmetric matrices. $\mathbf{1}_n$ is the all-one column vector of length n. $\mathbf{I}$ denotes the identity matrix. $\mathrm{tr}(\cdot)$ is the matrix trace. In addition, we define $\|\nabla \mathbf{Z}\|_{\ell_k} = \sum_{i,j} w_{ij} \|\mathbf{z}_i - \mathbf{z}_j\|_k^k$ $(k = 1, 2)$, where $\nabla \mathbf{Z}$ represents the gradient of $\mathbf{Z}$ and $\mathbf{W} = \{w_{ij}\}$ is the weight matrix of the data adjacency graph [6,7,12]. Specifically, provided that $\mathbf{X}$ is the data matrix and $\mathbf{Z}$ is the representation coefficient matrix, $\mathbf{Z}$ has sparse gradients if there is almost no disparity between $\mathbf{z}_i$ and $\mathbf{z}_j$ for most $(i,j) \in \{(i,j)|w_{ij} \neq 0\}$, i.e., most columns of $\mathbf{Z}$ corresponding to the adjacent points in the data graph are similar. Moreover, $\mathbf{z}_i$ and $\mathbf{z}_j$ are adjacent columns of $\mathbf{Z}$ if the corresponding $\mathbf{x}_i$ and $\mathbf{x}_j$ are adjacent points in the data graph, i.e., $w_{ij} \neq 0$. Note that even though $\mathbf{x}_i$ and $\mathbf{x}_j$ are adjacent points, they can belong to either the same class or different classes.

1.2 Motivation

In representation-based approaches, the representation matrix $\mathbf{Z}$ reflects the data structure.[1] Thus, how to describe the properties of $\mathbf{Z}$ plays a critical role in many applications, e.g., clustering and classification. We note that Laplacian-based methods [3,8] preserve the locally within-class data structure by minimizing a Laplacian regularization, i.e., $\|\nabla \mathbf{Z}\|_{\ell_2}$. However, minimizing an objective function with this penalty inevitably results in isotropic smoothing on the data graph and restrains the discrimination among different clusters (See Fig. 1(b)). Taking clustering as an example, under the penalty $\|\nabla \mathbf{Z}\|_{\ell_2}$, the representation coefficients corresponding to different clusters still contain obvious coincidences (See Fig. 1(d)) and do not distinguish different clusters as shown in Fig. 1(b).

As the input data not only contains homogeneous samples but also heterogeneous samples, only considering the structure of homogeneous samples is not enough (e.g., the Laplacian regularization). To keep the structures of not only within-class data points but also between-class samples, we propose a novel criterion. For adjacent homogeneous samples in the data graph, their reconstructed coefficients under the same dictionary should be similar with each other,

---

[1] For simplicity, we adopt the self-representation in this paper, but the proposed method is still valid with other dictionaries [13,14].

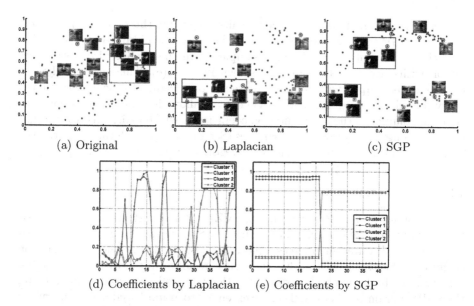

(a) Original   (b) Laplacian   (c) SGP

(d) Coefficients by Laplacian   (e) Coefficients by SGP

Fig. 1. The effect of the SGP criterion on clustering. (a)–(c) Face images displayed after projecting their features to the 2D space using PCA, where the features are: (a) the original image, (b)–(c) the representation matrices derived by Laplacian and SGP. (Faces in the orangered and darkgreen boxes in (a)–(c) belong to cluster 1 and cluster 2, respectively.) (d)–(e) The reconstruction coefficient distribution obtained by Laplacian and SGP, respectively. Each orangered or darkgreen line (shown in (d)–(e)) corresponds to the representation of one data sample in the cluster 1 or cluster 2. The horizontal axes in (d)–(e) denote the index number of the bases (e.g., the input data itself), the first half of which belongs to the cluster 1 and the second half belongs to the cluster 2. The vertical axes in (d)–(e) indicate the absolute values of the representation coefficients (normalized to [0 1] for display). Note that the representation coefficients generated by the SGP criterion look regular which accordingly leads to better clustering results (See (c)). (Images in this paper are best viewed on screen.) (Color figure online)

i.e., there is almost no disparity among the columns of $\mathbf{Z}$ corresponding to adjacent within-class samples. For heterogeneous samples, their corresponding coefficients should be different, i.e., discontinuity. In most real applications, we note that the input data usually contains lots of samples but consists of a few clusters, which indicates that the gradient of the coefficient matrix $\mathbf{Z}$ should be sparse. As the ℓ_1 norm is usually used to model the sparsity, we propose the Sparse Gradient Pursuit (SGP) criterion, i.e., $\|\nabla \mathbf{Z}\|_{\ell_1}$, to describe those aforementioned properties. Note that the new SGP criterion allows a few obvious discontinuities of the adjacent columns of $\mathbf{Z}$ (for heterogeneous samples) while keeps the most disparities of the adjacent columns of $\mathbf{Z}$ sparse (for homogeneous samples). One clustering example is shown in Fig. 1. In particular, we consider the faces marked by large orangered and darkgreen boxes in Fig. 1(a). Though belonging to different clusters, they look the same with a large part shadowed, making the clustering challenging. Most methods are likely to fail in this case (See Fig. 1(b)). Note that SGP pursues sparse gradients. Therefore, the obtained representation coefficients of adjacent within-class samples have the similar piecewise distribution with only one large jump as shown in Fig. 1(e). This ensures adjacent homogeneous samples with similar representations, leading to the preservation of the within-class homogeneity. Meanwhile, SGP allows few differences among the adjacent columns of the representation coefficients, which maintains the discrimination among the heterogeneous samples (See Fig. 1(e)). With this property, the samples from different clusters can be categorized well as shown in Fig. 1(c).

1.3   Our Contribution

We develop a novel SGP criterion, i.e., $\|\nabla \mathbf{Z}\|_{\ell_1}$, for robust visual data analysis to optimally capture the main data structure. Specifically, for general data clustering, the SGP criterion enforces a piecewise distribution on the representation coefficients (See Fig. 1(e)), so that the structures of both the within-class homogeneity and the between-class discrimination are well preserved. Furthermore, the SGP criterion is generic and can be extended to graph-based Semi-Supervised Learning (SSL), which integrates the graph learning and the label propagation in a unified framework. In addition, we design an algorithm to solve the proposed nonconvex SSL model, which is demonstrated to be effective in practice. We analyze the convergence property of this algorithm under some mild conditions. It is worthwhile to highlight main contributions of the proposed approach:

- We propose a novel SGP criterion to model the main data structure. The proposed criterion is able to model the structures of both the within-class homogeneity and the between-class discrimination.
- We show that the proposed SGP criterion is generic and can be extended to graph-based SSL. As a non-trivial extension, our SGP based SSL method can simultaneously learn an informative graph and data labels by integrating the graph learning and the label propagation in a unified framework, which is different from conventional graph-based SSL methods.
- We develop an effective numerical optimization method to solve the proposed nonconvex SSL model. Furthermore, we present the convergence property of the algorithm in both theoretical and practical analyses.

2   Related Work

In this section, we briefly review the most relevant algorithms and put this work in the proper context.

The Laplacian prior, as an important criterion in the construction of a data graph, has been widely used in high-dimensional data analysis. In dimensionality reduction methods, Locally Linear Embedding (LLE) [7] constructs the graph weight matrix by minimizing the reconstruction error of the coefficients. Instead of solving such a system of linear equations for the graph weights,

Laplacian Eigenmaps (LE) [6] computes the weight matrix directly by a heat kernel. To some extent, both methods can be reduced to study the spectral property of the Laplacian on the data graph.

Recently, the Laplacian criterion has emerged in many graph-oriented algorithms designed for the purposes of data clustering [1–5] and semi-supervised learning [8,9]. Among them, the representation-based methods [3,8] are the most popular approaches. One kind of representative methods construct the data graph using the Laplacian criterion as the regularization. In [3], Hu et al. propose a smooth representation model with a Laplacian regularization term for subspace clustering. As the Laplacian regularization diminishes the discrimination across clusters, an additional affinity measure of the graph weight is proposed to remedy this limitation in the final results. In [4], Liu et al. introduce a regularized Markov random walk learning model, which can be reformulated as a Laplacian regularized model. Yin et al. [8] develop a Laplacian regularized low-rank method to represent the locality and similarity information within each cluster. To alleviate the limitation of the Laplacian regularization, both methods use a low-rank penalty to model the global data structure. Instead of introducing additional processing and extra penalties, we propose a novel criterion to preserve both the within-class homogeneity and the between-class discrimination.

Graph-based SSL methods have attracted considerable attention in recent years [15–22]. Most of these SSL methods focus on how to construct an informative affinity graph of the data points without considering the label information. Therefore, the graph learning and the label propagation are usually treated as two separate stages. Note that the label information is able to facilitate the construction of the affinity graph. Li et al. [23] formulate these two stages of SSL into a unified framework. Different from [23], we present a new SSL model based on our proposed criterion, where the graph learning and the label propagation are solved in a unified framework. Moreover, we present an effective optimization method for the proposed SSL model.

3 Unsupervised Sparse Gradient Pursuit

In this section, we first provide a novel clustering model with the proposed SGP criterion and then present how to effectively solve this proposed model. Furthermore, we analyze the effectiveness of the SGP criterion and discuss its relationship with the most related methods.

3.1 Unsupervised Learning Model

In this section, we present how to apply our SGP criterion to solve clustering problems. We assume that the observed data $\mathbf{X}$ can be decomposed as $\mathbf{X} = \mathbf{X}\mathbf{Z} + \mathbf{E}$, where $\mathbf{E}$ denotes the noise [24]. Based on this representation and the proposed SGP criterion, our Unsupervised Sparse Gradient Pursuit (USGP) clustering model is defined as

$$\min_{\mathbf{Z},\mathbf{E}} \|\nabla \mathbf{Z}\|_{\ell_1} + \gamma \|\mathbf{E}\|_{2,1}, \text{ s.t. } \mathbf{X} = \mathbf{X}\mathbf{Z} + \mathbf{E} \qquad (1)$$

Here we adopt the $\ell_{2,1}$ norm to characterize the noise $\mathbf{E}$, since we aim at modeling sample-specific corruptions which can be arbitrary in magnitude but affect only a fraction of the entries [25,26]. Most clustering methods assume that the similarity graph is symmetric and use $(\mathbf{Z} + \mathbf{Z}^\top)/2$ to perform NCut [2] to obtain the final results. Therefore, without loss of generality, we symmetrize $\mathbf{Z}$ in each iteration. For computing simplicity, we rewrite $\|\nabla \mathbf{Z}\|_{\ell_1} = \|\mathbf{KZ}\|_1$, where $\mathbf{K}$ denotes the gradient matrix of the adjacency KNN graph [12,27]. Each row of $\mathbf{K}$ corresponds to an edge and each column corresponds to a data point. For any edge $(i,j) \in \{(i,j)|w_{ij} \neq 0\}$ in the graph, the corresponding k-th row in the matrix $\mathbf{K}$ satisfies that $\mathbf{K}_{ki} = -\mathbf{K}_{kj} = w_{ij}$. Then we reformulate model (1) as

$$\min_{\mathbf{Z},\mathbf{E}} \|\mathbf{KZ}\|_1 + \gamma\|\mathbf{E}\|_{2,1}, \text{ s.t. } \mathbf{X} = \mathbf{XZ} + \mathbf{E} \tag{2}$$

The effectiveness of the proposed SGP criterion has been illustrated in Fig. 1. The coefficients of the representation matrix show that the SGP criterion is able to preserve both the homogeneity of samples within class and the discrimination across clusters. Further analyses on SGP will be provided in Sect. 3.3.

3.2   Optimization Method for Solving USGP

To solve problem (2), we introduce an auxiliary variable $\mathbf{J}$ to reformulate (2) as

$$\min_{\mathbf{J},\mathbf{Z},\mathbf{E}} \|\mathbf{J}\|_1 + \gamma\|\mathbf{E}\|_{2,1}, \text{ s.t. } \mathbf{X} = \mathbf{XZ} + \mathbf{E}, \ \mathbf{KZ} = \mathbf{J} \tag{3}$$

Since the problem (3) is a multi-variable separable convex problem with linear constraints, we use the method [28] to solve it by minimizing the augmented Lagrangian function $\mathcal{L}_u$ with penalty μ:

$$\begin{aligned}
\mathcal{L}_u(\mathbf{J},\mathbf{Z},\mathbf{E},\mathbf{P}_1,\mathbf{P}_2) &= \|\mathbf{J}\|_1 + \gamma\|\mathbf{E}\|_{2,1} + \langle \mathbf{P}_1, \mathbf{X} - \mathbf{XZ} - \mathbf{E}\rangle + \\
&\quad \langle \mathbf{P}_2, \mathbf{KZ} - \mathbf{J}\rangle + \frac{\mu}{2}\left(\|\mathbf{X} - \mathbf{XZ} - \mathbf{E}\|^2 + \|\mathbf{KZ} - \mathbf{J}\|^2\right)
\end{aligned} \tag{4}$$

3.3   Discussion

In this section, we present more insights on the effectiveness of the proposed SGP criterion. We also discuss its relationship with the most related methods.

Effectiveness of the SGP Criterion. For representation/graph-based data analysis, the regularization for the representation coefficients is important. The smoothing representation method [3] constrains the coefficient matrix $\mathbf{Z}$ by a Laplacian regularization, i.e., $\|\nabla \mathbf{Z}\|_{\ell_2}$ based on our definition (Eq. (5) in [3]). As illustrated in Sect. 1.2, the Laplacian criterion over-smooths the discrimination among heterogeneous samples. However, the SGP criterion pursues sparse gradients and allows obvious differences among heterogeneous data rather than smooths them. Such property ensures that the SGP criterion is able to model the

Table 1. Quantitative comparisons of the average clustering accuracy with the SGP criterion and other related criteria on digital images. The SGP criterion compares favorably with other related criteria. The Sparsity and Laplacian regularization terms refer to $\|\mathbf{Z}\|_1$ and $\|\nabla\mathbf{Z}\|_{\ell_2}$, respectively.

Regularization	Sparsity	Laplacian	SGP
Average clustering accuracy (%)	74.53	69.78	**80.25**

data structure, where both the within-class homogeneity and the between-class discrimination can be well preserved. Specifically, for clustering and classification problems, data samples in the same class usually have similar primary structures. Therefore, they can be effectively reconstructed by only one or few clusters of data points, under the same dictionary (e.g., the input data matrix). The coefficients, corresponding to adjacent homogeneous samples, are similar, which leads to sparse gradients of the representation matrix. To further evaluate the effectiveness of the proposed SGP criterion, we compare it with the method based on the Laplacian regularization, i.e.,

$$\min_{\mathbf{Z},\mathbf{E}} \|\nabla\mathbf{Z}\|_{\ell_2} + \gamma\|\mathbf{E}\|_{2,1}, \text{ s.t. } \mathbf{X} = \mathbf{X}\mathbf{Z} + \mathbf{E} \tag{5}$$

using the same setting as Eq. (1). The SGP criterion imposes the sparsity on the gradient of $\mathbf{Z}$, and we note that another way to model $\mathbf{Z}$ is to enforce the sparsity on $\mathbf{Z}$ itself [29]. However, this penalty only considers the local data structure and may possibly lead to over-segmentation [30,31].

We quantitatively compare the SGP criterion with those aforementioned criteria for ten times. The average accuracies are shown in Table 1. For each time, we randomly select 1,000 images from the USPS database with 100 images for each digit (0–9). Note that graph-based clustering using the Laplacian or sparsity regularization on $\mathbf{Z}$ is less effective according to previous analyses. In contrast, as the SGP criterion is able to preserve both the within-class homogeneity and the between-class discrimination, the average accuracy is much higher. More comparisons are provided in Sect. 5.1.

Relation with the Multiclass Total Variation Clustering Method. We note that the recent method [32] proposes a Multiclass Total Variation (MTV) method based on total variation techniques to solve the multiclass clustering problem. However, our method is quite different from MTV [32]. First, MTV is based on the classical balanced-cut and derives a relaxation of the graph cut. To eliminate the unnatural bias for partitioning out small sets of samples, it computes the energy as a fraction of the total variation term to a balance term that favors equal sized partition. This formation is complex and difficult to be solved. Second, MTV is less robust to noise and outliers. The model used in [32] cannot be easily extended to handle these issues as this will make the problem intractable. Third, due to the complicated form of MTV and its special solution, this approach cannot be easily extended to other applications. In contrast, our

Table 2. The average clustering accuracy with USGP and MTV [32] on corrupted images. The proposed USGP is more robust to MTV.

Method	MTV	USGP
Average clustering accuracy (%)	54.90	**61.50**

method is a regularized representation-based approach with the characterization of noise, which is simple yet effective and much more robust for real applications, e.g., SSL problems (See Sect. 4). For comparison, we use the same data as before, but randomly corrupt 30% of them with random noise. Results shown in Table 2 demonstrate that MTV is less robust to noise.

4   Semi-supervised Sparse Gradient Pursuit

In this section, we show how to apply the proposed SGP criterion to SSL. Furthermore, we note that the label information is able to facilitate the graph construction. Conventional graph-based SSL methods usually treat the graph learning and the label propagation as two independent processes. In contrast, we further propose a new model for SSL, where the graph learning and the label propagation are solved in a unified framework. In the following sections, we first present our unified SSL model together with the proposed SGP criterion and then propose an efficient numerical algorithm to solve the proposed model.

4.1   Semi-supervised Learning Model

Definition 1 (Semi-supervised Learning). *Given a data set, denoted by a matrix* $\mathbf{X} = (\mathbf{x}_1, \ldots, \mathbf{x}_l, \mathbf{x}_{l+1}, \ldots, \mathbf{x}_n)$, *which consists of c classes, and* $l \ (\ll n)$ *given labels indicated by a matrix* $\mathbf{Y}_l = (\mathbf{y}_1, \ldots, \mathbf{y}_l)^\top$. *Each* $\mathbf{y}_i$ *is a c-dimensional indicator vector, where each component* $\mathbf{y}_{i,j} = 1$ *if data* $\mathbf{x}_i$ *belongs to class j and* $\mathbf{y}_{i,j} = 0$, *otherwise. SSL aims at learning a classification function* $\mathbf{H} = [\mathbf{H}_l; \mathbf{H}_u]$ *(i.e.,* $\mathbf{H}$ *is the concatenation of* $\mathbf{H}_l$ *and* $\mathbf{H}_u$) *to estimate the unknown labels for the remaining* $n - l$ *unlabeled data points. Here* $\mathbf{H}_l$ *and* $\mathbf{H}_u$ *denote the class probability matrices for labeled and unlabeled data points, respectively.*

Similar to existing SSL methods [18,19], we assume that $\mathbf{Z} \geq 0$ to ensure that the coefficients can be directly converted to graph weights, $\mathbf{Z} \in \mathscr{S}^n$ to make $\mathbf{Z}$ a similarity metric over the data space, and $\mathbf{1}_n^\top \mathbf{Z} = \mathbf{1}_n^\top$. To summarize, the solution of the representation matrix $\mathbf{Z}$ in SSL should be in the following set:

$$\mathcal{K} = \{\mathbf{Z} | \mathbf{1}_n^\top \mathbf{Z} = \mathbf{1}_n^\top, \ \mathbf{Z} \in \mathscr{S}^n, \ \mathbf{Z} \geq 0\} \qquad (6)$$

To simultaneously learn the graph with the weight matrix $\mathbf{Z} \in \mathcal{K}$ and the classification function $\mathbf{H}$, we impose $\mathbf{H}_l = \mathbf{Y}_l$ for the labeled data and assume that the labels are smoothly propagated over the graph [17]. Thus, our unified Semi-supervised Sparse Gradient Pursuit (SSGP) model is defined as

$$\begin{cases} \min_{\mathbf{Z},\mathbf{E},\mathbf{H}} \|\nabla \mathbf{Z}\|_{\ell_1} + \gamma \|\mathbf{E}\|_{2,1} + \delta G(\mathbf{H}), \\ \text{s.t. } \mathbf{X} = \mathbf{X}\mathbf{Z} + \mathbf{E}, \ \mathbf{Z} \in \mathcal{K}, \ \mathbf{H}_l = \mathbf{Y}_l \end{cases} \qquad (7)$$

where $G(\mathbf{H})$ controls the smoothness of label propagation. And we define

$$G(\mathbf{H}) = \frac{1}{2}\sum_{i,j} z_{ij}\|\mathbf{H}_{i,:} - \mathbf{H}_{j,:}\|^2 = \mathrm{tr}(\mathbf{H}^\top(\mathbf{D}-\mathbf{Z})\mathbf{H}) \qquad (8)$$

where $\mathbf{H}_{i,:}$ denotes the i-th row of $\mathbf{H}$. The degree matrix $\mathbf{D}$ is the diagonal matrix with d_i on the diagonal and $d_i = \sum_j z_{ij}$. Then, $\mathbf{D} = \mathrm{diag}(\mathbf{Z}\mathbf{1}_n) = \mathrm{diag}(\mathbf{1}_n) = \mathbf{I}$.

4.2 Optimization Method for Solving SSGP

Different from model (2), the proposed SSGP model is a nonconvex and nonsmooth optimization problem. As the Alternating Direction Method (ADM) is mainly designed for solving convex optimization problems [33], we propose an effective ADM based algorithm, termed Proximal ADM (PADM) to solve (7). Then we analyze the convergence property of PADM in both theory and practice.

Firstly we introduce two auxiliary variables $\mathbf{J}$ and $\mathbf{L}$, corresponding to $\mathbf{KZ}$ and $\mathbf{Z}$, respectively in Eq. (7). For convenience, we denote $\mathcal{R} = (\mathbf{J};\mathbf{L};\mathbf{Z};\mathbf{E};\mathbf{H})$ and $\mathcal{P} = (\mathbf{P}_1;\mathbf{P}_2;\mathbf{P}_3;\mathbf{P}_4)$. The augmented Lagrangian function of Eq. (7) is

$$\begin{aligned}
\mathcal{L}(\mathcal{R},\mathcal{P}) &= \|\mathbf{J}\|_1 + \gamma\|\mathbf{E}\|_{2,1} + \delta\mathrm{tr}(\mathbf{H}^\top(\mathbf{I}-\mathbf{L})\mathbf{H}) + \langle\mathbf{P}_1, \mathbf{X}-\mathbf{XZ}-\mathbf{E}\rangle \\
&\quad + \langle\mathbf{P}_2, \mathbf{1}_n^\top\mathbf{L}-\mathbf{1}_n^\top\rangle + \langle\mathbf{P}_3, \mathbf{KZ}-\mathbf{J}\rangle + \langle\mathbf{P}_4, \mathbf{Z}-\mathbf{L}\rangle \\
&\quad + \frac{\mu}{2}(\|\mathbf{X}-\mathbf{XZ}-\mathbf{E}\|^2 + \|\mathbf{1}_n^\top\mathbf{L}-\mathbf{1}_n^\top\|^2 + \|\mathbf{KZ}-\mathbf{J}\|^2 + \|\mathbf{Z}-\mathbf{L}\|^2)
\end{aligned} \qquad (9)$$

To solve this model, we propose to use the following iterative scheme:

$$\begin{cases}
\mathbf{J}_{k+1} = \underset{\mathbf{J}}{\mathrm{argmin}}\; \|\mathbf{J}\|_1 + \dfrac{\mu_k}{2}\left\|\mathbf{J}-(\mathbf{KZ}_k+\mathbf{P}_{3,k}/\mu_k)\right\|^2 + \dfrac{\beta}{2}\|\mathbf{J}-\mathbf{J}_k\|^2, \\[2mm]
\mathbf{L}_{k+1} = \underset{\mathbf{L}\in\mathscr{S}^n, \mathbf{L}\geq 0}{\mathrm{argmin}}\; \delta\mathrm{tr}(\mathbf{H}_k^\top(\mathbf{I}-\mathbf{L})\mathbf{H}_k) + \dfrac{\mu_k}{2}(\|\mathbf{1}_n^\top\mathbf{L}-\mathbf{1}_n^\top+\mathbf{P}_{2,k}/\mu_k\|^2 \\[2mm]
\qquad\qquad + \|\mathbf{L}-(\mathbf{Z}_k+\mathbf{P}_{4,k}/\mu_k)\|^2) + \dfrac{\beta}{2}\|\mathbf{L}-\mathbf{L}_k\|^2, \\[2mm]
\mathbf{Z}_{k+1} = \underset{\mathbf{Z}}{\mathrm{argmin}}\; \dfrac{\mu_k}{2}(\|\mathbf{X}-\mathbf{XZ}-\mathbf{E}_k+\mathbf{P}_{1,k}/\mu_k\|^2 + \|\mathbf{KZ}-\mathbf{J}_{k+1}+\mathbf{P}_{3,k}/\mu_k\|^2 \\[2mm]
\qquad\qquad + \|\mathbf{Z}-\mathbf{L}_{k+1}+\mathbf{P}_{4,k}/\mu_k\|^2) + \dfrac{\beta}{2}\|\mathbf{Z}-\mathbf{Z}_k\|^2, \\[2mm]
\mathbf{E}_{k+1} = \underset{\mathbf{E}}{\mathrm{argmin}}\; \gamma\|\mathbf{E}\|_{2,1} + \dfrac{\mu_k}{2}\left\|\mathbf{E}-(\mathbf{X}-\mathbf{XZ}_{k+1}+\mathbf{P}_{1,k}/\mu_k)\right\|^2 + \dfrac{\beta}{2}\|\mathbf{E}-\mathbf{E}_k\|^2, \\[2mm]
\mathbf{H}_{k+1} = \underset{\mathbf{H}_l=\mathbf{Y}_l}{\mathrm{argmin}}\; \delta\mathrm{tr}(\mathbf{H}^\top(\mathbf{I}-\mathbf{L}_{k+1})\mathbf{H}) + \dfrac{\beta}{2}\|\mathbf{H}-\mathbf{H}_k\|^2, \\[2mm]
\mathbf{P}_{1,k+1} = \mathbf{P}_{1,k} + \mu_k(\mathbf{X}-\mathbf{XZ}_{k+1}-\mathbf{E}_{k+1}), \\[1mm]
\mathbf{P}_{2,k+1} = \mathbf{P}_{2,k} + \mu_k(\mathbf{1}_n^\top\mathbf{L}_{k+1}-\mathbf{1}_n^\top), \\[1mm]
\mathbf{P}_{3,k+1} = \mathbf{P}_{3,k} + \mu_k(\mathbf{KZ}_{k+1}-\mathbf{J}_{k+1}), \\[1mm]
\mathbf{P}_{4,k+1} = \mathbf{P}_{4,k} + \mu_k(\mathbf{Z}_{k+1}-\mathbf{L}_{k+1}), \\[1mm]
\mu_{k+1} = \min(\rho\mu_k, \mu_{max})
\end{cases} \qquad (10)$$

Algorithm 1. Solving SSGP via PADM

Input: Data matrix $\mathbf{X}$, gradient matrix $\mathbf{K}$, parameters γ, δ.
1: **Initialize:** $\mathbf{J} = \mathbf{K}, \mathbf{L} = \mathbf{Z} = \mathbf{I}, \mathbf{E} = \mathbf{H} = 0, \mathbf{P}_1 = \mathbf{P}_2 = \mathbf{P}_3 = \mathbf{P}_4 = 0$.
2: **while** not converge **do**
3:    Update $(\mathbf{J}_{k+1}, \mathbf{L}_{k+1}, \mathbf{Z}_{k+1}, \mathbf{E}_{k+1}, \mathbf{H}_{k+1}, \{\mathbf{P}_{i,k+1}\}_{i=1}^4, \mu_{k+1})$ by Eq. (10).
4:    Check the convergence conditions
       $\|\mathbf{X} - \mathbf{XZ}_{k+1} - \mathbf{E}_{k+1}\|_\infty < \epsilon$, $\|\mathbf{1}_n^\top \mathbf{L}_{k+1} - \mathbf{1}_n^\top\|_\infty < \epsilon$,
       $\|\mathbf{KZ}_{k+1} - \mathbf{J}_{k+1}\|_\infty < \epsilon$, and $\|\mathbf{Z}_{k+1} - \mathbf{L}_{k+1}\|_\infty < \epsilon$.
5: **end while**
Output: $\mathbf{J}^*, \mathbf{L}^*, \mathbf{Z}^*, \mathbf{E}^*, \mathbf{H}^*$.

The complete algorithm is summarized in Algorithm 1. We set $\mu = 4$, $\mu_{\max} = 10^{10}$, $\rho = 1.1$, $\epsilon = 10^{-8}$, and $\beta = 10^{-4}$. Note that the subproblems with respect to $\mathbf{J}, \mathbf{L}, \mathbf{Z}, \mathbf{E}$, and $\mathbf{H}$ have closed-form solutions. $\mathbf{J}$ and $\mathbf{E}$ are solved by the soft thresholding operator [34] and the $\ell_{2,1}$ optimization operator [25], respectively. $\mathbf{H}$ is obtained in a similar way as [17]. In practice, we adopt a simple procedure called class mass normalization as stated in [17] to adjust the class distributions to match the label priors. In every iteration, we could obtain the indexes of unlabeled samples and their corresponding values in $\mathbf{H}$. Then we could utilize learned reliable indexes which have high values in $\mathbf{H}$ to update the adjacency graph after every certain iteration.

4.3 Convergence Analysis of PADM

We prove that PADM is able to converge to a KKT point of Eq. (7), while ADM does not have this property according to [35].

Theorem 1. *Assume that the sequence of $\{\mathcal{R}_k, \mathcal{P}_k\}_{k \in \mathbb{N}}$ is bounded and there exists a nonnegative sequence $\{\varepsilon_k\}_{k \in \mathbb{N}}$ that satisfies $\Sigma_k \varepsilon_k < \infty$ and $\|\mathcal{P}_{k+1} - \mathcal{P}_k\| \leq \varepsilon_k$ for all k. Then there exists a subsequence of $\{\mathcal{R}_k, \mathcal{P}_k\}$ which converges to a KKT point of the problem (7).*

Although we prove that the proposed algorithm is able to converge to a KKT point, as model (7) is nonconvex and nonsmooth, a natural question is whether our proposed PADM algorithm converges well in practice. We quantitatively evaluate the convergence property of PADM on the USPS database. We randomly select 100 images for each digit and randomly label 10% of them. Figure 2 shows that the proposed algorithm converges after less than 40 iterations, in terms of the objective function values.

Remark. For the nonconvex optimization problems, converging to a KKT point is the best result as far as we know. In addition, our numerical experiments show that the proposed algorithm converges well (Fig. 2). Furthermore, although the proposed algorithm is designed for solving (7), it can be applied to some other related optimization problems in image processing and visual analysis.

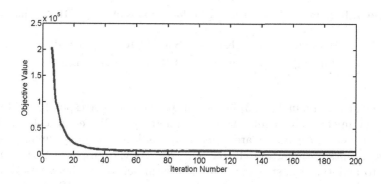

Fig. 2. Convergence of our proposed PADM.

5 Experiments

We evaluate the proposed methods on unsupervised clustering and semi-supervised classification tasks and compare it with several corresponding state-of-the-art methods. In addition, we verify the robustness of our algorithm on the data with large corruptions.

5.1 Unsupervised Face Clustering

We evaluate the effectiveness of the proposed method on clustering tasks and compare it with the state-of-the-art clustering methods: (1) K-means; (2) Sparse Subspace Clustering (SSC) [29]; (3) Low-Rank Representation (LRR) [36]; (4) SMooth Representation (SMR) [3]. In this experiment, the CMU-PIE [37] face image database is utilized to evaluate the performance of our method. This face database contains 42,368 images of 68 subjects with different poses, illumination conditions and expressions. Each sample is manually cropped to a size of 32×32 pixels. We only select a part of their images by fixing the pose and expression. For computing convenience, we preprocess the data with PCA whose dimension is set as 64 to keep most information. The clustering experiments are directed with various cluster numbers. That is, we use the first k classes

Table 3. Clustering accuracy (%) on CMU-PIE face images.

Clusters	K-means	SSC	LRR	SMR	USGP
4	**100**	**100**	98.81	**100**	**100**
12	91.67	91.67	93.65	99.60	**100**
28	84.34	85.54	82.82	88.19	**88.95**
44	76.33	79.04	74.92	79.26	**83.50**
60	73.99	72.55	74.22	73.67	**80.19**
68	74.25	70.46	70.81	75.09	**76.63**

Table 4. Running time of clustering for four groups of CMU-PIE face images.

Methods	K-means	SSC	LRR	SMR	USGP
Running time (/s)	0.28	0.61	1.31	**0.26**	0.42

(See the first column in Table 3) in the database for the corresponding data clustering experiments. The regularization parameter γ in (2) is empirically set to 50.

The clustering accuracies are reported in Table 3. The bold numbers denote the best results with certain cluster numbers. As the proposed SGP criterion is able to maintain the structures of both within-class and between-class data samples, the clustering results show that the proposed USGP algorithm outperforms other state-of-the-art methods. Moreover, we test the running time of the proposed method. Table 4 shows the average running time from five different clustering methods. We note that SMR needs less running time as it uses the Laplacian criterion to regularize the data and the Frobenius norm to model noise, which can be efficiently solved by a standard Sylvester equation [38]. In addition to the accuracy, our method achieves competitive results.

5.2   Semi-supervised Image Classification

We then evaluate the proposed method on semi-supervised image classification tasks. Three public databases are selected for our experiments: CMU-PIE, USPS [39], and Textures[2]. We select the first 15 subjects of the CMU-PIE face database and only use their images in five near frontal poses and under different illuminations. The USPS database consists of 9,298 handwritten digit images of 10 digits (0–9) in total, where the size of each image is 16×16 pixels. The Textures database includes 13 texture categories, where each category contains 112 samples. The size of each image is 128×128 pixels. We first normalize all the images so that they have a unit norm. For the CMU-PIE and Textures databases, 50 images of each subject are randomly selected as the data sets in each run. For the USPS database, we randomly select 100 images from each digit category. These selected images are randomly labeled. The percentage of labeled samples for each individual varies from 5% to 30% on all the databases, because the goal of SSL is to deal with practical tasks that have very limited labeled samples [15].

We compare our algorithm with the following widely used SSL methods including Laplacian Support Vector Machines (Lap-SVM) [9], Non-Negative Low-Rank and Sparse (NNLRS) [18], Low-Rank Coding based Balanced (LRCB) graph [19], Nonnegative Sparse Laplacian regularized Low-Rank Representation (NSLLRR) [8], and Structure-Constrained Low Rank Representation (SCLRR) [40]. Lap-SVM is a natural extension of SVM. The four latter methods are all graph-based SSL methods, which combine different graphs with existing SSL frameworks. After obtaining the graphs, we choose the Gaussian harmonic function to compare the effectiveness of different graphs.

---

[2] The Textures can be downloaded from: http://sipi.usc.edu/database/database.cgi? volume=textures.

Table 5. Classification accuracy (%) of various methods with standard deviations (shown in the parenthesis) under different percentages of labeled samples on three databases. (Face, Digit, and Tectures denote the CMU-PIE, USPS, and Textures database, respectively.)

Methods	Lap-SVM	NNLRS	LRCB	NSLLRR	SCLRR	SSGP
Face (5%)	83.31 ± 2.53	78.13 ± 3.01	73.05 ± 3.67	81.52 ± 3.67	80.02 ± 3.09	**84.55 ± 3.07**
Face (15%)	92.86 ± 1.44	92.34 ± 1.46	88.89 ± 1.61	93.41 ± 1.69	93.49 ± 1.92	**95.08 ± 2.23**
Face (30%)	96.91 ± 0.94	97.77 ± 0.93	95.81 ± 1.76	98.05 ± 0.90	97.82 ± 1.21	**98.86 ± 1.27**
Digit (5%)	80.47 ± 2.05	86.50 ± 1.25	85.48 ± 1.10	83.26 ± 2.78	89.53 ± 0.82	**91.50 ± 1.21**
Digit (15%)	88.07 ± 1.11	90.76 ± 1.05	88.34 ± 1.15	90.06 ± 1.47	92.35 ± 0.86	**94.00 ± 0.96**
Digit (30%)	90.54 ± 0.99	92.64 ± 0.75	92.86 ± 0.87	93.26 ± 0.66	93.61 ± 0.50	**95.57 ± 0.77**
Textures (5%)	89.04 ± 2.31	88.82 ± 2.85	88.05 ± 2.84	74.63 ± 4.21	86.90 ± 2.53	**90.67 ± 2.71**
Textures (15%)	94.17 ± 1.44	92.29 ± 1.10	89.19 ± 2.78	87.11 ± 5.02	92.16 ± 1.50	**94.51 ± 2.14**
Textures (30%)	96.81 ± 0.94	93.24 ± 0.59	91.43 ± 1.99	88.46 ± 2.35	93.77 ± 1.13	**97.60 ± 3.64**

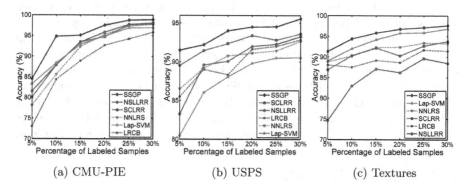

(a) CMU-PIE   (b) USPS   (c) Textures

Fig. 3. Classification accuracy of SSGP and compared methods with different percentages of labeled samples on three databases. (a)–(c) Accuracy plots of various methods on the CMU-PIE, USPS, and Textures database, respectively.

For all the three databases, we set $\gamma = 6$ and $\delta = 0.1$. Table 5 and Fig. 3 show the classification accuracies of various methods on three databases and the average accuracies versus varying percentages of labeled samples, respectively, where SSGP achieves the highest classification accuracy compared to all the other methods. On one hand, the better results are thanks to the fact that the SGP criterion is able to keep the structures of both the within-class homogeneity and the between-class discrimination. On the other hand, the SSGP framework utilizes the label prior to construct the adjacency graph and updates the graph learning with reliable learned label information, which improves the performance.

5.3   Robustness to Data Corruptions

The proposed method is robust to data corruptions. In this section, we further verify its classification capability in SSL tasks using corrupted data from the

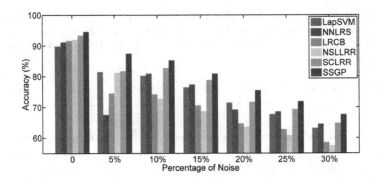

Fig. 4. Accuracy of different methods on USPS with corruptions.

USPS database, where we use the same data and experimental settings as the previous section with following changes. In each trial of the experiment, we randomly select both labeled and unlabeled data samples $\mathbf{x}$ corrupted by Gaussian noise with zero means and standard deviation $0.2\|\mathbf{x}\|_2$. We show the comparison results in Fig. 4, where the percentage of corrupted data is varied from 0 to 30%. The immediate observation is that our method achieves better performance than other methods. The robustness of our SSGP algorithm is not only due to the robust representation, but also benefits from the new SSL framework and the update strategy. We only select reliable learned labels with high confidence to update our graph, which helps graph learning from the effect of the corruptions.

6   Concluding Remarks

This paper introduces a novel criterion for robust visual analysis in high dimensional data space. Different from Laplacian-based methods which are not able to effectively preserve the structure of data across clusters, the proposed SGP based method aims at pursuing sparse gradients to preserve both the within-class homogeneity and the between-class discrimination. We show that the proposed SGP criterion is generic which can be applied to clustering and SSL problems. In addition, we develop a unified framework for SSL problems, where the graph construction and the label propagation are solved simultaneously. As the proposed SSL model is nonconvex and nonsmooth, we present an efficient optimization method based on ADM and provide analyses on the convergence properties both in theory and practice. Extensive experiments show that the proposed SGP criterion is effective for both unsupervised and semi-supervised learning problems.

Acknowledgement. Risheng Liu is supported by National Natural Science Foundation of China (NSFC) (Nos. 61300086, 61432003, 61672125), Fundamental Research Funds for the Central Universities (No. DUT15QY15), and the Hong Kong Scholar Program (No. XJ2015008). Zhixun Su is supported by NSFC (No. 61572099) and National Science and Technology Major Project (Nos. ZX20140419, 2014ZX04001011).

References

1. Von Luxburg, U.: A tutorial on spectral clustering. Stat. Comput. **17**, 395–416 (2007)
2. Shi, J., Malik, J.: Normalized cuts and image segmentation. TPAMI **22**, 888–905 (2000)
3. Hu, H., Lin, Z., Feng, J., Zhou, J.: Smooth representation clustering. In: CVPR, pp. 3834–3841 (2014)
4. Liu, R., Lin, Z., Su, Z.: Learning markov random walks for robust subspace clustering and estimation. Neural Netw. **59**, 1–15 (2014)
5. He, X., Cai, D., Shao, Y., Bao, H., Han, J.: Laplacian regularized Gaussian mixture model for data clustering. TKDE **23**, 1406–1418 (2011)
6. Belkin, M., Niyogi, P.: Laplacian eigenmaps for dimensionality reduction and data representation. Neural Comput. **15**, 1373–1396 (2003)
7. Roweis, S.T., Saul, L.K.: Nonlinear dimensionality reduction by locally linear embedding. Science **290**, 2323–2326 (2000)
8. Yin, M., Gao, J., Lin, Z.: Laplacian regularized low-rank representation and its applications. TPAMI **38**, 504–517 (2016)
9. Belkin, M., Niyogi, P., Sindhwani, V.: Manifold regularization: a geometric framework for learning from labeled and unlabeled examples. JMLR **7**, 2399–2434 (2006)
10. Gao, S., Tsang, I.W.H., Chia, L.T.: Laplacian sparse coding, hypergraph laplacian sparse coding, and applications. TPAMI **35**, 92–104 (2013)
11. He, X., Yan, S., Hu, Y., Niyogi, P., Zhang, H.J.: Face recognition using Laplacianfaces. TPAMI **27**, 328–340 (2005)
12. Yang, Y., Wang, Z., Yang, J., Han, J., Huang, T.S.: Regularized l1-graph for data clustering. In: BMVC (2014)
13. Protter, M., Elad, M.: Image sequence denoising via sparse and redundant representations. TIP **18**, 27–35 (2009)
14. Mairal, J., Bach, F., Ponce, J., Sapiro, G.: Online dictionary learning for sparse coding. In: ICML, pp. 689–696 (2009)
15. Chapelle, O., Schölkopf, B., Zien, A.: Semi-supervised Learning. MIT Press, Cambridge (2006)
16. Zhu, X.: Semi-supervised learning. In: Encyclopedia of Machine Learning, pp. 892–897 (2011)
17. Zhu, X., Ghahramani, Z., Lafferty, J., et al.: Semi-supervised learning using Gaussian fields and harmonic functions. In: ICML, vol. 3, pp. 912–919 (2003)
18. Zhuang, L., Gao, H., Lin, Z., Ma, Y., Zhang, X., Yu, N.: Non-negative low rank and sparse graph for semi-supervised learning. In: CVPR, pp. 2328–2335 (2012)
19. Li, S., Fu, Y.: Low-rank coding with b-matching constraint for semi-supervised classification. In: IJCAI (2013)
20. Joachims, T., et al.: Transductive learning via spectral graph partitioning. In: ICML, vol. 3, pp. 290–297 (2003)
21. Belkin, M., Matveeva, I., Niyogi, P.: Regularization and semi-supervised learning on large graphs. In: Shawe-Taylor, J., Singer, Y. (eds.) COLT 2004. LNCS, vol. 3120, pp. 624–638. Springer, Heidelberg (2004). doi:10.1007/978-3-540-27819-1_43
22. Yan, S., Wang, H.: Semi-supervised learning by sparse representation. In: SDM, pp. 792–801 (2009)
23. Li, C.G., Lin, Z., Zhang, H., Guo, J.: Learning semi-supervised representation towards a unified optimization framework for semi-supervised learning. In: ICCV, pp. 2767–2775 (2015)

24. Wright, J., Ganesh, A., Rao, S., Peng, Y., Ma, Y.: Robust principal component analysis: exact recovery of corrupted low-rank matrices via convex optimization. In: NIPS, pp. 2080–2088 (2009)
25. Liu, G., Lin, Z., Yan, S., Sun, J., Yu, Y., Ma, Y.: Robust recovery of subspace structures by low-rank representation. TPAMI **35**, 171–184 (2013)
26. Li, Z., Liu, J., Tang, J., Lu, H.: Robust structured subspace learning for data representation. TPAMI **37**, 2085–2098 (2015)
27. Zheng, M., Bu, J., Chen, C., Wang, C., Zhang, L., Qiu, G., Cai, D.: Graph regularized sparse coding for image representation. TIP **20**, 1327–1336 (2011)
28. Liu, R., Lin, Z., Su, Z.: Linearized alternating direction method with parallel splitting and adaptive penalty for separable convex programs in machine learning. In: ACML, pp. 116–132 (2013)
29. Elhamifar, E., Vidal, R.: Sparse subspace clustering: algorithm, theory, and applications. TPAMI **35**, 2765–2781 (2013)
30. Nasihatkon, B., Hartley, R.: Graph connectivity in sparse subspace clustering. In: CVPR, pp. 2137–2144 (2011)
31. Tang, K., Dunson, D.B., Su, Z., Liu, R., Zhang, J., Dong, J.: Subspace segmentation by dense block and sparse representation. Neural Netw. **75**, 66–76 (2016)
32. Bresson, X., Laurent, T., Uminsky, D., von Brecht, J.: Multiclass total variation clustering. In: NIPS, pp. 1421–1429 (2013)
33. Boyd, S., Parikh, N., Chu, E., Peleato, B., Eckstein, J.: Distributed optimization and statistical learning via the alternating direction method of multipliers. Found. Trends Mach. Learn. **3**, 1–122 (2011)
34. Lin, Z., Liu, R., Su, Z.: Linearized alternating direction method with adaptive penalty for low-rank representation. In: NIPS, pp. 612–620 (2011)
35. Yuan, G., Ghanem, B.: l0tv: a new method for image restoration in the presence of impulse noise. In: CVPR, pp. 5369–5377 (2015)
36. Liu, G., Lin, Z., Yu, Y.: Robust subspace segmentation by low-rank representation. In: ICML, pp. 663–670 (2010)
37. Sim, T., Baker, S., Bsat, M.: The CMU pose, illumination, and expression (PIE) database. In: AFG, pp. 46–51 (2002)
38. Bartels, R.H., Stewart, G.W.: Solution of the matrix equation AX+XB=C. CACM **15**, 820–826 (1972)
39. Hull, J.J.: A database for handwritten text recognition research. TPAMI **16**, 550–554 (1994)
40. Tang, K., Liu, R., Su, Z., Zhang, J.: Structure-constrained low-rank representation. TNNLS **25**, 2167–2179 (2014)
41. Bolte, J., Sabach, S., Teboulle, M.: Proximal alternating linearized minimization for nonconvex and nonsmooth problems. Math. Program. **146**, 459–494 (2014)

F-SORT: An Alternative for Faster Geometric Verification

Jacob Chan[1]([✉]), Jimmy Addison Lee[2], and Kemao Qian[1]

[1] School of Computer Engineering (SCE), Nanyang Technological University,
Block N4 Nanyang Avenue, Singapore 639798, Singapore
{jchan015,MKMQian}@ntu.edu.sg
[2] Institute for Infocomm Research (I2R), Agency for Science,
Technology and Research (A*STAR), 1 Fusionopolis Way,
Connexis (South Tower), Singapore 138632, Singapore
jalee@i2r.a-star.edu.sg

Abstract. This paper presents a novel geometric verification approach coined Fast Sequence Order Re-sorting Technique (F-SORT), capable of rapidly validating matches between images under arbitrary viewing conditions. By using a fundamental framework of re-sorting image features into local sequence groups for geometric validation along different orientations, we simulate the enforcement of geometric constraints within each sequence group in various views and rotations. While conventional geometric verification (e.g. RANSAC) and state-of-the-art fully affine invariant image matching approaches (e.g. ASIFT) are high in computational cost, our approach is multiple times less computational expensive. We evaluate F-SORT on the Stanford Mobile Visual Search (SMVS) and the Zurich Buildings (ZuBuD) image databases comprising an overall of 9 image categories, and report competitive performance with respect to PROSAC, RANSAC and ASIFT. Out of the 9 categories, F-SORT wins PROSAC in 9 categories, RANSAC in 8 categories and ASIFT in 7 categories, with a significant reduction in computational cost of over nine-fold, thirty-fold and hundred-fold respectively.

1 Introduction

The local descriptor SIFT [1] has proven to be remarkably successful in many applications. Some notable examples are object recognition [1], image stitching [2,3], visual mapping [4], etc. Nevertheless, it imposes a high computational burden, especially for real-time systems such as visual odometry, or for low-power devices such as smart phones. This has led to an intensive search for replacements with lower computational cost; arguably the best of these to date are SURF [5] and ORB [6].

The discriminative power of these local descriptors, however, degrades rapidly with the growth of the image data set due to increased matching ambiguity. Thus, geometric verification [1,7–11] becomes an important post-processing step for obtaining a reasonable retrieval precision, especially for low-resolution images

© Springer International Publishing AG 2017
S.-H. Lai et al. (Eds.): ACCV 2016, Part I, LNCS 10111, pp. 385–399, 2017.
DOI: 10.1007/978-3-319-54181-5_25

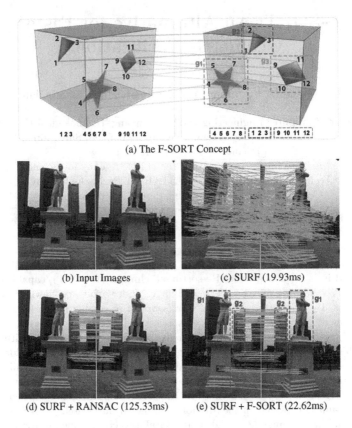

(a) The F-SORT Concept

(b) Input Images

(c) SURF (19.93ms)

(d) SURF + RANSAC (125.33ms)   (e) SURF + F-SORT (22.62ms)

Fig. 1. Our F-SORT framework and comparison to conventional method. (a) The F-SORT concept of bundling image features into local sequence groups (g_1, g_2 and g_3 in this case) for further geometric verification. (b) Two input images obtained from different viewing angles. (c) SURF result shows a large number of matches heavily contaminated by outliers with computational time of 19.93 ms. (d) Conventional method using RANSAC shows filtered matches falling mainly on a single building facade with computational time of 125.33 ms. (e) SURF + F-SORT result shows filtered matches falling on the non-planar statue and the building facade, grouped as g_1 and g_2 respectively, and with computational time of 22.62 ms.

commonly used in mobile image matching applications. But again, many existing geometric verification methods are computationally expensive and infeasible for real-time performance. Therefore, in practice, geometric verification is only applied to a subset of the top-ranked candidate images, which may not be sufficient in larger scale image retrieval systems. Another crucial deficiency of many existing methods [7,8,12] lies in the possibility that the inliers do not necessarily cluster together on a single dominant plane. A more computationally expensive multi-model fitting is required to identify all inliers on different planes.

In this paper, we aim to design a computationally-efficient replacement to conventional geometric verification methods (i.e. RANSAC [3,8]) that is similar in performance accuracy, less affected by geometric changes, not restricted to a single plane, and is capable of being used in low-latency image matching applications. The proposed fast verification method is called F-SORT (Fast Sequence Order Re-sorting Technique). As illustrated in Fig. 1, we bundle image features into local sequence groups to increase their discriminative power and robustness to image variations induced by 3D viewpoint changes. While a single 2D projective transformation is unable to deal with multiple planes, our approach provides a less rigid representation that allows fast and robust geometric constraints to be enforced onto individual sequence group of features on planar and non-planar surfaces on multiple planes. As the algorithm is fast, the feature sequence order can simply be re-sorted along different orientations to handle larger rotations. The effectiveness of the approach is substantiated by a pairwise image matching experiment where the results favor F-SORT for its improved accuracy and lower computational cost.

The organization of this paper is as follows. Section 2 discusses the related work. Section 3 presents the F-SORT methodology. Section 4 showcases the experimental results. Lastly, Sect. 5 concludes the work.

2   Related Work

In computer vision applications, false matches or more commonly known as outliers, frequently occur after the execution of a detection algorithm. Therefore, any system which aims to fulfill visual tasks must address this problem. Conventional methods, such as RANSAC [8], verify matches by estimating a model from a minimal number of points with algorithms for instance the 8-point [13], the 7-point [14], or the more recent 5-point [15]. The main drawback of RANSAC is its iterative process of computing affine models, which makes the algorithm slow and therefore unattractive for real-time image matching applications. This robust method has been modified many times with examples such as MLE-SAC [16], LO-RANSAC [17], WaldSAC [18], QDEGSAC [19], GroupSAC [20], and EVSAC [21]. The proposed changes are made either to the cost function, the sampling technique, or detecting the degeneracies in data. However, none of the above are directly applicable in a real-time setting. Nistér [22] proposes a radically different approach in which multiple hypotheses are scored in parallel, with the least promising hypotheses being dropped at successive stages. It is named Preemptive RANSAC, intended for use in real-time applications. One of the primary limitations of preemptive RANSAC is that only a fixed number of hypotheses are evaluated, which is equivalent to a priori assumption that a lower bound on the fraction of inliers is known. This limits the applicability of preemptive RANSAC in wide baseline stereo where the fraction of inliers varies widely. For low contamination situations, preemptive RANSAC can be even slower than standard RANSAC, since it evaluates many more hypotheses than necessary. Another algorithm, named PROSAC [7], modifies RANSAC to use a measure of the quality of the data points in order to preferentially generate hypotheses that are more likely to be valid. In practice, PROSAC often achieves significant computational savings, since good hypotheses are generated early on in the sampling process. Nevertheless, when the quality scores are less helpful, for instance in the case of highly repetitive nature of the man-made environments, the performance of PROSAC reduces to that of RANSAC. It can be prone to degeneracies, since when there is no dominant plane in the scene, the top ranked points often lie on the same surface. A series of works by Raguram et al. [23–26] also made substantial impact in RANSAC related iterations. Out of which, ARRSAC [23], an extension of Preemptive RANSAC, enables real-time performance by estimating inliers with only a fraction of hypothesized observations. However, it only averages about 3 times faster than PROSAC and relies heavily on its derived hypothesis for good performance.

In recent years, methods [10, 27] that verify spatial consistency of features in local areas instead of the entire image are also gaining popularity in the computer vision community. Local spatial consistency from k (=15) spatial nearest neighbors is used in [27] to filter out false matches. Although this geometric constraint is computationally more feasible, it is sensitive to resolution changes and image noise from background clutter. In [10], MSER regions [28] are used to bundle SIFT features [1] which allow simple geometric constraints to be enforced at the bundle level. This approach improves robustness to occlusion and geometric changes. However, due to the complexity of computing MSER and SIFT, the time improvement ($\approx$ twofold) over conventional geometric verification method is not very significant. We have also come across an impressive work by Morel and Yu [29], which post-processes very well using epipolar filtering with the Moisan and Stival's Optimized Random Sampling Algorithm (ORSA) [12]. This approach, coined Affine-SIFT or ASIFT, simulates all image views obtainable by varying the camera optical axis direction and is capable of finding large numbers of correct matches on multiple planes. Nevertheless, this comes at an extra computational cost.

3   F-SORT Verification

This section presents the methodology of F-SORT, where we introduce the notion of sorting matched keypoints in an axial manner to search for ascending clustered sequences for the validation of matches within each independent group. These local bundled features are able to effectively refine matches while being insensitive to occlusion and image variations caused by viewpoint changes. Furthermore, by carefully considering the possibility of rotation and clutter scenarios, the feature sequence order is re-sorted along multiple orientations for re-bundling to verify the best sequence score.

3.1   Matching Features

Let $\mathbf{P} = \{p_i\}$ and $\mathbf{Q} = \{q_j\}$ be the features (e.g. SIFT [1] or SURF [5] features) detected in two images. We find matches between the two sets of features by Euclidean distance, using the k-d data structure and search algorithm [30]. The algorithm is highly time-efficient especially when the number of reference images in the database is large. It reduces the time to locate nearest neighbors from the brute-force $O(N)$ to $O(\log N)$, with N being the number of reference images. Although we suggest using the k-d tree algorithm in this paper, other matching algorithms can also be applied without compromising F-SORT's performance.

3.2   Sorting and Grouping Features

After feature matching, we discard any $q_j \in \mathbf{Q}$ that does not have a matched $p_i \in \mathbf{P}$, and remove any duplicates with a lower match score. Next, we sort the remaining $\{p_i\}$ in a geometric order according to their X-coordinates and denote the order for each of their corresponding matched feature $q'p_i$ in $\mathbf{Q}$ as $O_q[p_i]$. Ideally, $O_q[p_i]$ should follow the same sorted sequence order in $\{p_i\}$ where $O_q[p_i] < O_q[p_{i+1}]$ as depicted in Fig. 2 if there are no significant orientation and viewing angle differences between the two images. However, due to geometric changes the sequence orders between them may have changed as shown in Fig. 3. As such, penalizing the geometric inconsistency of the matches between the

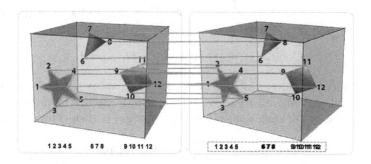

Fig. 2. An ideal case where the sequence orders of the matching features between two images are exactly the same (order numbers 1 to 12).

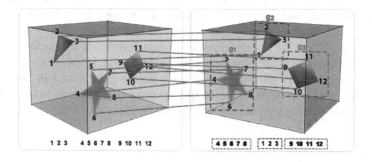

Fig. 3. A case where the sequence orders of the matching features between two images are different due to geometric change. F-SORT algorithm bundles the features into three independent sequence groups ($\mathbf{g}_1$, $\mathbf{g}_2$ and $\mathbf{g}_3$ in this case) for further geometric verification.

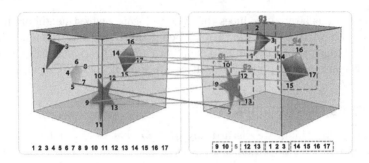

Fig. 4. A case where there exists a false match (order number "5") among the matching features that does not fulfill the ordering constraint and is therefore not allocated to any groups. F-SORT bundles the remaining features into four independent sequence groups ($\mathbf{g}_1$, $\mathbf{g}_2$, $\mathbf{g}_3$ and $\mathbf{g}_4$ in this case) for further geometric verification.

entire two images is not appropriate. This motivates us to bundle the features into independent sequence groups $\mathbf{G} = \{\mathbf{g}_k\}$ where k is the group number index.

We allocate a feature p_i and its next neighbor p_{i+1} to the same group if the following condition is satisfied:

$$Cond_{\mathbf{g}_k}(p_i, p_{i+1}) : O_q[p_{i+1}] - O_q[p_i] = \tilde{1}, \tag{1}$$

where $\tilde{1}$ denotes a positive integer value ≈ 1. This value is usually fixed at 1, but can be relaxed with caution to grant a little flexibility for noise intrusion sensitivity. The size of each sequence group $Size(\mathbf{g}_k)$ is then computed as:

$$Size(\mathbf{g}_k) = 1 + \sum_i \vartheta[Cond_{\mathbf{g}_k}(p_i, p_{i+1})], \tag{2}$$

where $\vartheta[.]$ is an indicator function and a minimum of $Size(\mathbf{g}_k) = 2$ is required in order to form a group. Note that there may be some features not allocated to any groups, and will be discarded. These stand-alone features do not fulfill the ordering constraint and are therefore not reliable to be taken as inliers. Figure 4 shows an example.

Equations (1) and (2) were inferred from our observation that upon viewpoint changes, the sorted features break up into groups which may have their positions interchanged with each other. This scenario can be found in both Figs. 3 and 4. Our algorithm catches hold of this transposition pattern among the feature groups, simulating the viewing of the feature groups in different angle views.

3.3 Angle-Variance Constraint

Based on observation, there may be a small subset of outliers which fortuitously fulfill the ordering constraint in each local sequence group. As the matching features are already in sorted order within their local groups, the angle variance between their adjacent neighbors should not be dramatic. We perform a

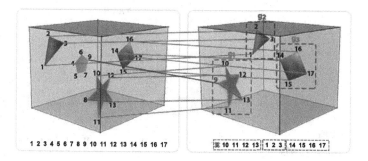

Fig. 5. A case where there exists a false match (order number "9") which fortuitously fulfills the ordering constraint and is therefore allocated to a group ($\mathbf{g}_1$). Nevertheless, it does not fulfill the angle-variance constraint and is being removed from the group.

geometric verification on each individual group of bundled features using angle-variance constraint between adjacent neighbors. Those features that do not fulfill the angle-variance constraint are most likely outliers and will be removed from the group. Figure 5 illustrates the above.

Let $\mathbf{g}_A \in \mathbf{G}$ be a group of bundled features with $p_i \in \mathbf{g}_A$ and $p_{i+1} \in \mathbf{g}_A$ be a feature point and its subsequent neighbor respectively. The angle consistency check between them is represented by:

$$Consist_\angle(p_i, p_{i+1}) = |\angle q[p_{i+1}] - \angle q[p_i]| < \tau_c, \ \forall p_i, p_{i+1} \in \mathbf{g}_A, \tag{3}$$

where $Consist_\angle(p_i, p_{i+1})$ expresses a consistency check between the angles formed by each corresponding feature pair $(p_i, q'p_i)$ and its subsequent neighboring feature pair $(p_{i+1}, q'p_{i+1})$, defined as $\angle q[p_i]$ and $\angle q[p_{i+1}]$ respectively. τ_c is an adjustable angle-variance constraint parameter to enforce a non-dramatic angle variance between adjacent neighbors. It defaults to $\tau_c = 15°$ as derived and used in our experiments. The angle $\angle q[p_i]$ is formed by the line made with $(p_i, q'p_i)$ and the X-axis given by:

$$\angle q[p_i] = \tan^{-1}\left(\frac{y^{q'p_i} - y^{p_i}}{x^{q'p_i} - x^{p_i}}\right), \angle q[p_i] \in [0°, 360°), \tag{4}$$

where x^{p_i} and y^{p_i} denote the X- and Y-coordinates of p_i respectively.

3.4   Handling Rotations and Scores

The generalization to handle different rotations is straightforward, an efficient way is by re-sorting features along different orientations instead of rotating the whole image. As our algorithm is considerably fast, we can afford to iterate it multiple times in different orientations. For instance, the difference between 1 rotation and 10 rotations (our default setting) on a query verification is only 1 to 2 ms, depending on the number of detected features. The rotation angle which is in closest proximity to the targeted image gives the best score as depicted in Fig. 6, where the score is computed as $\sum_k Size(\mathbf{g}_k)$ for each rotation.

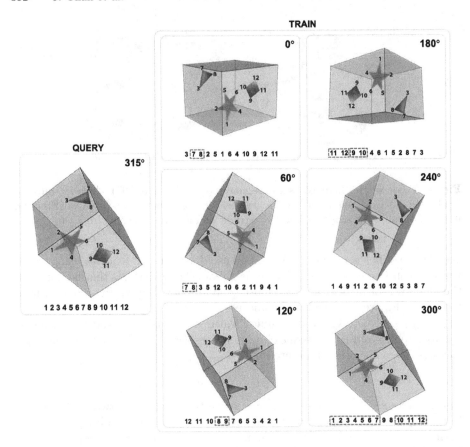

Fig. 6. An example of a train image computed along six different rotation angles $(0°, 60°, 120°, 180°, 240°, 300°)$, and the angle $(300°)$ which is in closest proximity with the query image $(315°)$ obtains the highest score of 10 (determined by the sum of features in all sequence groups after F-SORT verification).

4   Experimental Results

This section presents the experimental results that validate the effectiveness and efficiency of F-SORT. For the experiments reported here, we adopt the Fast-Hessian detector and SURF descriptor in [5], however other options [1,6] could be applied as well. We compare F-SORT with other related approaches in a pairwise image matching experiment to demonstrate the verification performance between them.

Data Sets. We validate empirically our approach on the Stanford Mobile Visual Search (SMVS) [31] and the Zurich Buildings (ZuBuD) [32] image databases. The number of database and query images for a total of 9 image categories used in our evaluation are summarized in Table 1.

Table 1. Number of query and database images in SMVS and ZuBuD with a total of 9 image categories.

Data set	Category	Database	Query
SMVS	CDs	100	400
SMVS	DVDs	100	400
SMVS	Books	100	400
SMVS	Video clips	100	400
SMVS	Landmarks	500	500
SMVS	Business cards	100	400
SMVS	Text documents	100	400
SMVS	Paintings	100	400
ZuBuD	ZuBuD buildings	201	919
Total	9	1401	4219

The SMVS data set provides a total of 3300 query images for 1200 distinct classes across 8 image categories. The ZuBuD image database consists of 201 classes, with 5 images per class (5 different views of each building) hence adding up to a total of 1005 database images. There are 115 query images which are not contained in the database. With the above setting, it is fairly easy to achieve close to 100 % accuracy in image matching experiments as reported in several papers [33,34]. We make this data set more challenging by reducing the class size from 5 images to 1 image per class, with the other 4 images (randomly selected) added to the existing queries.

No. of Rotations. As discussed in Sect. 3.4, we have to determine the number of rotations for F-SORT. We deduce the best choice experimentally by studying a range of values (1–24) in a pairwise image matching experiment across the 9 image categories. The best choice is defined as the least number of rotations that produces highest matching accuracy. The results are recorded in Fig. 7. As the graph shows, the rotation value of 10 gives the best image matching results in 6 categories, and this is the number of rotations used for the experiments throughout the paper. We also observe that the number of rotations above 11 does not obtain higher accuracy in any of the 9 categories.

Comparisons. We compare combination schemes as follows: (1) SURF standalone, (2) SURF + RANSAC [3,8] homography, (3) SURF + PROSAC [7] homography, (4) ASIFT [29], (5) SURF + F-SORT. The evaluation measures are straightforward. The candidate with the most inlier matches is chosen as the matched image. We report the percentage of correctly matched images and the average query time for each of the 9 image categories. We perform all our experiments with a single CPU on a 2.8 GHz Intel Xeon desktop with 12 GB memory.

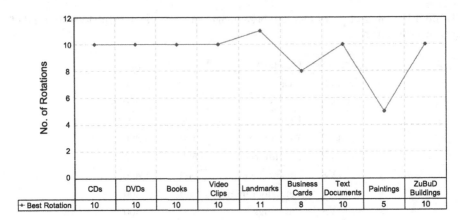

Fig. 7. A graph shows the least number of rotations between 1 and 36 for F-SORT to achieve best pairwise image matching result in each of the 9 image categories. From the results, it shows that the most effective number of rotations for F-SORT is 10, which repetitively attains best results (6 out of 9) in the image categories.

Table 2. Pairwise image matching results (correctly matched images in %) of different combination schemes in each of the 9 image categories. Top score in each category is indicated in bold type. SURF + F-SORT leads in 7 categories, while SURF + RANSAC and ASIFT each leads in 1 category. SURF stand-alone and SURF + PROSAC do not lead in any category.

Category	SURF	SURF + RANSAC	SURF + PROSAC	ASIFT	SURF + F-SORT
CDs	86.75	89.75	87.75	90.75	**93**
DVDs	87.75	89.75	88.75	94	**94.75**
Books	94	94	91.25	90.25	**96**
Video clips	98	**99**	98.25	98.75	98.5
Landmarks	39.40	39	37.80	**43.80**	42.80
Business cards	77.25	81	78.75	84	**84.5**
Text documents	68.25	69.25	68.75	69.75	**73.75**
Paintings	62.75	77.50	72.75	82.75	**87.25**
ZuBuD buildings	87.59	85.74	83.57	85.09	**87.92**

Table 2 compares pairwise image matching results between the above five combination schemes in each of the 9 image categories, and Fig. 8 shows their average query time for one image query. Out of the 9 categories, SURF + F-SORT dominates 7 categories, winning over SURF stand-alone and SURF + PROSAC in all categories, and SURF + RANSAC and ASIFT in 8 and 7 categories respectively. As reported in the figure, SURF + F-SORT is also close to 5 times, 2 times and 136 times faster than SURF + RANSAC, SURF + PROSAC and ASIFT respectively.

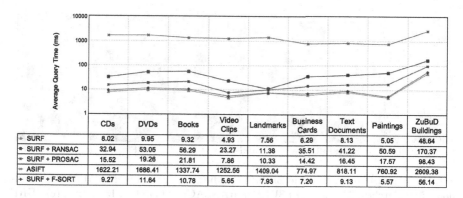

	CDs	DVDs	Books	Video Clips	Landmarks	Business Cards	Text Documents	Paintings	ZuBuD Buildings
SURF	8.02	9.95	9.32	4.93	7.56	6.29	8.13	5.05	48.64
SURF + RANSAC	32.94	53.05	56.29	23.27	11.38	35.51	41.22	50.59	170.37
SURF + PROSAC	15.52	19.26	21.81	7.86	10.33	14.42	16.45	17.57	98.43
ASIFT	1622.21	1686.41	1337.74	1252.56	1409.04	774.97	818.11	760.92	2609.38
SURF + F-SORT	9.27	11.64	10.78	5.65	7.93	7.20	9.13	5.57	56.14

Fig. 8. Average query time (in milliseconds) of different combination schemes in each of the 9 image categories. On average, SURF + F-SORT is 4.5 times faster than SURF + RANSAC, 1.86 times faster than SURF + PROSAC and 135.93 times faster than ASIFT.

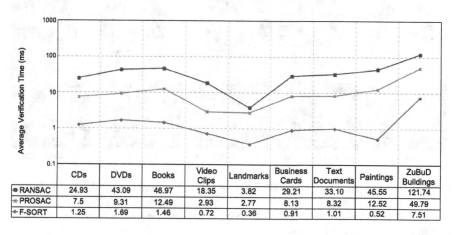

	CDs	DVDs	Books	Video Clips	Landmarks	Business Cards	Text Documents	Paintings	ZuBuD Buildings
RANSAC	24.93	43.09	46.97	18.35	3.82	29.21	33.10	45.55	121.74
PROSAC	7.5	9.31	12.49	2.93	2.77	8.13	8.32	12.52	49.79
F-SORT	1.25	1.69	1.46	0.72	0.36	0.91	1.01	0.52	7.51

Fig. 9. Average verification time (in milliseconds - without including SURF feature extraction and matching time) of RANSAC, PROSAC and F-SORT in each of the 9 image categories. On average, F-SORT is 31.33 times faster than RANSAC and 9.72 times faster than PROSAC.

The results clearly indicate that upon applying F-SORT verification to SURF, it translates into better accuracy and faster computation time.

We also compare the computation time just between the three geometric verification methods RANSAC, PROSAC and F-SORT without including the feature extraction and matching time in each of the 9 image categories. The results as reported in Fig. 9 clearly reveals that F-SORT significantly outperforms the two methods in all categories. On average, F-SORT is above 31 times and 9 times faster than RANSAC and PROSAC respectively.

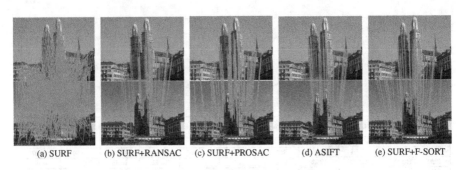

(a) SURF	(b) SURF+RANSAC	(c) SURF+PROSAC	(d) ASIFT	(e) SURF+F-SORT

Fig. 10. Example results on the ZuBuD Buildings image database. (a) SURF stand-alone: 540 matches, 23.57 ms. (b) SURF + RANSAC: 72 matches, 135.32 ms. (c) SURF + PROSAC: 65 matches, 37.58 ms. (d) ASIFT: 95 matches, 1783.13 ms. (e) SURF + F-SORT: 132 matches, 26.11 ms.

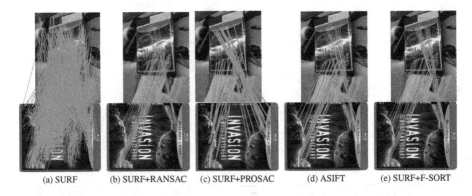

(a) SURF	(b) SURF+RANSAC	(c) SURF+PROSAC	(d) ASIFT	(e) SURF+F-SORT

Fig. 11. Example results on the SMVS data set, DVDs category. (a) SURF stand-alone: 656 matches, 20.45 ms. (b) SURF + RANSAC: 49 matches, 122.69 ms. (c) SURF + PROSAC: 44 matches, 40.29 ms. (d) ASIFT: 66 matches, 1524.78 ms. (e) SURF + F-SORT: 51 matches, 21.42 ms.

Some sample results of the five different combination schemes are displayed in Figs. 10, 11 and 12. The number of matches and computational time for each scheme are included in the figures, and corresponding lines between matching feature pairs are deliberately drawn for method comparison purposes. In comparison, e.g. in Fig. 10, we see that SURF + F-SORT is less rigid compared to other schemes in attaining matches on different planes. The computational time is also drastically reduced. In the other two figures, we see comparable results with ASIFT which has the fully affine invariant advantage and can handle larger transition tilts than SURF.

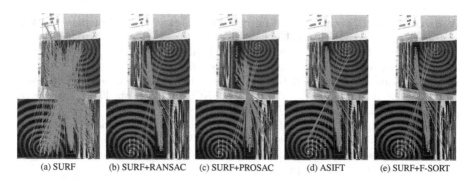

(a) SURF   (b) SURF+RANSAC   (c) SURF+PROSAC   (d) ASIFT   (e) SURF+F-SORT

Fig. 12. Example results on the SMVS data set, Books category. (a) SURF stand-alone: 542 matches, 16.58 ms. (b) SURF + RANSAC: 142 matches, 116.86 ms. (c) SURF + PROSAC: 144 matches, 35.43 ms. (d) ASIFT: 166 matches, 1357.27 ms. (e) SURF + F-SORT: 151 matches, 17.02 ms.

5   Conclusion

We have proposed an efficient alternative to conventional geometric verification methods called F-SORT, which shows high potential in low-latency image matching applications. We introduced our sorting and grouping methods which are not only fast, but more crucially the methods provide a representation that allows robust geometric constraints to be enforced at the group level. The bundled features are more discriminative than individual features, and more robust to image variations induced by viewpoint changes. Our proposed geometric verification to each group involves the ordering and angle-variance constraints which are less rigid than the robust fitting methods, and experimental results have shown that F-SORT is able to obtain more correct matches on different planes on planar and non-planar objects after verification. Results over 9 different image categories indicated that application of F-SORT verification to the SURF algorithm translates into higher accuracy and lower computational cost which easily outperforms other approaches in the pairwise image matching experiment. Out of the 9 categories, F-SORT dominated in 7 categories, and on average F-SORT achieved superior efficiency of over thirty-fold faster than RANSAC, over nine-fold faster than PROSAC and over hundred-fold faster than ASIFT when applied to SURF.

Acknowledgements. This research was partially supported by National Research Foundation, Prime Minister's Office, Singapore under its IDM Futures Funding Initiative and AcRF Tier 1 (RG28/15).

References

1. Lowe, D.G.: Distinctive image features from scale-invariant keypoints. IJCV **60**, 91–110 (2004)
2. Brown, M., Lowe, D.G.: Automatic panoramic image stitching using invariant features. IJCV **74**, 59–73 (2007)

3. Hartley, R.I., Zisserman, A.: Multiple View Geometry in Computer Vision. Cambridge University Press, UK (2000)
4. Se, S., Lowe, D.G., Little, J.: Mobile robot localization and mapping with uncertainty using scale-invariant visual landmarks. IJRR **21**, 735–758 (2002)
5. Bay, H., Tuytelaars, T., Gool, L.: SURF: speeded up robust features. In: Leonardis, A., Bischof, H., Pinz, A. (eds.) ECCV 2006. LNCS, vol. 3951, pp. 404–417. Springer, Heidelberg (2006). doi:10.1007/11744023_32
6. Rublee, E., Rabaud, V., Konolige, K., Bradski, G.: ORB: an efficient alternative to SIFT and SURF. In: Proceedings of the ICCV, pp. 2564–2571 (2011)
7. Chum, O., Matas, J.: Matching with PROSAC - progressive sample consensus. In: Proceedings of the CVPR, pp. 220–226 (2005)
8. Fischler, M.A., Bolles, R.C.: Random sample consensus: a paradigm for model fitting with applications to image analysis and automated cartography. Commun. ACM **24**, 381–395 (1981)
9. Philbin, J., Chum, O., Isard, M., Sivic, J., Zisserman, A.: Object retrieval with large vocabularies and fast spatial matching. In: Proceedings of the CVPR, pp. 1–8 (2007)
10. Wu, Z., Ke, Q., Isard, M., Sun, J.: Bundling features for large scale partial-duplicate web image search. In: Proceedings of the CVPR, pp. 25–32 (2009)
11. Zhang, W., Košecká, J.: Image based localization in urban environments. In: Proceedings of the 3DPVT, pp. 33–40 (2006)
12. Moisan, L., Stival, B.: A probabilistic criterion to detect rigid point matches between two images and estimate the fundamental matrix. Int. J. Comput. Vis. **57**, 201–218 (2004)
13. Hartley, R.I.: In defense of the eight-point algorithm. IEEE TPAMI **19**, 580–593 (1997)
14. Zhang, Z.: Determining the epipolar geometry and its uncertainty: a review. IJCV **27**, 161–195 (1998)
15. Stewénius, H., Engels, C., Nistér, D.: Recent developments on direct relative orientation. ISPRS J. Photogrammetry Remote Sens. **60**, 284–294 (2006)
16. Torr, P.H.S., Zisserman, A.: MLESAC: a new robust estimator with application to estimating image geometry. CVIU **78**, 138–156 (2000)
17. Chum, O., Matas, J., Kittler, J.: Locally optimized RANSAC. In: Michaelis, B., Krell, G. (eds.) DAGM 2003. LNCS, vol. 2781, pp. 236–243. Springer, Heidelberg (2003). doi:10.1007/978-3-540-45243-0_31
18. Matas, J., Chum, O.: Randomized RANSAC with sequential probability ratio test. In: Proceedings of the ICCV, pp. 1727–1732 (2005)
19. Frahm, J.M., Pollefeys, M.: RANSAC for (quasi-)degenerate data (QDEGSAC). In: Proceedings of the CVPR, pp. 453–460 (2006)
20. Ni, K., Jin, H., Dellaert, F.: GroupSAC: efficient consensus in the presence of groupings. In: Proceedings of the ICCV, pp. 2193–2200 (2009)
21. Fragoso, V., Sen, P., Rodriguez, S., Turk, M.: EVSAC: accelerating hypotheses generation by modeling matching scores with extreme value theory. In: Proceedings of the ICCV, pp. 2472–2479 (2013)
22. Nistér, D.: Preemptive RANSAC for live structure and motion estimation. In: Proceedings of the ICCV, pp. 199–206 (2003)
23. Raguram, R., Frahm, J.-M., Pollefeys, M.: A comparative analysis of RANSAC techniques leading to adaptive real-time random sample consensus. In: Forsyth, D., Torr, P., Zisserman, A. (eds.) ECCV 2008. LNCS, vol. 5303, pp. 500–513. Springer, Heidelberg (2008). doi:10.1007/978-3-540-88688-4_37

24. Raguram, R., Frahm, J., Pollefeys, M.: Exploiting uncertainty in random sample consensus. In: ICCV, pp. 2074–2081 (2009)
25. Raguramand, R., Frahm, J.: RECON: Scale-adaptive robust estimation via Residual Consensus. In: ICCV, pp. 1299–1306 (2011)
26. Raguram, R., Chum, O., Pollefeys, M., Matas, J., Frahm, J.: USAC: a universal framework for random sample consensus. IEEE TPAMI **35**, 2022–2038 (2013)
27. Sivic, J., Zisserman, A.: Video Google: a text retrieval approach to object matching in videos. In: Proceedings of the ICCV, pp. 1470–1477 (2003)
28. Matas, J., Chum, O., Urban, M., Pajdla, T.: Robust wide baseline stereo from maximally stable extremal regions. In: Proceedings of the BMVC, pp. 384–393 (2002)
29. Morel, J.M., Yu, G.: ASIFT: a new framework for fully affine invariant image comparison. SIAM J. Imaging Sci. **2**, 438–469 (2009)
30. Bentley, J.L.: Multidimensional binary search trees used for associative searching. Commun. ACM **18**, 509–517 (1975)
31. Chandrasekhar, V.R., Chen, D.M., Tsai, S.S., Cheung, N.M., Chen, H., Takacs, G., Reznik, Y., Vedantham, R., Grzeszczuk, R., Bach, J., Girod, B.: The stanford mobile visual search data set. In: Proceedings of the MMSys, pp. 117–122 (2011)
32. Shao, H., Svoboda, T., Van-Gool, L.: Zubud-zurich buildings database for image based recognition. Technical report 260 (2003)
33. Obdržálek, Š., Matas, J.: Image retrieval using local compact DCT-based representation. In: Michaelis, B., Krell, G. (eds.) DAGM 2003. LNCS, vol. 2781, pp. 490–497. Springer, Heidelberg (2003). doi:10.1007/978-3-540-45243-0_63
34. Obdržálek, S., Matas, J.: Sub-linear indexing for large scale object recognition. In: Proceedings of the BMVC, pp. 1–10 (2005)

Clustering Symmetric Positive Definite Matrices on the Riemannian Manifolds

Ligang Zheng[1]([✉]), Guoping Qiu[2], and Jiwu Huang[3]

[1] Guangzhou University, Guangzhou, China
zlg@gzhu.edu.cn
[2] University of Nottingham, Nottingham, UK
guoping.qiu@nottingham.ac.uk
[3] Shenzhen University, Shenzhen, China
jwhuang@szu.edu.cn

Abstract. Using structured features such as symmetric positive definite (SPD) matrices to encode visual information has been found to be effective in computer vision. Traditional pattern recognition methods developed in the Euclidean space are not suitable for directly processing SPD matrices because they lie in Riemannian manifolds of negative curvature. The main contribution of this paper is the development of a novel framework, termed Riemannian Competitive Learning (RCL), for SPD matrices clustering. In this framework, we introduce a conscious competition mechanism and develop a robust algorithm termed Riemannian Frequency Sensitive Competitive Learning (rFSCL). Compared with existing methods, rFSCL has three distinctive advantages. Firstly, rFSCL inherits the online nature of competitive learning making it capable of handling very large data sets. Secondly, rFSCL inherits the advantage of conscious competitive learning which means that it is less sensitive to the initial values of the cluster centers and that all clusters are fully utilized without the "dead unit" problem associated with many clustering algorithms. Thirdly, as an intrinsic Riemannian clustering method, rFSCL operates along the geodesic on the manifold and the algorithms is completely independent of the choice of local coordinate systems. Extensive experiments show its superior performance compared with other state of the art SPD matrices clustering methods.

1 Introduction

Recent times have seen a steep rise of data which are encoded as symmetric positive definite (SPD) matrices. Examples of SPD matrices in computer vision applications include diffusion tensors, structure tensors and region covariance descriptors [1–3]. Compared with vectors, SPD matrices offer new means of capturing intrinsic geometric structures. Benefiting from the additional structure, matrices (also known as *tensors*) are often more informative and have been empirically found to be more effective feature representations [2, 4–10].

There is an urgent demand for a rigorous framework to deal with SPD matrices. In this paper, we investigate the clustering of SPD matrices - one of the fundamental operations for the SPD matrices, grouping intrinsically *similar* SPD

© Springer International Publishing AG 2017
S.-H. Lai et al. (Eds.): ACCV 2016, Part I, LNCS 10111, pp. 400–415, 2017.
DOI: 10.1007/978-3-319-54181-5_26

matrices into the same cluster. However, the SPD matrices lie on a Riemannian manifold that constitutes a convex half-cone in the vector space of matrices [11,12]. In other words, the space of SPD matrices, although a subset of vector space, is not a vector space, e.g., the negation of a positive definite matrix is not positive definite. As a result of lacking usual vector operations like subtraction, addition and mean, directly adopting the traditional clustering algorithms often leads to unsatisfactory performance [1,7,13–15].

In recent years, several SPD matrices clustering methods taking into account the Riemannian geometry have appeared in the literature. Using matrix logarithm, the authors in [16] first mapped the manifold onto the tangent space which is Euclidean at the mean tensors and then used standard techniques for image segmentation. Taking the same strategy, the authors in [17] developed a semi-supervised framework for clustering covariance descriptors. Unfortunately, the above mappings just approximate the manifold by Euclidean space but theoretically there is not such a mapping that could globally preserve the manifold structure.

In [6], the authors used affine-invariant Riemannian metric (AIRM) [18,19] as similarity measure to implement SPD matrices K-means for human activity analysis. The authors in [5] used Jensen-Bregman LogDet Divergence (JBLD) as similarity measure for clustering covariance matrices. In the above K-means type Riemannian clustering methods, there is a need to compute the Karcher mean [13,20] which usually incurs heavy computational overheads (there is no closed form solution). In order to decrease the computation time of the K-means type Riemannian clustering algorithm, the authors in [15] presented a recursive estimation of Stein metric [21] center of SPD matrices. However, this method will sacrifice the accuracy to some extent.

In [14], the authors proposed to map SPD matrices to high dimensional Reproducing Kernel Hilbert Space (RKHS) and then extend the kernel-based algorithm developed for Euclidean space to Riemannian manifold of SPD matrices. The authors in [1] clustered submanifolds of Riemannian space using the basic concepts from Riemannian geometry and nonlinear dimension reduction. The clustering is performed in a low dimensional space after dimension reduction, which doesn't necessarily preserve all the information in original data. Unfortunately, similar to the kernel method [14] that needs to calculate a kernel matrix, this method have to compute a similarity matrix (or affinity matrix). This hinders the method's application to problems where the sample sizes are large. In [22], the authors proposed a kernelised random projection approach which achieves significant speed increase while maintaining clustering performance. However, experiments show that the method is not stable.

An EM algorithm for clustering SPD matrices using a mixture of Wishart distribution is suggested in [23]. In order to make the clustering algorithm to scale to real-world problems, the authors in [24] presented the Dirichlet Process Mixture Model (DPMM) framework for clustering SPD matrices. DPMM can dynamically update the number of clusters according to the complexity of the data. However, in some experiments DPMM is slow and sometimes can fail to converge.

1.1 Contributions

Drawing inspiration from the principle of competitive learning in the Euclidean space, we develop a new framework, termed Riemannian Competitive Learning (RCL), for SPD matrices clustering. Different from the traditional competitive learning, the new framework RCL considered the intrinsic geometry of the Riemannian manifold of SPD matrices by operating along the geodesics on the manifold. Similar to other winner-take-all clustering algorithm, RCL is sensitive to initial states. In order to overcome this drawback, we further developed rFSCL by improve RCL with a conscious mechanism. Experiments demonstrate the effectiveness of rFSCL algorithm.

Compared with existing state of the art SPD matrices clustering algorithms, the new framework has several distinctive advantages. Firstly, rFSCL inherits the online nature of competitive learning making it very effective for handling very large data sets with moderate computer memory resources. In contrast, existing methods are mainly batch algorithms and there is a need to compute and store the similarity matrix or kernel matrix of the samples making them difficult to cope with large data sets. Secondly, rFSCL inherits the advantage of conscious competitive learning which means that it is less sensitive to the initial values of the cluster centers and that all clusters are fully utilized without the "dead unit" problem associated with many clustering algorithms. Thirdly, as a intrinsic Riemannian clustering method, rFSCL operates along the geodesic on the manifold and the algorithms is completely independent of the choice of local coordinate systems. Extensive experiment results show that the clustering performance of rFSCL is superior to state of the art clustering methods for SPD matrices.

2 Preliminaries

2.1 Riemannian Manifold

A manifold is a topological space that locally looks like an Euclidean space. For every point of a m-dimensional manifold, there exists a neighborhood that is homeomorphic (one-to-one, onto and continuous mapping in both directions) to some Euclidean space $\mathbb{R}^m$. For differential manifolds, the derivatives at a point X on the manifold lies in a vector space which is also called the tangent space at X. A tangent space of $\mathbf{M}$ at point $X \in \mathbf{M}$, denoted as $T_X\mathbf{M}$, is defined as the span of the tangent vectors of all curves passing through the point X. In differential geometry, a *Riemannian manifold* $\mathbf{M}$ is a real Riemannian space in which the inner product $\langle \cdot, \cdot \rangle_X$ in each tangent space $T_X\mathbf{M}$ varies smoothly from point to point. This family of inner product is called *Riemannian metric* on $\mathbf{M}$.

Riemannian Distance and Geodesics. Let $\gamma(t) : [0,1] \to \mathbf{M}$ be a sufficiently smooth curve in the Riemannian manifold $\mathbf{M}$. For a particular $t \in [0,1]$, $\gamma(t)$ lies on the manifold and $\dot{\gamma}(t)$ is the tangent which points along the curve $\gamma(t)$.

To compute the length of the curve, we can proceed by integrating this value along the curve

$$\mathcal{L}(\gamma(t)) = \int_0^1 \sqrt{\langle \dot{\gamma}(t), \dot{\gamma}(t) \rangle_{\gamma(t)}}\, dt \tag{1}$$

Let $\mathcal{C}$ denote the space of piecewise smooth curves joining X and Y with $\gamma(0) = X$ and $\gamma(1) = Y$. The distance $dist(X, Y)$ associated to a given Riemannian metric is the minimum length among the smooth curves joining X and Y

$$dist(X, Y) = \inf\{\mathcal{L}(\gamma(t)) | \gamma(t) \in \mathcal{C}\} \tag{2}$$

Accordingly, the curve realizing this minimum for any two points on the manifold are called *geodesic*.

Riemannian Exponential and Logarithm Map. For Riemannian manifolds, tangents (on the tangent space) and geodesics (on the manifold) are closely related. For a tangent vector $v \in T_x\mathbf{M}$, locally there exist a unique geodesic $\gamma_v(t)$ starting at $X \in \mathbf{M}$ with initial velocity v. The exponential mapping $\exp_X : T_X\mathbf{M} \to \mathbf{M}$ maps tangent vector v to the point on the manifold reached by the geodesic.

For each point $X \in \mathbf{M}$, there exist a neighbourhood of $\mathcal{U}$ of the origin in the tangent space $T_X\mathbf{M}$, such that $\exp_X$ is diffeomorphism from $\mathcal{U}$ onto a neighbourhood $\mathbf{U}$ of X. Over the neighbourhood $\mathcal{U}$, we can define the inverse of exponential map from $\mathbf{U}$ to $\mathcal{U}$. This mapping is known as *logarithm map*, $\log_X() = \exp_X(-1)$. Note that both exponential and logarithm operators are point dependent where the dependance is made explicit with subscript. An toy illustration of basic notions for a two-dimensional manifold is shown in Fig. 1.

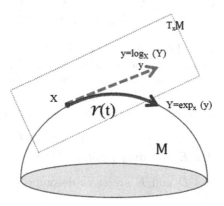

Fig. 1. A two-dimensional manifold M, T_X is tangent space of point X, y is the tangent vector in the tangent space T_X and $\gamma(t)$ is the geodesic between X and Y.

2.2   The Space of SPD Matrices

The inner product between two vectors x and y is written as $\langle x, y \rangle = x^T y$. A matrix $P \in \mathbb{R}^{n \times n}$ is called positive if

$$\langle x, Px \rangle > 0, for \quad all \quad x \in \mathbb{R}^n \backslash \{\mathbf{0}\} \tag{3}$$

which we also denote by $P \succ 0$. Denote $\mathcal{S}(n) = \{S \in \mathbb{R}^{n \times n}, S^T = S\}$ as the symmetric space of all $n \times n$ symmetric matrix and denote $\mathcal{P}(n) = \{P \in \mathcal{S}(n), P \succ 0\}$. Thus, $\forall P \in \mathcal{P}(n)$ is a SPD matrix. The space $\mathcal{P}(n)$ forms a convex subset of $\mathbb{R}^{n^2}$.

Suppose $\alpha \in \mathbb{R}, \alpha > 0$ and $A \in \mathcal{P}(n)$ is a SPD matrix, thus

$$\langle x, -\alpha Ax \rangle = -\alpha \cdot \langle x, Ax \rangle < 0 \tag{4}$$

Equation (4) tells us that $(-\alpha A) \notin \mathcal{P}(n)$, which means that $\mathcal{P}(n)$ is not closed under multiplication with a negative scalar. Therefore, the space of SPD matrices, although a subset of vector space, is not a vector space. Instead, the set of SPD matrices forms a differential Riemannian manifold of nonpositive curvature [11,19,25].

According to [18], the affine-invariant Riemannian metric (AIRM)[1] for two tangent vectors $y, z \in T_X P(n), X \in P(n)$ is as follows,

$$\begin{aligned} \langle y, z \rangle_X &= \left\langle X^{-\frac{1}{2}} y X^{-\frac{1}{2}}, X^{-\frac{1}{2}} z X^{-\frac{1}{2}} \right\rangle \\ &= tr(X^{-\frac{1}{2}} y X^{-1} z X^{-\frac{1}{2}}) \end{aligned} \tag{5}$$

This metric varies smoothly as X moves. The Riemannian exponential and logarithm map associated with the metric have the expression

$$\begin{aligned} \exp_X(v) &= X^{\frac{1}{2}} exp(X^{-\frac{1}{2}} v X^{-\frac{1}{2}}) X^{\frac{1}{2}} \\ \log_X(Y) &= X^{\frac{1}{2}} \log(X^{-\frac{1}{2}} Y X^{-\frac{1}{2}}) X^{\frac{1}{2}} \end{aligned} \tag{6}$$

where, manifold exponential operator, $\exp_X : T_X P(n) \to P(n)$, maps the tangent vector v to the location on the manifold reached by geodesic starting at X in the tangent direction. Its inverse, the Riemannian logarithm operator, $\log_X : P(n) \to T_X P(n)$, gives the vectors in $T_X P(n)$ corresponding to the geodesic from X to Y. The matrix logarithm $\log(\cdot)$ and matrix exponential $\exp(\cdot)$ are calculated as

$$\begin{aligned} \exp(D) &= U diag(\exp(diag(S))) U^T \\ \log(E) &= V diag(\log(diag(T))) V^T \end{aligned} \tag{7}$$

where two SPD matrices D and E are eigen-decomposed as $D = USU^T$ and $E = VTV^T$ respectively.

---

[1] Please note that an equivalent form of affine-invariant Riemannian metric (AIRM) is given in [19]. Affine-invariant Riemannian metric was first used to calculate the geodesic distance of two SPD matrices in [26] and Pennec *et al.* promoted AIRM as a computing framework in [18].

3   Conscious Competitive Learning over Riemannian Manifold

3.1   Simple Competitive Learning

Competitive learning [27–29] is a form of unsupervised learning which performs online clustering over the input data. It is sometimes convenient to implement this as a two-layer neural network [27] as shown in Fig. 2. The input layer and the output layer are fully connected, each input neuron corresponds to one of the elements of the input vector $X \in \mathbb{R}^d$ and every competitive neuron (output neuron) is described by a vector of weights $W \in \mathbb{R}^d$.

Given a set of observations $X_1, X_2, \cdots, X_n$, where each observation is a d-dimensional real vector, competitive learning aims to partition the n observations into K sets ($K \leq n$), $S = S_1, S_2, \cdots, S_K$ so as to minimize the mean square error (MSE) function [27]

$$\min_{S} \arg \frac{1}{n} \sum_{i=1}^{K} \sum_{j=1}^{n} y_i \times Dist(X_j, W_i) \tag{8}$$

where $Dist(X_j, W_i)$ is the Euclidean distance between two vectors X_j and W_i, y_i is the state of output neuron i indicating the membership of X_j into the prototype W_i. Mathematically, y_i is defined as

$$\begin{aligned} y_i &= 1, \quad if \ W_i \ win \ X_j \\ y_i &= 0, \quad else \end{aligned} \tag{9}$$

For each competitive neuron $W_i, i = 1, 2, \cdots, K$, the neuron with a minimum distance is declared a winner.

$$k = \arg \min Dist(X_j, W_i), i = 1, 2, \cdots, K \tag{10}$$

Based on the MSE criteria (8) and the gradient-descent method, assuming W_k wins X, the winning neuron's weight vector is updated as

$$W_k(t+1) = W_k(t) + \eta(t)[X(t) - W_k(t)] \tag{11}$$

where t is the iteration index, $\eta(t)$ is the learning rate. All other competitive neurons' weight vectors remain unchanged.

$$W_i(t+1) = W_i(t), i \neq k \tag{12}$$

A Key Insight into the Competitive Learning Algorithm: We here present a key observation of the competitive learning algorithm of (11) and (12) which inspires the development of the new data clustering algorithm in the Riemannian manifold of SPD matrices. As illustrated in Fig. 3, at each iteration of the training process, when a data point X is presented to the network for training, the algorithm moves the winning weight vector $W_k(t)$ towards the input vector X along

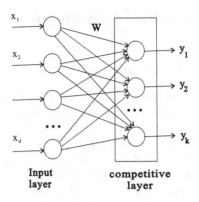

Fig. 2. Competitive neural network architecture.

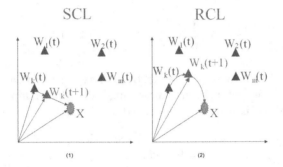

Fig. 3. An illustration of the learning step of competitive learning. In the Euclidean space, the winner $W_k(t)$ moves forward X along a straight line. In the Riemannian framework, the winner moves along the geodesic.

the direction **e** from $W_k(t)$ to X. The amount of movement is proportional to the difference between $W_k(t)$ and X and is controlled by the learning rate $\eta(t)$. The larger the learning rate, the closer $W_k(t)$ will be moved towards X. Eventually, when the learning process is completed, the weight vectors will be stabled, *i.e.*, further iterations will not move the weights, and these stable weight vectors form the cluster centres of data clusters. Based on this important insight, we introduce competitive learning in the Riemannian manifold in the next subsection.

3.2   Riemannian Competitive Learning

The simple competitive learning (SCL) framework described above operates in the Euclidean space, the algorithm moves the winning weight vector (cluster centre) towards the input data point along the linear direction from the weight to the input. However, as stated previously the space of SPD matrices is a Riemannian manifold of nonpositive curvature rather than a vector space, *i.e.*, don't adhere to Euclidean geometry, the SCL algorithm cannot be directly applied to SPD matrices for clustering.

Based on the key insight into the operation of the SCL, we adapt the SCL algorithm for the clustering of SPD matrices by taking into account the intrinsic geometry of the Riemannian manifold. We name this Riemannian extension of SCL algorithm Riemannian competitive learning (RCL). Suppose a set of observations X_1, X_2, $\cdots$ are SPD matrices, we define the mean geodesic error (MGE) as

$$MGE = \frac{1}{n}\sum_{i=1}^{K}\sum_{j=1}^{n} y_i \times GeoDist(X_j, W_i) \tag{13}$$

where $GeoDist(X_j, W_i)$ is the geodesic distance between two SPD matrices X_j and W_i and can be calculated using Riemannian metric.

In order to optimize the MGE, RCL also has two similar steps with SCL, (1) find the winner and (2) move the winning weight (cluster center) towards the input. However, different from SCL, RCL takes a Riemannian framework. Specifically, instead of using Euclidean distance to find the winner and moving the wining weight along the linear direction between the weight and the input, RCL uses geodesic distance to find the winner and moves the wining cluster centre along the geodesic between the wining weight and the input SPD matrix. An illustration of this process is given in Fig. 3.

In RCL, the winner can be determined using formula (10) by replacing Euclidean distance with geodesic distance. In the Riemannian manifold, common operations like addition and subtraction in vector space can be reinterpreted through introducing the notion of Riemannian exponential and logarithmic mappings [18,19]. Accordingly, in Riemannian framework, the winning neuron's weight vector is updated as

$$W_k(t+1) = \exp_{W_k(t)}(\eta(t)\overrightarrow{W_k(t)X(t)}) \tag{14}$$

where $\eta(t)$ is the learning rate, $\overrightarrow{W_k(t)X(t)}$ is the tangent vector of geodesic from $W_k(t)$ to $X(t)$. This tangent vector can be expressed as Riemannian logarithm mapping (15).

$$\overrightarrow{W_k(t)X(t)} = \log_{W_k(t)}(X(t)) \tag{15}$$

In order to use tools from differential geometry to explicitly solve (14), we choose a suitable Riemannian metric for the space of SPD matrices [18,19]. Replacing formula (14) with Riemannian exponential (6), Riemannian logarithm (6) and the tangent vector at $W_k(t)$ (15), the learning equation of RCL (14) can be reformulated as (16).

$$W_k(t+1) = W_k(t)^{\frac{1}{2}}(W_k(t)^{-\frac{1}{2}}X(t)W_k(t)^{-\frac{1}{2}})^{\eta(t)}W_k(t)^{\frac{1}{2}} \tag{16}$$

The formula (16) update the winner's weight intrinsically and no local coordinate system is involved in the process. The RCL is summarized in Algorithm 1.

Algorithm 1. Riemannian Competitive Learning (RCL)

1: Initialized the cluster centers, $W_i(0)$, $i = 1, 2, ..., K$, set $t = 0$, where t is the sequence index.
2: **repeat**
3:    Present the samples, $X(t)$, calculate the geodesic distance $GeoDist_i(t)$ between $X(t)$ and the ith cluster centers, $i = 1, 2, ..., K$.
4:    Find the wining cluster k according to (17)

$$k = \arg\min_i\{GeoDist_i(t)\} \tag{17}$$

5:    Update the wining cluster center W_k according to (16).
6: **until** Convergence

3.3   Adding a Conscious Mechanism

Like many other clustering techniques such as the well-known k-means algorithm, the simple winner takes all competitive learning algorithm described above will be sensitive to the initial clusters which are normally set randomly. These simple algorithms only adjust the wining cluster and leave the loosing cluster centers unchanged, this may result in some of the clusters wining all the time while others may never win thus leading to poor clustering performance, *e.g.*, some clusters are over used whilst others are under-utilized. To overcome this drawback, the conscious mechanism can be used. The idea is that if a cluster wins the competition in the current iteration, the chance of it wining again in the next iteration is reduced, or equivalently, if a cluster looses the competition in the current iteration, the chance of it wining the competition the next time is increased. Eventually, all clusters get equal chance of wining the competition and all clusters are fully utilized. The frequency sensitive competitive learning (FSCL) [28] introduced a very simple conscious mechanism and we adapt this to RCL and introduce the Riemannian Frequency Competitive Learning (rFSCL) algorithm which is summarized in Algorithm 2.

It is seen from Algorithm 2 that in rFSCL, if a cluster wins a competition, its counter is increased, because the competition is based on a modified distance, $GeoDist_i^*(t) = C_i(t) \times GeoDist_i(t)$, the chance of it wining the completion next time is decreased. Because only the winner's counter is increased, this effectively gives the losers a higher chance to win the next time around.

We give a performance comparison between RCL and rFSCL in Table 1. From the table, we can see that rFSCL gets better mean and variance performance than RCL[2]. Especially, the mean clustering results of rFSCL are bigger than RCL while the variance of rFSCL are far smaller than RCL, meaning that rFSCL not only has a better performance but also is far more consistent for different runs or equivalently much less sensitive to initial clusters center positions. The experiment demonstrates the effectiveness of introducing a conscious mechanism in designing competitive learning based clustering algorithms in Riemannian framework.

---

[2] The experiments are conducted on the simulated dataset.

Algorithm 2. Riemannian Frequency Sensitive Competitive Learning (rFSCL)

1: Initialized the cluster centers, $W_i(0)$, $i = 1, 2, ..., K$.

2: For each cluster, introduce an associated counter C_i, $i = 1, 2, ..., K$, initialize the counters as $C_i(0) = 1$, set $t = 0$, where t is the sequence index.

3: **repeat**

4:   Present the samples, $X(t)$, calculate the geodesic distance $GeoDist_i(t)$ between $X(t)$ and the ith cluster centers, $i = 1, 2, ..., K$.

5:   Modify the distance,

$$GeoDist_i^*(t) = C_i(t) \times GeoDist_i(t) \tag{18}$$

6:   Find the wining cluster k according to (19)

$$k = \arg\min_i \{GeoDist_i^*(t)\} \tag{19}$$

7:   Update the wining cluster center W_k according to (16).

8:   Update the sequence index $t = t + 1$, update the wining counter $C_k(t) = C_k(t - 1) + 1$.

9: **until** Convergence

Table 1. Performance comparison between RCL and rFSCL

	5×5		10×10		15×15		20×20	
	μ	σ	μ	σ	μ	σ	μ	σ
RCL	0.9415	0.0654	0.9069	0.0521	0.8485	0.049	0.8546	0.059
rFSCL	0.9926	0.0003	0.9894	0.0041	0.9852	0.0069	0.9927	0.0073

3.4   Learning Rate and Convergence

The convergence property of competitive learning algorithms has been studied extensively in the neural network community [30,31]. As a conscious type competitive learning algorithm, the convergence properties of FSCL algorithm was analysed by Aristides *et al.* in the paper [30] through approximating the final phase of FSCL learning by a diffusion process with a Fokker-Plank equation (FPE) that describes the evolution of the process. Aristides *et al.*'s research results showed that the convergence conditions involve a small learning rate.

The proposed rFSCL can be seen as a Riemannian extension of FSCL. The rFSCL algorithm should inherit the converge property of FSCL, and will reach a local equilibrium for suitably chosen small learning rates. We have performed extensive experiments which confirmed that rFSCL always converged within a finite number of iterations.

From the experiments, we also observed that in general, a larger training rate will converge faster but the clustering quality may not be as good, on the other hand, a small training rate will result in higher quality clustering but the convergence rate may be slower. The reason is that the algorithm is a gradient

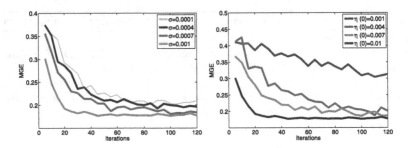

Fig. 4. Typical convergence behaviors of rFSCL for different learning rate settings

descent algorithm optimising a cost function which has many local minima. A large training rate moves the weights in a larger step, which could miss the global minimum leading the algorithm converge to a local minimum. Similar to [31] and others, we adopt a strategy that starts with an initial learning rate and then gradually attenuate the learning rate value as training progresses, as follows

$$\eta(t) = \eta(0)e^{-\sigma t} \tag{20}$$

where $\eta(0)$ is the initial learning rate, σ controls the descending rate of the learning rate as the training progresses. We have done extensive experiments by varying the the two parameters of the training rate and found that rFSCL always converge within finite number of iterations. Figure 4 shows the typical convergence behaviours for the data used in this paper with different learning rate parameters (see Footnote 2). Here it is seen that the algorithm converges between 60 and 120 iterations. Empirically, we found that setting $\eta(0)$ between 0.01 and 0.001 and σ between 0.0001 and 0.01 always worked well.

4 Applications and Experiments

This section details experiments on three applications. We employ purity [24] to evaluate the clustering performance. We compare rFSCL against geodesic K-means (gKmeans) [6], kernel K-means (knKmeans)[3] [14], Riemannian locally linear embedding (rLLE) [1], recursive Stein mean method (RSM)[4] [15], kernelised orthogonal random projection (KORP) [22] and log-Euclidean K-means [4] (leKmeans) which are the state of the art algorithms for solving SPD matrices clustering problem. We also cluster the SPD matrices using FSCL, termed eFSCL, without considering the Riemannian geometry.

4.1 DTI Image Segmentation

Diffusion tensor imaging (DTI) is a magnetic resonance imaging technique that enables the measurement of the diffusion of water in tissue by modeling the

---

[3] The code is downloaded from http://www.robots.ox.ac.uk/~sadeep/.

[4] The code is downloaded from http://www.cise.ufl.edu/~salehian/Softwares.html.

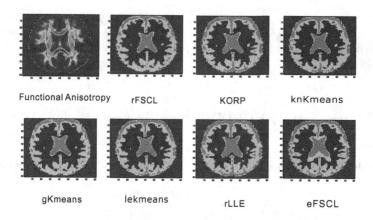

Functional Anisotropy   rFSCL   KORP   knKmeans

gKmeans   lekmeans   rLLE   eFSCL

Fig. 5. DTI segmentation results for different methods.

local diffusivity in each voxel as a 3×3 symmetric positive definite matrix [32]. It provides useful structural information about anatomical tissue such as white matter fibers in the brain and muscle fibers in the heart. Many algorithms for DTI analysis have been proposed. DTI segmentation plays an indispensable role for diffusion tensor image analysis. In this part, we apply rFSCL algorithm to segment a brain DTI slice along with the ellipsoid and fractional anisotropy representation of the original DTI image. We cluster the voxels of an "image" into three classes. In this experiment, we run each algorithm 10 times. The best segmentation results of a 91×81 DTI image using different methods are shown in Fig. 5. Three colors in each figure show the three clusters respectively. We can see from the figure that rFSCL yields a clearer segmentation result that is at least comparable to other state of the art Riemannian clustering methods.

It is worth mentioning that we have performed experiments on a larger (256×256) DTI image. Whilst rFSCL, leKemans and gKmeans achieved similar results as those shown in Fig. 5, rLLE and knKmeans crashed on the same PC because both rLLE and knKmeans require the computation of large similarity or kernel matrix. Although a machine with larger memory can cope, the point is that rFSCL is an on online clustering algorithm and can handle large problems with moderate computational resources.

4.2 Categorization

Material Categorization [33] - KTH-TIPS2b material dataset is used to evaluate our proposal with the problem of unsupervised material categorization. The dataset contains images of 11 materials and each material totally has 108 samples (4 illuminations $\times$ 3 poses $\times$ 9 scales). The variation in appearance between the samples in each category is larger for some categories than others. Some sample images are given in Fig. 6(a). A 15-dimensional feature vector for each pixel is defined as $\mathcal{F}(x, y) = [I(x, y), Y(x, y), C_b(x, y), C_r(x, y), F^1_{(x,y)}(Y), \cdots, F^8_{(x,y)}(Y)]$,

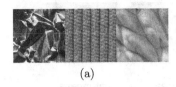

(a)   (b)

Fig. 6. Examples from (a) KTH-TIPS2b material dataset [33], (b) ETH-80 object dataset [35].

Table 2. Performance comparison for different algorithms in two datasets

	KTH-TIPS2b	ETH-80
rFSCL	48.17 ± 0.39	87.48 ± 0.45
knKmeans	47.52 ± 1.05	85.43 ± 1.53
leKmeans	45.66 ± 1.21	85.26 ± 1.46
gKmeans	46.25 ± 1.32	85.36 ± 1.35
eFSCL	38.5 ± 1.52	71.83 ± 1.73
RSM	46.75 ± 0.98	83.96 ± 1.06
rLLE	41.25 ± 1.26	74.33 ± 1.68
KORP	47.13 ± 1.13	85.3 ± 1.5

where $I(x, y)$ is the image grey value at pixel (x, y); Y, C_b, C_r are color values converted from RGB color space; $F^1_{(x,y)}(Y), \cdots, F^8_{(x,y)}(Y)$ are the filter banks, consisting of scaled symmetric difference of offset Gaussian (DOOG) [34] applied on the luminance channel Y. Then the covariance matrix is calculated using $\mathcal{F}(x, y)$ extracted from each image. There are totally $11 \times 108 \times 4 = 4752$ SPD matrices in the experiment. SPD clustering algorithms are employed to cluster 4752 SPD matrices into 11 material categories.

Object Categorization [35] - The ETH-80 dataset [35] contains 8 categories with 10 objects. Each object is represented by 41 images of views from the upper hemisphere. Sample images of the objects are give in Fig. 6(b). For each image, we construct a 9×9 covariance matrix using 9-dimensional feature vectors $[x, y, R, G, B, \frac{\partial I}{\partial x}, \frac{\partial I}{\partial y}, \frac{\partial^2 I}{\partial x^2}, \frac{\partial^2 I}{\partial y^2}]$, where x, y are pixel locations and RGB are three color channels. Therefore, there are totally $41 \times 10 \times 8 = 3280$ SPD matrices in this experiment. SPD clustering algorithms are employed to cluster 3280 SPD matrices into 8 categories.

Results - Table 2 reports the average results over 10 runs with different random initialization in the KTH-TIPS2b and ETH-80 dataset. From the table, we can see that (1) the mean clustering results of rFSCL are bigger than other clustering methods; (2) the variances of rFSCL algorithm are far smaller than other methods; (3) without considering the intrinsic geometry of SPD manifold, eFSCL has the worst performance compared to other Riemannian clustering methods.

Therefore, taking Riemannian geometry into consideration is very important for SPD matrices clustering. rFSCL algorithm not only has a better performance but also is far more consistent for different runs or equivalently much less sensitive to initial clusters center positions, which may be partly due to the introduction of conscious mechanism.

5   Conclusion and Future Work

This paper successfully extends the framework of conscious competitive learning to the Riemannian manifold of the SPD matrices through operating on the intrinsic geometric structure. We have shown that our new technique has several advantages over existing SPD matrices clustering methods. We have presented experimental results in real world applications and have demonstrated that our new method outperformed state of the art techniques. In the future, I will adapt the RCL framework to other kinds of Riemannian manifolds such as Lie group, Grassmannian manifold, $SO(n), etc.$

Acknowledgement. Part of this paper is supported by NSFC (61332012, 61300205), Shenzhen R&D Program (JCYJ20160328144421330, GJHZ20140418191518323).

References

1. Goh, A., Vidal, R.: Clustering and dimensionality reduction on Riemannian manifolds. In: IEEE Conference on Computer Vision and Pattern Recognition (CVPR) (2008)
2. Chiang, M.C., Dutton, R.A., Hayashi, K.M., Lopez, O.L., Aizenstein, H.J., Toga, A.W., Becker, J.T., Thompson, P.M.: 3d pattern of brain atrophy in HIV/AIDS visualized using tensor-based morphometry. Neuroimage **34**, 44–60 (2007)
3. Tuzel, O., Porikli, F., Meer, P.: Region covariance: a fast descriptor for detection and classification. In: Leonardis, A., Bischof, H., Pinz, A. (eds.) ECCV 2006. LNCS, vol. 3952, pp. 589–600. Springer, Heidelberg (2006). doi:10.1007/11744047_45
4. Arsigny, V., Fillard, P., Pennec, X., Ayache, N.: Fast and simple calculus on tensors in the log-Euclidean framework. In: Duncan, J.S., Gerig, G. (eds.) MICCAI 2005. LNCS, vol. 3749, pp. 115–122. Springer, Heidelberg (2005). doi:10.1007/11566465_15
5. Cherian, A., Sra, S., Banerjee, A., Papanikolopoulos, N.: Jensen-bregman logdet divergence with application to efficient similarity search for covariance matrices. IEEE Trans. Pattern Anal. Mach. Intell. **35**, 2161–2174 (2013)
6. Chaudhry, R., Ivanov, Y.: Fast approximate nearest neighbor methods for non-euclidean manifolds with applications to human activity analysis in videos. In: Daniilidis, K., Maragos, P., Paragios, N. (eds.) ECCV 2010. LNCS, vol. 6312, pp. 735–748. Springer, Heidelberg (2010). doi:10.1007/978-3-642-15552-9_53
7. Malcolm, J., Rathi, Y., Tannenbaum, A.: A graph cut approach to image segmentation in tensor space. In: IEEE Conference on Computer Vision and Pattern Recognition (CVPR), pp. 1–8 (2007)
8. Zheng, L., Lei, Y., Qiu, G., Huang, J.: Near-duplicate image detection in a visually salient Riemannian space. IEEE Trans. Inf. Forensics Secur. **7**, 1578–1593 (2012)

9. Alavi, A., Wiliem, A., Zhao, K., Lovell, B., Sanderson, C.: Random projections on manifolds of symmetric positive definite matrices for image classification. In: 2014 IEEE Winter Conference on Applications of Computer Vision (WACV), pp. 301–308 (2014)
10. Zhang, S., Kasiviswanathan, S., Yuen, P., Harandi, M.: Online dictionary learning on symmetric positive definite manifolds with vision applications. In: Twenty-Ninth AAAI Conference on Artificial Intelligence (AAAI 2015), pp. 3165–3173 (2015)
11. Bridson, M.R.: Metric Spaces of Non-Positive Curvature. Springer, Heidelberg (1999)
12. Tyagi, A., Davis, J.W.: A recursive filter for linear systems on riemannian manifolds. In: IEEE Conference on Computer Vision and Pattern Recognition, pp. 1–8 (2008)
13. Dryden, I., Koloydenko, A., Zhou, D.: Non-euclidean statistics for covariance matrices, with applications to diffusion tensor imaging. Ann. Appl. Stat. **3**, 1102–1123 (2009)
14. Jayasumana, S., Hartley, R., Salzmann, M., Li, H., Harandi, M.: Kernel methods on the Riemannian manifold of symmetric positive definite matrices. In: IEEE Conference on Computer Vision and Pattern Recognition (CVPR), pp. 73–80 (2013)
15. Salehian, H., Cheng, G., Vemuri, B., Ho, J.: Recursive estimation of the stein center of SPD matrices and its applications. In: IEEE International Conference on Computer Vision (ICCV), pp. 1793–1800 (2013)
16. Rathi, Y., Tannenbaum, A., Michailovich, O.: Segmenting images on the tensor manifold. In: IEEE Conference on Computer Vision and Pattern Recognition (CVPR), pp. 1–8 (2007)
17. Sivalingam, R., Morellas, V., Boley, D., Papanikolopoulos, N.: Metric learning for semi-supervised clustering of region covariance descriptors. In: Third ACM/IEEE International Conference on Distributed Smart Cameras, pp. 1–8 (2009)
18. Pennec, X., Fillard, P., Ayache, N.: A Riemannian framework for tensor computing. Int. J. Comput. Vis. **66**, 41–66 (2006)
19. Bhatia, R.: Positive Definite Matrices. Princeton University Press, Princeton (2007)
20. Karcher, H.: Riemannian center of mass and mollifier smoothing. Commun. Pure Appl. Math. **30**, 509–541 (1977)
21. Sra, S.: Positive definite matrices and the s-divergence (2013). http://people.kyb.tuebingen.mpg.de/suvrit/
22. Zhao, K., Alavi, A., Wiliem, A., Lovell, B.C.: Efficient clustering on Riemannian manifolds: a kernelised random projection approach. Pattern Recogn. **51**, 333–345 (2016)
23. Hidot, S., Saint-Jean, C.: An expectation-maximization algorithm for the Wishart mixture model: application to movement clustering. Pattern Recogn. Lett. **31**, 2318–2324 (2010)
24. Cherian, A., Morellas, V., Papanikolopoulos, N., Bedros, S.J.: Dirichlet process mixture models on symmetric positive definite matrices for appearance clustering in video surveillance applications. In: IEEE Conference on Computer Vision and Pattern Recognition, pp. 3417–3424 (2011)
25. Hiai, F., Petz, D.: Riemannian metrics on positive definite matrices related to means. Linear Algebra Appl. **430**, 3105–3130 (2009)
26. Frstner, W., Moonen, B.: A metric for covariance matrices. Technical report, Stuttgart University (1999)
27. Du, K.L.: Clustering: a neural network approach. Neural Netw. **23**, 89–107 (2010)

28. Ahalt, S.C., Krishnamurthy, A.K., Chen, P., Melton, D.E.: Competitive learning algorithms for vector quantization. Neural Netw. **3**, 277–290 (1990)
29. Banerjee, A., Ghosh, J.: Frequency-sensitive competitive learning for scalable balanced clustering on high-dimensional hyperspheres. IEEE Trans. Neural Netw. **15**, 702–719 (2004)
30. Galanopoulos, A.S., Moses, R.L., Ahalt, S.C.: Diffusion approximation of frequency sensitive competitive learning. IEEE Trans. Neural Netw. **8**, 1026–1030 (1997)
31. Qiu, G., Duana, J., Finlaysonb, G.D.: Learning to display high dynamic range images. Pattern Recogn. **40**, 2641–2655 (2007)
32. Basser, P.J., Mattiello, J., LeBihan, D.: Estimation of the effective self-diffusion tensor from the NMR spin echo. J. Magn. Reson. Ser. B **103**(3), 247–254 (1994)
33. Caputo, B., Hayman, E., Mallikarjuna, P.: Class-specific material categorisation. In: Proceedings of the Tenth IEEE International Conference on Computer Vision-ICCV 2005, vol. 2, pp. 1597–1604. IEEE Computer Society, Washington, DC (2005)
34. Dollár, P.: Piotr's image and video matlab toolbox (pmt) (2013). http://vision.ucsd.edu/~pdollar/toolbox/doc/index.html
35. Leibe, B., Schiele, B.: Analyzing appearance and contour based methods for object categorization. In: IEEE Computer Society Conference on Computer Vision and Pattern Recognition, vol. 2, pp. II-409–II-415 (2003)

Subspace Learning
Based Low-Rank Representation

Kewei Tang[1(✉)], Xiaodong Liu[2], Zhixun Su[2(✉)], Wei Jiang[1],
and Jiangxin Dong[2]

[1] School of Mathematics, Liaoning Normal University,
Dalian, People's Republic of China
tkwliaoning@gmail.com, swxxjw@aliyun.com
[2] School of Mathematical Sciences, Dalian University of Technology,
Dalian, People's Republic of China
liuxdxiaodong@gmail.com, zxsu@dlut.edu.cn, dongjxjx@gmail.com

Abstract. Subspace segmentation has been a hot topic in the past decades. Recently, spectral-clustering based methods arouse broad interests, however, they usually consider the similarity extraction in the original space. In this paper, we propose subspace learning based low-rank representation to learn a subspace favoring the similarity extraction for the low-rank representation. The process of learning the subspace and achieving the representation is conducted simultaneously and thus they can benefit from each other. After extending the linear projection to nonlinear mapping, our method can handle manifold clustering problem which is a general case of subspace segmentation. Moreover, our method can also be applied in the problem of recognition by adding suitable penalty on the learned subspace. Extensive experimental results confirm the effectiveness of our method.

1 Introduction

Many kinds of vision data have the structure of multiple subspaces, leading to the heated discussions on the problem of subspace segmentation. With respect to a collection of data points drawn from a union of subspaces, the goal of subspace segmentation is to segment these data points according to their underlying subspaces. Independence and disjointness are two classes of formally defined subspace arrangements in existing work [1]. Independent subspaces mean that the sum of their dimension is equal to the dimension of the whole space whereas the subspaces are said to be disjoint if the only intersection of every two subspace is the origin. Disjoint subspaces are more general according to the definition. Numerous subspace segmentation methods have been proposed in the past two decades, and can be classified into four categories including algebraic

Electronic supplementary material The online version of this chapter (doi:10.1007/978-3-319-54181-5_27) contains supplementary material, which is available to authorized users.

S.-H. Lai et al. (Eds.): ACCV 2016, Part I, LNCS 10111, pp. 416–431, 2017.
DOI: 10.1007/978-3-319-54181-5_27

approaches, iterative approaches, statistical approaches, and spectral clustering-based approaches, according to the review [2].

Spectral clustering-based approaches are very popular in recent years. This kind of methods first construct an affinity matrix representing the similarity between the data points, and then perform spectral clustering [3] algorithm such as Normalized Cuts (NC) [4] on the affinity matrix to produce the segmentation results. Hence, the construction of the affinity matrix is a key step. As two typical spectral clustering-based methods, Low-Rank Representation (LRR) [5] and Sparse Subspace Clustering (SSC) [6] employ the self-representation of the data points to extract the similarity. In the hope that the data points are only represented by those in the same subspace, LRR puts the low-rank penalty on the affinity matrix whereas SSC uses the sparse criterion. In the past few years, LRR and SSC have generated a number of literatures, such as Subspace Segmentation via Quadratic Programming (SSQP) [7], Least Squares Regression (LSR) [8], Correlation Adaptive Subspace Segmentation (CASS) [9], Smooth Representation (SMR) [10], Structure-Constrained Low-Rank Representation (SC-LRR) [11], Dense Block and Sparse Representation (DBSR) [12], etc., which can be summarized in the following framework

$$\min_{\mathbf{Z},\mathbf{E}} f(\mathbf{Z}) + \alpha l(\mathbf{E}) \quad s.t. \quad \mathbf{X} = \mathbf{XZ} + \mathbf{E}, \tag{1}$$

where $\mathbf{X}$ denotes the sample set with each column denoting a data point, $\mathbf{Z}$ is the affinity matrix with its element located in i-th row and j-th column representing the similarity of data points located in i-th column of $\mathbf{X}$ and j-th column of $\mathbf{X}$, $\mathbf{E}$ denotes the noise, $f(\mathbf{Z})$ and $l(\mathbf{E})$ are the penalty on the affinity matrix and the noise, respectively, and α is a tradeoff balancing these two terms. Because $l(\mathbf{E})$ is usually selected as $\ell_{2,1}$-norm, i.e. the sum of ℓ_2-norm of each column vector, and F-norm, i.e. the square root of the sum of all the elements' square, the main difference of these approaches is $f(\mathbf{Z})$ summarized in Table 1, where $\| \bullet \|_*$ denotes the nuclear norm, i.e. the sum of all the singular values, $\| \bullet \|_{1,1}$ denotes the 1, 1-norm, i.e. the sum of the absolute value of each element. $\| \bullet \|_2$ denotes the 2-norm of the matrix, i.e. the maximal singular value, $[\bullet]$ with subscripts denotes the element of the matrix, for example, $[\mathbf{Z}]_{:i}$ denotes the i-th column of $\mathbf{Z}$, $diag(\mathbf{x})$ is a diagonal matrix with its diagonal elements being $\mathbf{x}$ when $\mathbf{x}$ is a vector, $\odot$ denotes the Hadmard product, i.e. $[\mathbf{W} \odot \mathbf{Z}]_{ij} = [\mathbf{W}]_{ij}[\mathbf{Z}]_{ij}$ [13]. The properties of different $f(\mathbf{Z})$ result in different characteristics, for example, the algorithm of SSQP and LSR is fast; LRR, LSR, CASS, and SMR exhibit grouping effect; SSC and DBSR can induce sparsity and dense block, respectively; SC-LRR can provide theoretical guarantees on disjoint subspace segmentation whereas other methods can only provide theoretical guarantees on independent subspace segmentation. However, these methods only consider the self-representation of the data points in the original space. Actually, using self-representation in other spaces to explore the affinity between the data points may be more effective. Although other extension of LRR such as Latent Low-Rank Representation [14]

Table 1. $f(\mathbf{Z})$ of different methods

Methods	LRR	SSC	SSQP	LSR
$f(\mathbf{Z})$	$\|\mathbf{Z}\|_*$	$\|\mathbf{Z}\|_{1,1}$	$\|\mathbf{Z}^T\mathbf{Z}\|_{1,1}$	$\|\mathbf{Z}\|_F^2$
Methods	CASS	SMR	SC-LRR	DBSR
$f(\mathbf{Z})$	$\sum_i \|\mathbf{X}diag([\mathbf{Z}]_{:i})\|_*$	$tr(\mathbf{ZLZ}^T)$	$\|\mathbf{Z}\|_* + \beta\|\mathbf{W}\odot\mathbf{Z}\|_{1,1}$	$\|\mathbf{Z}\|_2 + \beta\|\mathbf{Z}\|_{1,1}$

(LLRR) exploits $\mathbf{X} = \mathbf{XZ} + \mathbf{LX} + \mathbf{E}$ instead of the commonly used $\mathbf{X} = \mathbf{XZ} + \mathbf{E}$ by introducing hidden effect, it also conducts the representation in the original space.

In fact, those self-representation based methods usually project the data into a low-dimensional space by Principle Component Analysis (PCA) to obtain a more meaningful representation in their experiments. This preprocessing step can often reduce the computation time and improve the results to some extent. However, this strategy separates the process of finding the subspace from obtaining the representation, leading to the result that the subspace is only determined by the dimensionality reduction methods and thus may not favor the extraction of the similarity. Learning a subspace and representing the data at the same time seems more reasonable. In this way, the learned subspace is likely to greatly facilitate similarity extraction by self-representation. Latent Space Sparse Subspace Clustering [15] (LS3C) is an example of learning a subspace to obtain the representation. It considers the self-representation in the low-dimensional latent space instead of the original space by seeking a subspace minimizing the reconstruction of the data points to simultaneously preserve the information and reduce the dimensionality as well as possible. Although it has the idea of learning a subspace a bit, this idea is only from the perspective of reducing the computation time of SSC not from the view of favoring similarity extraction.

In this paper, we propose Subspace Learning based Low-Rank Representation (SLLRR) to extract the similarity between the data points by the self-representation in a learned subspace. Since we simultaneously learn the subspace and obtain the low-rank representation, the learned subspace is decided not only by the term describing its property but also by the low-rank penalty. Hence, it has significant difference from the strategy that first project the data in a subspace by dimensionality reduction methods and then obtain the segmentation results by LRR. Different from LS3C incorporating PCA in the model, our method specially constructs a penalty on subspace to favor similarity extraction. After extending the linear projection in the model to nonlinear mapping, our method can achieve state-of-the-art results on both subspace segmentation and manifold clustering. Furthermore, referring to the conclusion in graph embedding framework [16], our method can also be applied in recognition problem such as recognition based on image set. In summary, our contributions are as follows:

- We present subspace learning based low-rank representation. It can learn the subspace favoring similarity extraction by self-representation.

– Our method can handle both subspace segmentation and manifold clustering by considering the nonlinear mapping.
– As a non-trivial byproduct, our proposed method can also be applied into recognition according to its characteristic.

The reminder of the paper is organized in the following: In Sect. 2, we propose subspace learning based low-rank representation, extend our method by considering the nonlinear mapping, illustrate how to apply our method for manifold clustering, and discuss the term describing the property of the subspace in details. In Sect. 3, we conduct several experiments to test our method on the problem of manifold clustering. In Sect. 4, we show that our method can be applied into recognition based on image set. In Sect. 5, we conclude this paper.

2   Subspace Learning Based Low-Rank Representation

In this section, we focus on the manifold clustering problem which is a general case of subspace segmentation. Given a set of data points drawn from a union of manifolds, the task of manifold clustering is to segment the data according to their underlying manifolds they are drawn from [17]. Manifold clustering includes two branches: linear and nonlinear [18]. The linear case is subspace segmentation, which arouses widely discussions recently. First, we will develop our model only devoted to subspace segmentation. Second, by extending the mapping from linear to nonlinear, our model can also be applied for nonlinear manifold clustering. Then, we make a detailed discussion on the learning of the subspace for manifold clustering, and provide the numerical solution of our proposed method. In the last, we take an example to illustrate our motivation.

2.1   The Basic Model

Since the data points may exhibit more meaningful linear structure in another subspace instead of the original subspace, we present the Subspace Learning based Low-Rank Representation model (SLLRR) to learn a subspace for low-rank representation. First, we consider to learn the linear subspace by a linear projection $\mathbf{W}$. In this case, the model of SLLRR can be formulated as

$$\min_{\mathbf{Z},\mathbf{W}} \|\mathbf{Z}\|_* + \alpha\|\mathbf{WX} - \mathbf{WXZ}\|_F^2 + \beta f(\mathbf{W}) \\ s.t. \quad \mathbf{WW}^T = \mathbf{I}_d \tag{2}$$

where $\mathbf{X}$, $\mathbf{W}$, and $\mathbf{Z}$ denotes the data set, linear projection, and representation matrix, respectively. Assuming the size of $\mathbf{X}$ is $D \times n$ implies that there are n data points in the D-dimensional space. Denoting the dimensionality of the learned subspace as d, the size of $\mathbf{W}$ will be $d \times D$. The constraint $\mathbf{WW}^T = \mathbf{I}_d$ is used to keep $\mathbf{W}$ as the orthogonal basis where $\mathbf{I}_d$ denotes the identity matrix of the size $d \times d$. The property of the learned subspace is described by the term $f(\mathbf{W})$ whose definition will be discussed in details in Sects. 2.3 and 4.1. With respect to different problems, we will construct different $f(\mathbf{W})$ to enforce the appropriate properties. The term $\|\mathbf{Z}\|_*$ and $\|\mathbf{WX} - \mathbf{WXZ}\|_F^2$ are used to enforce the low-rank property on the representation matrix and keep the projected data points represented by itself as well as possible, respectively. Both α and β are the trade-off parameters balancing these terms.

2.2   Nonlinear Extension

With respect to the data points drawn from a mixture of subspaces, the model (2) makes sense. However, in some real-world problems, we usually encounter with the nonlinear manifold instead of the linear subspace. In order to make SLLRR applicable to manifold clustering, we extend our model in the following.

$$\min_{\mathbf{Z},\mathbf{W}} \|\mathbf{Z}\|_* + \alpha\|\mathbf{W}\phi(\mathbf{X}) - \mathbf{W}\phi(\mathbf{X})\mathbf{Z}\|_F^2 + \beta f(\mathbf{W})$$
$$s.t. \quad \mathbf{WW}^T = \mathbf{I}_d \tag{3}$$

With regard to manifold clustering, we assume the data points can be linearly represented by the data points drawn from the same manifold after a nonlinear mapping ϕ. Then, we also learn a subspace favoring the similarity extraction in order that we can get the effective segmentation results by spectral clustering algorithm. Hence, we change $\mathbf{X}$ in the model (2) into $\phi(\mathbf{X})$ in the model (3). The new model can be understood in the perspective that we learn a nonlinear subspace for the low-rank representation.

Since the bases can be linearly represented by the mapped data points, we have $\mathbf{W} = \mathbf{U}\phi(\mathbf{X})^T$. Then the model (3) can be reformulated in the following

$$\min_{\mathbf{Z},\mathbf{U}} \|\mathbf{Z}\|_* + \alpha\|\mathbf{UK} - \mathbf{UKZ}\|_F^2 + \beta f(\mathbf{U})$$
$$s.t. \quad \mathbf{UKU}^T = \mathbf{I}_d \tag{4}$$

where $\mathbf{K} = \phi(\mathbf{X})^T\phi(\mathbf{X})$ is the kernel. We develop different kernels for different problems with detailed discussion provided in the experiments.

Similar with the spectral clustering-based method such as LRR and SSC for subspace segmentation, SLLRR can handle manifold clustering by Algorithm 1 after getting $\mathbf{Z}$ by Algorithm 2.

Algorithm 1. Manifold Clustering by SLLRR

---

1. Obtain $\mathbf{Z}$ from Algorithm 2.
2. Let $\mathbf{L} = \frac{1}{2}(|\mathbf{Z}| + |\mathbf{Z}^T|)$.
3. Apply NC to $\mathbf{L}$ to conduct a segmentation.

2.3   Learning Subspace for Manifold Clustering

Since linear subspace is a special manifold, we first consider the case that all the mixing manifolds are subspaces to construct $f(\mathbf{W})$ in the model (2), and then make the extension to obtain $f(\mathbf{U})$ in the model (4) for manifold clustering.

Referring to the result in the work of graph embedding framework [16], we can incorporate the objective functions of some dimensionality reduction methods in our model to describe the property of the subspace. However, their objective functions are usually constructed for discrimination since the recognition is their main application. This kind of rule may not be suitable for the self-representation extracting the similarity between data points. If the data points in different subspace are projected into one subspace in spite of desired separation in each class, it will bring in more difficulty for clustering. In addition, the label information required by the supervised approaches can not be obtained in this unsupervised problem. Our idea is to learn a subspace by keeping the local structure of the data points drawn from the same subspace. We pull the data points both near to each other and in the same subspace closer to make them represented by each other better.

Although the neighbors of the data points near the intersection of the subspaces include the data points drawn from different subspaces, numerous data points with small Euclidean distance are in the same subspace. Hence, we first choose the neighbors of each data point by Euclidean distance. After denoting n and c as the number of the data points and the subspaces, respectively, we set the number of the neighbors as $n/2c$. Then we compute the angles between every two data points in the neighbor to establish $\mathbf{G}$ in the following

$$[\mathbf{G}]_{ij} = \begin{cases} 1 & \mathbf{x}_i \in knn(\mathbf{x}_j) \quad or \quad \mathbf{x}_j \in knn(\mathbf{x}_i) \\ 0 & otherwise \end{cases}, \tag{5}$$

where $knn(\mathbf{x}_i)$ denotes the data points in the neighbor of $\mathbf{x}_i$ with the k smallest angles. Because the data points in the same subspaces usually have smaller angle than those in different subspaces, we can select the nearby data points from the same subspace in the above way. Next, we pull them close to each other in the learned subspace by minimizing the following objective function.

$$\sum_{i,j} \|\mathbf{W}\mathbf{x}_i - \mathbf{W}\mathbf{x}_j\|_2^2 [\mathbf{G}]_{ij}. \tag{6}$$

It can be reformulated as

$$tr(\mathbf{W}\mathbf{X}\mathbf{S}\mathbf{X}^T\mathbf{W}^T), \tag{7}$$

where $\mathbf{S} = \mathbf{H} - \mathbf{G}$ is the Laplacian matrix, in which $\mathbf{H}$ is the diagonal matrix with $[\mathbf{H}]_{ii} = \sum_{j=1}^{n}[\mathbf{G}]_{ij}$. Hence, $f(\mathbf{W}) = tr(\mathbf{W}\mathbf{X}\mathbf{S}\mathbf{X}^T\mathbf{W}^T)$ in the model (2).

With respect to the general case of manifold clustering, we assume the data points transformed by the mapping ϕ have the structure of mixing subspaces and thus the Eq. (6) should be changed into the following form.

$$\sum_{i,j} \|\mathbf{W}\phi(\mathbf{x}_i) - \mathbf{W}\phi(\mathbf{x}_j)\|_2^2 [\mathbf{G}]_{ij}. \tag{8}$$

By means of $\mathbf{W} = \mathbf{U}\phi(\mathbf{X})^T$ and $\mathbf{K} = \phi(\mathbf{X})^T\phi(\mathbf{X})$, we can reformulate it as

$$tr(\mathbf{U}\mathbf{K}\mathbf{S}\mathbf{K}^T\mathbf{U}^T), \tag{9}$$

Hence, we establish $f(\mathbf{U}) = tr(\mathbf{U}\mathbf{K}\mathbf{S}\mathbf{K}^T\mathbf{U}^T)$ for the model (4).

2.4   Numerical Solution

First, we further extend our model to a more general case by considering a dictionary $\mathbf{A}$.

$$\min_{\mathbf{Z},\mathbf{W}} \|\mathbf{Z}\|_* + \alpha\|\mathbf{WX} - \mathbf{WAZ}\|_F^2 + \beta tr(\mathbf{WASA}^T\mathbf{W}^T) \\ s.t. \quad \mathbf{WW}^T = \mathbf{I}_d \tag{10}$$

If the dictionary includes N atoms, the size of $\mathbf{A}$ will be $D \times N$. When we choose $\mathbf{A} = \mathbf{X}$, the model (10) becomes the model (2). Its nonlinear extension can be formulated as follows

$$\min_{\mathbf{Z},\mathbf{U}} \|\mathbf{Z}\|_* + \alpha\|\mathbf{UK}_B - \mathbf{UK}_A\mathbf{Z}\|_F^2 + \beta tr(\mathbf{UK}_A\mathbf{SK}_A^T\mathbf{U}^T) \\ s.t. \quad \mathbf{UK}_A\mathbf{U}^T = \mathbf{I}_d \tag{11}$$

where $\mathbf{K}_B = \phi(\mathbf{A})^T\phi(\mathbf{X})$, and $\mathbf{K}_A = \phi(\mathbf{A})^T\phi(\mathbf{A})$. When $\mathbf{K}_B = \mathbf{K}_A = \mathbf{K}$, the model (11) becomes the model (4). In the following, we will provide the numerical solution for this general model.

We can utilize the Alternating Direction Method (ADM) [19] to solve the model (11) by introducing one auxiliary variable $\mathbf{J}$ to make the objective function separable.

$$\min_{\mathbf{Z},\mathbf{J},\mathbf{U}} \|\mathbf{J}\|_* + \alpha\|\mathbf{UK}_B - \mathbf{UK}_A\mathbf{Z}\|_F^2 + \beta tr(\mathbf{UK}_A\mathbf{SK}_A^T\mathbf{U}^T) \\ s.t. \quad \mathbf{UK}_A\mathbf{U}^T = \mathbf{I}_d \quad \mathbf{Z} = \mathbf{J} \tag{12}$$

Then, we get the following augmented Lagrange function

$$\|\mathbf{J}\|_* + \alpha\|\mathbf{UK}_B - \mathbf{UK}_A\mathbf{Z}\|_F^2 + \beta tr(\mathbf{UK}_A\mathbf{SK}_A^T\mathbf{U}^T) \\ + \frac{\mu}{2}(\|\mathbf{Z} - \mathbf{J}\|_F^2) + \langle \mathbf{\Lambda}, \mathbf{Z} - \mathbf{J} \rangle \tag{13}$$

We iteratively update each variable by fixing the others, fortunately the closed form solution exists for each optimization problem. The numerical algorithm of the model (11) is provided in the Algorithm 2 where the convex optimization in step 2 can be solved by Singular Value Thresholding (SVT) operator [20].

2.5   An Example Illustrating Our Motivation

In this section, we will take an example to illustrate our motivation. As shown in Fig. 1(a), there are some data points sampled from two spirals. Since one spiral is a nonlinear manifold instead of a linear subspace, the self-representation in the original space can not extract the similarity well. We take LRR as an example to test the performance of self-representation based approaches on this problem. The result of LRR is shown in Fig. 1(b) where the data points near the intersection are misclassified. Then, we use PCA as a preprocessing step for LRR and provide the result in Fig. 1(c). Since PCA can not project the data points drawn from one spiral into a linear subspace, LRR still achieves the wrong segmentation result.

Algorithm 2. Solving Problem (11) by ADM

Input: $\mathbf{K}_B$, $\mathbf{K}_A$, $\mathbf{S}$, α, β, d.

Initialize: $\mathbf{Z} = \mathbf{I}_{n \times N}$, $\mathbf{U} = \mathbf{I}_{d \times m}$, $\mathbf{\Lambda} = \mathbf{0}$, $\mu = 10^{-6}$, $max_\mu = 10^{30}$, $\rho = 1.1$, $\varepsilon = 10^{-4}$.

   while not converged **do**

      1. Fix the others and update $\mathbf{U}$ by computing the generalized eigenvalue decomposition problem:

      $(\beta \mathbf{K}_A \mathbf{S} \mathbf{K}_A^T + \alpha(\mathbf{K}_B - \mathbf{K}_A \mathbf{Z})(\mathbf{K}_B - \mathbf{K}_A \mathbf{Z})^T)\mathbf{U}^T = \lambda \mathbf{K}_A \mathbf{U}^T$

      2. Fix the others and update $\mathbf{J}$

      $\mathbf{J} = argmin \frac{1}{\mu}\|\mathbf{J}\|_* + \frac{1}{2}\|\mathbf{J} - (\mathbf{Z} + \frac{\mathbf{\Lambda}}{\mu})\|_F^2$.

      3. Fix the others and update $\mathbf{Z}$

      $\mathbf{Z} = (\mu \mathbf{I} + 2\alpha \mathbf{K}_A^T \mathbf{W}^T \mathbf{W} \mathbf{K}_A)^{-1}(2\alpha \mathbf{K}_A^T \mathbf{W}^T \mathbf{W} \mathbf{K}_B + \mu \mathbf{J} - \mathbf{\Lambda})$

      4. Update the multipliers

      $\mathbf{\Lambda} = \mathbf{\Lambda} + \mu(\mathbf{Z} - \mathbf{J})$.

      5. Update the parameter μ

      $\mu = min(\rho\mu, max_\mu)$.

      6. Check the convergence conditions

      $\|\mathbf{Z} - \mathbf{J}\|_\infty < \varepsilon$.

   end while

Next, we test the performance of LS3C on this synthetic data. First, we try the polynomial kernel frequently used in its experiment. As shown in Fig. 1(d), it also can not handle this case well. Second, we construct a kernel for LS3C in the following:

$$[\mathbf{K}]_{ij} = \exp(-\frac{1 - (\hat{\mathbf{y}}_i^T \hat{\mathbf{y}}_j)^2}{\sigma^2}), \tag{14}$$

where $\hat{\mathbf{y}}_i = \frac{\mathbf{y}_i}{\|\mathbf{y}_i\|_2}$, $i = 1, \cdots, N$, σ^2 is empirically set as $\frac{\sum_{i,j}(1-(\hat{\mathbf{y}}_i^T \hat{\mathbf{y}}_j)^h)}{N^2}$. $\mathbf{Y} = [\mathbf{y}_1, \cdots, \mathbf{y}_i, \cdots, \mathbf{y}_n]$ is defined as follows

$$\mathbf{Y} = \mathbf{M} \odot \mathbf{Q}, \tag{15}$$

where the construction of $\mathbf{Q}$ is similar with $\mathbf{K}$ in Eq. (14) by replacing $\mathbf{y}_i$ with $\mathbf{x}_i$, $\mathbf{M}$ is defined as a binary matrix in the way that if $\mathbf{x}_i$ is among the k-nearest neighbors of $\mathbf{x}_j$ measured by Euclidean distance or vice versa, $[\mathbf{M}]_{ij} = 1$, otherwise, $[\mathbf{M}]_{ij} = 0$. The construction of $\mathbf{K}$ is inspired by the work [21]. According to its analysis, $[\mathbf{K}]_{ij}$ represents the angle between the $\mathbf{x}_i$ and $\mathbf{x}_j$ transformed by the Veronese map of degree 2, which can transform some disjoint but not independent subspaces into independent subspaces. Since independent subspace segmentation is an easier task than disjoint subspace segmentation, the $\mathbf{K}$ can improve the result for subspace segmentation. Our idea is applying the transformed data points approximately with the structure of mixing subspace for $\mathbf{K}$. Hence, we construct $\mathbf{Y}$ which can be viewed as the transformed data points. In order to enable $\mathbf{Y}$ to approximately have the appreciated structure, we try to let the transformed data points drawn from the same spiral have small angle between them. The local information collected in $\mathbf{M}$ and the angle measure represented by $\mathbf{Q}$ are used to achieved this aim. Although we deliberately devise the kernel for LS3C, it can not achieve the right results too. The reason may

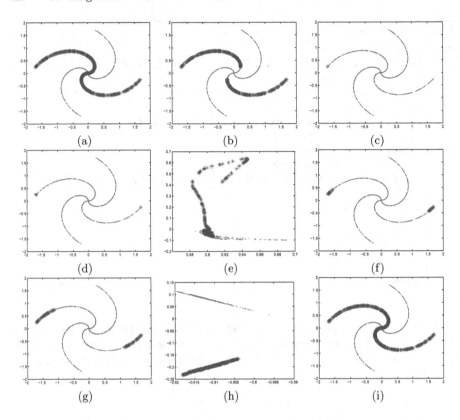

Fig. 1. The results of different methods on two spirals: (a) shows the original data. There are some data points sampled on two spirals. (b) is the segmentation result of LRR. (c) is the segmentation result of LRR by using PCA as a preprocessing step. (d) is the segmentation result of LS3C with the polynomial kernel. (e) is the projected data points in the latent space of LS3C with the kernel in Eq. 14. (f) is the segmentation result of LS3C with the kernel in Eq. 14. (g) is the segmentation result of SLLRR with the kernel in Eq. 14 and the graph of PCA. (h) is the transformed data points in the learned subspace of SLLRR with the kernel constructed in Eq. 14 and the graph established in Sect. 2.3. (i) is the segmentation result of SLLRR with the kernel constructed in Eq. 14 and the graph established in Sect. 2.3.

be the criterion of the subspace does not favor the self-representation extracting the similarity. Figure 1(e) shows the transformed data points in the latent space by setting its dimensionality as 2. Since these projected data points do not have the structure of mixing subspaces, it achieves the wrong segmentation results shown in Fig. 1(f).

In the last, we try the **K** in Eq.(14) for SLLRR on this synthetic data by setting the dimensionality of the learned subspace as 2. As shown in Fig. 1(i), our method achieves the right segmentation results with $\alpha = 1, \beta = 100$. The transformed data points in the learned subspace are shown in Fig. 1(h) where the data points drawn from one spiral distribute on the same line. This result indicates that SLLRR learns a subspace favoring the self-representation to extract similarity and thus performs well. According to the results in the work of graph embedding framework [16], the objective function of the dimensionality reduction methods can be summarized as

$$\min_{\mathbf{W}} \frac{tr(\mathbf{W}\mathbf{X}\mathbf{S}_W\mathbf{X}^T\mathbf{W}^T)}{tr(\mathbf{W}\mathbf{X}\mathbf{S}_B\mathbf{X}^T\mathbf{W}^T)}, \tag{16}$$

where $\mathbf{S}_W$ and $\mathbf{S}_B$ represent the similarity and dissimilarity information among the data set, respectively. Inspired by [22], we reformulate it as

$$\min_{\mathbf{W}}(tr(\mathbf{W}\mathbf{X}\mathbf{S}_W\mathbf{X}^T\mathbf{W}^T) - tr(\mathbf{W}\mathbf{X}\mathbf{S}_B\mathbf{X}^T\mathbf{W}^T)), \tag{17}$$

meaning that we can incorporate previous dimensionality reduction methods into our model by setting $f(\mathbf{W}) = tr(\mathbf{W}\mathbf{X}(\mathbf{S}_W - \mathbf{S}_B)\mathbf{X}^T\mathbf{W}^T)$. With respect to the manifold clustering, $f(\mathbf{U}) = tr(\mathbf{U}\mathbf{K}(\mathbf{S}_W - \mathbf{S}_B)\mathbf{K}^T\mathbf{U}^T)$. According to the result in Table 1 of the Ref. [16], we apply the graph of PCA for SLLRR and provide the result in Fig. 1(g). Under this condition, SLLRR achieves the wrong segmentation result. This phenomenon implies the importance of criterion for the learned subspace.

3   Experimental Results

In this section, we test the performance of SLLRR on manifold clustering by comparing it with the state-of-the-art on the benchmark databases. Since SLLRR can be viewed as an extension of LRR, we choose LRR and its three different kinds of extensions LSR, LLRR, SC-LRR in our comparison. In addition, we also compare SLLRR with LS3C and its basic method SSC. For fair comparison, the important parameters of each method are empirically tuned according to the recommendations in the original references as well as the source codes provided by the original authors. Segmentation error is used to evaluate the performance of the state-of-the-art, which is defined as the ratio of the number of the misclassified data points to the number of all the data points. Clearly, lower clustering error means better performance.

3.1   Hopkins 155

Motion segmentation refers to the problem of separating different object motions in a video sequence. Existing work usually is based on the extraction of the feature point trajectories from the video. It has been proved that the feature point trajectories on one motion are in a subspace whose dimension is not more than 4 [23]. Hence, segmenting the feature point trajectories on different object motions can be cast as the subspace segmentation problem. With regard to this problem, we choose the kernel defined in Eq. 14 by replacing $\mathbf{y}_i$ with the original data points. Hopkins 155 [24] is a database including 155 motion segmentation

Table 2. Mean segmentation error of all the compared methods on Hopkins 155.

Methods	LRR	SSC	LLRR	LSR	SC-LRR	LS3C	SLLRR
2 motions	3.14%	1.83%	2.08%	1.80%	1.60%	1.62%	1.63%
3 motions	4.86%	4.40%	3.74%	4.14%	5.31%	4.38%	3.15%
All the motions	3.74%	2.41%	2.45%	2.33%	2.43%	2.31%	1.97%

problem, each of which includes two or three motions. Hence, the number of the subspace in each subspace segmentation problem of this experiment is not more than 3. This may be the reason that all the compared methods achieve low segmentation error on Hopkins 155. The mean of the segmentation error of 2 motions, 3 motions and all the motions has been shown in Table 2. Under the setting of $\alpha = 20, \beta = 1$, SLLRR achieves 1.63% on 2 motions, not far from the lowest result 1.60%. With respect to the 3 motions and all the motions, SLLRR both achieves the best results. This experiment confirms that SLLRR can achieve the state-of-the-art result.

3.2 Extended Yale B

The reference [25] implies that the images taken under different light conditions can be well approximated by a low-dimensional space. Hence, the problem of face clustering can also be cast as a subspace segmentation problem. In this situation, we still choose the kernel defined in Eq. 14 by replacing $\mathbf{y}_i$ with the original data points. Since the number of the subspace in Hopkins 155 is small, we test the performance of all the compared methods on 38 subspaces in this experiment. Extended Yale B [26] includes 2432 frontal face images of 38 subjects under different light conditions, with each subjects including 64 face images. We choose the first 20 images of each subject and resize them as 42×48 to conduct this experiment. Table 3 provides the results of all the compared methods, among which SLLRR achieves the lowest segmentation error 21.32% with $\alpha = 50, \beta = 1$. LRR and its extensions achieve low segmentation error than SSC and its extensions. The reason may be that the sparse affinity matrix can result in over-segmentation [27] easily occurred with many categories.

Table 3. Segmentation error of all the compared methods on Extended Yale B.

Methods	LRR	SSC	LLRR	LSR	SC-LRR	LS3C	SLLRR
Error	24.34%	36.71%	26.18%	25.00%	23.55%	26.58%	21.32%

3.3 COIL-20

The processed COIL-20 [28] has been used to test the state-of-the-art on the non-linear manifold clustering in previous work [29]. It is a database containing 1440

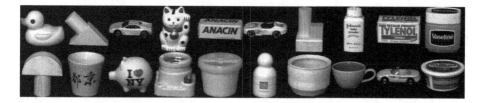

Fig. 2. The first image of each object in COIL-20

gray-scale images of 20 objects, with each object including 72 images. Figure 2 shows the first image of each object in this database. The size of each image is 128 × 128. We extract the spatial pyramid bag of words (SPBOW) [30] by setting the size of dictionary at 80, and explore these descriptors as the original data points. Due to the effectiveness of SPBOW, the results in Table 4 are much better than those in Table 3 of the reference [29]. Since this experiment is not the case of linear subspace, we use Gaussian kernel with the Chi-square distance. From Table 4, it can be found that SLLRR performs best with $\alpha = 20, \beta = 1$. This result confirms the effectiveness of SLLRR again.

Table 4. Segmentation errors of all the compared methods on COIL-20.

Methods	LRR	SSC	LLRR	LSR	SC-LRR	LS3C	SLLRR
Error	27.78%	26.18%	30.35%	25.42%	25.69%	25.42%	20.63%

4   A Non-trivial Byproduct

With respect to the recognition problem, Nearest Neighbors (NN) and Sparse Representation-based Classification (SRC) [31] are two kinds of well-known classifiers. The dimensionality reduction approaches are usually combined with NN for recognition while the representation based methods such as LRR can also be applied for recognition by adopting the similar strategy of SRC outlined in Algorithm 3. Since SLLRR includes the terms associated with both dimensionality reduction and LRR, it is natural to apply SLLRR for recognition.

Algorithm 3. Conduct recognition by the representation-based methods

---

Input: Let $\mathbf{T} = [\mathbf{T}_1, \cdots, \mathbf{T}_i, \cdots, \mathbf{T}_c]$ be the training samples from c classes, and $\mathbf{X} = [\mathbf{x}_1, \cdots, \mathbf{x}_i, \cdots, \mathbf{x}_n]$ be the testing set including n samples.

1. Solving the model to obtain the representation matrix $\mathbf{Z} = [\mathbf{Z}_1; \cdots; \mathbf{Z}_i; \cdots; \mathbf{Z}_c]$, where $\mathbf{Z}_i$ denotes the representation coefficients associated with $\mathbf{T}_i$.

2. Compute the residuals $[\mathbf{R}]_{i:} =$ the ℓ_2 norm of each column of the matrix $\mathbf{X} - \mathbf{T}_i \mathbf{Z}_i$.

3. Identity each $\mathbf{x}_j, j = 1, \cdots, n$ by minimizing its residuals with respect to each class, i.e. min$[\mathbf{R}]_{:j}$.

4.1   Learning Subspace for Recognition

By denoting the $\mathbf{X}$ and $\mathbf{T}$ as the testing data set and the training data set, respectively, SLLRR for recognition can be formulated as

$$\min_{\mathbf{Z},\mathbf{W}} \|\mathbf{Z}\|_* + \alpha\|\mathbf{W}\mathbf{X} - \mathbf{W}\mathbf{T}\mathbf{Z}\|_F^2 + \beta f(tr(\mathbf{W}\mathbf{T}(\mathbf{S}_W - \mathbf{S}_B)\mathbf{T}^T\mathbf{W}^T))$$
$$s.t.\quad \mathbf{W}\mathbf{W}^T = \mathbf{I}_d \tag{18}$$

It means that we learn a subspace by the training data set and simultaneously obtain the low-rank representation of the testing data set in terms of the training data set. With respect to the nonlinear case, we should replace the $\mathbf{T}$ with $\phi(\mathbf{T})$, meaning that we choose $\mathbf{K}_B = \phi(\mathbf{T})^T\phi(\mathbf{X})$, $\mathbf{K}_A = \phi(\mathbf{T})^T\phi(\mathbf{T})$, and $\mathbf{S} = \mathbf{S}_W - \mathbf{S}_B$ for the model (11). This result indicates that SLLRR can obtain $\mathbf{U}$ and $\mathbf{Z}$ from Algorithm 2 to conduct recognition by NN or Algorithm 3. Because the last term in the model (18) is only an approximation to the dimensionality reduction methods including both the graph describing the similarities and dissimilarities such as Linear Discriminant Analysis (LDA) [32], Marginal Fisher Analysis (MFA) [16], etc., due to Eq. (17), we prefer to incorporate the graph of Lorentzian Discriminant Projection (LDP) [33] in our model.

4.2   Experimental Results of Recognition Based on Image Set

Compared with the traditional recognition based on single-shot image, the recognition based on image set attracts increasing interest in computer vision. The image set collected from consecutive video sequences or unordered photo album will cover large variations of one subject. The task of recognition based on image set is matching the probe image set against all the gallery image sets with each image set corresponding to one subject. Some methods [34–36] conduct recognition by the kernel extension of the dimensionality reduction methods such as Kernel Discriminant Analysis (KDA) [37] after constructing the kernel between the image sets. Because SLLRR includes both the kernelization of the representation based methods and dimensionality reduction methods, we can improve the previous work by replacing the step of the kernel dimensionality reduction methods with SLLRR. We take Covariance Discriminative Learning (CDL) [36] as an example to conduct ten-fold experiments on two benchmark databases YouTube Celebrities and ETH-80, and then report the average recognition accuracy. YouTube Celebrities (YTC) [38] contains 1910 videos of 47 celebrities collected from YouTube while ETH-80 [39] contains images of 8 categories. With respect to these two databases, we randomly select 3 image sets of each category for training and 7 for testing. The preprocessing of histogram equalization has been performed on all the images after resizing them as 20×20.

The experimental results are summarized in Table 5 where CDLLDA and CDLPLS denotes the method in [36] whose step of recognition is performed by KDA and Kernel Partial Least Squares (KPLS) [40], respectively, SLLRRNN and SLLRRRC denotes the method replacing KDA in CDLLDA with the part of dimensionality reduction method and representation based method in SLLRR,

Table 5. Average recognition accuracy of the compared methods on YTC and ETH-80.

Methods	CDLLDA	CDLPLS	SLLRRNN	SLLRRRC
Accuracy on ETH-80	85.54%	89.64%	88.21%	93.04%
Accuracy on YTC	55.96%	62.77%	56.29%	63.05%

respectively. Compared with CDLLDA, SLLRRNN achieves a little improvement on both databases due to the use of kernel LDP [33]. In contrast to the dimensionality reduction methods, the representation based methods operates in a different manner for recognition. This kind of method usually outperforms NN combined with dimensionality reduction methods in previous references. In this experiment, utilizing the kernel extension of the representation based methods for recognition, SLLRRRC performs best with $\alpha = 20, \beta = 1$.

5   Conclusion and Future Work

This paper presents SLLRR to obtain a more meaningful representation in a learned subspace for manifold clustering. Moreover, it can also be applied for recognition based on image set. Extensive experimental results confirm its effectiveness. However, there remain several directions for future work. First, we will try our method on other graph based application such as graph based semi-supervised learning. Second, we will consider the combination of the dimensionality reduction methods and representation based methods in our model to form a strong classifiers for the recognition problem. In addition, we will discuss the construction of the kernel suitable for SLLRR in the domain of recognition based on image set.

Acknowledgement. The work of K. Tang was supported by the Educational Commission of Liaoning Province, China (No. L201683662). The work of Z. Su was supported by the National Natural Science Foundation of China (No. 61572099), National Science and Technology Major Project (No. 2014ZX04001011, ZX20140419). The work of W. Jiang was supported by the Natural Science Foundation of Liaoning Province, China (No. 60875029).

References

1. Soltanolkotabi, M., Candès, E.J.: A geometric analysis of subspace clustering with outliers. Ann. Stat. **40**, 2195–2238 (2011)
2. Vidal, R.: Subspace clustering. IEEE Sig. Process. Mag. **28**, 52–68 (2011)
3. Luxburg, U.: A tutorial on spectral clustering. Stat. Comput. **17**, 395–416 (2007)
4. Shi, J., Malik, J.: Normalized cuts and image segmentation. IEEE Trans. Pattern Anal. Mach. Intell. **22**, 888–905 (2000)
5. Liu, G., Lin, Z., Yan, S., Sun, J., Yu, Y., Ma, Y.: Robust recovery of subspace structures by low-rank representation. IEEE Trans. Pattern Anal. Mach. Intell. **35**, 171–184 (2013)

6. Elhamifar, E., Vidal, R.: Sparse subspace clustering: algorithm, theory, and applications. IEEE Trans. Pattern Anal. Mach. Intell. **35**, 2765–2781 (2013)
7. Wang, S., Yuan, X., Yao, T., Yan, S., Shen, J.: Efficient subspace segmentation via quadratic programming. In: AAAI (2011)
8. Lu, C.-Y., Min, H., Zhao, Z.-Q., Zhu, L., Huang, D.-S., Yan, S.: Robust and efficient subspace segmentation via least squares regression. In: Fitzgibbon, A., Lazebnik, S., Perona, P., Sato, Y., Schmid, C. (eds.) ECCV 2012. LNCS, vol. 7578, pp. 347–360. Springer, Heidelberg (2012). doi:10.1007/978-3-642-33786-4_26
9. Lu, C., Feng, J., Lin, Z., Yan, S.: Correlation adaptive subspace segmentation by trace lasso. In: ICCV (2013)
10. Hu, H., Lin, Z., Feng, J., Zhou, J.: Smooth representation clustering. In: CVPR (2014)
11. Tang, K., Liu, R., Su, Z., Zhang, J.: Structure-constrained low-rank representation. IEEE Trans. Neural Netw. Learn. Syst. **25**, 2167–2179 (2014)
12. Tang, K., Dunson, D.B., Su, Z., Liu, R., Zhang, J., Dong, J.: Subspace segmentation by dense block and sparse representation. Neural Netw. **75**, 66–76 (2016)
13. Zhang, X.: Matrix Analysis and Applications. Tsinghua University Press, Beijing (2004)
14. Liu, G., Yan, S.: Latent low-rank representation for subspace segmentation and feature extraction. In: ICCV (2011)
15. Patel, V.M., Nguyen, H.V., Vidal, R.: Latent space sparse subspace clustering. In: ICCV (2013)
16. Yan, S., Xu, D., Zhang, B., Zhang, H., Yang, Q., Lin, S.: Graph embedding and extensions: a general framework for dimensionality reduction. IEEE Trans. Pattern Anal. Mach. Intell. **29**, 40–51 (2007)
17. Souvenir, R., Pless, R.: Manifold clustering. In: ICCV (2005)
18. Wang, Y., Jiang, Y., Wu, Y., Zhou, Z.-H.: Multi-manifold clustering. In: Zhang, B.-T., Orgun, M.A. (eds.) PRICAI 2010. LNCS (LNAI), vol. 6230, pp. 280–291. Springer, Heidelberg (2010). doi:10.1007/978-3-642-15246-7_27
19. Lin, Z., Chen, M., Wu, L., Ma, Y.: The augmented Lagrange multiplier method for exact recovery of corrupted low-rank matrices. UIUC Technical report, UILU-ENG-09-2215 (2009)
20. Cai, J., Candès, E.J., Shen, Z.: A singular value thresholding algorithm for matrix completion. SIAM J. Optim. **20**, 1956–1982 (2010)
21. Tang, K., Zhang, J., Su, Z., Dong, J.: Bayesian low-rank and sparse nonlinear rerpesentation for manifold clustering. Neural Process. Lett. **44**(3), 1–15 (2015)
22. Zhao, D., Lin, Z., Tang, X.: Classification via semi-Riemannian spaces. In: CVPR, pp. 1–8 (2008)
23. Vidal, R., Ma, Y., Sastry, S.: Generalized principal component analysis (GPCA). IEEE Trans. Pattern Anal. Mach. Intell. **27**, 1945–1959 (2005)
24. Tron, R., Vidal, R.: A benchmark for the comparison of 3-d motion segmentation algorithms. In: CVPR (2007)
25. Ho, J., Yang, M.H., Lim, J., Lee, K.C., Kriegman, D.J.: Clustering appearances of objects under varying illumination conditions. In: CVPR (2003)
26. Lee, K.C., Ho, J., Kriegman, D.J.: Acquiring linear subspaces for face recognition under variable lighting. IEEE Trans. Pattern Anal. Mach. Intell. **27**, 684–698 (2005)
27. Nasihatkon, B., Hartley, R.I.: Graph connectivity in sparse subspace clustering. In: CVPR, pp. 2137–2144 (2011)
28. Nene, S.A., Nayar, S.K., Murase, H.: Columbia object image library (coil-20). Technical report, CUCS-005-96 (1996)

29. Yong, W., Yuan, J., Yi, W., Zhou, Z.: Spectral clustering on multiple manifolds. IEEE Trans. Neural Netw. Learn. Syst. **22**, 1149–1161 (2011)
30. Lazebnik, S., Schmid, C., Ponce, J.: Beyond bags of features: spatial pyramid matching for recognizing natural scene categories. In: CVPR (2006)
31. Wright, J., Yang, A.Y., Ganesh, A., Sastry, S.S., Ma, Y.: Robust face recognition via sparse representation. IEEE Trans. Pattern Anal. Mach. Intell. **31**, 210–227 (2009)
32. Belhumeur, P.N., Hespanha, J.P., Kriegman, D.J.: Eigenfaces vs. fisherfaces: recognition using class specific linear projection. IEEE Trans. Pattern Anal. Mach. Intell. **19**, 711–720 (1997)
33. Liu, R., Lin, Z., Su, Z., Tang, K.: Feature extraction by learning lorentzian metric tensor and its extensions. Pattern Recogn. **43**, 3298–3306 (2010)
34. Harandi, M.T., Sanderson, C., Shirazi, S.A., Lovell, B.C.: Graph embedding discriminant analysis on Grassmannian manifolds for improved image set matching. In: CVPR (2011)
35. Wang, W., Wang, R., Huang, Z., Shan, S., Chen, X.: Discriminant analysis on Riemannian manifold of Gaussian distributions for face recognition with image sets. In: CVPR (2015)
36. Wang, R., Guo, H., Davis, L.S., Dai, Q.: Covariance discriminative learning: a natural and efficient approach to image set classification. In: CVPR (2012)
37. Baudat, G., Anouar, F.: Generalized discriminant analysis using a kernel approach. Neural Comput. **12**, 2385–2404 (2000)
38. Kim, M., Kumar, S., Pavlovic, V., Rowley, H.A.: Face tracking and recognition with visual constraints in real-world videos. In: CVPR (2008)
39. Leibe, B., Schiele, B.: Analyzing appearance and contour based methods for object categorization. In: CVPR (2003)
40. Rosipal, R., Trejo, L.J.: Kernel partial least squares regression in reproducing Kernel Hilbert space. J. Mach. Learn. Res. **2**, 97–123 (2001)

Author Index

Printed in the United States
By Bookmasters